BIOMASS VOLUME ESTIMATION AND VALORIZATION FOR ENERGY

Edited by **Jaya Shankar Tumuluru**

Biomass Volume Estimation and Valorization for Energy
http://dx.doi.org/10.5772/62678
Edited by Jaya Shankar Tumuluru

Contributors
Tran Dang Xuan, WeiSheng Zeng, Sibel Irmak, Hua Song, Aiguo Wang, Danielle Austin, Valter Silva, Nuno Couto, Lei Shi, Shirong Liu, C. Luke Williams, Jaya Shankar Tumuluru, Rachel M. Emerson, Adejumoke Inyinbor, Abimbola Oluyori, Tabitha Adelani-Akande, Borja Velazquez-Marti, Isabel López Cortés, Domingo M. Salazar Hernández, Ángel-Jesús Callejón-Ferre, Luis Fernandez Linares, Adélia Sousa, Ana Cristina Gonçalves, José Marques Da Silva, Olga Anne, Thorsten Ahrens, Silvia Drescher-Hartung, Martin Bosak, Allison Ray, Chenlin Li, Vicki Thompson, Dayna Daubaras, Nick Nagle, Claudia Ortiz, Hector Cid, Daniel Barros, Carlos Escudero-Oñate, Isabel Villaescusa, Jordi Poch, Nuria Fiol, Robert Clubb, Grace Huang, Frank Hedlund, Øssur J Hilduberg, Leticia Saiz-Rodriguez, Jean-Philippe Faure, Adrien Zambon, José María Bermejo, Chuan-Fu Liu, Hui-Hui Wang, Xue-Qin Zhang, Yi Wei, Mohammed Abed Fattah Hamad

Published by InTech
Janeza Trdine 9, 51000 Rijeka, Croatia

Publishing Process Manager Maja Bozicevic
Technical Editor SPi Global
Cover InTech Design team

First published February, 2017
Printed in Croatia
Legal deposit, Croatia: National and University Library in Zagreb

Additional hard copies can be obtained from orders@intechopen.com

Biomass Volume Estimation and Valorization for Energy, Edited by Jaya Shankar Tumuluru
p. cm.
Print ISBN 978-953-51-2937-0
Online ISBN 978-953-51-2938-7

For EU product safety concerns:
IN TECH d.o.o., Prolaz Marije Krucifikse Kozulić 3, 51000 Rijeka, Croatia,
info@intechopen.com or visit our website at intechopen.com.

Meet the editor

Dr. Jaya Shankar Tumuluru works as a research scientist in the Biofuels Department at the Idaho National Laboratory. His major research areas include biomass pretreatment, densification, storage, food processing, data analysis and modeling using statistical methods, evolutionary and back propagation algorithms. Dr. Tumuluru has acted as a reviewer for national and international grant agencies and high-impact factor peer-reviewed journals. He has published more than 60 peer-reviewed journal papers, 10 book chapters, and numerous conference proceeding papers. He has received numerous honors, including outstanding reviewer award from the American Society of Agricultural and Biological Engineers and the Institution of Chemical Engineers, and INL's Exceptional Contributions Award Program for the Outstanding Achievement in Scientific and Technical Publication. He received his PhD degree from the Agricultural and Food Engineering Department at the Indian Institute of Technology, Kharagpur, India.

Contents

Preface

The United Nations Paris Framework Convention on Climate Change has agreed to reduce greenhouse gas emission, in an effort to reduce the global average temperature increase, to below 2°C. Energy produced from renewable energy resources will play a key role in reducing the greenhouse gas emissions. Among the various renewable energy sources, biomass is considered as carbon neutral because the carbon dioxide released during its conversion is already part of the carbon cycle. Biomass energy is also known to be relatively sustainable.

By definition, biomass means any plant-derived organic matter which is available on a renewable basis. Examples of biomass which are used for energy generation are agricultural crop wastes and residues, wood waste and residues, aquatic plants, animal waste, municipal solid waste, and other waste materials from food and feed processing industries. Biomass resources can be used to produce a number of products including electricity; liquid, solid, and gaseous fuels; heat; chemicals; and other high value-added materials. Currently, second-generation biofuel crops such as the perennial grasses (e.g., miscanthus, switchgrass, and others) and dedicated forest plantations are grown for bioenergy production. These dedicated herbaceous and woody biomasses can be grown in underutilized or marginal lands without significantly displacing the land which is typically used for food and feed crops. The growing and processing of dedicated energy crops and forest plantations for energy and other value-added bio-based products will help to improve economic activity in rural regions.

This book is the outcome of contributions by many experts in the field from different disciplines, various backgrounds, and diverse expertise. This book provides information on biomass volume calculation methods and biomass valorization for energy production. The chapters presented in this book include original research and review articles. I hope the research presented in this book will help to advance the use of biomass for bioenergy production and valorization. I want to thank all the authors who have contributed to this book, without whom it would have been impossible to get this book published. I also want to thank the Department of Energy, Office of Energy Efficiency and Renewable Energy under the Department of Energy, and Idaho Operations Office Contract DE-AC07-05ID14517 for supporting me in completing this work.

The key features of the book are:

- Providing information on biomass volume estimation using direct, nondestructive and remote-sensing methods
- Biomass valorization for energy using thermochemical (gasification and pyrolysis) and biochemical (fermentation) conversion processes

Jaya Shankar Tumuluru, PhD
Research Scientist, Biofuels Department
Energy Systems Laboratory
Energy and Environment Directorate
Idaho National Laboratory, Idaho Falls, Idaho, USA

Section 1

Biomass Volume Estimation

Chapter 1

Developing Tree Biomass Models for Eight Major Tree Species in China

WeiSheng Zeng

Additional information is available at the end of the chapter

http://dx.doi.org/10.5772/65664

Abstract

In the context of climate change, estimating forest biomass for large regions is key to national carbon stocks, but few models have been developed at regional level. Based on mensuration data from large samples (4818 and 1626 trees for above- and belowground biomass, respectively) of eight major tree species in China, the author developed one- and two-variable compatible integrated model systems for aboveground and belowground biomass, biomass conversion factor (*BCF*) and root-to-shoot ratio (*RSR*), using the error-in-variable simultaneous equations. Furthermore, the differences of aboveground and belowground biomass among various species were analyzed using the dummy variable approach. The results indicated that (1) two-variable models were almost better than one-variable models for aboveground biomass estimation, while the two model systems were not significantly different for belowground biomass estimation; (2) the eight species can be ranked in terms of aboveground biomass from *Quercus* (largest), *Betula, Populus, Pinus massoniana, Picea, Larix, Abies* to *Cunninghamia lanceolata* and in terms of belowground biomass from *Quercus* (largest), *Betula, Larix, Picea, Populus, P. massoniana, C. lanceolata* to *Abies*; (3) mean prediction errors (*MPEs*) of aboveground biomass models for the species were less than 5%, whereas *MPEs* of belowground biomass equations were less than 10%, except for *Abies*.

Keywords: aboveground biomass, belowground biomass, biomass conversion factor, root-to-shoot ratio, error-in-variable simultaneous equations

1. Introduction

Increasingly, governments worldwide attach considerable importance to estimating biomass and carbon storage of forest ecosystems in the context of global climate change. To help countries conduct national greenhouse gas inventories, forest biomass estimation and carbon stock assessment, the Intergovernmental Panel on Climate Change (IPCC) provided such carbon-accounting parameters as biomass expansion factors (*BEF*) and root-to-shoot ratios (*RSR*) for estimating different geographic zones in 2003 [1]. However, it probably has great uncertainty to apply these parameters for biomass estimation. Developing individual tree biomass models and parameters for national monitoring and assessment of biomass and carbon storage of forest ecosystems has become fundamentally important.

The earliest research on forest biomass abroad can be traced to the 1870s [2]. In recent years, biomass models for major tree species in America, Canada and some European countries have been developed or improved [3–11]. Their purpose was to assess and monitor forest biomass and carbon storage and to provide a basis for evaluating the contribution of forest ecosystems to the global carbon cycle. Studies on forest biomass in China have only been implemented since the late 1970s when some related articles were published [12, 13], i.e., a century after the earliest study abroad. Due to special historical reasons, China did not participate in the International Biological Program (IBP), initiated by the International Union of Forest Research Organizations (IUFRO), during the period of 1964–1974 and thus missed the golden development stage of forest biomass research [14].

Reviewing the development of forest biomass modeling near 40 years in China, three stages could be classified: the first is estimating biomass and productivity of major forest types toward the end of the twentieth century [13, 15–30]; the second is assessing carbon storage in Chinese forest ecosystems since the beginning of the current century [31–37]; and the third is the new development stage for monitoring and assessing forest biomass and carbon storage at provincial and national levels [14, 38]. To monitor forest biomass and carbon storage in the National Forest Inventory (NFI) system, the National Forest Biomass Modeling Program has been implemented since early 2009. Up to now, many papers on modeling individual tree biomass have been published [39–51], which classified 70 modeling populations for developing individual tree biomass models, determined the sample structure of each population and studied the modeling methods including nonlinear error-in-variable simultaneous equations, mixed-effects modeling, dummy variable modeling and segmented modeling approaches. Also, logarithmic regression and weighted regression were analyzed [52] and goodness evaluation and precision analysis of biomass models were studied [53]. Based on the studying achievements, two ministerial standards on technical regulations and five ministerial standards on biomass models have been approved for application [54–60]. In the near future, more ministerial standards on biomass models for other tree species would be published.

From the published papers and ministerial standards, we could find that the aboveground and belowground biomass models were developed separately owing to the unequal sample sizes and most of the studies were only based on sample trees of one tree species. In this chapter, the author will use the mensuration data of aboveground and belowground biomass from 4818 to 1626 destructive sample trees of eight major tree species, respectively. The main purpose was to develop an integrated individual tree model system for aboveground and belowground biomass, biomass conversion factor (*BCF*) and root-to-shoot ratio (*RSR*), using the approach of nonlinear error-in-variable simultaneous equations with dummy variable. The system could assure aboveground biomass models compatible with stem volume models and *BCF* models and belowground biomass models compatible with aboveground biomass models and *RSR* models. Secondly, the generalized dummy-variable models of aboveground and belowground biomass for eight major tree species were established and compared and the ranks of eight species for aboveground and belowground biomass estimation were provided respectively from the species-specific parameter estimates.

2. Materials and methods

2.1. Data

During the 5 years between 2009 and 2013, a total amount of 4818 sample trees for 31 modeling populations of eight major tree species or species groups, namely, *Picea* spp., *Abies* spp., *Betula* spp., *Quercus* spp., *Populus* spp., *Larix* spp., *Cunninghamia lanceolata* and *Pinus massoniana*, which occupied more than 60% of forest volume in China [39], were felled for aboveground biomass mensuration. The sample trees were evenly distributed in ten diameter classes of 2, 4, 6, 8, 12, 16, 20, 26, 32 and more than 38 cm for each modeling population, and about 15 sample trees in each diameter class were selected by height class as evenly as possible. For example, if three height classes were defined, i.e., low, intermediate and high, then five sample trees should be selected in each height class. For each sample tree, the diameter at breast height of stem was measured in the field. After the tree was felled, total trunk length (tree height, from ground level to the top) and live crown length were also measured. The trunk was divided into 11 sections at points corresponding to 0, 0.05, 0.1, 0.2, 0.3, 0.4, 0.5, 0.6, 0.7, 0.8 and 0.9 of tree height. Base diameters of all sections were measured and the tree volume was computed using Smalian's formula [61], which referred to total volume over bark. Specifically, the formula was written as $V = (A_1 + A_2)/2 \times L$ with V as the volume of a section of tree trunk, A_1 and A_2 as two areas of the small and large ends of the section and L as the section length. The fresh weights of stem, branch and foliage were also measured; subsamples were selected and weighed in the field [54]. Among all sample trees, about one third (1626 trees) were selected for measuring both aboveground and belowground biomass. The whole roots were excavated out, fresh weights of stump, coarse roots (more than 10 mm) and small roots (2–10 mm, not including fine roots less than 2 mm) were measured, respectively and subsamples were selected. After being taken into the laboratory, all subsamples were oven-dried at 85°C until a constant weight was reached. According to the ratio of dry weight to fresh weight, each component biomass was computed and the aboveground biomass of the tree was obtained by summation [54]. **Table 1** shows the general situation for biomass samples of eight major tree species or groups.

2.2 Model construction

The general form of individual tree biomass and stem volume models is as follows [45, 62]:

$$y = \beta_0 x_1^{\beta_1} x_2^{\beta_2} \cdots x_j^{\beta_j} + \varepsilon \tag{1}$$

where y is biomass (kg), x_j are predictive biometric variables, which reflect the dimensions of a tree, such as diameter at breast height D (cm) and tree height H (m), β_j are parameters and ε is the error term. Because the biomass data are significantly heteroscedastic, some measures should be taken to eliminate heteroscedasticity prior to parameter estimation. In this paper, weighted regression was applied and the specific weight functions were derived from the

Species	Samples	Variables	Mean	Min	Max	S.D.	CV (%)
Picea spp.	900/295	Diameter D (cm)	17.0	1.0	65.5	12.8	75.6
		Height H (m)	12.3	1.4	46.9	8.1	66.4
		Stem volume V (dm^3)	343.0	0.6	6770.7	609.9	177.8
		Aboveground biomass M_a (kg)	174.5	0.4	1668.9	251.3	143.9
		Belowground biomass M_b (kg)	41.2	0.1	289.1	61.3	148.8
Abies spp.	751/249	Diameter D (cm)	17.1	1.1	68.0	13.0	76.6
		Height H (m)	11.9	1.5	39.0	7.4	62.7
		Stem volume V (dm^3)	352.4	0.5	4525.0	589.5	167.3
		Aboveground biomass M_a (kg)	168.9	0.3	1817.0	262.7	155.6
		Belowground biomass M_b (kg)	29.0	0.1	393.4	52.4	180.7
Betula spp.	690/236	Diameter D (cm)	15.9	1.0	60.8	11.8	73.7
		Height H (m)	11.3	1.9	33.0	6.2	55.1
		Stem volume V (dm^3)	235.0	0.3	2782.7	345.9	147.2
		Aboveground biomass M_a (kg)	167.4	0.2	1671.0	240.6	143.7
		Belowground biomass M_b (kg)	45.0	0.1	343.6	67.0	148.8
Quercus spp.	670/228	Diameter D (cm)	16.1	1.5	54.0	11.6	72.1
		Height H (m)	10.9	1.4	28.6	6.3	57.6
		Stem volume V (dm^3)	253.2	0.2	2487.1	370.9	146.5
		Aboveground biomass M_a (kg)	208.2	0.3	1664.1	295.2	141.8
		Belowground biomass M_b (kg)	51.4	0.1	385.9	71.6	139.4
Populus spp.	602/207	Diameter D (cm)	16.4	1.2	48.9	11.9	72.3
		Height H (m)	12.9	2.4	31.1	6.9	53.6
		Stem volume V (dm^3)	281.4	0.3	2228.4	385.3	136.9
		Aboveground biomass M_a (kg)	174.1	0.2	1065.1	241.3	138.6
		Belowground biomass M_b (kg)	35.6	0.1	384.5	54.3	152.7
Larix spp.	602/199	Diameter D (cm)	16.7	1.5	54.2	12.3	73.7
		Height H (m)	12.6	1.4	37.5	7.6	60.0
		Stem volume V (dm^3)	316.6	0.6	3016.6	471.7	149.0
		Aboveground biomass M_a (kg)	160.4	0.2	1301.9	231.1	144.1
		Belowground biomass M_b (kg)	41.0	0.1	300.0	61.8	150.9
Cunninghamia lanceolata	302/108	Diameter D (cm)	16.4	1.8	42.0	11.8	71.8
		Height H (m)	11.5	1.9	33.0	7.1	61.7
		Stem volume V (dm^3)	293.7	0.6	1815.2	409.7	139.5
		Aboveground biomass M_a (kg)	75.6	0.3	644.9	105.5	139.5
		Belowground biomass M_b (kg)	25.9	0.1	174.9	37.7	145.8

Species	Samples	Variables	Mean	Min	Max	S.D.	CV (%)
Pinus massoniana	301/104	Diameter D (cm)	16.5	1.2	47.2	11.9	72.4
		Height H (m)	12.1	1.6	30.3	7.2	59.4
		Stem volume V (dm^3)	300.8	0.3	1825.4	405.7	134.9
		Aboveground biomass M_a (kg)	125.1	0.1	1079.3	171.6	137.2
		Belowground biomass M_b (kg)	35.5	0.1	285.0	53.7	151.6

Min—minimum, Max—maximum, S.D.—standard deviation, and CV—coefficient of variation.
The sample sizes are for aboveground and belowground biomass mensuration, respectively.

Table 1. General situation of biomass samples for eight major tree species.

residuals of independently fitted models by ordinary least squares regression [62, 63]. Since models based on one (D) or two variables (D and H) have been commonly used, this paper develops both one- and two-variable models. The aboveground biomass, belowground biomass and stem volume models based on two variables can be expressed respectively as:

$$M_a = a_0 D^{a_1} H^{a_2} + \varepsilon \tag{2}$$

$$M_b = b_0 D^{b_1} H^{b_2} + \varepsilon \tag{3}$$

$$V = c_0 D^{c_1} H^{c_2} + \varepsilon \tag{4}$$

where M_a and M_b are aboveground and belowground biomass (kg), respectively; V is stem volume (dm^3); a_i, b_i and c_i are parameters; and other symbols are the same as above.

2.2.1 Integrated compatible model systems

The aboveground biomass is correlated to stem volume through biomass conversion factor (*BCF*), which is equal to biomass expansion factor (*BEF*) multiplied by basic wood density following the IPCC's approach [64]. Because the *BCF* is an important parameter for forest biomass estimation [65], it is very common to develop both an aboveground biomass model and a *BCF* model that are compatible with stem volume model [45, 51]. Similarly, belowground biomass is connected with aboveground biomass model through root-to-shoot ratio (*RSR*) [66, 67]. Because the *RSR* model is also an important parameter for forest biomass estimation, generally both belowground biomass model and *RSR* model compatible with aboveground biomass model are developed simultaneously [44]. Therefore, we can develop an integrated aboveground and belowground biomass model system through using the nonlinear error-in-variable simultaneous equation approach [51, 68]. Because the belowground biomass observations were only 1/3 of the aboveground biomass observations, a dummy variable (x) was required for those trees for which no belowground biomass observation was available, i.e., 1 for the trees with belowground biomass observation and 0 for the trees with no belowground biomass observation [69]. The system can ensure the compatibility between aboveground biomass, belowground biomass, stem volume, *BCF* and *RSR*. The one- and two-variable integrated systems are as follows, respectively:

$$\begin{cases} M_a = a_0 D^{a_1} + \varepsilon \\ M_b = b_0 D^{b_1} x + \varepsilon \\ V = c_0 D^{c_1} + \varepsilon \\ BCF = a_0 D^{a_1} / c_0 D^{c_1} + \varepsilon \\ RSR = b_0 D^{b_1} x / a_0 D^{a_1} + \varepsilon \end{cases} \quad (5)$$

$$\begin{cases} M_a = a_0 D^{a_1} H^{a_2} + \varepsilon \\ M_b = b_0 D^{b_1} H^{b_2} x + \varepsilon \\ V = c_0 D^{c_1} H^{c_2} + \varepsilon \\ BCF = a_0 D^{a_1} H^{a_2} / c_0 D^{c_1} H^{c_2} + \varepsilon \\ RSR = b_0 D^{b_1} H^{b_2} x / a_0 D^{a_1} H^{a_2} + \varepsilon \end{cases} \quad (6)$$

where, M_a, M_b, V, BCF and RSR are aboveground biomass, belowground biomass, stem volume, biomass conversion factor and root-to-shoot ratio, respectively, which are regarded as error-in-variables; D and H are diameter at breast height and tree height, which are regarded as error-free variables; x is a dummy variable to distinguish if belowground biomass is available; and a_i, b_i and c_i are parameters.

Various methods have been attempted to estimate the parameters of the simultaneous equations. Parresol [63] used the seemingly unrelated regression (SUR) for solving the additivity of simultaneous biomass equations. Tang et al. [70] further developed an error-in-variable modeling approach to estimate the parameters of simultaneous equations, which has been widely used in recent years [40, 45, 49, 51]. In this study, the error-in-variable simultaneous equation approach was used to estimate the parameters of the integrated systems based on maximum likelihood estimation through ForStat software (statistical software with analytical tools for forestry as well as general statistical procedures, developed in the Chinese Academy of Forestry, Beijing, China) [68].

In addition, the weighted regression method was used to eliminate the heteroscedasticity commonly exhibited in biomass and volume data by using specific weight functions, which were derived from the residuals of biomass or volume equations fitted through the ordinary least square (OLS) technique [52, 62]. For biomass conversion factor and root-to-shoot ratio modeling, the OLS regression technique was directly used to estimate the parameters because the *BCF* and *RSR* data mostly exhibited homoscedasticity.

2.2.2 Generalized dummy variable models

The one-variable biomass equation was the most widely used model in estimating individual tree biomass [3, 7]. The power function of one-variable aboveground biomass equation was based on the WBE theory for the origin of allometric scaling laws [71, 72]. According to the results from Zeng and Tang [73], the generalized one-variable aboveground biomass model can be expressed as:

$$M_a = aD^{7/3} + \varepsilon \quad (7)$$

That is, the power parameter of the allometric model is constantly equal to 7/3 (≈2.33), only the parameter a depends on tree species. If a variable vector $\boldsymbol{z}$ was defined as dummy variable to indicate tree species, then the generalized model (7) could be expressed as:

$$M_a = (a + v_a z)D^{7/3} + \varepsilon \tag{8}$$

where a is the global parameter and $\boldsymbol{v_a}$ is tree species-specific parameter vector. The dummy variable vector $\boldsymbol{z}$ includes seven elements, indicating the eight tree species by the following combinations:

$z_1 = 1$, $z_2 = 0$, $z_3 = 0$, $z_4 = 0$, $z_5 = 0$, $z_6 = 0$ and $z_7 = 0$ for *Picea* spp.

$z_1 = 0$, $z_2 = 1$, $z_3 = 0$, $z_4 = 0$, $z_5 = 0$, $z_6 = 0$ and $z_7 = 0$ for *Abies* spp.

$z_1 = 0$, $z_2 = 0$, $z_3 = 1$, $z_4 = 0$, $z_5 = 0$, $z_6 = 0$ and $z_7 = 0$ for *Betula* spp.

$z_1 = 0$, $z_2 = 0$, $z_3 = 0$, $z_4 = 1$, $z_5 = 0$, $z_6 = 0$ and $z_7 = 0$ for *Quercus* spp.

$z_1 = 0$, $z_2 = 0$, $z_3 = 0$, $z_4 = 0$, $z_5 = 1$, $z_6 = 0$ and $z_7 = 0$ for *Populus* spp.

$z_1 = 0$, $z_2 = 0$, $z_3 = 0$, $z_4 = 0$, $z_5 = 0$, $z_6 = 1$ and $z_7 = 0$ for *Larix* spp.

$z_1 = 0$, $z_2 = 0$, $z_3 = 0$, $z_4 = 0$, $z_5 = 0$, $z_6 = 0$ and $z_7 = 1$ for *C. lanceolata*

$z_1 = 0$, $z_2 = 0$, $z_3 = 0$, $z_4 = 0$, $z_5 = 0$, $z_6 = 0$ and $z_7 = 0$ for *P. massoniana*

Consequently, from comparing the estimated values of species-specific parameter vector $\boldsymbol{v_a}$, the differences among various tree species could be analyzed.

2.3 Model evaluation

Many statistical indices could be used to evaluate individual tree biomass models [63]. According to the study results from Zeng and Tang [53], the following six statistical indices, namely, the coefficient of determination (R^2), standard error of estimate (*SEE*), mean prediction error (*MPE*), total relative error (*TRE*), average systematic error (*ASE*) and mean percent standard error (*MPSE*), were very important for assessing biomass models. In this study, the same six statistical indices were used for model evaluation [50, 51]:

$$R^2 = 1 - \sum\left(y_i - \hat{y}_i\right)^2 / \sum\left(y_i - \bar{y}\right)^2 \tag{9}$$

$$SEE = \sqrt{\sum\left(y_i - \hat{y}_i\right)^2 / (n - p)} \tag{10}$$

$$TRE = \sum\left(y_i - \hat{y}_i\right) / \sum\hat{y}_i \times 100 \tag{11}$$

$$ASE = \sum\left(y_i - \hat{y}_i\right) / \hat{y}_i / n \times 100 \tag{12}$$

$$MPE = t_\alpha \cdot \left(SEE / \bar{y}\right) / \sqrt{n} \times 100 \tag{13}$$

$$MPSE = \sum\left|\left(y_i - \hat{y}_i\right) / \hat{y}_i\right| / n \times 100 \tag{14}$$

where y_i are observed values, $\hat{y}_i$ are estimated values, $\overline{y}$ is mean value of samples, n is the number of samples, p is the number of parameters and t_α is the t-value at confidence level α with n-p degrees of freedom.

3. Results and analysis

The one- and two-variable integrated systems (Eqs. (5) and (6)) for eight tree species or groups were estimated using the error-in-variable simultaneous equation approach through ForStat (**Tables 2** and **3**). The six fitting statistics, R^2, *SEE, TRE, ASE, MPE* and *MPSE*, were calculated and could be used for evaluating the goodness-of-fit of the three models (**Table 4**). From the

	Aboveground biomass models		Belowground biomass models		Stem volume models	
Species	a_0	a_1	b_0	b_1	c_0	c_1
Pi	0.17417	2.2270	0.04853	2.1954	0.1528	2.4548
Ab	0.10195	2.3676	0.02873	2.2452	0.1297	2.5106
Be	0.13392	2.3401	0.05767	2.2039	0.1712	2.3653
Qu	0.16592	2.3409	0.10619	2.0373	0.1448	2.4351
Po	0.09198	2.4490	0.02958	2.3200	0.1410	2.4702
La	0.12473	2.3190	0.03154	2.3355	0.1464	2.4737
Cl	0.09782	2.3099	0.02853	2.2500	0.1144	2.5421
Pm	0.13771	2.3243	0.01959	2.4400	0.1514	2.4655

Pi—*Picea* spp., Ab—*Abies* spp., Be—*Betula* spp., Qu—*Quercus* spp., Po—*Populus* spp., La—*Larix* spp., Cl—*Cunninghamia lanceolata,* and Pm—*Pinus massoniana.* Same in **Tables 3–7**.

Table 2. The parameter estimates of the one-variable integrated system (Eq. (5)).

	Aboveground biomass models			Belowground biomass models			Stem volume models		
Species	a_0	a_1	a_2	b_0	b_1	b_2	c_0	c_1	c_2
Pi	0.11007	2.1369	0.2615	0.03284	2.3516	−0.0527	0.07763	1.7758	1.0122
Ab	0.06720	2.0221	0.5442	0.02412	2.5974	−0.3600	0.07429	1.8135	0.9975
Be	0.08322	2.0749	0.4844	0.04531	2.1630	0.1401	0.08383	1.8246	0.8965
Qu	0.10520	1.9808	0.5939	0.09338	2.1694	−0.1091	0.07796	1.8607	0.9115
Po	0.06304	2.2460	0.3588	0.03216	2.5313	−0.2697	0.07611	1.9503	0.7927
La	0.07437	2.0003	0.5438	0.02195	2.2354	0.2369	0.07610	1.8067	0.9827
Cl	0.06740	1.9253	0.5765	0.02252	2.5080	−0.2072	0.07417	1.7949	1.0121
Pm	0.10462	2.1591	0.2857	0.01744	2.5697	−0.1028	0.09393	1.8696	0.8451

Table 3. The parameter estimates of the two-variable integrated system (Eq. (6)).

Species	Systems	Items	R^2	*SEE*	*MPE* (%)	*TRE* (%)	*ASE* (%)	*MPSE* (%)
Pi	(5)	AB	0.9109	75.05	2.81	1.31	−2.44	24.21
		BB	0.7842	28.55	5.51	−0.25	−3.75	40.29
		SV	0.8380	245.64	4.68	2.63	0.19	27.65
	(6)	AB	0.9061	77.10	2.89	0.37	7.68	25.26
		BB	0.7751	29.19	5.63	0.65	9.28	43.70
		SV	0.9744	97.70	1.86	0.44	6.01	15.87
Ab	(5)	AB	0.9223	73.29	3.10	0.28	−1.24	22.19
		BB	0.5474	35.29	11.97	−1.15	−2.74	49.06
		SV	0.9222	164.52	3.34	−0.60	−0.29	20.79
	(6)	AB	0.9434	62.61	2.65	0.24	4.97	22.50
		BB	0.5547	35.08	11.89	0.37	7.64	52.51
		SV	0.9800	83.41	1.69	0.25	4.34	13.05
Be	(5)	AB	0.9139	70.63	3.15	2.02	0.11	22.52
		BB	0.7734	31.95	6.16	2.47	−1.15	37.64
		SV	0.9118	102.74	3.26	2.87	7.19	26.85
	(6)	AB	0.9332	62.27	2.78	1.05	2.47	21.30
		BB	0.7741	31.97	6.17	0.18	1.16	38.38
		SV	0.9566	72.13	2.29	0.49	6.53	19.60
Qu	(5)	AB	0.9030	91.99	3.35	2.67	0.93	27.82
		BB	0.8168	30.73	5.23	2.21	−1.83	40.37
		SV	0.9262	100.81	3.01	4.11	4.36	27.33
	(6)	AB	0.9285	79.04	2.87	1.67	2.81	24.66
		BB	0.8133	31.09	5.29	1.11	2.35	41.08
		SV	0.9790	53.88	1.61	1.07	3.71	16.72
Po	(5)	AB	0.9379	60.17	2.76	1.51	−0.86	17.55
		BB	0.8440	21.51	5.47	2.21	−2.71	32.15
		SV	0.9539	82.81	2.35	−0.41	1.51	15.87
	(6)	AB	0.9506	53.72	2.47	1.17	1.25	17.11
		BB	0.8618	20.29	5.15	1.77	1.15	32.11
		SV	0.9842	48.49	1.38	0.83	1.04	9.48
La	(5)	AB	0.9123	68.47	3.41	1.26	1.15	23.18
		BB	0.6773	35.22	8.65	−0.52	0.14	37.37
		SV	0.9016	148.10	3.74	0.10	4.26	25.04
	(6)	AB	0.9432	55.14	2.75	0.22	6.91	24.68
		BB	0.6675	35.83	8.81	−0.83	6.78	40.08
		SV	0.9770	71.63	1.81	0.21	5.58	15.05

Species	Systems	Items	R^2	*SEE*	*MPE* (%)	*TRE* (%)	*ASE* (%)	*MPSE* (%)
Cl	(5)	AB	0.9614	30.62	2.98	3.24	1.03	22.99
		BB	0.8414	15.08	7.43	1.66	1.75	38.39
		SV	0.9474	94.14	3.62	1.99	2.28	17.25
	(6)	AB	0.9774	23.47	2.28	1.47	6.10	23.66
		BB	0.8342	15.50	7.64	−0.88	9.87	41.43
		SV	0.9931	34.10	1.31	0.50	5.21	11.30
Pm	(5)	AB	0.9542	48.45	3.21	0.91	0.25	17.29
		BB	0.8509	20.85	7.64	1.55	−2.12	39.21
		SV	0.9503	90.62	3.40	0.17	0.34	20.87
	(6)	AB	0.9572	46.89	3.10	0.41	3.76	16.30
		BB	0.8546	20.69	7.59	−0.63	1.44	39.73
		SV	0.9846	50.59	1.90	0.64	1.13	12.51

AB—aboveground biomass, BB—belowground biomass, SV—stem volume, R^2—coefficient of determination, *SEE*—standard error of estimate, *MPE*—mean prediction error, *TRE*—total relative error, *ASE*—average systematic error, and *MPSE*—mean percent standard error.
Units of *SEE*: dm^3 for volume and kg for biomass.

Table 4. The fitting statistics of two integrated systems (Eqs. (5) and (6)).

fitting results of integrated systems (Eqs. (5) and (6)), the parameter estimates of the *BCF* and *RSR* models could be obtained (**Table 5**).

From comparison of the fitting statistics of two integrated systems (Eqs. (5) and (6)) in **Table 4**, we can found that for aboveground biomass estimation, two-variable models were better than one-variable models except *Picea*. For belowground biomass estimation, one- and two-variable models were not significantly different, even some of one-variable models were slightly better than two-variable models, such as *Picea, Quercus, Larix* and *C. lanceolata*. Considering that tree height measurement is time consuming and two-variable biomass models are not significantly different from one-variable models, especially for belowground biomass estimation, it was commended to apply one-variable models in forestry practice such as National Forest Inventory.

From **Table 2**, it was found that the estimates of parameter a_1 were approximately equal to 7/3, confirming the results of an earlier study [73]. To analyze the difference among various tree species, the dummy model (8) was fitted using the aboveground biomass data of all eight species (**Table 6**).

According to the parameter estimates in **Table 6**, we could rank the eight tree species by aboveground biomass estimates in descending order as *Quercus, Betula, Populus, P. massoniana, Picea, Larix, Abies* and *C. lanceolata*. That is, *Quercus* had the largest aboveground biomass, whereas *C. lanceolata* had the smallest one for the same diameter trees. The aboveground biomass estimates of the dummy model (Eq. (8)) for *Quercus, Betula, Populus, P. massoniana,*

Species	Systems	*BCF* models	*RSR* models
Pi	(5)	$BCF = 1.1401\ D^{-0.2278}$	$RSR = 0.2786\ D^{-0.0316}$
	(6)	$BCF = 1.4183\ D^{0.3611}\ H^{-0.7507}$	$RSR = 0.2983\ D^{0.2147}\ H^{-0.3142}$
Ab	(5)	$BCF = 0.7858\ D^{-0.1430}$	$RSR = 0.2818\ D^{-0.1224}$
	(6)	$BCF = 0.9046\ D^{0.2086}\ H^{-0.4533}$	$RSR = 0.3589\ D^{0.5753}\ H^{-0.9041}$
Be	(5)	$BCF = 0.7821\ D^{-0.0252}$	$RSR = 0.4307\ D^{-0.1362}$
	(6)	$BCF = 0.9928\ D^{0.2504}\ H^{-0.4121}$	$RSR = 0.5445\ D^{0.0880}\ H^{-0.3443}$
Qu	(5)	$BCF = 1.1456\ D^{-0.0942}$	$RSR = 0.6400\ D^{-0.3036}$
	(6)	$BCF = 1.3494\ D^{0.1201}\ H^{-0.3177}$	$RSR = 0.8877\ D^{0.1887}\ H^{-0.7030}$
Po	(5)	$BCF = 0.6522\ D^{-0.0212}$	$RSR = 0.3216\ D^{-0.1290}$
	(6)	$BCF = 0.8283\ D^{0.2958}\ H^{-0.4339}$	$RSR = 0.5102\ D^{0.2853}\ H^{-0.6285}$
La	(5)	$BCF = 0.8522\ D^{-0.1547}$	$RSR = 0.2528\ D^{0.0165}$
	(6)	$BCF = 0.9773\ D^{0.1936}\ H^{-0.4389}$	$RSR = 0.2951\ D^{0.2351}\ H^{-0.3068}$
Cl	(5)	$BCF = 0.8554\ D^{-0.2321}$	$RSR = 0.2917\ D^{-0.0599}$
	(6)	$BCF = 0.9087\ D^{0.1303}\ H^{-0.4356}$	$RSR = 0.3341\ D^{0.5828}\ H^{-0.7836}$
Pm	(5)	$BCF = 0.9096\ D^{-0.1412}$	$RSR = 0.1422\ D^{0.1157}$
	(6)	$BCF = 1.1138\ D^{0.2894}\ H^{-0.5593}$	$RSR = 0.1667\ D^{0.4106}\ H^{-0.3885}$

Table 5. The simultaneously estimated *BCF* and *RSR* models.

Species	Global parameter (a)	Species-specific parameters (v_a)
Pi	0.13485	−0.01084
Ab		−0.01959
Be		0.00443
Qu		0.03908
Po		0.00106
La		−0.01441
Cl		−0.04254
Pm		0.00000

Table 6. The parameter estimates of dummy aboveground biomass model (Eq. (8)).

Picea, Larix and *Abies* were 88%, 51%, 47%, 46%, 34%, 30% and 25% larger, respectively, than that for *C. lanceolata* (see **Figure 1**).

Similarly, for one-variable belowground biomass models, it was found that the estimates of parameter b_1 for eight species were not significantly different. To analyze the difference of belowground biomass estimation among various tree species, we fitted the following dummy model:

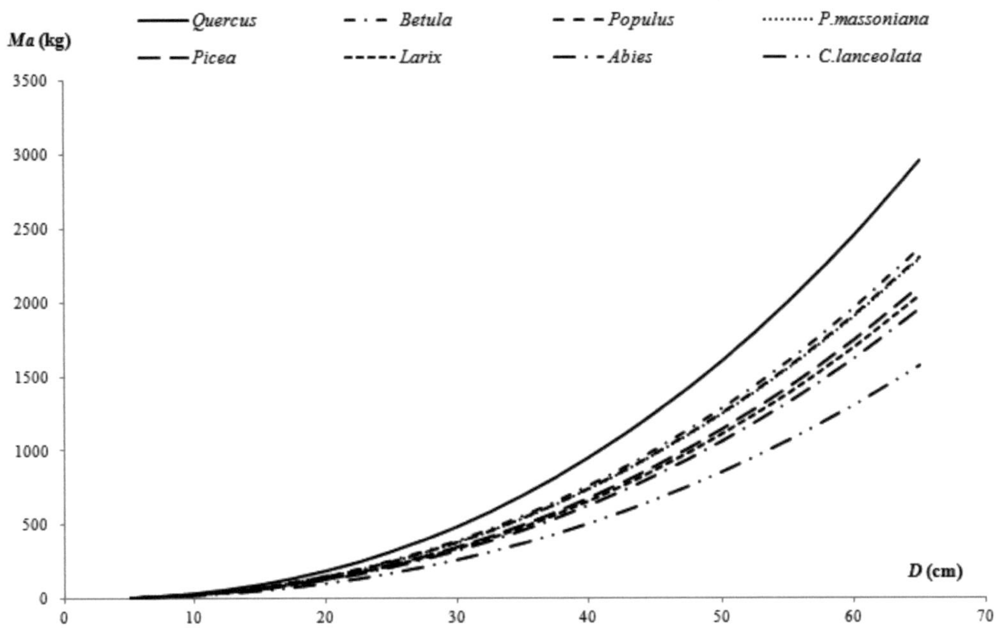

Figure 1. Comparison of aboveground biomass models for eight tree species.

	Global parameters		
Species	b_0	b_1	**Species-specific parameters (v_b)**
Pi	0.03551	2.2544	0.00424
Ab			−0.00792
Be			0.01437
Qu			0.01835
Po			0.00091
La			0.00583
Cl			−0.00761
Pm			0.00000

Table 7. The parameter estimates of dummy belowground biomass model (Eq. (15)).

$$M_b = (b_0 + v_b z)D^{b_1} + \varepsilon \tag{15}$$

where b_0 and b_1 are global parameters and $\boldsymbol{v_b}$ is species-specific parameter vector. The parameter estimates of dummy model (Eq. (15)) are listed in **Table 7**.

According to the parameter estimates in **Table 7**, we could rank the eight tree species by belowground biomass estimates in descending order as *Quercus, Betula, Larix, Picea, Populus, P.*

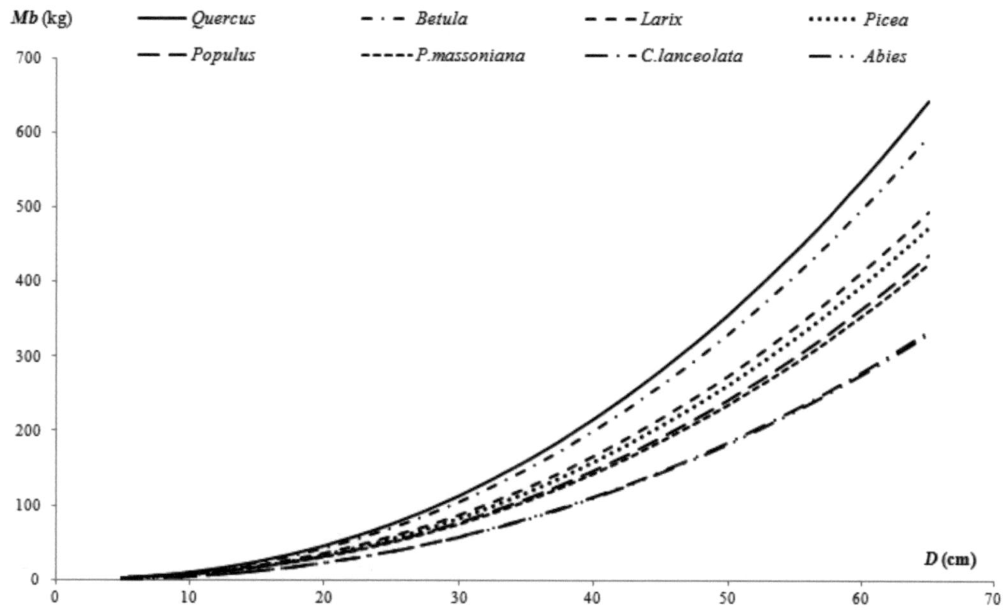

Figure 2. Comparison of belowground biomass models for eight tree species.

massoniana, C. lanceolata and *Abies*. That is, *Quercus* had the largest belowground biomass, while *Abies* had the smallest one for the same diameter trees. The belowground biomass estimates of the dummy model (Eq. (15)) for *Quercus, Betula, Larix, Picea, Populus, P. massoniana* and *C. lanceolata* were 95%, 81%, 50%, 44%, 32%, 29% and 1% larger, respectively, than that for *Abies* (see **Figure 2**).

4. Discussion and conclusion

In this study, data on above- and belowground biomass from 4818 to 1626 sample trees, respectively, for eight major tree species in China were used to develop compatible individual tree biomass models. The models included aboveground biomass equations and *BCF* equations compatible with stem volume equations and belowground biomass equations and *RSR* models compatible with aboveground biomass equations. To solve compatibility of the biomass models, the nonlinear error-in-variable simultaneous equations were applied and to solve the issue of unequal sample sizes for above- and belowground biomass, the dummy-variable model approach was used. In the technical regulation on methodology for tree biomass modeling [55], the segmented modeling approach was recommended when the biomass estimate of small trees was obviously biased [43, 46]. Furthermore, for the tree species distributed in various regions, it was generally needed to develop biomass models for different regions. For example, according to the population classification on modeling of single-tree biomass equations [39], it was necessary to establish five sets of biomass models for both *Abies* and *Picea*. But in this study, the segmented modeling approach was not used to develop biomass models for large and small trees, respectively and the differences among various regions were not taken into account, only one set of biomass models, including one- and two-variable models, was developed for each tree species.

The data of three tree species, i.e., *C. lanceolata, P. massoniana* and *Larix* spp., were used or partly used to develop biomass models, which were published as original papers [40–51] or ministerial standards [56, 57]. Comparing with the study results by Zeng et al. [47], the parameter estimates and fitness indices of aboveground biomass and volume models are very close to those for *C. lanceolata* in this study. From the achievements by Zeng and Tang [45], we can find that the parameter estimates of aboveground biomass and volume models are not significantly different from those for *P. massoniana* in this chapter, but this study provided better models considering the statistical indices of goodness-of-fit. Comparing with the biomass models published as ministerial standards [56, 57], the developed models in this study are more generalized and simpler for application in national and regional biomass estimation. There are four sets of biomass models in total for trees (dbh $\geq$ 5 cm) and saplings (dbh $<$ 5 cm) for two modelling populations of each tree species in the ministerial standards [56, 57] and here we have only one set of biomass models which are suitable for both trees and saplings and for the whole country.

The results indicated that two-variable models were almost better than one-variable models for aboveground biomass estimation, while the two model systems were not significantly different for belowground biomass estimation. The mean prediction errors (*MPEs*) of aboveground biomass models for the eight species were less than 5%, whereas *MPEs* of belowground biomass equations were less than 10%, except for *Abies*. The models developed in this study can provide a basis for estimating biomass for the eight major tree species in China and will fill in the lack for China on the web platform GlobAllomeTree [74]. Also, they will have the potential to support the implementation of policies and mechanisms designed to mitigate climate change (e.g., CDM and REDD+) and to calculate costs and benefits associated with forest carbon projects. In addition, the overall modeling methodology presented in this study can be taken into consideration in any case that involves individual tree biomass modeling.

Acknowledgements

The author acknowledges the National Biomass Modeling Program in Continuous Forest Inventory (NBMP-CFI), which was funded by the State Forestry Administration of China, for providing the mensuration biomass data of eight tree species. This study was financially supported by the National Natural Science Foundation of China (Grant No. 31370634).

Author details

WeiSheng Zeng

Address all correspondence to: zengweisheng@forestry.gov.cn

Academy of Forest Inventory and Planning, State Forestry Administration, Beijing, China

References

[1] IPCC. Good practice guidance for land use, land use change and forestry. The Institute for Global Environmental Strategies for the IPCC, Japan; 2003

[2] Ebermeyr E. A comprehensive study of forest litter with regard to silvicultural effects on soil chemistry. Springer, Berlin; 1876

[3] Jenkins JC, Chojnacky DC, Heath LS, Birdsey RA. National-scale biomass estimators for United States tree species. For Sci. 2003; 49(1):12–35

[4] Bi H, Turner J, Lambert MJ. Additive biomass equations for native eucalypt forest trees of temperate Australia. Trees. 2004; 18:467–479

[5] Lambert MC, Ung CH, Raulier F. Canadian national tree aboveground biomass equations. Can J For Res. 2005; 35:1996–2018

[6] Snorrason A, Einarsson SF. Single-tree biomass and stem volume functions for eleven tree species used in Icelandic forestry. Icelandic Agric Sci. 2006; 19:15–24

[7] Muukkonen P. Generalized allometric volume and biomass equations for some tree species in Europe. Eur J For Res. 2007; 126:157–166

[8] Návar J. Allometric equations for tree species and carbon stocks for forests of northwestern Mexico. For Ecol Manage. 2009; 257: 427–434

[9] Blujdea VNB, Pilli R, Dutca I, Ciuvat L, Abrudan IV. Allometric biomass equations for young broadleaved trees in plantations in Romania. For Ecol Manage. 2012; 264:172–184

[10] Fayolle A, Doucet JL, Gillet JF, Bourland N, Lejeune P. Tree allometry in central Africa: testing the validity of pantropical multi-species allometric equations for estimating biomass and carbon stocks. For Ecol Manage. 2013; 305:29–37

[11] Sileshi GW. A critical review of forest biomass estimation models, common mistakes and corrective measures. For Ecol Manage. 2014; 3: 237–254

[12] Li WH. Concept of forest biomass productivity and its basic studying approach. Nat Res. 1978; 1:71–92

[13] Pan WC, Li LC, Gao ZH. Biomass and nutrient elements distribution of two different forest types of Chinese fir. Hunan For Sci Technol. 1980; 4:1–14

[14] Zeng WS. Development of monitoring and assessment of forest biomass and carbon storage in China. For Ecosyst. 2014; 1:20. DOI: 10.1186/s40663-014-0020-5

[15] Li WH, Deng KM, Li F. Research on biomass productivity of major ecosystems in the Changbai Mountain. For Ecosyst Res. 1981; (Test issue):34–50

[16] Feng ZW, Chen CY, Zhang JW, Wang KP, Zhao JL, Gao H. Determination of biomass of *Pinus massoniana* stand in Huitong county, Hunan province. Sci Silv Sin. 1982; 18(2):127–134

[17] Ye JZ. Annual dynamic of the biomass of Chinese fir forests on Yangkou forestry farm, J Nanjing For Univ. 1984; 4:1–9

[18] Chen LZ, Chen QL, Bao XC, Ren JK, Miu YG, Hu YH. Studies on Chinese arborvitae (*Platycladus orientalis*) forest and its biomass in Beijing. J Plant Ecol. 1986; 10(1):17–25

[19] Xu ZB. Biomass productivity of major forest types in Daxinganling. Chin J Ecol. 1988; 7 (Sp):49–54

[20] Ma QY. A study on the biomass of Chinese pine forests. J Beijing For Univ. 1989; 11(4):1–10

[21] Chen CG, Zhu JF. Manual of tree biomass for main species in northern China. Chinese Forestry Press, Beijing; 1989.

[22] Liu SR, Chai YX, Cai TJ, Peng CH. Research on biomass and net primary productivity of *Larix gmelini* plantations. J Northeast For Univ. 1990; 18(2): 40–46

[23] Liu ZG. Research on biomass and productivity of *Larix principis-rupprechtii* plantations. J Beijing For Univ. 1992; 14(Sp.1):114–123

[24] Liu XZ. Study on biomass of Masson pine forests at different ages. For Res Manage. 1993; 2: 77–80

[25] Fang JY, Liu GH, Xu SL. Biomass and net production of forest vegetation in China. Acta Ecol Sin. 1996; 16(4): 497–508

[26] Luo TX. Patterns of net primary productivity for Chinese major forest types and their mathematical models. Dissertation, Commission for Integrated Survey of Natural Resources, the Chinese Academy of Sciences and State Planning Commission, Beijing; 1996.

[27] Tian DL, Pan HH, Kang WX, Fang HB Study on biomass of second generation of Chinese fir plantations. J Cent S Univ For Technol. 1998; 18(3):14–19

[28] Feng ZW, Wang XK, Wu G. The biomass and productivity of forest ecosystem in China. Science Press, Beijing; 1999

[29] Zeng WS, Luo QB, He DB. Study on compatible nonlinear tree biomass models. Chin J Ecol. 1999; 18(4):19–24

[30] Tang SZ, Zhang HR, Xu H. Study on establishing and estimating method of compatible biomass model. Sci Silv Sin. 2000; 36(Sp.1):19–27

[31] Liu GH, Fu BJ, Fang JY. Carbon dynamics of Chinese forests and its contribution to global carbon balance. Acta Ecol Sin. 2000; 20(5):733–740

[32] Zhou YR, Yu ZL, Zhao SD. Carbon storage and budget of major Chinese forest types. Acta Phytoecologica Sinica. 2000; 24(5):518–522

[33] Fang JY, Chen AP, Peng CH, Zhao SQ, Ci LJ. Changes in forest biomass carbon storage in China between 1949 and 1998. Science. 2001; 292:2320–2322

[34] Xu B, Guo ZD, Piao SL, Fang JY. Biomass carbon stocks in China's forests between 2000 and 2050: a prediction based on forest biomass–age relationships. Sci China Life Sci. 2010; 53(7):776–783. DOI: 10.1007/s11427-010-4030-4

[35] Li HK, Lei YC. Estimation and evaluation of forest biomass carbon storage in China. Chinese Forestry Press, Beijing; 2010, 60 p.

[36] Li HK, Lei YC, Zeng WS. Forest carbon storage in China estimated using forest inventory data. Sci Silv Sin. 2011; 47(7):7–12

[37] Li HK, Zhao PX, Lei YC, Zeng WS. Comparison on estimation of wood biomass using forest inventory data. Sci Silv Sin. 2012; 48(5):44–52

[38] Zheng DX, Liao XL, Li CW, Ye QL, Chen PL. Estimation and dynamic change analysis of forest carbon storage in Fujian province. Acta Agric Univ Jiangxiensis. 2013; 35(1):112–116

[39] Zeng WS, Tang SZ, Huang GS, Zhang M. Population classification and sample structure on modeling of single-tree biomass equations for national biomass estimation in China. For Res Manage. 2010; 3:16–23

[40] Zeng WS, Tang SZ. Using measurement error modeling method to establish compatible single-tree biomass equations system. For Res. 2010; 23(6):797–803

[41] Zeng WS, Zhang HR, Tang SZ. Using the dummy variable model approach to construct compatible single-tree biomass equations at different scales—a case study for Masson pine (*Pinus massoniana*) in southern China. Can J For Res. 2011; 41(7): 1547–1554. DOI: 10.1139/X11-068

[42] Zeng WS. Methodology on modeling of single-tree biomass equations for national biomass estimation in China [thesis]. Chinese Academy of Forestry, Beijing; 2011.

[43] Zhang LJ, Zeng WS, Tang SZ. Comparison of nonlinear regression equation with intercept and segmented modeling approach for estimation of single tree biomass. For Res. 2011; 24(4): 453–457

[44] Zeng WS, Tang SZ. Establishment of belowground biomass equations for larch in northeastern and Masson pine in southern China. J Beijing For Univ. 2011; 33(2):1–6

[45] Zeng WS, Tang SZ. Modeling compatible single-tree biomass equations of Masson pine (*Pinus massoniana*) in southern China. J For Res. 2012; 23(4): 593–598. DOI: 10.1007/s11676-012-0299-4

[46] Dang YF, Wang XJ, Zeng WS. Using segmented modeling approach to construct tree volume and biomass equations for larch in northeastern China. For Res. 2012; 25(5): 558–563

[47] Zeng M, Nie XY, Zeng WS. Compatible tree volume and aboveground biomass equations of Chinese fir in China. Sci Silv Sin. 2013; 49(10):74–79. DOI: 10.11707/j.1001-7488.20131012

[48] Fu LY, Zeng WS, Zhang HR, Wang GX, Lei YC, Tang SZ. Generic linear mixed-effects individual-tree biomass models for *Pinus massoniana* Lamb. in southern China. South For. 2014; 76(1):47–56. DOI: 10.2989/20702620.2013.870389

[49] Zou WT, Zeng WS, Zhang LJ, Zeng M. Modeling crown biomass for four pine species in China. Forests. 2015; 6(2):433–449. DOI: 10.3390/f6020433

[50] Zeng WS. Using nonlinear mixed model and dummy variable model approaches to construct origin-based single tree biomass equations. Trees Struct Funct. 2015; 29(1): 275–283. DOI: 10.1007/s00468-014-1112-0

[51] Zeng WS. Integrated individual tree biomass simultaneous equations for two larch species in northeastern and northern China. Scand J For Res. 2015; 30(7): 594–604. DOI: 10.1080/02827581.2015.1046481

[52] Zeng WS, Tang SZ. Bias correction in logarithmic regression and comparison with weighted regression for non-linear models. Nat Prec. 2011; Available from http://dx.doi.org/10.1038/npre.2011.6708.1

[53] Zeng WS, Tang SZ. Goodness evaluation and precision analysis of tree biomass equations. Sci Silvae Sin. 2011; 47(11): 106–113

[54] State Forestry Administration of China. Technical regulation on sample collections for tree biomass modeling. China Standard Press, Beijing; 2015; 11 p.

[55] State Forestry Administration of China. Technical regulation on methodology for tree biomass modeling. China Standard Press, Beijing; 2015; 7 p.

[56] State Forestry Administration of China. Tree biomass models and related parameters to carbon accounting for *Cunninghamia lanceolata*. China Standard Press, Beijing; 2015; 14 p.

[57] State Forestry Administration of China. Tree biomass models and related parameters to carbon accounting for *Pinus massoniana*. China Standard Press, Beijing; 2015; 14 p.

[58] State Forestry Administration of China. Tree biomass models and related parameters to carbon accounting for *Pinus yunnanensis*. China Standard Press, Beijing; 2015; 9 p.

[59] State Forestry Administration of China. Tree biomass models and related parameters to carbon accounting for *Pinus tabulaeformis*. China Standard Press, Beijing; 2015; 9 p.

[60] State Forestry Administration of China. Tree biomass models and related parameters to carbon accounting for *Pinus elliottii*. China Standard Press, Beijing; 2015; 9 p.

[61] Jayaraman K. A statistical manual for forestry research. FAO Regional Office for Asia and the Pacific, Bangkok, 1999; 231 p.

[62] Parresol BR. Assessing tree and stand biomass: a review with examples and critical comparisons. For Sci. 1999; 45: 573–593

[63] Parresol BR. Additivity of nonlinear biomass equations. Can J For Res. 2001; 31: 865–878

[64] IPCC. IPCC guidelines for national greenhouse gas inventories—agriculture, forestry and other land use (Volume 4). Published: IGES, Japan, 2006.

[65] Pajtík J, Konôpka B, Lukac M. Biomass functions and expansion factors in young Norway spruce (*Picea abies* L. Karst) trees. For Ecol Manage. 2008; 256: 1096–1103

[66] Wang XP, Fang JY, Zhu BA. Forest biomass and root–shoot allocation in northeast China. For Ecol Manage. 2008; 255: 4007–4020

[67] Mugasha WA, Eid T, Bollandsas OM, Malimbwi RE, Chamshama SAO, Zahabu E, Katani JZ. Allometric models for prediction of above- and belowground biomass of trees in the miombo woodlands of Tanzania. For Ecol Manage. 2013; 310: 87–101

[68] Tang SZ, Lang KJ, Li HK. Statistics and computation of biomathematical models. Science Press, Beijing; 2008. 584 p.

[69] Crecente-Campo F, Soares P, Tomé M, Diéguez-Aranda U. Modelling annual individual-tree growth and mortality of Scots pine with data obtained at irregular measurement intervals and containing missing observations. For Ecol Manage. 2010; 260: 1965–1974

[70] Tang SZ, Li Y, Wang YH. Simultaneous equations, error-in-variable models and model integration in systems ecology. Ecol Model. 2001; 142: 285–294

[71] West GB, Brown JH, Enquist BJ. A general model for the origin of allometric scaling laws in biology. Science. 1997; 276(5309): 122–126

[72] West GB, Brown JH, Enquist BJ. A general model for the structure and allometry of plant vascular systems. Nature. 1999; 400: 664–667

[73] Zeng WS, Tang SZ. A new general allometric biomass model. Nat Prec. 2011; Available from http://dx.doi.org/10.1038/npre.2011.6704.2

[74] Henry M, Bombelli A, Trotta C, et al. GlobAllomeTree: international platform for tree allometric equations to support volume, biomass and carbon assessment. iForest (early view): e1–e5 [Internet]. 2013. Available from: http://www.sisef.it/iforest/contents/?id=ifor0901-006 [Assessed 2016-09-04]

Chapter 2

Methods of Estimating Forest Biomass: A Review

Lei Shi and Shirong Liu

Additional information is available at the end of the chapter

http://dx.doi.org/10.5772/65733

Abstract

Forest plays a special role in carbon sequestration and thus mitigating climate change. However, the large uncertainty in biomass estimation is unable to meet the requirement of the accurate carbon accounting. The use of a suitable and rigor method to accurately estimate forest biomass is significant. Moreover, the world is increasingly facing the conflicting pressures of economic growth and environmental protection. Improving energy structure and vigorously developing biomass energy has become the development trend of energy utilization in the future. As energy plant is characterized by a large net accumulation of biomass. Therefore, the scientific evaluation of the size and potential of energy from plant also requires a suitable method for estimating biomass. Here, we reviewed the estimate methods, including allometric equation, mean biomass density, biomass expansion factor, geostatistics, etc. For each method, we will present background, rational, applicability, as well as estimation procedure by exemplifying a case. In this chapter, we argued that the new developed technique such as geo-statistics and remote sensing technique (e.g. LIDAR) would be the key tools to improve forest biomass estimation accuracy. However, prior to this, spatial variation of forest biomass at various levels should be explored using multi-source data and multi-approaches.

Keywords: carbon accounting, climate change, field survey, geostatistics, remote sensing technique, scale, uncertainty

1. Introduction

Currently, CO_2 and other greenhouse gas are inducing global warming, and vegetation is the only natural ecosystems to fix atmospheric CO_2. Forest is the main component of vegetation.

Accordingly, forest ecosystem is destined to be paid more attention by governments, academics, and the general public [1]. According to the Global Forest Resources Assessment 2010 [2], the global forest biomass (including above- and belowground) is 600 Pg, with a mean biomass density of 148.8 t/ha (**Table 1**). It is estimated that carbon sequestered in forest can account for about 77% of terrestrial ecosystem [3].

Region	**Biomass (×10^6^ t)**	**Biomass density (t/ha)**
Eastern and Southern Africa	33,385	124.8
Northern Africa	3711	47.1
Western and Central Africa	816.3	248.7
Total Africa	**118,700**	**176.0**
East Asia	18,429	72.4
South and Southeast Asia	51,933	176.4
Western and Central Asia	3502	80.5
Total Asia	**73,864**	**124.7**
Europe excl. Russian Federation	25,602	130.7
Total Europe	**90,602**	**90.2**
Caribbean	1092	157.5
Central America	3715	190.5
North America	76,929	113.3
Total North and Central America	**81,736**	**115.9**
Total Oceania	**21,302**	**111.3**
Total South America	**213,863**	**247.4**
World	**600,066**	**148.8**

Table 1. Forest biomass and its density by region, 2010 [2].

Global deforestation is undergoing seriously [2], which contributes to a quarter of carbon released into the atmosphere each year [4]. Land use/land use change (mainly deforestation) is considered to be an important approach to the release of CO_2, which affects the carbon cycle on various spatial and temporal scales, and then global climate change [5, 6]. Therefore, the scientific and real-time monitoring of forest cover change and more accurate estimates of forest biomass and its magnitude is of significance to clarify the contribution of forests in global climate change.

In addition, the current world is facing the dual pressures of economic growth and environmental protection. Adjusting and optimizing energy structure, vigorously developing biomass energy has become the main developing trend of energy in the future. As energy plant, the most important characteristic is to possess a large net accumulation of biomass. Therefore, the scientific evaluation of the size and potential of biomass needs a suitable method used to estimate its biomass potential.

However, a large uncertainty exists in biomass estimation, which is unable to meet the requirement of the accurate carbon accounting required by Kyoto Protocol. The use of a suitable and rigor method to accurately estimate the size and distribution of forest biomass is of significance and also urgently needed. And also, the method used to estimate forest biomass is more likely to vary frequently with scale.

Given this, we reviewed the commonly used methods to estimate forest biomass across the scale in this chapter, for the purpose of operation and guidance, which includes allometric equation, mean biomass density, biomass expansion factor, forest identity, remote sensing- and geostatistics-based estimation methods, etc. For each method, we will present background, rational, applicability, as well as estimation procedure by exemplifying a case. At the end of this chapter, we argued that the new developed techniques such as geostatistics and remote sensing technique (e.g., LIDAR) would be the key tools to improve forest biomass estimation with a high accuracy. However, prior to this, spatial variation in forest biomass at various levels should be first explored using multi-source data and multi-approaches.

2. Allometry and allometric equation

If one organ is correlated to another of a plant, or a certain attribute is to plant size, we can call it allometry [7, 8], which is frequently expressed with a power relationship below [9, 10]:

$$y = ax^b \tag{1}$$

or

$$lgy = lga + blgx \tag{2}$$

where y often represents an attribute of plant (such as metabolic rate, biomass, etc.), x shows the size of the plant body (such as diameter at breast height and/or height), and a and b are coefficients.

In botany, the allometric relationship is able to be used to calculate biomass and other ecological factors by measuring the easily measured diameter at breast height (and/or height). Theoretically, tree D (diameter at breast height) and H (height) can both affect tree biomass. Thus, tree biomass can be estimated by allometric equation (Eqs. (3) and (4)), which includes both D and H [11–13]:

$$w = aD^b H^c \; or \lg w = a' + b\lg D + c\lg H \tag{3}$$

$$w = a(D^2 H)^b \; or \lg w = a' + b\lg(D^2 H) \tag{4}$$

where *D* and *H* represent tree diameter at breast height (cm) and height (m) and *a*, *b*, *a'*, and *c* are regression coefficients.

Tree *H* is not easy to measure in field survey, so many researchers have used *H-D* model to estimate the *H* through easily measured *D* and then to estimate vegetation biomass [14–16]. In contrast to *H*, it is easier to measure *D*, and the measurement error is relatively small while measuring *D* [17, 18]. Furthermore, it is very common and efficient to use allometric relationship in scientific literatures, which includes *D* only, while estimating biomass [19, 20]. The allometric equation including *D* only can be reused by others and also make comparable among regions. The study performed by Wang [12] has also indicated that the including of *H* variable (i.e., including both *D* and *H*) in allometric relationship was unable to improve biomass estimate significantly (increased determination coefficient less than 4%). Therefore, many scholars contain *D* only as independent variable while fitting biomass allometric relationship (i.e., allometric equations (1) and (2) above) [12, 21, 22].

Such allometric relationship is based on the measured sample tree and aims to estimate vegetation biomass as the mathematical model (hereafter also referred to as a biomass allometric equation). Apparently, plant allometry is the theoretical basis of vegetation biomass estimation, which makes biomass estimate possible. Recently, remote sensing technique has been increasingly applied to estimate the biomass [23, 24]. However, data derived from allometric equation must be verified with field data in the method [25]. Generally, the use of allometric equation is indispensable to estimate biomass for both tree and forest.

3. Procedure of estimating multi-scale forest biomass

Multi-scale aboveground biomass estimation is demonstrated as an example to show the procedure (**Figure 1**). First, a number of plots are set, where field survey is performed (step 1); then several sample trees are cut to fit individual-level allometric equation (step 2); the use of developed allometric equation, together with filed survey data (mainly *D*), estimates biomass for each tree in plot and sums as stand-level biomass (step 3); finally, such upscaling methods as the mean biomass density, geostatistical technique, and others are used to upscale the regional forest biomass (step 4).

While estimating forest carbon stock, most scholars assumed that carbon content in plant biomass is constant (approximately 50%) [27–29]. Therefore, we estimate forest biomass first, multiplied by 50%, and can calculate the corresponding forest biomass carbon stock. It is not difficult to conclude that the method used to estimate forest carbon stock is almost entirely consistent with one used for biomass estimation; thus, the method of estimating forest biomass was addressed below, which can also be used to estimate forest carbon stock.

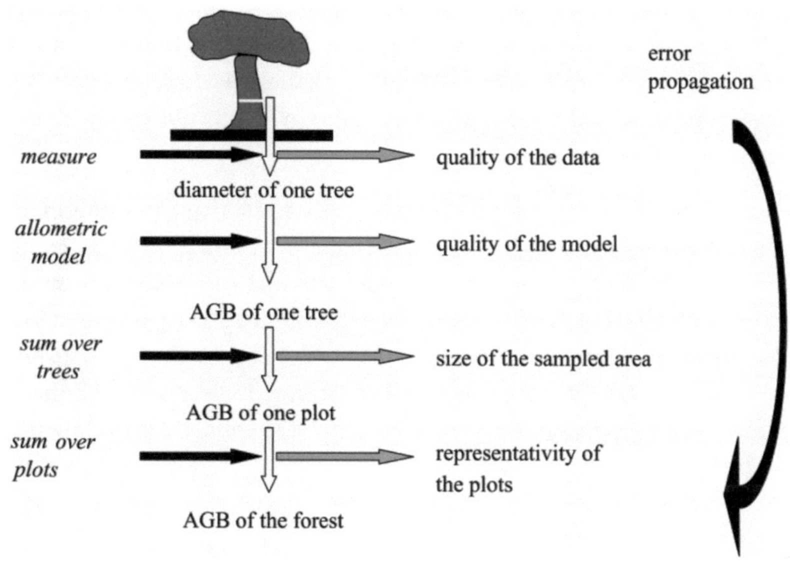

Figure 1. Procedure for biomass estimate and its error propagation [26].

4. Current methods of estimating forest biomass

4.1. Biomass estimate at individual level

Destructive sampling method (or harvesting method) and developed allometric equation can both be used to estimate individual-level biomass. For tree biomass estimate, destructive

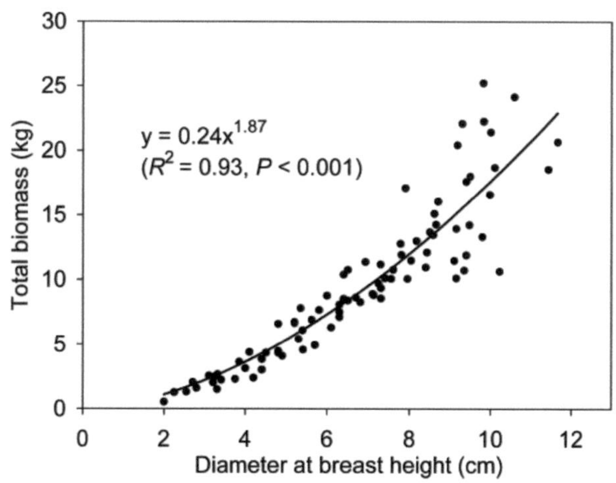

Figure 2. The relationship between the total biomass (*y*, kg) and *D* (*x*, cm) for the moso bamboo (*Phyllostachys edulis*) in South Anhui Province, China. The two variables exhibited a strong power function [30].

sampling method is more accurate than the use of developed allometric equation, because all the developed allometric equations are fitted (derived) from the biomass data based on the destructive sampling method. However, destructive method needs to cut down several sample trees and is thus expensive and time-consuming; moreover, it is not practical to weigh all the biomass for each tree in a stand or forest.

The general procedure for estimating biomass using destructive sampling method is to cut down several sample trees and weigh its different components (e.g., foliage, branch, stem, and root), respectively. After field survey, the components of the sample trees are collected and immediately taken to the laboratory to determine the water content. Subsequently, the (total) biomass can be determined by multiplying the fresh weight by the dry/fresh weight ratio. Then allometric equation can be fitted between the sampling biomass and D (and/or H) (e.g., **Figure 2**), and the developed equation can be employed to estimate individual-level biomass for each standing tree.

4.2. Biomass estimate at stand level

4.2.1. Mixed stand: simple allometric equation

The choice of stand-level biomass estimation is varied with the proportion of stand species composition (i.e., mixed or pure forests). Mixed stand-level biomass estimates may be estimated using allometric equation and then obtained by the addition of entire stands.

4.2.2. Pure stand: diameter-distribution model

This stand-level method is similar to the large-scale mean biomass density method described above, which does not take variations in biomass within a stand into account. In addition, the aforementioned simple allometric equation method is unable to fully reflect the developments and changes in stand structures. The corporation of the commonly used simple allometric equation and diameter-distribution functions (e.g., normal, lognormal, gamma, logistic, exponential, Richards, or Weibull functions) into a model (hereafter referred to as a diameter-distribution model) would likely improve the biomass estimation accuracy and strengthen the power of forest dynamics analyses.

The paper reported by Qi et al. [30] has exemplified the diameter-distribution model (Eq. (5)), which combined a three-parameter diameter-distribution function with an allometric equation to estimate the biomass of pure moso bamboo forests in China. The study found that a three-parameter Weibull distribution best characterized the diameter distribution of the moso bamboo stands. The biomass derived using the allometric equation was estimated 52.39 t/ha, smaller than 53.25 t/ha estimated using the Weibull distribution model (**Table 2**); this implied that the use of the common allometric equation alone to estimate forest biomass and carbon stocks may lead to an underestimate. It is concluded that using the diameter-distribution model to estimate forest biomass and carbon stock is expected to improve the accuracy.

Plot	BD_{WD} (t/ha)	BD_{AE} (t/ha)	RE (%)	Plot	BD_{WD} (t/ha)	BD_{AE} (t/ha)	RE (%)
PG1	45.54	46.29	−1.64	PH15	47.48	46.16	2.79
PG2	35.88	34.79	3.04	PH16	31.43	32.17	−2.34
PG3	43.00	42.19	1.87	PH17	34.99	34.59	1.14
PG4	41.46	38.87	6.25	PH18	24.87	23.78	4.37
PH5	18.96	18.02	4.94	PH19	61.27	59.91	2.23
PH6	51.25	50.72	1.03	PH20	13.07	12.68	3.00
PH7	40.39	39.02	3.38	PHS21	87.39	87.36	0.03
PH8	53.54	52.05	2.78	PHS22	63.389	63.28	0.16
PH9	64.17	62.37	2.81	PHS23	49.339	48.96	0.76
PH10	44.38	43.79	1.32	PHS24	61.22	60.13	1.78
PH11	52.57	50.57	3.80	PHS25	100.93	100.79	0.14
PH12	43.42	41.94	3.40	PHS26	102.58	102.10	0.47
PH13	62.56	60.81	2.79	PHS27	102.44	101.54	0.87
PH14	60.40	59.69	1.18	All	53.25	52.39	1.91

Note: The abbreviations PG and PH correspond to the plots at the Guangde Forest Station and the Huangshan Forest Station, respectively. BD_{WD}, BD_{AE}, and RE are estimated biomass density from Weibull-distribution model and allometric equation alone and relative error calculated using the equation RE (%) = (BD_{WD} − BD_{AE})/BD_{WD} × 100%.

Table 2. Comparison of biomass density (BD, t/ha) based on both the Weibull-distribution model and allometric equation alone for the 27 moso bamboo stands.

$$y = N\int_{D_{min}}^{D_{max}} g(x)f(x)dx \tag{5}$$

where x and y denote D and stand total biomass; $g(x)$ is the allometric relationship of stand biomass versus D; $f(x)$ is the probability density function of D for the given stand; $D_{\min}$ and $D_{\max}$ represent the minimum and maximum D for the stand, respectively; and N is the total culm number within the bamboo stand.

4.3. Large-scale biomass estimate

4.3.1. Mean biomass density method

Early in the International Biosphere Plan (IBP) period, Whittaker et al. [31, 32] have assessed forest biomass and carbon stock on the regional and global scales, via mean biomass density method, where one can estimate biomass for a stand or forest by the mean biomass density multiplied by the area.

Shi [33] selected 36 plots of moso bamboo forests to first calculate the mean D and biomass at the stand level using filed survey data and the developed allometric equation (**Figure 2**) and then estimate forest biomass and carbon stock via mean biomass density method (**Table 3**).

According to the sixth National Forest Inventory (NFI) data, bamboo forest area in Anhui Province is about 152,700 ha [34], so bamboo forest biomass of 1999–2003 period in the southern Anhui Province was estimated about 5.70 Tg (=37.33 t ha × 15.27 × 104 ha) (1 Tg = 10^{12} g), approximately 0.05% of the national forest biomass of the same period [35].

Plot	D_{mean} (cm)	Biomass density (t/ha)	Carbon density (t/ha)	Plot	D_{mean} (cm)	Biomass density (t/ha)	Carbon density (t/ha)
P1	9.56	45.94	22.97	P20	4.76	16.56	8.28
P2	7.82	33.03	16.52	P21	4.62	8.81	4.4
P3	9.31	37.99	18.99	P22	6.68	14.66	7.33
P4	6.4	34.26	17.13	P23	6.71	16.64	8.32
P5	6.7	23.33	11.66	P24	8.12	34.5	17.25
P6	8.79	47.79	23.89	P25	10.57	52.75	26.37
P7	7.73	60.46	30.23	P26	8.11	21.46	10.73
P8	8.54	64.65	32.33	P27	6.12	37.49	18.75
P9	8.89	72.69	36.35	P28	5.56	22.51	11.26
P10	7.82	38.89	19.45	P29	5.98	25.26	12.63
P11	8.09	46.84	23.42	P30	7.14	24.35	12.18
P12	7.18	43.36	21.68	P31	5.91	18.44	9.22
P13	7.81	30.61	15.31	P32	7.86	31.12	15.56
P14	9.81	52.73	26.37	P33	9.00	60.95	30.48
P15	7.73	45.80	22.9	P34	10.51	49.96	24.98
P16	8.02	27.09	13.54	P35	10.77	57.13	28.57
P17	8.72	29.62	14.81	P36	8.17	39.05	19.53
P18	6.33	22.09	11.05				
P19	9.07	54.91	27.46	Mean		37.33	18.66

Note: The biomass of *Cunninghamia lanceolata* in each bamboo stand was estimated with the one-parameter power equation developed by Li et al. [36] ($y = 0.1606D^{2.1203}$, $R^2 = 0.99$, $P < 0.001$).

Table 3. Mean biomass method-derived biomass and carbon stock at the stand level in South Anhui Province, China.

However, the measured biomass density (i.e., data or plot) is more limited in this method, and plot location is frequently in the well-growing stand, easily leading to an overestimation [37, 38].

4.3.2. Geostatistics in biomass estimation

As a spatial statistics, geostatistics has become an indispensable method used to study the nature with the dual characteristics of randomness and regularity over the past 50 years [39]. Forest is affected by physical, climate, and other natural disturbances with a high degree of heterogeneity and relevance; therefore geostatistics is also gradually used in forest ecology, including forest biomass estimate [40, 41].

Zhang et al. [42] confirmed the feasibility of geostatistical methods used in estimating bamboo forest biomass. They chose Huangkeng Town, southern Wuyishan Mountain, as the study area and cut 103 sample bamboo culms to develop an allometric equation for moso bamboo of the region, combined with data of field survey and leaf area index data to estimate the biomass of a total of 209 plots at stand level. By means of ARCGIS software, statistical technique was used to estimate bamboo forest biomass of the whole town, and spatial distribution of biomass was also visualized.

4.3.3. *National forest inventory and biomass expansion factor*

Many countries have implemented national forest inventory (NFI) regularly or irregularly in order to grasp the forest resources and their dynamics, as well as more scientific development of forest policies [43]. Subsequently, many scholars estimated forest biomass using NFI data at the region level, and then a biomass expansion factor (BEF) method came into being [28, 44]. This method assumes that there is a certain relationship between the forest growing stock and biomass; thus, biomass can be estimated based on the growing stock (derived from NFI) multiplied by the BEF conversion factor. Some researchers hold that BEF is a constant; we can call it mean BEF method. Actually, BEF varied frequently with forest age, site class, and stand density [45–47]. Hence, an improved method (called continuous BEF) [38, 45, 48, 49] is gradually being accepted by many scientists.

Current researches regarding forest biomass change are mainly based on NFI data [46, 47, 50–54]. But for the nation with a larger land area, the NFI points are limited, and remote areas are also difficult to reach, thereby often creating a bias in estimates. Moreover, forest resource assessment is incomparable; for example, the error of tropical deforestation rate estimated by FAO may be up to 50%, which is mainly due to the differences in national inventory methods and the definition of forest in tropical region [3]. Additionally, since NFI data have no information recording the spatial distribution of plots; therefore, spatial variation cannot be analyzed while using NFI data.

4.3.4. *Remote sensing technique and its application in biomass estimation*

Remote sensing technique developed rapidly in the late twentieth century, and remote sensing data with high spatiotemporal resolution, wide coverage, and timely updates has been widely used in the assessment of forest biomass and carbon stock on various scales [55–57]. Currently, remote sensing-derived biomass estimation has become the leading method of large-scale forest biomass and carbon stock estimation. The use of remote sensing technique to assess forest biomass is mainly based on the normalized difference vegetation index (NDVI) datasets. While using NDVI datasets, most researches frequently overlay the vegetation maps of a region and NDVI datasets to explore the spatiotemporal changes. But this method only pays attention to changes in productivity, without considering the change in area caused by land-use change [58, 59].

The method can be exemplified with the study conducted by Shi [24]. In this paper, National Forest Inventory and normalized difference vegetation index (NDVI, comes from Global Inventory Monitoring and Modeling Studies) datasets were integrated via matching forest type for Heilongjiang, Liaoning, and Jilin Provinces while fitting the inverse model (**Figure 3**). The developed inverse models were used to estimate forest growing stock and carbon stock, respectively. Consequently, changes in growing stock and biomass between 1982 and 2006 were analyzed.

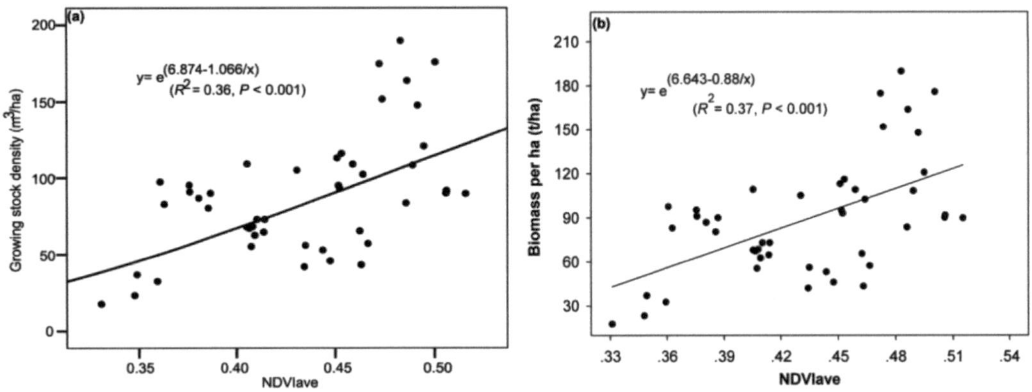

Figure 3. Relationship between forest timber volume density (m^3/ha) (a), biomass density (t/ha) (b), and annual mean NDVI (NDVIave) of northeast China (including Heilongjiang, Liaoning, and Jilin provinces), respectively; four forest types for each inventory period (four inventory periods of and a total of 48 data points) are used in the regression.

4.3.5. Forest identity

For any forest, area, growing stock, biomass, and carbon can be linked using Eqs. (6)–(8). The forest identity method defines these four valued attributes by using measurable variables, and it quantitatively and logically integrates their changes into a causal relationship (i.e., Eqs. (9)–(11)) [60, 61].

$$V\left(\mathrm{m}^3\right) = A\left(\mathrm{ha}\right) \times D\left(\mathrm{m}^3/\mathrm{ha}\right) \tag{6}$$

$$M\left(\mathrm{Mg}\right) = A\left(\mathrm{ha}\right) \times D\left(\mathrm{m}^3/\mathrm{ha}\right) \times B\left(\mathrm{Mg}/\mathrm{m}^3\right) = V\left(\mathrm{m}^3\right) \times B\left(\mathrm{Mg}/\mathrm{m}^3\right) \tag{7}$$

$$Q\left(\mathrm{Mg\ C}\right) = A\left(\mathrm{ha}\right) \times D\left(\mathrm{m}^3/\mathrm{ha}\right) \times B\left(\mathrm{Mg}/\mathrm{m}^3\right) \times C\left(\mathrm{Mg\ C}/\mathrm{Mg}\right) \tag{8}$$

Then, $d\ln(Q)/dt = d\ln(A)/dt + d\ln(D)/dt + d\ln(B)/dt + d\ln(C)/dt$.

Let $q \approx d\ln(Q)/dt$, $a \approx d\ln(A)/dt$, $d\ b \approx d\ln(B)/dt$, $d \approx d\ln(B)/dt$, and $c \approx d\ln(C)/dt$.

Then,

$$q = a + d + b + c \tag{9}$$

Similarly,

$$v = a + d \tag{10}$$

And also,

$$m = a + d + b \tag{11}$$

where *V, M, Q, A, D, B,* and *C* represent growing stock (m^3), biomass (Mg), forest carbon stock (Mg C), area (ha), growing stock density (m^3/ha), the conversion ratio of biomass to growing stock (Mg/m^3), and carbon concentration in biomass (Mg C/Mg) at the provincial or national level, respectively, and *v, m, q, a, d, b,* and *c* represent the corresponding derivatives of these attributes with respect to time.

Shi et al. [35] applied both forest identity and regression approaches to explore the temporal changes in China's forest. This paper showed that China's forest area and growing stock density increased by 0.51 and 0.44% annually over the past three decades, while the conversion ratio of forest biomass to growing stock declined by 0.10% annually. These developments resulted in a net annual increase in 0.85% in forest carbon sequestration, which is equivalent to a net biomass carbon uptake of 43.8 Tg per year (1 Tg = 10^{12} g).

In the paper, two regression equations below (i.e., Eqs. (12) and (13)) were used to obtain the derivatives of the forest attributes with respect to time:

$$y = \text{slope} \times x + \text{intercept} \tag{12}$$

$$RR(\text{yr}/\%) = \{\text{slope}/[(y_1 + y_2 + y_3 + y_4 + y_5)/5]\} \times 100 \tag{13}$$

where *y* represents the forest attributes (i.e., area, growing stock density, the conversion ratio, or carbon content) at the provincial or national level; y_1, y_2, y_3, y_4, and y_5 denote the corresponding forest attributes for the inventories of 1977–1981, 1984–1988, 1989–1993, 1994–1998, and 1999–2003, respectively; *RR* (%/yr) is the relative annual change rate of the forest attributes; slope denotes the amplitude and direction of annual absolute change for each forest attribute; and *x* represents the corresponding periods of NFI. In other words, the relative annual change rate (*RR*, %) defined here is equivalent to *q, a, d,* or *b* mentioned above.

5. Method comparison

Each estimation method has its advantages and disadvantages. None of these methods mentioned in this chapter is the best from individual to large scale (**Table 4**). Fitting allometric equation requires some down sampling trees, but the number is small relative to all the standing trees. The equation (coefficients) varies frequently with species, terrain, temperature, and rainfall. To improve its prediction power, combining the field survey with LIDAR and incorporating the variation into allometric coefficients are the two key elements. It is fully convinced that allometric equation, together with LIDAR, is increasing widely used in estimate forest biomass at individual and stand levels in the future.

Method	Scale	Major limitation	Improvement practices
Allometric equation	Individual or stand	Varying frequently with species, terrain, temperature, and rainfall; less sampling trees	Incorporating these factors into allometric coefficients; combine with LIDAR
Mean biomass density	Stand or region	Easily leading to an overestimation	Randomly set more plots
Biomass expansion factor	Stand	Varying frequently with species, terrain, temperature, and rainfall	Incorporating these into conversion factor
Forest identity	Region	Comprehensive analysis	–
Remote sensing	Stand or region	Saturation and bidirectional reflectance of surface features	Higher spatiotemporal resolution, advanced algorithm and technology
Geostatistics	Region	More field data	Constructing the biomass database

Table 4. Comparison of the forest biomass estimation methods.

Mean biomass density easily leads to an overestimation. However, the uncertainty will be reduced if we randomly select some plots across the study region as more as possible.

Like allometric equation, biomass expansion factor also varies frequently with species, terrain, temperature, and rainfall. It is of significance to explore the variation of conversion factor with species and environment; then the conversion factor can be estimated as possible as we can.

Forest identity is a tool used to comprehensively analyze the forest change, which is more likely be employed only by a small number of scientists.

Remote sensing technique has been booming since the 1980s, but saturation of vegetation index (e.g., widely used NDVI) and bidirectional reflectance of surface features still occur. Exploring the higher spatiotemporal resolution, advanced algorithm and technology are needed in the future.

Geostatistics is an important technique to upscale forest biomass but needs more field data. More field data implies more cost and time-consuming. It is, therefore, important to construct the biomass database.

6. Uncertainty

Many scholars have conducted researches to estimate forest biomass and carbon stock on various scales, but estimates have a large uncertainty [62–65]. Dixon et al. [37] used mean biomass density method to estimate China's forest, whose carbon storage and carbon density were 7.0 Pg C and 114 t/ha, while biomass was estimated 6.2 Pg C and 57.07 t/ha based on the 720 plots conducted by Zhou et al. [66]. Estimation error of tropical deforestation rates may be as high as 50% due to the difference by national forest inventory method in tropical regions and different definitions of forest [3]. With regard to carbon stock for the Amazon rain forest, the estimation results varied from 58 [67] to 134 Pg C [68], as well as 89 Pg C estimated by FAO [2]. Given the large uncertainty, Houghton et al. [69] conducted a study regarding the comparison of different methods by different scholars and concluded that carbon storage estimates are rarely considerable, even using similar data and the same estimation method.

As addressed above, the procedure of estimating forest biomass includes four steps (**Figure 1**). Each step can produce uncertainty: step 1 induces measure error, step 2 causes model error, step 3 involves sampling error, and step 4 often ignores biomass spatial variability and increases uncertainty. Uncertainty resulting from the previous steps will affect (propagate) the following biomass estimation process (result). Compared to measure and sampling errors, model and spatial variability errors are much larger but easier to solve. Taking the inverse model (see **Figure 3**), for example, the densities of growing stock and biomass both have significant relationships with NDVI, but determination coefficients (R^2) are very low, implying that the current remote sensing data is unable to fully retrieve forest biomass. This is mainly caused by the complicated soil background, bidirectional reflectance of surface features, and saturation of NDVI. Higher spatiotemporal remote sensing data in the future can improve the prediction accuracy while using inverse model.

In ecological researches, a number of parameters (such as biomass) are not easy to be measured directly; thus, the indirect measurement is available, namely, measuring certain indirect variables first and calculating direct one based on the relationship between indirect and direct variables. Direct measurement error will be transformed and propagated to indirect measurement error by a certain function, namely, error propagation [70]. Overall, the uncertainty is mainly due to the spatial variability of forest carbon stock and error propagated from the estimation process [26, 71, 72]. In general, the total uncertainty can be obtained by summing the error produced during the procedure [73].

Currently, little research has been conducted on the uncertainty of biomass estimation [73], and only a few studies have focused on one or several errors. For example, Ketterings et al. [74] have focused on the error induced by step 2 and presented a simple method to reduce error based on the tree *H* and stand density. Phillips et al. [73] analyzed the uncertainty using NFI data of five states of the USA and found that the uncertainty of large-scale forest volume estimation was mainly from the sampling error. Chave et al. [26] pointed out that sampling, identifying allometric equation, measuring tree *D*, and other stand variables can propagate error to the aboveground biomass estimates and the developed allometric model accounts for the largest source of uncertainty; thus, fitting allometric model should be given adequate attention. Holdaway et al. [75] used Monte Carlo method to analyze the uncertainty while estimating carbon stock in temperate rain forest, New Zealand. Su et al. [76] pointed out that non-geo-referenced plot location would bring biomass estimation error and also estimated the aboveground biomass of China's forest via incorporating sampling error model.

With the increasing value of climate change by world dignitaries and growing global carbon market, the accuracy problems in estimating forest carbon stock on various scales have become increasingly prominent. The uncertainty in forest carbon reserve estimation can involve the climate negotiations and the development of carbon trading market to some extent, thereby affecting the development of government policies and international negotiations regarding the forest response to climate change. Therefore, the scientific community has become increasingly concerned about the uncertainty in the estimation of forest carbon stock. Changes in publication papers from 1990 to date regarding biomass estimation uncertainty can be evident (**Figure 4**); prior to 1997, the number of papers is relatively stable, but article number increases exponentially since 1997 (when Kyoto Protocol effects).

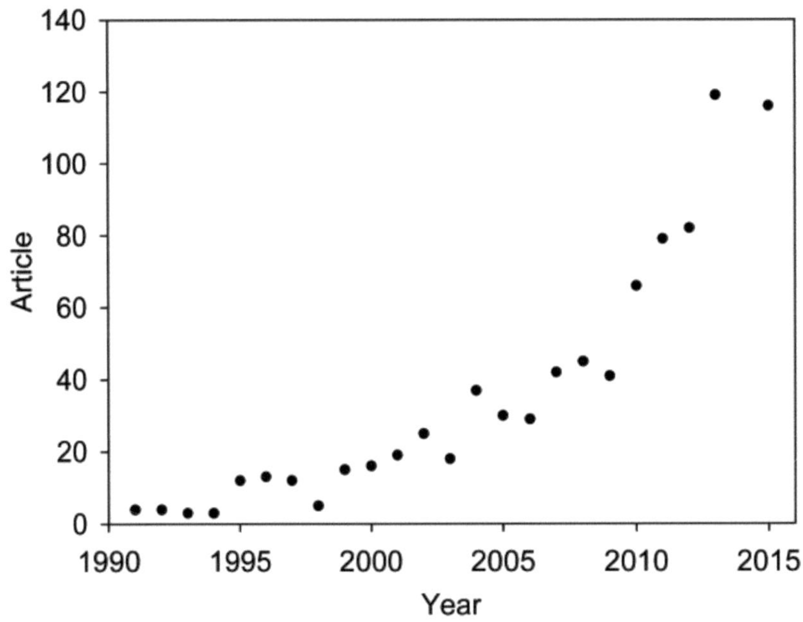

Figure 4. Articles regarding the uncertainty analysis of forest carbon stock versus time. The number is obtained via searching forest, biomass, and uncertainty in core journal of ISI Web of Knowledge.

7. Challenges and prospects

7.1. Global field biomass dataset

Although remote sensing, NFI, geographic information system, and other advanced techniques have been applied to estimate forest biomass, but the accuracy of estimate of forest biomass needs to be improved. In recent years, with the advent of LIDAR, multispectral data, and geostatistical technique, most scholars paid more attention to the application of such advanced technology in biomass estimates but ignored the acquisition and compiling field biomass data. Although acquisition of field biomass data using harvesting method is time-consuming and laborious, the fitted inverse model based on remote sensing data must be verified using the measured data on biomass. Matching remote sensing data and ground-truth data is often a difficult problem during biomass estimation. In various biomass estimation methods, most researchers assume that the ground-truth biomass data are the most accurate. It is thus important to carry out field biomass measurement globally using a unified investigation specification, as well as the establishment of a unified biomass database.

7.2. Upscaling in biomass estimation

Scale has been one of the core focuses of ecology. Scale and upscale affect the accuracy of forest biomass estimation and are also an important issue in biomass estimation. Biomass estimation method varied with scale. The mean biomass density method, BEF method, remote sensing, and geostatistic method are used to estimate large-scale forest biomass and carbon stock, which involves scale biomass (upscaling); allometric equation is used for small-scale biomass estimation. However, large-scale biomass estimate must be based on the allometric equation. If forest biomass estimation method you use is different from another, estimate may be different even if the same data is used. The rapid development of remote sensing data, geostatistical technique, and others make scaling possible while estimating biomass. Therefore, upscaling forest biomass from stand scale to regional and even global scale should be one of the key elements for future biomass researches. Only by addressing the scaling issue and clarifying the spatial variation of forest biomass can a unified framework or system for estimating forest biomass be presented. Then national and international standards of estimating forest biomass can be developed to improve biomass estimation with a high accuracy.

7.3. High-accuracy biomass estimate needs remote sensing

Remote sensing data with a high spatiotemporal resolution, wide coverage, and timely updates have been widely used in the assessment of forest biomass and carbon stock on various scales, which play an important role in improving estimation accuracy.

In addition, scaling biomass cannot do without application of remote sensing data. However, remote sensing data have no continuity of the time series, the resolution is not high enough, and the fitted inverse model is often saturated. Moreover, the quality of the current multispectral data is not high, and LIDAR data and UAV technology applied in forest biomass estimate are not yet mature. Improving the quality of multispectral data, LIDAR and UAV technology are two of the most active frontiers in the future.

7.4. Geographical variation of allometric equation

As mentioned above, allometric equation is indispensable in biomass estimation. However, the coefficients of the equation varied frequently with species, terrain, temperature, and rainfall [77–79]. Small variation in the allometric coefficients is likely to result in a larger biomass estimation error [80]. Therefore, the clarification of the variation in the allometric coefficients along different gradients is also the element to improve the accuracy of biomass estimation.

Biomass estimate may be different using different allometric equations while estimating vegetation biomass. Thus, it is likely to come to inconsistent results, even when different scholars estimate the same forest. In other words, the choice of the appropriate predictors and optimization of allometric equation and its parameters can contribute to reduce biomass estimation uncertainty and improve the ability to predict the biomass and carbon stock. It is worth noting that recently developed metabolic rate theory (fractal networks) is contrary to conventional wisdom. The theory inference indicated that the forest biomass (M) and its D on the species and biota scale both showed a constant relationship, which can be written as $M \propto D^{8/3}$ [81–83]. Whether the coefficients of biomass allometric equation varied with the species, site, and climatic factors or not is a hot debate in recent years [84]. If the relationship between forest biomass and the corresponding D remains to be constantly exponential on various scales, a unified allometric biomass estimation model is more likely to be developed, which helps to save costs in in situ measurements of biomass, but also can contribute to promote the in-depth study of biomass estimation methods.

8. Conclusions

Plant allometry is the theoretical basis of vegetation biomass estimation. Generally, the use of allometric equation is indispensable to estimate biomass for both tree and forest. By reviewing the methods used for estimating forest biomass, we can conclude that each estimation method has its advantages and disadvantages, and none of these methods mentioned is always the best from individual to large scales. A large uncertainty exists in biomass estimation, which is unable to meet the requirement of the accurate carbon accounting required by Kyoto Protocol. It is of significance and urgently needed to develop the more suitable and rigor method to accurately estimate the size and distribution of forest biomass. To achieve this goal, we argued that the new developed technique such as geostatistics and remote sensing technique (e.g., LIDAR) would be the key tools to improve forest biomass estimation with a high accuracy. However, prior to this, spatial variation in forest biomass at various levels should be explored using multi-source data and multi-approaches.

Acknowledgements

This study was supported by the Special Fund for Basic Scientific Research of International Center for Bamboo and Rattan (1632015005) and the National Natural Science Foundation of China (31300177).

Author details

Lei Shi[1*] and Shirong Liu[1,2]

*Address all correspondence to: leishi@icbr.ac.cn

1 State Forestry Administration Key Laboratory for Science & Technology of Bamboo & Rattan, International Center for Bamboo and Rattan, Beijing, China

2 Institute of Forest Ecology, Environment and Protection, Chinese Academy of Forestry, Beijing, China

References

[1] FAO (Food and Agriculture Organization, the United Nations). State of the world's forests 2011. Rome: UN Food and Agriculture Organization; 2011.

[2] FAO (Food and Agriculture Organization, the United Nations). Global Forests Resources Assessment 2010: Main Report. Rome: UN Food and Agriculture Organization; 2010.

[3] Watson RT, Noble A, Bolin B, Ravindranath NH, Verado D. Land use, land-use change, and forestry. Intergovernmental Panel on Climate Change Special Report. Cambridge: Cambridge University Press; 2000.

[4] IPCC (Intergovernmental Panel on Climate Change) Climate Change 2007: Synthesis Report. Core Writing Team P, Pachauri RK, Reisinger A, Editor. Geneva, Switzerland: Contribution of Working Groups I, II and III to the Fourth Assessment Report of the Intergovernmental Panel on Climate Change; 2007.

[5] Jackson RB, Jobbagy EG, Avissar R, Roy SB, Barrett DJ, Cook CW, Farley KA, le Maitre DC, McCarl BA, Murray BC. Trading water for carbon with biological sequestration. Science. 2005; 310: 1944–1947. DOI: 10.1126/science.1119282.

[6] Piao SL, Friedlingstein P, Ciais P, de Noblet-Ducoudre N, Labat D, Zaehle S. Changes in climate and land use have a larger direct impact than rising CO_2 on global river runoff trends. Proceedings of the National Academy of Sciences of the United States of America. 2007; 104: 15242–15247. DOI: 10.1073/pnas.0707213104.

[7] Niklas KJ, Owens T, Reich PB, Cobb ED. Nitrogen/phosphorus leaf stoichiometry and the scaling of plant growth. Ecology Letters. 2005; 8: 636–642. DOI: 10.1111/j. 1461-0248.2005.0075 9.x.

[8] Huxley JS. Problems of relative growth. London: Methuen. 1932. 276.

[9] Enquist BJ, Niklas KJ. Invariant scaling relations across tree-dominated communities. Nature. 2001; 410: 655–660. DOI: 10.1038/35070500.

[10] Cheng DL. Study on the correlation between plant biomass allocation model and growth rate [Thesis]. Lanzhou: Lanzhou University; 2007.

[11] Shanmughavel P and Francis K. Above ground biomass production and nutrient distribution in growing bamboo (*Bambusa bambos* (L) Voss). Biomass & Bioenergy. 1996; 10: 383–391.

[12] Wang CK. Biomass allometric equations for 10 co-occurring tree species in Chinese temperate forests. Forest Ecology and Management. 2006; 222: 9–16. DOI: 10.1016 / j.foreco.2005.10.074.

[13] Gao X, Jiang ZH, Guo QR, Zhang Y, Yin H, He F, Qi LH, Shi L. Allometry and biomass production of *Phyllostachys edulis* across the whole lifespan. Polish Journal of Environmental Studies. 2015; 24: 511–517.

[14] Feldpausch TR, Banin L, Phillips OL, Baker TR, Lewis SL, Quesada CA, Affum-Baffoe K, Arets E, Berry NJ, Bird M, Brondizio ES, de Camargo P, Chave J, Djagbletey G, Domingues TF, Drescher M, Fearnside PM, Franca MB, Fyllas NM, Lopez-Gonzalez G, Hladik A, Higuchi N, Hunter MO, Iida Y, Salim KA, Kassim AR, Keller M, Kemp J, King DA, Lovett JC, Marimon BS, Marimon BH, Lenza E, Marshall AR, Metcalfe DJ, Mitchard ETA, Moran EF, Nelson BW, Nilus R, Nogueira EM, Palace M, Patino S, Peh KSH, Raventos MT, Reitsma JM, Saiz G, Schrodt F, Sonke B, Taedoumg HE, Tan S, White L, Woll H, Lloyd J. Height-diameter allometry of tropical forest trees. Biogeosciences. 2011; 8: 1081–1106. DOI: 10.5194/bg-8-1081-2011.

[15] Peng C, Zhang L, Liu J. Developing and validating nonlinear height-diameter models for major tree species of Ontario's boreal forests. Northern Journal of Applied Forestry. 2001; 18: 87–94.

[16] Gao X, Li ZD, Yu HM, Jiang ZH, Wang C, Zhang Y, Qi LH, Shi L. Modeling of the height-diameter relationship using an allometric equation model: a case study of stands of *Phyllostachys edulis*. Journal of Forestry Research. 2016; 27: 339–347. DOI: 10.1007/s11676-015-0145-6.

[17] Gower ST, Kucharik CJ, Norman JM. Direct and indirect estimation of leaf area index, f(APAR), and net primary production of terrestrial ecosystems. Remote Sensing of Environment. 1999; 70: 29–51. DOI: 10.1016/s0034-4257(99)00056-5.

[18] Mowrer HT and Frayer WE. Variance propagation in growth and yield projections. Canadian Journal of Forest Research-Revue Canadienne De Recherche Forestiere. 1986; 16: 1196–1200. DOI: 10.1139/x86-213.

[19] Pilli R, Anfodillo T, Carrer M. Towards a functional and simplified allometry for estimating forest biomass. Forest Ecology and Management. 2006; 237: 583–593. DOI: 10.1016/j.foreco.2006.10.004.

[20] Andrade HA, Campos RO. Allometry coefficient variations of the length-weight relationship of skipjack tuna (*Katsuwonus pelamis*) caught in the southwest South Atlantic. Fisheries Research. 2002; 55: 307–312. DOI: 10.1016/s0165-7836(01)00305-8.

[21] Singh AN, Singh JS. Biomass, net primary production and impact of bamboo plantation on soil redevelopment in a dry tropical region. Forest Ecology and Management. 1999; 119: 195–207.

[22] Isagi Y, Kawahara T, Kamo K. Biomass and net production in a bamboo *Phyllostachys-Bambusoides* stand. Ecological Research. 1993; 8: 123–133.

[23] Shi L. Changes in China's forests over the past 25 years [Thesis]. Beijing: Peking University; 2011.

[24] Shi L. Changes of forest in Northeast China over the past 25 years: an analysis based on remote sensing technique. Proceedings of SPIE. 2011; 8203: The Society of Photo-Optical Instrumentation Engineers (SPIE); International Society for Digital Earth; UNDESA GAID e-SDDC; Asian Association on Remote Sensing; IEEE Beijing Section.

[25] Mitchard ETA, Saatchi SS, Lewis SL, Feldpausch TR, Gerard FF, Woodhouse IH, Meir P. Comment on 'A first map of tropical Africa's above-ground biomass derived from satellite imagery'. Environmental Research Letters. 2011; 6: 6. DOI: 10.1088/1748-9326/6/4/04 9 001.

[26] Chave J, Condit R, Aguilar S, Hernandez A, Lao S, Perez R. Error propagation and scaling for tropical forest biomass estimates. Philosophical Transactions of the Royal Society of London Series B-Biological Sciences. 2004; 359: 409–420. DOI: 10.1098/rstb.2003.1425.

[27] Matthews G. The carbon content of trees. Edinburgh, Scotland: Forestry Commission; 1993.

[28] Birdsey RA. Carbon storage and accumulation in United States forest ecosystems. Washington D.C.: U.S. Department of Agriculture, Forest Service, Washington Office; 1992.

[29] Brown S. Estimating biomass and biomass change of tropical forests: a primer. Rome: Food and Agriculture Organization of the United Nations; 1997.

[30] Qi LH, Liu XJ, Jiang ZH, Yue XH, Li ZD, Fu JH, Liu GL, Guo BH, Shi L. Combining diameter-distribution function with allometric equation in biomass estimates: a case study of *Phyllostachys edulis* forests in South Anhui, China. Agroforestry Systems. 2015. DOI: 10.1007/s 10457-015-9887-6.

[31] Whittaker RH. Net production of heath balds and forest heaths in the Great Smoky Mountains. Ecology. 1963; 44: 176–182.

[32] Woodwell GM and Whittaker RH. Primary production in terrestrial ecosystems. American Zoologist. 1968; 8: 19–30.

[33] Shi L. Carbon stock estimation of *Phyllostachys edulis* forest: a case study in South Anhui Province, China. Beijing: International Center for Bamboo and Rattan (ICBR). 2014; p. 55.

[34] SFA (State Forestry Administration). National Forest Resources Statistics (1999–2003). Beijing: China Forestry Publishing House; 2004.

[35] Shi L, Zhao SQ, Tang ZY, Fang JY. The changes in China's forests: an analysis using the forest identity. PLoS One. 2011; 6: e20778. DOI: 10.1371/journal.pone.0020778.

[36] Li Y, Zhang JG, Duan AJ, Xiang CW. Selection of biomass estimation models for Chinese fir plantation. Chinese Journal of Applied Ecology. 2010; 21: 3036–3046.

[37] Dixon RK, Brown S, Houghton RA, Solomon AM, Trexler MC, Wisniewski J. Carbon pools and flux of global forest ecosystems. Science. 1994; 263: 185–190.

[38] Fang JY, Oikawa T, Kato T, Mo WH, Wang ZH. Biomass carbon accumulation by Japan's forests from 1947 to 1995. Global Biogeochemical Cycles. 2005; 19: Gb2004. DOI: 10.1029/2004gb002253.

[39] Feng YM, Tang SZ, Li ZY. Application of spatial statistic analysis in forestry. Scientia Silvae Sinicae. 2004; 40: 149–155.

[40] Yadav BKV, Nandy S. Mapping aboveground woody biomass using forest inventory, remote sensing and geostatistical techniques. Environmental Monitoring and Assessment. 2015; 187: DOI: 10.1007/s10661-015-4551-1.

[41] Barra M, Petitgas P, Bonanno A, Somarakis S, Woillez M, Machias A, Mazzola S, Basilone G, Giannoulaki M. Interannual changes in biomass affect the spatial aggregations of anchovy and sardine as evidenced by geostatistical and spatial indicators. Plos One. 2015; 10: e0135808. DOI: 10.1371/journal.pone.0135808.

[42] Zhang Y, Yue XH, Qi LH, Jiang ZH, Shi L. Estimation of *Phyllostachys edulis* forest biomass in southern Wuyishan Mountain using allometric equation and geostatistical technique. Chinese Journal of Ecology. 2016; 35: 1957–1962.

[43] Kangas A, Maltamo M. Forest inventory: methodology and applications. Dordrecht, the Netherlands: Springer Verlag. 2006.

[44] Turner DP, Koerper GJ, Harmon ME, Lee JJ. A carbon budget for forests of the conterminous United-States. Ecological Applications. 1995; 5: 421–436.

[45] Brown S, Lugo AE. Aboveground biomass estimates for tropical moist forests of the Brazilian Amazon. Interciencia. 1992; 17: 8–18.

[46] Fang JY, Chen AP, Peng CH, Zhao SQ, Ci L. Changes in forest biomass carbon storage in China between 1949 and 1998. Science. 2001; 292: 2320–2322.

[47] Fang JY, Wang GG, Liu GH, Xu SL. Forest biomass of China: an estimate based on the biomass-volume relationship. Ecological Applications. 1998; 8: 1084–1091.

[48] Fang JY, Guo ZD, Piao SL, Chen AP. Terrestrial vegetation carbon sinks in China, 1981–2000. Science in China Series D-Earth Sciences. 2007; 50: 1341–1350. DOI: 10.1007/s11430-007-0049-1.

[49] Schroeder P, Brown S, Mo JM, Birdsey R, Cieszewski C. Biomass estimation for temperate broadleaf forests of the United States using inventory data. Forest Science. 1997; 43: 424–434.

[50] Corona P, Fattorini L, Franceschi S. Estimating the volume of forest growing stock using auxiliary information derived from relascope or ocular assessments. Forest Ecology and Management. 2009; 257: 2108–2114. DOI: 10.1016/j.foreco.2009.02.017.

[51] Fazakas Z, Nilsson M, Olsson H. Regional forest biomass and wood volume estimation using satellite data and ancillary data. Agricultural and Forest Meteorology. 1999; 98-9: 417–425.

[52] Franco-Lopez H, Ek AR, Bauer ME. Estimation and mapping of forest stand density, volume, and cover type using the k-nearest neighbors method. Remote Sensing of Environment. 2001; 77: 251–274.

[53] Al-Bakri JT and Taylor JC. Application of NOAA AVHRR for monitoring vegetation conditions and biomass in Jordan. Journal of Arid Environments. 2003; 54: 579–593. DOI: 10.1006/jare.2002.1081.

[54] Peng CH, Zhou XL, Zhao SQ, Wang XP, Zhu B, Piao SL, Fang JY. Quantifying the response of forest carbon balance to future climate change in Northeastern China: model validation and prediction. Global and Planetary Change. 2009; 66: 179–194. DOI: 10.1016/ j.gloplacha.2008.12.001.

[55] Lu DS. The potential and challenge of remote sensing-based biomass estimation. International Journal of Remote Sensing. 2006; 27: 1297–1328. DOI: 10.1080/01431 160500486732.

[56] Nelson R, Krabill W, Tonelli J. Estimating forest biomass and volume using airborne laser data. Remote Sensing of Environment. 1988; 24: 247–267.

[57] Timothy D, Onisimo M, Cletah S, Adelabu S, Tsitsi B. Remote sensing of aboveground forest biomass: a review. Tropical Ecology. 2016; 57: 125–132.

[58] Piao SL, Fang JY, Zhou LM, Ciais P, Zhu B. Variations in satellite-derived phenology in China's temperate vegetation. Global Change Biology. 2006; 12: 672–685. DOI: 10.1111/ j.1365-2486.2006.01123.x.

[59] Myneni RB, Dong J, Tucker CJ, Kaufmann RK, Kauppi PE, Liski J, Zhou L, Alexeyev V, Hughes MK. A large carbon sink in the woody biomass of northern forests. Proceedings of the National Academy of Sciences of the United States of America. 2001; 98: 14784–14789.

[60] Waggoner PE. Using the Forest Identity to grasp and comprehend the swelling mass of forest statistics. International Forestry Review. 2008; 10: 689–694.

[61] Kauppi PE, Ausubel JH, Fang JY, Mather AS, Sedjo RA, Waggoner PE. Returning forests analyzed with the forest identity. Proceedings of the National Academy of Sciences of the United States of America. 2006; 103: 17574–17579. DOI: 10.1073/pnas.0608343103.

[62] Mitchard ETA, Feldpausch TR, Brienen RJW, Lopez-Gonzalez G, Monteagudo A, Baker TR, Lewis SL, Lloyd J, Quesada CA, Gloor M, ter Steege H, Meir P, Alvarez E, Araujo-Murakami A, Aragao L, Arroyo L, Aymard G, Banki O, Bonal D, Brown S, Brown FI, Ceron CE, Moscoso VC, Chave J, Comiskey JA, Cornejo F, Medina MC, Da Costa L, Costa FRC, Di Fiore A, Domingues TF, Erwin TL, Frederickson T, Higuchi N, Coronado ENH, Killeen TJ, Laurance WF, Levis C, Magnusson WE, Marimon BS, Marimon BH, Polo IM, Mishra P, Nascimento MT, Neill D, Vargas MPN, Palacios WA, Parada A, Molina GP, Pena-Claros M, Pitman N, Peres CA, Poorter L, Prieto A, Ramirez-Angulo H, Correa ZR, Roopsind A, Roucoux KH, Rudas A, Salomao RP, Schietti J, Silveira M, de Souza PF, Steininger MK, Stropp J, Terborgh J, Thomas R, Toledo M, Torres-Lezama A, van Andel TR, van der Heijden GMF, Vieira ICG, Vieira S, Vilanova-Torre E, Vos VA, Wang O, Zartman CE, Malhi Y, Phillips OL. Markedly divergent estimates of Amazon forest carbon density from ground plots and satellites. Global Ecology and Biogeography. 2014; 23: 935–946. DOI: 10.1111/ geb.12168.

[63] Yuen JQ, Fung T, Ziegler AD. Review of allometric equations for major land covers in SE Asia: uncertainty and implications for above- and below-ground carbon estimates. Forest Ecology and Management. 2016; 360: 323–340. DOI: 10.1016/j.foreco.2015.09.016.

[64] Chen Q, Laurin GV, Valentini R. Uncertainty of remotely sensed aboveground biomass over an African tropical forest: propagating errors from trees to plots to pixels. Remote Sensing of Environment. 2015; 160: 134–143. DOI: 10.1016/j.rse.2015.01.009.

[65] Ahmed R, Siqueira P, Hensley S, Bergen K. Uncertainty of forest biomass estimates in north temperate forests due to allometry: implications for remote sensing. Remote Sensing. 2013; 5: 3007–3036. DOI: 10.3390/rs5063007.

[66] Zhou YR, Yv ZL, Zhao SD. Carbon storage and budget of major Chinese forest types. Chinese Journal of Plant Ecology. 2000; 24: 518–522.

[67] Olson JS, Watts JA, Allison LJ. Carbon in live vegetation of major world ecosystems. Oak Ridge National Lab., Oak Ridge, TN (USA). 1983.

[68] Fearnside PM. Greenhouse gases from deforestation in Brazilian Amazonia: net committed emissions. Climatic Change. 1997; 35: 321–360. DOI: 10.1023/a:1005336 724350.

[69] Houghton RA, Lawrence KT, Hackler JL, Brown S. The spatial distribution of forest biomass in the Brazilian Amazon: a comparison of estimates. Global Change Biology. 2001; 7: 731–746. DOI: 10.1046/j.1365-2486.2001.00426.x.

[70] Yang ZC. Error Theory. Changsha: Central South University of Technology Press. 1987.

[71] Rejou-Mechain M, Muller-Landau HC, Detto M, Thomas SC, Le Toan T, Saatchi SS, Barreto-Silva JS, Bourg NA, Bunyavejchewin S, Butt N, Brockelman WY, Cao M,

Cardenas D, Chiang JM, Chuyong GB, Clay K, Condit R, Dattaraja HS, Davies SJ, Duque A, Esufali S, Ewango C, Fernando RHS, Fletcher CD, Gunatilleke I, Hao Z, Harms KE, Hart TB, Herault B, Howe RW, Hubbell SP, Johnson DJ, Kenfack D, Larson AJ, Lin L, Lin Y, Lutz JA, Makana JR, Malhi Y, Marthews TR, McEwan RW, McMahon SM, McShea WJ, Muscarella R, Nathalang A, Noor NSM, Nytch CJ, Oliveira AA, Phillips RP, Pongpattananurak N, Punchi-Manage R, Salim R, Schurman J, Sukumar R, Suresh HS, Suwanvecho U, Thomas DW, Thompson J, Uriarte M, Valencia R, Vicentini A, Wolf AT, Yap S, Yuan Z, Zartman CE, Zimmerman JK, Chave J. Local spatial structure of forest biomass and its consequences for remote sensing of carbon stocks. Biogeosciences. 2014; 11: 6827–6840. DOI: 10.5194/bg-11-6827-2014.

[72] Gregoire TG, Naesset E, McRoberts RE, Stahl G, Andersen HE, Gobakken T, Ene L, Nelson R. Statistical rigor in LiDAR-assisted estimation of aboveground forest biomass. Remote Sensing of Environment. 2016; 173: 98–108. DOI: 10.1016/j.rse.2015.11.012.

[73] Phillips DL, Brown SL, Schroeder PE, Birdsey RA. Toward error analysis of large-scale forest carbon budgets. Global Ecology and Biogeography. 2000; 9: 305–313. DOI: 10.1046/j.1365-2699.2000.00197.x.

[74] Ketterings QM, Coe R, van Noordwijk M, Ambagau Y, Palm CA. Reducing uncertainty in the use of allometric biomass equations for predicting above-ground tree biomass in mixed secondary forests. Forest Ecology and Management. 2001; 146: 199–209. DOI: 10.1016/s0378-1127(00)00460-6.

[75] Holdaway RJ, McNeill SJ, Mason NWH, Carswell FE. Propagating uncertainty in plot-based estimates of forest carbon stock and carbon stock change. Ecosystems. 2014; 17: 627–640. DOI: 10.1007/s10021-014-9749-5.

[76] Su YJ, Guo QH, Xue BL, Hu TY, Alvarez O, Tao SL, Fang JY. Spatial distribution of forest aboveground biomass in China: estimation through combination of spaceborne lidar, optical imagery, and forest inventory data. Remote Sensing of Environment. 2016; 173: 187–199. DOI: 10.1016/j.rse.2015.12.002.

[77] Wang XP, Fang JY, Tang ZY, Zhu B. Climatic control of primary forest structure and DBH-height allometry in Northeast China. Forest Ecology and Management. 2006; 234: 264–274. DOI: 10.1016/j.foreco.2006.07.007.

[78] Lopez-Serrano FR, Garcia-Morote A, Andres-Abellan M, Tendero A, del Cerro A. Site and weather effects in allometries: a simple approach to climate change effect on pines. Forest Ecology and Management. 2005; 215: 251–270. DOI: 10.1016/j.foreco.2005.05.014.

[79] Berninger F and Nikinmaa E. Implications of varying pipe model relationships on Scots Pine growth in different climates. Functional Ecology. 1997; 11: 146–156. DOI: 10.1046/j.1365-2435.1997.00067.x.

[80] Geron CD, Ruark GA. Comparison of constant and variable allometric ratios for predicting foliar biomass of various tree Genera. Canadian Journal of Forest Research-Revue Canadienne De Recherche Forestiere. 1988; 18: 1298–1304. DOI: 10.1139/x88-200.

[81] West GB, Brown JH, Enquist BJ. A general model for the structure and allometry of plant vascular systems. Nature. 1999; 400: 664–667.

[82] Enquist BJ, Brown JH, West GB. Allometric scaling of plant energetics and population density. Nature. 1998; 395: 163–165. DOI: 10.1038/25977.

[83] West GB, Brown JH, Enquist BJ. A general model for the origin of allometric scaling laws in biology. Science. 1997; 276: 122–126. DOI: 10.1126/science.276.5309.122.

[84] Marshall AR, Willcock S, Platts PJ, Lovett JC, Balmford A, Burgess ND, Latham JE, Munishi PKT, Salter R, Shirima DD, Lewis SL. Measuring and modelling above-ground carbon and tree allometry along a tropical elevation gradient. Biological Conservation. 2012; 154: 20–33. DOI: 10.1016/j.biocon.2012.03.017.

Chapter 3

Above-Ground Biomass Estimation with High Spatial Resolution Satellite Images

Adélia M. O. Sousa, Ana Cristina Gonçalves and
José R. Marques da Silva

Additional information is available at the end of the chapter

http://dx.doi.org/10.5772/65665

Abstract

Assessment and monitoring of forest biomass are frequently done with allometric functions per species for inventory plots. The estimation per area unit is carried out with an extrapolation method. In this chapter, a review of the recent methods to estimate forest above-ground biomass (AGB) using remote sensing data is presented. A case study is given with an innovative methodology to estimate above-ground biomass based on crown horizontal projection obtained with high spatial resolution satellite images for two evergreen oak species. The linear functions fitted for pure, mixed and both compositions showed a good performance. Also, the functions with dummy variables to distinguish species and compositions adjusted had the best performance. An error threshold of 5% corresponds to stand areas of 8.7 and 5.5 ha for the functions of all species and compositions without and with dummy variables. This method enables the overall area evaluation, and it is easily implemented in a geographic information system environment.

Keywords: QuickBird, multi-resolution segmentation, crown horizontal projection, forest inventory, regressions

1. Introduction

1.1. Role of forest inventories in biomass estimation

Forest covers a larger area on Earth and provides many products and services. Forest evaluation inventories were initiated when wood shortages arose. It can be said that they are the driving force to acquiring information on forest areas, stand composition and products. In the beginning

(Middle Ages), it was focused on timber volume and forest planning. With time demands for products and services changed, shifting their focus [1].

Forest inventories are based on a sampling design for a given threshold error, with ground plot assessment, providing data sets, which enable the forest evaluation. The need for measuring an increasing number of variables to evaluate products and services expected from forests turns them increasingly expensive and labour intensive. Remote sensing enabling the evaluation of some of those variables (*e.g.* areas, stand composition, crown cover) has been included in forests inventories. Though it cannot replace field work, it can rationalise it, allowing sample efficiency and reducing costs, error and labour intensity [1].

Biomass, both above and below ground, estimation, distribution and dynamics have been acquiring an increasing importance, especially in the last couple of decades. It was powered by the use of wood for bioenergy and by the evaluation of carbon stocks, sequestration and losses. The estimation of biomass can be grouped in two main methods: direct and indirect. The *direct method*, thought the most accurate, requires cutting trees, separating their different components (*e.g.* wood, bark, branches and leaves) and determining their dry weight. It is an expensive and labour demanding method, especially in what roots are concerned, which is feasible for small samples and prohibitive for large ones. The *indirect method* is traditionally based on mathematical relations between biomass (dependent variable) and one or a few easy-to-measure tree variables (independent variables). The biomass estimates attained with the former method are used to fit allometric functions at tree level, whose most frequent explanatory variables are the diameter at breast height and/or total height. Up to 2005 for Europe, North America and Australia, many functions were compiled by several authors [2–6]. More recent ones were found [7–12]. This large number of allometric functions stresses their species-specific and site-specific domain, reason why the selection of the best-suited function is of the utmost importance.

The tree-level biomass functions are frequently used to make estimations at plot scale (usually summing the biomass of all stems). Their estimation for a given area is based on the inventory plots (which depend on the sampling design and intensity for a certain error threshold) and an extrapolation method. In the literature, different extrapolation methods are described [1] with accuracy decreasing with the increase in the forest area, the variability of stand composition and structure, topography, soil and climate [13].

The forest inventories have cycles of 5 or 10 years. In between remote sensing can be used to evaluate the forest dynamics, in general, and biomass in particular. On the other hand, advantages can be gained as it can work at different scales and time frames; all area is under evaluation no requiring extrapolation methods; and maps can be produced in a geographical information system (GIS) environment.

A wide range of studies of biomass estimation derived from remote sensing can be found in the literature. In the next subchapter (2), a brief description of the methods and techniques is given. A case study (subchapter 3) will be used to illustrate the development of a methodology which was aimed to be simple and to be used either by researches or by technicians. The challenge was to produce data from remote sensing images for two evergreen oaks (holm and cork oak), native of the Mediterranean basin, and develop accurate above-ground biomass (AGB) allometric functions whose explanatory variable is crown horizontal projection, for pure and mixed stands of both species, in the latter analysing also the influence of the predictive ability of independent variables that differentiate species and composition.

The innovation of this study is the estimation of above-ground biomass considering or not stand composition, which is identified by the processing of high spatial resolution of satellite image data. In earlier studies, AGB estimation functions using satellite images did not consider the composition as independent variable. In the studies in which medium and low spatial resolution images are used, the pixel size frequently does not allow the separation of the species. High spatial resolution satellite images data enable the identification and delimitation of crowns of different tree species. Nonetheless, the studies with these satellite images seem to be focused in pure stands. The case study presented demonstrates that stand composition improves the accuracy of AGB estimation functions.

1.2. Contribution of remote sensing data to biomass estimation

Remote sensing was under a strong development in the last three decades due to the rapid advancement of remote sensing technology, increasing the availability of satellite imagery with different spectral, spatial, radioactive and temporal resolutions. Their sensor technologies enable a wide range of Earth surface monitoring scales [14] of forest areas distribution, species, and physic and biochemical properties [15]. The data can be derived from two sensors: the passive and the active.

In the ***passive sensor***, the optical spectral reflectance of Earth surface is registered. This multi-spectral information is sensitive to vegetation characteristics such as their shadows and textures, tree density, leaf area index (LAI) and crown size [16]. Good correlations are found between the former and the latter, namely with biomass [17] for a wide range of geographical areas [18]. To develop biomass functions, several methods are used. The most frequent is the regression analysis using forest inventory biomass data and satellite generated data [19], such as spectral reflectance, crown diameter and crown horizontal projection [20–24], original bands and/or vegetation indices [25–28]. The green vegetation canopy has different interactions with specific electromagnetic spectrum ranges, especially the red (R) and the near infrared (NIR) regions, due to the leaves photosynthetic activity and thus with LAI and biomass. The middle infrared (MIR) region, because of its ability to normalise the canopy cover variability, plays also an important contribution [15]. High relations are established between spectral reflectance and/or vegetation indices, and canopy structure, LAI or biomass [16] from tropical [29, 30] to Mediterranean evergreen oaks and shrubland [17, 26, 31, 32] landscapes. Vegetation indices, condensing satellite imagery data in a quantitative numeric form, are related to some forest parameter's estimation, such as the number of trees per hectare [33], canopy cover [26, 34] and volume, basal area and biomass [35]. The aforementioned indices show a better sensitivity when compared with the spectral reflectance [36]. Apart from regression, other methods are described in the literature, such as k-nearest neighbour, neural network, regression tree, random forest and support vector machine [29, 37–40].

Passive sensor, satellite imagery data, depends on its spatial resolution to determine directly the working scale and can be divided into three groups: low, medium and high.

The ***low spatial resolution*** satellite imagery data are characterised by scenes covering large geographical areas. The first studies of biomass estimation at local, regional and global scales used the images from the National Oceanic and Atmosphere Administration (NOAA) with the Advanced Very High Resolution Radiometer (AVHRR) sensor, with 1.1 km of spatial resolution; Moderate Resolution Imaging Spectroradiometer (MODIS), with 500 m [23]; and Satellite *Pour l'Observation de la Terre* (SPOT) Vegetation with 1 km of spatial resolution [41]. The advantage that can be pointed out in the biomass estimation with this spatial resolution is that one scene covers large area allowing working at country and global scales. Conversely, the pixel size makes the image classification more difficult due to the different land cover types encountered in it, resulting frequently in low accuracy biomass estimations [42]. To surpass this disadvantage in some studies, spectral mixture analysis was employed, where each pixel is a physical mixture of the multiple components weighted by the dominant area and the mixture spectrum is a linear combination of the endmember reflectance spectra [43].

The ***medium spatial resolution*** (30 m) satellite imagery data more frequently used result from the processing of Landsat Thematic Mapper (TM), Enhanced Thematic Mapper Plus (ETM +), Advanced Spaceborne Thermal Emission and Reflection radiometer (ASTER) and Wide Field Sensors (WiFS) [23, 25, 44, 45]. The data set is usually combined with forest inventory sample plots data and low spatial resolution imagery. Examples of the first are the volume and biomass estimation of conifer species of boreal forest in northern Europe, with Landsat images and NOAA-AVHRR [46] and Landsat and WiFS [25]; example of the second is AGB estimation for local and regional scales [25, 39, 46].

The data of ***high spatial resolution*** (IKONOS, QuickBird, WordView 2) have had, more recently, a high availability increase. Improved accuracy in biomass estimation is reached when compared with the former two spatial resolutions. The main disadvantage derives from their spatial resolution, which makes the data processing more time-consuming, thus better suited for local or regional scales [27]. In the literature, some references were found for these data sets, but few for biomass estimation, most of which with functions developed with regression analysis methods. Examples with IKONOS image data are the estimation of wet and dry biomass of oil palm plantations in Africa, considering vegetation indices and individual reflectance bands [47] and allometric equations of diameter at breast height and AGB as a function of crown width in tropical forests [48]. Examples with QuickBird are a methodology to estimate and map AGB for black spruce stands in Canada with shadow fraction as explanatory variable [49], AGB functions whose independent variable is crown horizontal projection, applying multi-resolution segmentation and object-oriented classification methods, with regression analysis for *Eucalyptus* sp. and *Tectona grandis* L. [50] and *Quercus rotundifolia* [51].

The three referred spatial resolutions of the optical remote sensing imagery for the same region and their image processing techniques allow the monitoring of AGB with different degrees of detail which are able to decrease the processing time and costs [19, 27, 52].

The ***active sensor*** is based on a mechanism that transmits and receives a portion of energy (microwaves region), interacting with Earth surface, and whose relevance to modelling forest variables depends on how deep the signal penetrates in the canopy. The mostly used data are derived from the Space-borne synthetic Aperture Radar (SAR) and L-band ALOS PALSAR. Two groups of data can be defined: RAdio Detection And Ranging (RADAR) and Light Detection and Ranging (LiDAR). The ***RADAR*** data's main advantage is their independence of atmospheric conditions as wavelength of radiation is much larger that the atmospheric particles [53], reason why it has acquired a marked importance in regions with frequent cloud cover [15]. Several examples of AGB estimation with RADAR were found in the literature [54–59]. The ***LiDAR*** is mainly obtained by airborne discrete-return [60], airborne waveform and from ground-based [61] LiDAR. Several authors developed nonlinear functions for AGB with its derived variables [15, 62] for individual trees, plot and stands, depending on the type of LiDAR system used [16]. LiDAR derives forest structural information (canopy cover), and RADAR backscatter seems to be able to estimate AGB with good accuracy [63].

In summary, the estimation of forest AGB with remote sensing data has advantages and disadvantages. To obtain very good accuracy, these approaches need a heavy field work in order to attain training data sets. However, many studies present methodologies with very good results at local, regional and global scale for AGB estimation. There is a trend towards the combination of several types of remote sensing imagery data to generate vegetation parameters and their relation with forest biomass.

2. Case study

2.1. Introduction

The data resulting from passive sensors, with low and medium spatial resolution, are usually used for the country or global scales. Due to their pixel size, it is not possible to identify and delimitate the tree crowns. Conversely, high spatial resolution satellite images enable it with good accuracy. The active sensors processing is more difficult than the passive sensors. Thus, when developing a methodology to be used both by researches and by technicians, the passive sensors data processing is more straightforward than that of the active sensors. The objectives of this study are the development of allometric functions for the estimation of AGB with crown horizontal projection, obtained with high spatial resolution satellite image, as independent variable with linear regression for (i) cork oak pure stands, (ii) mixed cork oak and holm oak stands (from this point forward referred as evergreen oaks) and (iii) both pure and mixed stands (from this point forward referred as all). Though with similar stand parameters there are some differences between pure stands of cork oak or holm oak and mixed stands of both species. Thus it was considered that the model for both species might be improved with the inclusion of dummy variables for species and composition as independent variables.

Cork oak (*Quercus suber* L.) and holm oak (*Quercus rotundifolia* Lam.) are native of the Mediterranean basin, distributed from Portugal to Greece and Portugal to Syria [64]. These species occur usually in low-density stands in a silvopastoral system, called *montado*, with open heterogeneous canopies, where the main products are bark and fruit, associated with extensive grazing, both in pure and in mixed stands. According to the Portuguese National Forest Inventory, cork oak and holm oak stands account for 22.5 and 13.7% of the forest area, respectively. The biomass for pure cork oak stands is 49.5 t/ha with an error of 5.2% and for holm oak 23.2 t/ha with an error of 7.2%. Mixed stands with cork oak or holm oak as dominant species correspond to 17% of their total area. The biomass estimation for mixed stands is 35.6 t/ha with an error of 13.7% [65]. They are similar in terms of habit, especially due to pruning to enhance fruit production and growth; both are slow growing [64].

2.2. Materials and methods

2.2.1. *Study area*

The study area is in southern Portugal, region of Mora, with approximately 80 km^2 (**Figure 1**). The area is mainly occupied by forest, composed predominantly by cork and holm oak in both pure and mixed stands. This region has a Mediterranean climate with a hot and dry summer and rainy winter with lower temperature. The terrain is characterised by small variations, with a mean elevation of about 200 m. The used QuickBird satellite image (August 2006) has a spatial resolution of 0.70 m, resulting from a fusion of panchromatic band with the four multi-spectral bands, blue (B), green (G), red (R) and near infrared (NIR).

2.2.2. *Processing satellite image*

The image was geometric and radiometric corrected using ENVI4.8 [66]. The geometric correction was based on ground control points obtained with Global Navigation Satellite System (GNSS) and geodetic vertices, identified on the ground and in the images, with a Root Mean Square Error (RMSE) of 0.49 m. The pixel values that are the digital numbers of the image were converted to top-of-atmosphere reflectance and finally to soil reflectance, using the dark object subtraction method [67]. The Normalized Difference Vegetation Index (NDVI) [68] was calculated with the latter and used to generate a vegetation mask, isolating tree crowns from other land cover types. For this propose, the multi-resolution segmentation method with the contrast split segmentation algorithm on eCognition, version 8.0.1 was applied [69]. Over this vegetation mask, forest species were identified using the nearest neighbour classification method, producing a map with the delimitation of trees crowns identified by specie [70].

The area was divided into a square grid of 45.5 m× 45.5 m (2070.25 m^2). Grids were classified according to the species present. The crown cover obtained with satellite data (CC_s) was defined as the percentage of area occupied by the crown horizontal projection per grid (CHP_s) in relation to the grid area. It was calculated with the vegetation mask per species using a geographic information system, ArcGIS, version 10 [71]. A grid was considered pure when crown cover of one species was equal or larger than 75% and mixed otherwise. The design of the forest inventory was a random stratified sampling by proportional allocation. Three strata were defined as function of crown cover: (i) 10–30%, (ii) 30–50% and (iii) >50%.

http://dx.doi.org/10.5772/65665

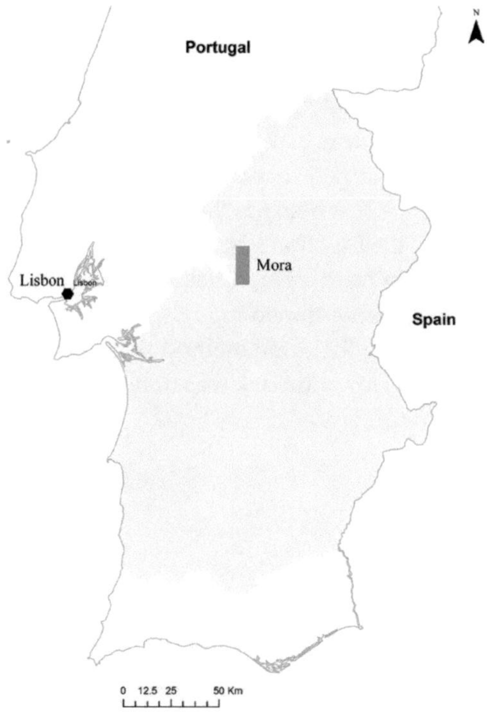

Figure 1. Location of the study area (dark grey) and Alentejo region (light grey).

2.2.3. Forest inventory data and biomass estimation

Forest inventory data set is composed of 17 plots of pure holm oak, 11 plots of pure cork oak and 23 mixed plots of holm oak and cork oak, with a total sampled area of 10.6 ha. In these plots, for all individuals with a breast height diameter ≥6 cm, the diameter at breast height, total height and crown radii in four directions (North, South, East and West) were measured [72]. Each tree geographical location was recorded by GNSS. Tree crown horizontal projection (CHP_p) was calculated as the circle whose radius is the arithmetic mean of the four crown radii measured. The relation (in %) of sum of the tree crown horizontal projections to plot area was used to estimate plot crown cover (CC_p). Ref. [8] developed a biomass allometric function per tree with simultaneous fitting for cork and holm oak. As the stands referred by the aforementioned authors were similar to those of this study, their functions were used to estimate above-ground biomass (*AGB*, Eq. (1)) as the sum of wood (*ww*), bark (*wb*) and crown (*wc*) biomass. Above-ground biomass per plot (AGB_p) was defined as the sum of *AGB* of all the trees in the plot.

$$AGB_t = ww + wb + wc = 0.164185 \times dbh^{2.011002} + 0.600169 \times dbh^{1.355957} + 1.909152 \times dbh^{1.200354} \quad (1)$$

2.2.4. Statistical analysis

Statistical analysis included correlation analysis and linear and multiple regression, implemented in R statistical software [73]. As AGB_p, CHP_p and CHP_s did not have normal distribution (evaluated with Shapiro-Wilk normality test), the Spearman correlation test was used. The linear function was selected as it is especially suited for one layer or without crown closure stands or forests [26]. As it was assumed that null *CHP* corresponded to null *AGB*, the linear regression through the origin (Eq. (2), where β is the slope) was used. The ordinary least square linear method was selected to fit the functions, both for the individual and for their cumulative values (**Figure 2.**) with CHP_s as independent variable. Also, in order to understand whether AGB could be estimated with aforementioned models, a preliminary analysis was carried out for plot inventory data (CHP_{pi}) with the same methodology for cork oak, evergreen oaks and all plots. The preliminary analysis for holm oak was done by [51], and the data were included in the evaluation for all plots in this study.

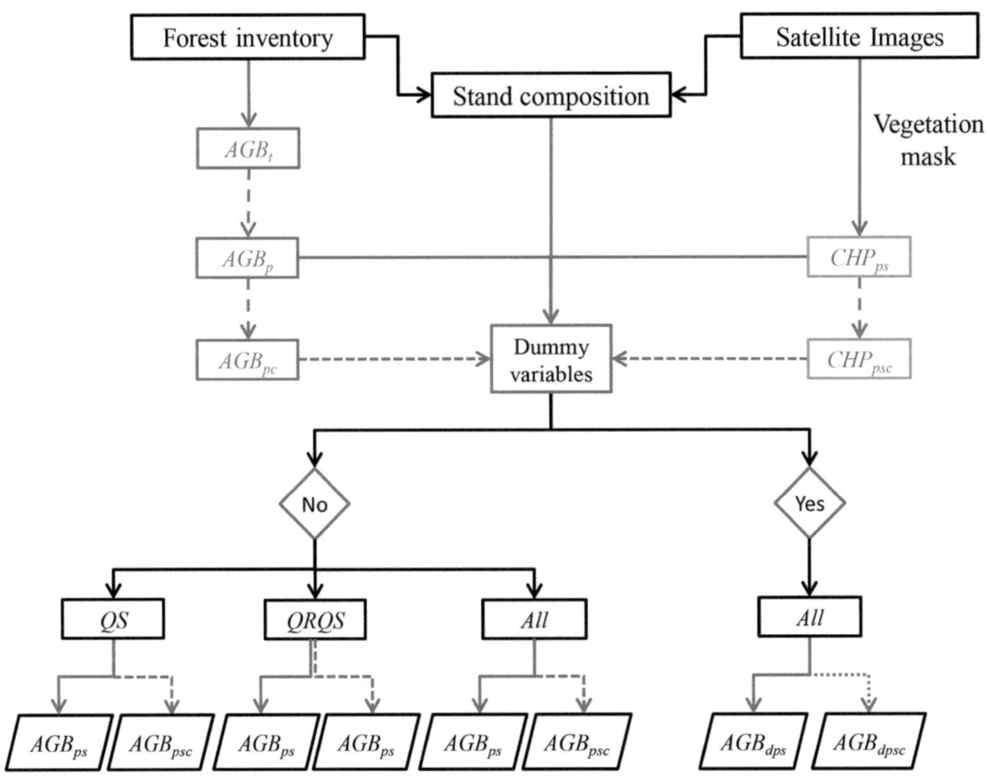

Figure 2. Flow diagram of the models developed (where AGB is the above-ground biomass in kg, CHP is the crown horizontal projection in m^2, t is the tree, p is the plot values, i is the inventory data, c is the cumulative values, s is satellite image data and d is dummy variables).

As the goal was to develop functions able to estimate AGB as function of *CHP* for the two evergreen oaks, in pure and mixed stands, dummy variables were defined to test the species composition contribution in the function predictive ability (**Figure 2**). Three dummy variables were defined. For each dummy variable, 1 indicates that the plot is pure holm oak (*dQR*), pure cork oak (*dQS*) or mixed holm oak and cork oak (*dQRQS*) and 0 otherwise. Multiple linear regression with stepwise method, using Akaike information criterion (AIC) as selection criteria, was used to fit the allometric functions (Eq. (3), where β_i are the regression coefficients, i = 0, 1, 2, 3). Multicollinearity among explanatory variables was analysed with variance inflation factor (VIF), considering that values exceeding 10 are sign of serious multicollinearity [74, 75].

As suggested by several authors [76, 77], the models were studied by the sum of squares of the residuals (SQR), the coefficient of determination (R^2) and the adjusted coefficient of determination (R^2_{aj}). Validation tests entail an independent data set. To overtake the inexistence of an independent data set, Refs. [78–80] suggest using predicted residual error. The sum of its square values, PRESS (Eq. (4)), and the sum of its absolute values, APRESS (Eq. (5)), were used as the validation test. The closer to the null value of residuals, the better is the model. The per cent value of the estimated and calculated AGB defined the error.

$$AGB = \beta \times CHP \tag{2}$$

$$AGB = \beta_0 \times CHP + \beta_1 \times dQS + \beta_2 \times dQR + \beta_3 \times dQRQS \tag{3}$$

$$PRESS = \sum_{i=1}^{n} \left(y_i - \hat{y}_{i,-i} \right)^2 \tag{4}$$

$$APRESS = \sum_{i=1}^{n} \left| y_i - \hat{y}_{i,-i} \right|^2 \tag{5}$$

2.3. Results and discussion

2.3.1. Multi-resolution segmentation and object-oriented classification

Figure 3 presents the vegetation mask resulting from the multi-resolution segmentation and object-oriented classification process (yellow line), over the QuickBird image with false colour composite for an area of Mora. The vegetation index NDVI enables a spectral signature sufficiently different to obtain a good distinction of the two forest species (cork and holm oak), with an agreement between the classification and ground truth using Kappa statistic [81, 82], of 78% and a global precision of 89%. Thus, the simultaneous analysis of these two evergreen oak species in pure and mixed stands can be useful for the estimation of AGB and also because, though they are visually similar, the methodology allows a mapping per species (cork and holm oak) with a good classification accuracy.

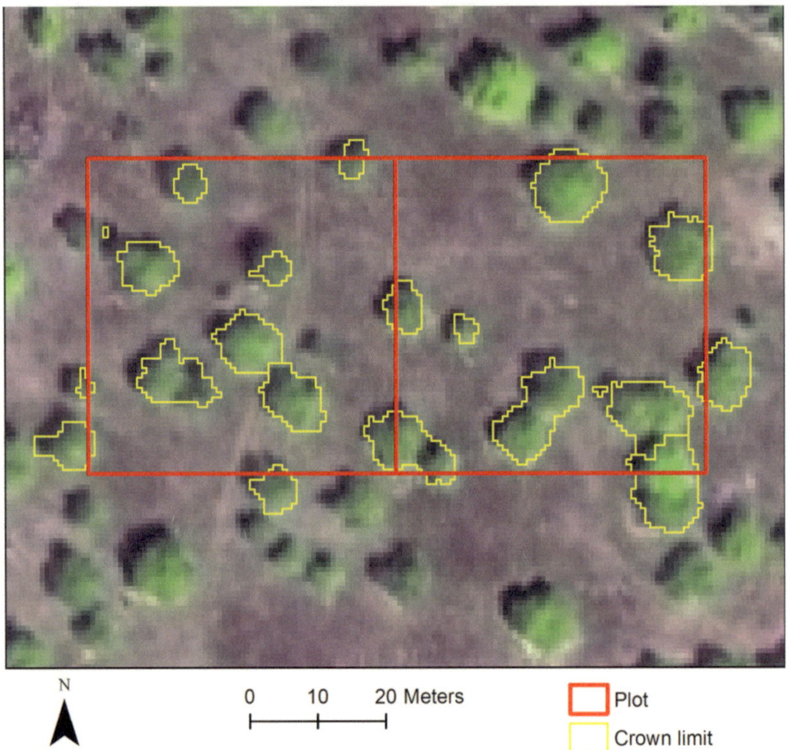

Figure 3. QuickBird image with false colour composite (RGB = red, NIR and blue) and illustration of the results of multi-resolution segmentation process (yellow line).

2.3.2. Individual trees and plot characteristics and above-ground biomass functions with inventory data

The descriptive statistics: minimum (*min*), maximum (*max*), arithmetic mean (*mean*), standard variation (*SD*) and coefficient of variation (*CV*) for all plots (*All*), evergreen oaks mixed plots (*QRQS*) and cork pure plots (*QS*) of the data are presented in **Table 1**. The density measures, number of trees per hectare (*N*), basal area per hectare (*G*), crown cover (CC_s) and above-ground biomass per hectare (*AGB*) were considered for the inventory data, while crown horizontal projection per plot (CHP_{ps}) and crown cover (CC_s) were considered for satellite data. In **Table 1**, the differences between the three groups of plots can be observed. Comparing the descriptive statistics of cork oak pure plots with those of evergreen oak plots (**Table 1**), higher densities and *AGB* are observed for the former. A more detailed analysis at tree level reveals that in the mixed plots about 54% of the trees are holm oak and 46% are cork oak. The differences are, at least partially, explained by the individual tree dimensions. While in the mixed plots mean crown radius is equal for both species (3.1 m) and similar to the cork oak pure plots (3.0 m), the mean total height and the mean quadratic diameter (d_g) are different:

6.8 m, 7.1 m, 35.8 cm and 37.6 cm, respectively. This is also reflected in the AGB_t with a mean larger for the cork oak pure plots (463.9 kgtree^{-1}) than for the mixed plots (424.0 kgtree^{-1}). When the two species are analysed in the evergreen oak plots, the cork oak individuals have, in average, larger dimensions (7.1 m of total height and 38.4 cm of d_g) than the holm oak ones (6.3 and 33.6 cm), which is also reflected in the mean AGB_t (481.7 and 382.3 kgtree^{-1}, respectively). The diameter distribution of the evergreen oak plots covers a wide spam indicative of stands of several cohorts (or age classes). The species spatial distribution visual analysis point outs towards an individual distribution both by species and by cohort. This might be probably due to the mutualistic and competitive interactions [83] and also by the sociability and compatibility [84] of these oaks, with similar growth rates, both slow, and rotation length. The holm oak pure plots [51] have values between the former two compositions. Also, trees with similar crown sizes may have quite different AGB_t. This is expectable considering two factors. First, these species have weak epinastic control, but when in group tend to reduce their lateral development due to branch abrasion or to the effects of high and low shading [83], especially because they are shade intolerant. As in *montado* tree spatial distribution is sparse, though irregular, trees can be found isolated or in clusters and in the latter is the competition phenomenon that explains, at least partially, the variability in crown dimensions. Second, in the *montado* system pruning to enhance fruit production is frequent [64] in mature stands and this reverses temporarily the rate between CHP_t and AGB_t.

	All plots (*All*)					**Evergreen oaks mixed plots (*QRQS*)**					**Cork oak pure plots (*QS*)**				
Variables	**min**	**max**	**mean**	**SD**	**CV**	**min**	**max**	**mean**	**SD**	**CV**	**min**	**max**	**mean**	**SD**	**CV**
N (treesha^{-1})	19	140	70	28	40.5	19	140	61	28	46.5	53	135	88	24	27.7
G (m^{2}ha^{-1})	2.6	15.4	7.1	2.7	38.1	2.6	13.1	6.2	2.1	34.3	5.2	15.4	9.8	3.0	31.2
CCp (%)	8.7	49.3	25.3	9.2	36.5	8.7	49.3	20.9	8.2	39.3	18.7	42.3	27.5	7.4	26.8
CCs (%)	13.7	70.5	35.0	14.7	41.9	14.3	58.8	28.5	11.2	39.2	23.6	70.5	44.9	13.5	30.0
CHP_{ps} (m^2)	284.2	1460.2	725.3	303.7	41.9	296.9	1216.7	589.3	230.9	39.2	489.0	1460.2	928.7	278.7	30.0
AGB (tha^{-1})	10.4	62.7	30.1	11.1	36.8	10.4	51.5	26.0	8.6	33.1	23.8	62.7	40.7	11.9	29.2

N: number of trees per hectare, G: basal area per hectare, CC_s: crown cover calculated with satellite data, CHP_{ps}: plot crown horizontal projection evaluated with satellite data, AGB: above-ground biomass per hectare, min: minimum, max: maximum, mean: arithmetic mean, SD: standard variation, CV: coefficient of variation.

Table 1. Descriptive statistics.

The strongest positive correlations are found for pure plots of cork oak between AGB_p and CHP_p (0.827), expected due to the smaller variability (**Table 1**). Strong positive correction for pure cork oak plots between AGB_p and CHP_s was attained (0.627). The difference between AGB_p and CHP_s and CHP_p can be, at least partially, explained by the stems spatial distribution, the shade casted by trees and understory. In tree clusters, multi-resolution segmentation process is not able to isolate tree crowns, including mixed pixels of vegetation, soil and/or shadow inside the polygon that delimitates the cluster. Some types of understory vegetation, near tree crowns, have similar spectral signature, and when delimiting the crown those pixels are included [85, 86]. The weakest correlations were observed for the evergreen oak plots between AGB_p and CHP_p (0.482) and CHP_s (0.552), which is expected due to the variability increase in those parameters. When all the plots are considered, strong positive correlations are found between AGB_p and CHP_p (0.671) and CHP_s (0.746).

Model	Plots	Equation	SQR	R^2	R^2_{aj}	PRESS	APRESS
		Inventory data					
1	QS	$AGB_{pi\_QS} = 14.7087 \times CHP_{pi\_QS}$	23576825	0.972	0.969	0.00000019	0.00101252
2	QRQS	$AGB_{pi\_QRQS} = 11.5889 \times CHP_{pi\_QRQS}$	74401387	0.899	0.894	0.00000065	0.00313966
3	All	$AGB_{pi\_All} = 11.4139 \times CHP_{pi\_All}$	187602462	0.916	0.915	0.00000113	0.00631091
4	QS	$AGB_{pic\_QS} = 14.8889 \times CHP_{pic\_QS}$	16220130	0.9995	0.999	0.00000002	0.00034940
5	QRQS	$AGB_{pic\_QRQS} = 11.75374 \times CHP_{pic\_QRQS}$	292087434	0.997	0.997	0.00000004	0.00049897
6	All	$AGB_{pic\_All} = 11.75374 \times CHP_{pic\_All}$	3066829502	0.998	0.998	0.00000001	0.00049770
		Satellite image data					
7	QS	$AGB_{ps\_QS} = 8.847 \times CHP_{ps\_QS}$	40091932	0.953	0.948	0.00000016	0.00106716
8	QRQS	$AGB_{ps\_QRQS} = 8.5093 \times CHP_{ps\_QRQS}$	72931734	0.901	0.896	0.00000061	0.00317184
9	All	$AGB_{ps\_All} = 8.1309 \times CHP_{ps\_All}$	163946315	0.927	0.925	0.00000120	0.00656152
10	QS	$AGB_{psc\_QS} = 9.2671 \times CHP_{psc\_QS}$	43379144	0.999	0.998	0.00000001	0.00025905
11	QRQS	$AGB_{psc\_QRQS} = 9.037 \times CHP_{psc\_QRQS}$	104841889	0.999	0.999	0.00000005	0.00068011
12	All	$AGB_{psc\_All} = 8.85869 \times CHP_{psc\_All}$	1722352935	0.999	0.999	0.00000001	0.00050963
		Satellite image data with dummy variables					
13	All	$AGB_{dps} = 4.836 \times CHP_{ps} + 2191.3997 \times dQR + 3944.2222 \times dQS + 2532.8074 \times dQSQR$	104975100	0.953	0.949	0.00000151	0.00661458
14	All	$AGB_{dpsc} = 8.6029 \times CHP_{psc} + 6425.9261 \times dQR + 6105.1531 \times dQS + 4827.1112 \times dQSQR$	1235713930	0.991	0.999	0.00000007	0.00070861

AGB: above-ground biomass in kg, CHP: crown horizontal projection in m^2, p: plot, i: inventory data, c: cumulative values, s: satellite data, QS: cork oak, QSQR: evergreen oaks, All: all plots, d: dummy variable.

Table 2. Properties of the fitted models.

The statistical properties and the PRESS and APRESS statistics of the linear models for plot CHP_{pi} (models 1–3) and CHP_{pic} (models 4–6) are indicative of their good performance (**Table 2**).

When comparing the models for the four sets of plots, the statistical properties of the pure plots are better than those of the evergreen oak plots and all plots (**Table 2**), due to the increase in variability in the latter two. The results attained by [51] for holm oak pure plots are also better when compared with the evergreen oak and all plots of this study. Nonetheless, smaller differences in model performance are found for models 4–6 when compared with models 1–3. The differences might be partially due to the small differences in habit of the two species, the spatial distribution pattern of the individuals of both species and silvicultural practices, especially pruning where focus is put in the crown enlargement more for holm oak than for cork oak. This is also reflected in the overall model error. Larger overall errors are found for the models 2 and 3 (24.3% and 23.7%, respectively) when compared with model 1 (13.6%), and models 5 and 6 (9.8% and 9.2%, respectively) when compared with model 4 (4.4%).

2.3.3. Above-ground biomass allometric functions with satellite image data

The statistical properties and the validation statistics of the linear models for CHP_{ps} and CHP_{psc} for pure cork oak plots (models 7 and 10), for evergreen oak plots (models 8 and 11) and for all plots (models 9 and 12, **Figure 4a**, **b**) are indicative of their good performance (**Table 2**). These results are in accordance with [26] that states that linear regression behaves well in stands with one layer, while [47] refers that in multi-layered stands nonlinear functions are better suited. Noteworthy is that models 10–12 have better statistical properties than the models 7–9. When comparing the models for evergreen oaks with those for pure plots, the statistical properties of the former are worse than of latter (**Table 2**). This, as already referred, is due to the larger variability observed in the mixed plots. When comparing the models fitted with inventory and satellite image data, the performance is better for the former, which can be partially explained by the differences in the calculation of *CHP* with the sets of data. Models for cumulative values have slightly better performance for the inventory (models 4–6) than for the satellite image data (models 10–12), corroborated by the overall error of 3.7, 4.4, 9.8 and 9.2%, and 13.4, 4.8, 6.4 and 5.8%, respectively.

Model 10, for cork oak pure stands, has larger errors up to 4000 m^2 and an irregular trend afterwards. For error thresholds of 10 and 5% *CHP,* areas larger than 4000 and 5000 m^2, respectively, are needed. Considering a mean *CHP* area of 4486 m^2ha^{-1}, the corresponding stand area is 0.9 and 1.1 ha, respectively. Model 11, for evergreen oak stands, shows higher errors than the former models for *CHP* areas up to 1000 m^2, with a continuous decrease, in absolute value, up to 3000 m^2 and stabilisation afterwards. Errors smaller or equal than 10 and 5% are attained for *CHP* areas of 3000 and 7000 m^2, respectively. The mean *CHP* area for these stands is 2847 m^2ha^{-1}, which corresponds to the above-referred errors to stand areas of 1.1 and 2.5 ha, respectively. Model 12, for both pure and mixed stands, shows the same error trend as model 11. Errors below the aforementioned are attained for *CHP* areas of 2000 and 8000 m^2. As the mean *CHP* area is 3503 m^2ha^{-1}, the corresponding stand areas are 0.6 and 2.3 ha. As 92% of the holm oak stands and 89% of the cork oak stands have areas bigger than 2 ha [65], this allometric function can be used at local and regional level.

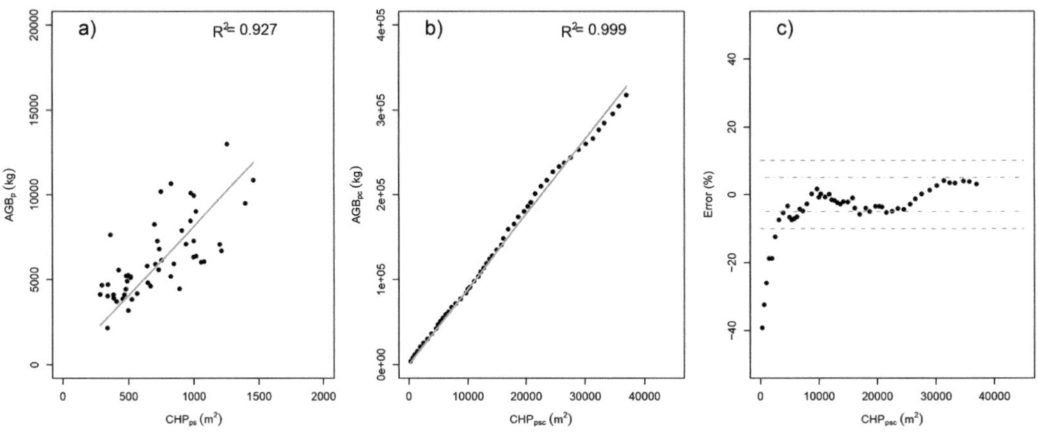

Figure 4. Crown horizontal projection derived from satellite image per plot (CHP_{ps}) versus above-ground biomass per plot (AGB_p) and equation (in grey) for model 9 (a) and their cumulative values, CHP_{psc} vs AGB_{pc}, and equation (in grey) (b) and error (c) for model 12.

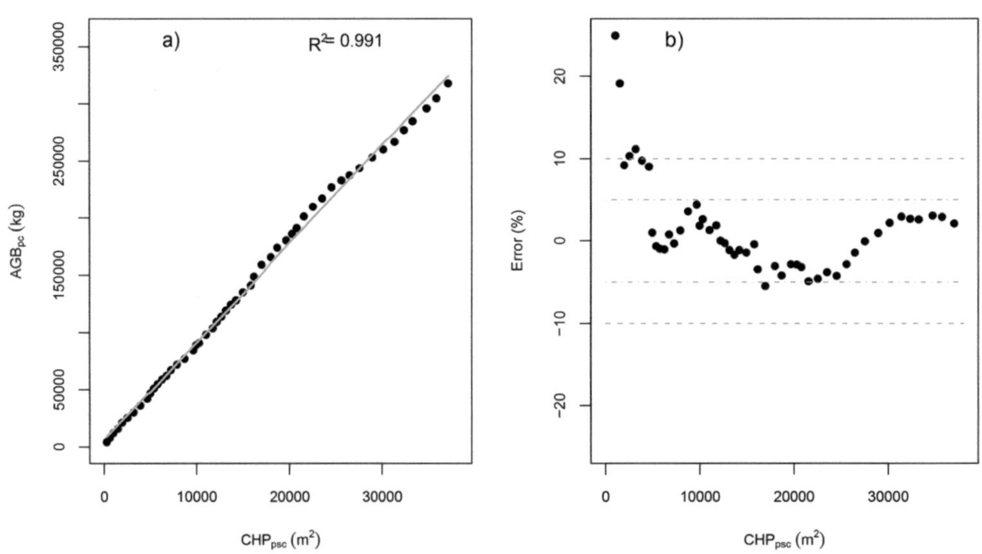

Figure 5. Cumulative values of crown horizontal projection derived from satellite image per plot (CHP_{psc}) versus cumulative values of above-ground biomass per plot (AGB_p) and equation (in grey, a), and error distribution (b) for model 14.

The inclusion of independent variables identifying the plot composition originated a generalised improvement in the statistical properties of the models (**Table 2**) for CHP_{ps} (model 13) and CHP_{psc} (model 14), when compared with those without composition variables (models 9 and 12). In fact, plot composition has revealed important as for both models the three dummy variables were included. Also, multicollinearity did not present any problem as VIF values were always lower than 10, being for *CHP*, *dQR*, *dQS* and *dQRQS*, 8.5, 3.8, 3.6 and 3.2 for model 13, and 4.6, 1. 8, 3.0 and 1.8 for model 14, respectively. The models accuracy increases from the models with only *CHP* (9 and 12) to those that include the dummy variables (models 13 and 14). Noteworthy is also the decrease in SQR of 36% from models 9 to 13, and of 39% from models 12 to 14, with the corresponding increase in the adjusted regression coefficients. There is a considerable improvement in the overall error from models 9 to 13, from 23.1 to 5.2% and from models 12 to 14, from 5.8 to 0.1%. The error for model 14 shows an irregular trend with error smaller than 10 and 5% for cumulative crown projection for areas larger than 1500 and 2000 m^2, corresponding to stand areas of 0.4 ha and 0.6 ha, respectively (**Figure 5**). Considering the threshold of 5% for model 14, -1.7 ha is needed than for model 12.

All the fitted models (models 7–14) have a high goodness of fit, with more than 90% of the variability explained. It seems that crown horizontal projection as better predictive abilities with linear functions than the shadow fraction processed from QuickBird images [49]. The models developed in this study show also better performances than multiple regression functions with crown diameter and total height as independent variables, derived from a panchromatic band of QuickBird satellite [50] or than exponential functions with reflectance of band 3 or NDVI for IKONOS image [47]. Likewise, biomass estimation functions derived from medium spatial resolution satellite image data do not show as good performance as those obtained in this study. This is probably related to dissimilarity between ground and satellite data [27]. The former was confirmed in a study with Landsat 5 TM and MODIS images for P*inus pinaster* stands, whose independent variable is vegetation indices [17].

2.4. Conclusions

Remote sensing gives a good contribution to estimate the above-ground biomass. The high spatial resolution satellite image allows mapping with high accuracy the above-ground biomass at local and regional scale.

The overestimation of crown horizontal projection per plot using high spatial resolution satellite image, when compared with that calculated using inventory data, is related to the inclusion of mixed pixels in the boundary of tree crown delimitation in the multi-resolution segmentation and trees' spatial distribution. Though the cork and holm oak are visually similar, their spectral signature is sufficiently different to obtain a very good classification by forest specie.

In general, above-ground biomass allometric functions for plot and their cumulative values showed a good performance, though better for pure stands than for mixed stands, which might be explained by the larger variability observed in the latter. However, the inclusion of dummy variables, which reflect the differences between the species and the stand structure, originates a generalised improvement in the functions performance. The same threshold errors of 10 and 5% are attained by the latter with less 33 and 70% of the stand areas. Considering that to a 10% error correspond stand areas of 1.1 ha, it includes about 90% of the area occupied by these species in Portugal [65]. This method has the advantage of enabling the overall area evaluation, not requiring forest inventory or extrapolation procedures. It can be used when species composition is not (models 9 and 12) and is (models 13 and 14) differentiated.

Acknowledgements

The authors would like to thank the forest producers for allowing plot installation and to Paulo Mesquita for satellite image processing. The work was financed by Programa Operativo de Cooperação Transfronteiriço Espanha—Portugal (POCTEP); project Altercexa—Medidas de Adaptación y Mitigación del Cambio Climático a Través del Impulso de las Energías Alternativas en Centro, Alentejo y Extremadura (Ref_a 0317_Altercexa_I_4_E and 0406_ALTERCEXA_II_4_E); TrustEE—innovative market based Trust for Energy Efficiency investments in industry (Project ID: H2020—696140); and National Funds through FCT—Foundation for Science and Technology under the Project UID/AGR/00115/2013.

Abbreviations

AGB	Above-Ground Biomass per tree (kg)
AGB_t	tree Above-Ground Biomass per tree (kg)
AIC	Akaike Information Criterion
All	all plots
ALOS	Advanced Land Observing Satellite
APRESS	Absolute Predicted Residual Error Sum
ASTER	Advanced Spaceborne Thermal Emission and Reflection Radiometer
AVHRR	Advanced Very High Resolution Radiometer
c	cumulative values
CC	Crown Cover (%)
CHP	Crown Horizontal Projection (m^2)
CV	Coefficient of Variation
d	dummy variable
dbh	diameter at breast height, measured at 1.30 m height (cm)
dg	mean quadratic diameter (cm)
G	basal area per hectare (m^2ha^{-1})
GNSS	Global Navigation Satellite System
i	inventory data
ETM +	Enhanced Thematic Mapper +
LAI	Leaf Area Index
max	maximum value
mean	arithmetic mean value
min	minimum value
MODIS	Moderate Resolution Imaging Spectroradiometer
N	number of trees per hectare (tree ha^{-1})
NDVI	Normalized Difference Vegetation Index

http://dx.doi.org/10.5772/65665

NOAA	National Oceanic and Atmospheric Administration
p	plot
PALSAR	Phased Array type L-band Synthetic Aperture Radar
PRESS	Predicted Residual Error Sum of Squares
QR	holm oak (*Quercus rotundifolia*) pure plots
QRQS	mixed plots of holm and cork oaks (evergreen oaks)
QS	cork oak (*Quercus suber*) pure plots
R^2	coefficient of determination
R^2_{aj}	adjusted coefficient of determination
RADAR	RAdio Detection And Ranging
RMSE	Root Mean Square Error
s	satellite data images
SAR	Space-borne synthetic Aperture Radar
SD	Standard Deviation
SQR	Sum of Squares of the Residuals
t	tree
TM	Thematic Mapper
VIF	Variance Inflation Factor
WiFS	Wide Field Sensor
ww	wood biomass (kg)
wb	bark biomass (kg)
wc	crown biomass (kg)

Author details

Adélia M. O. Sousa*, Ana Cristina Gonçalves and José R. Marques da Silva

*Address all correspondence to: asousa@uevora.pt

Department of Rural Engineering, School of Sciences and Technology, Institute of Mediterranean Agricultural and Environmental Sciences (ICAAM), Institute of Research and Advanced Information (IIFA), University of Évora, Évora, Portugal

References

[1] McRoberts R, Tomppo E, Naesset E. Advanced and emerging issues on national forest inventories. Scandinavian Journal of Forest Research. 2010;25:368–381.

[2] Ter-Mikaelian MT, Korzukhin MD. Biomass equations for sixty-five North American tree species. Forest Ecology and Management. 1997;97:1–24.

[3] Eamus D, McGuinness K, William B. editors. Review of Allometric Relationships for Estimating Woody Biomass for Queensland, the Northern Territory and Western Australia, Technical report no. 5a. Australia: Australian Greenhouse Office. 2000. 56p.

[4] Keith H, Barrett D, Keenan R. editors. Review of Allometric Relationships for estimating Woody Biomass for New South Wales, the Australian Capital Territory, Victoria, Tasmania and South Australia, Technical report no. 5b. Australia: Australian Greenhouse Office. 2000. 112p.

[5] Jenkins JC, Chojnacky DC, Heath LS, Birdsey RA. National-scale biomass estimators for United States tree species. Forest Science. 2003;49(1):12–35.

[6] Zianis D, Muukkonen P, Mäkipää R, Mencuccini M. Biomass and Stem Volume equations for tree species in Europe. The Finnish Society of Forest Science, The Finnish Forest Research Institute. Tampere, Finland; 2005. 63p.

[7] Fehrmann L, Kleinn C. General considerations about the use of allometric equations for biomass estimation on the example of Norway spruce in central Europe. Forest Ecology and Management. 2006;236:412–421.

[8] Paulo JA, Tomé M. Equações para estimação do volume e biomassa de duas espécies de carvalhos: *Quercus suber* e *Quercus ilex*. [Equations for estimation of the volume and the biomass of two oak species: *Quercus suber* and *Quercus ilex*]. Publicações do GIMREF; RC1. Instituto Superior de Agronomia, Departamento de Engenharia Florestal. Lisboa. 2006.

[9] Miksys V, Varnagiryte-Kabasinskiene I, Stupak I, Armolaitis K, Kukkola M, Wójcik J. Above-ground biomass functions for Scots pine in Lithuania. Biomass and Bioenergy. 2007;31:685–692.

[10] Correia AC, Faias S, Tomé M, Evangelista M, Freire J, Ochoa P. Ajustamento simultâneo de equações de biomassa de pinheiro manso no Sul de Portugal. [Simultaneous fitting of biomass equations for stone pine in southern Portugal]. Silva Lusitana. 2008;16:197–205.

[11] Djomo AN, Ibrahima A, Saborowski J, Gravenhorst G. Allometric equations for biomass estimations in Cameroon and pan moist tropical equations including biomass data from Africa. Forest Ecology and Management. 2010;260: 1873–1885.

[12] Arias D, Calvo-Alvarado J, Richter DB, Dohrenbusch A. Productivity, aboveground biomass, nutrient uptake and carbon content in fast-growing tree plantations of native and introduced species in the Southern Region of Costa Rica. Biomass and Bioenergy. 2011;35:1779–1788.

[13] Somogyi Z, Cienciala E, Mäkipää R, Muukkonrn P, Lehtonen A, Weiss P. Indirect methods of large-scale forest biomass estimation. European Journal of Forest Research. 2007;126:197–207.

[14] Gail WB. Remote sensing in the coming decade: the vision and the reality. Journal of Application of Remote Sensing. 2007;1(1):012505.

[15] Boyd, DS, Danson FM. Satellite remote sensing of forest resources: three decades of research development. Progress in Physical Geography. 2005;29(1):1–26.

[16] Zhang X, Ni-meister W. Remote sensing of forest biomass. In: Hanes J.M. editor. Biophysical Applications of Satellite Remote Sensing. Springer-Verlag: Berlin Heidelberg; 2014. p. 63–98.

[17] Viana H, Aranha J, Lopes D, Cohen WB. Estimation of crown biomass of *Pinus pinaster* stands and shrubland above-ground biomass using forest inventory data, remotely sensed imagery and spatial prediction models. Ecological Modelling. 2012;226:22–35.

[18] Ahmed T, Tian L, Zhang Y, Ting KC. A review of remote sensing methods for biomass feedstock production. Biomass and Bioenergy. 2011;35:2455–2469.

[19] Pizaña JMG, Hernández JMN, Romero NC. Remote sensing-based biomass estimation. In: Marghny M. editor. Earth and Planetary Sciences, Geology and Geophysics, Environmental Applications of Remote Sensing. Croatia: Intech; 2016. p. 3–40.

[20] Woodcock CE, Collins JB, Jakabhazy VD, Li X, Macomber S, Wu Y. Inversion of the Li-Strahler canopy reflectance model for mapping forest structure. IEEE Transactions on Geoscience and Remote Sensing. 1997;35:405–414.

[21] Phua M, Saito H. Estimation of biomass of a mountainous tropical forest using Landsat TM data. Canadian Journal of Remote Sensing. 2003;29:429–440.

[22] Popescu SC, Wynne RH, Nelson, RF. Measuring individual tree crown diameter with LiDAR assessing its influence on estimating forest volume and biomass. Canadian Journal of Remote Sensing. 2003;29:564–577.

[23] Baccini A, Friedl MA, Woodcock CE, Warbington R. Forest biomass estimation over regional scales using multisource data. Geophysical Research Letters. 2004;31:L10501.

[24] Powell, SL, Warren B, Cohen WB, Healey SP, Kennedy RE, Moisen GG, Pierce KB, Ohmann JL. Quantification of live aboveground forest dynamics with Landsat time-series and field inventory data: a comparison of empirical modelling approaches. Remote Sensing of Environmental. 2010;114:1053–1068.

[25] Tomppo E, Nilsson M, Rosengren M, Aalto P, Kennedy P. Simultaneous use of Landsat-TM and IRS-1C WiFS data in estimating large area tree stem volume and aboveground biomass. Remote Sensing of Environment. 2002;82:156–171.

[26] Carreiras JM, Pereira JMC, Pereira JS. Estimation of tree canopy cover in evergreen oak woodlands using remote sensing. Forest Ecology and Management. 2006;223:45–53.

[27] Lu D. The potential and challenge of remote sensing-based biomass estimation. International Journal of Remote Sensing. 2006;27:1297–1328.

[28] Muukkonen P, Heiskanen J. Biomass estimation over a large area based on standwise forest inventory data and ASTER and MODIS satellite data: a possibility to verify carbon inventories. Remote Sensing of Environment. 2007;107:617–624.

[29] Steininger MK, Steininger, M.K., 2000. Satellite estimation of tropical secondary forest above ground biomass data from Brazil and Bolivia. International Journal of Remote Sensing. 2000;21:1139–1157.

[30] Foody GM, Boyd DS, Cutler MEJ. Predictive relations of tropical forest biomass from Landsat TM data and their transferability between regions. Remote Sensing of Environment. 2003;85:463–474.

[31] Pereira JMC, Oliveira TM, Paul JCP. Satellite-based estimation of Mediterranean shrubland structural parameters. EARSel Advances in Remote Sensing. 1995;4(3–XII): 14–20.

[32] Calvão T, Palmeirim JM. Mapping Mediterranean scrub with satellite imagery: biomass estimation and spectral behaviour. International Journal of Remote Sensing. 2004;25(16):3113–3126.

[33] Joffre R, Lacaze B. Estimating tree density in oak savanna-like 'dehesa' of southern Spain from SPOT data. International Journal of Remote Sensing. 1993;14:685–697.

[34] Salvador R, Pons X. On the applicability of Landsat TM images to Mediterranean forest inventories. Forest Ecology and Management. 1998;104:193–208.

[35] Maciel MNM, Bastos PCO, Carvalho JOP, Watrin OS. Uso de imagens orbitais na estimativa de parâmetros estruturais de uma floresta primária no município de Paragominas, Estado do Pará. [Use of orbital images in the estimation of structural parameters of a primary forest in the municipality of Paragominas, Pará state]. Revista Ciência Agrária. 2009;52:159–178.

[36] Asrar G, Fuchs M, Kanemasu ET, Hatfield JL. Estimating absorbed photosynthetic radiation and leaf area index from spectral reflectance in wheat. Agronomy Journal. 1984;76:300–306.

[37] Nelson RF, Kimes DS, Salas WA, Routhier M. Secondary forest age and tropical forest biomass estimation using Thematic Mapper imagery. BioScience. 2000;50:419–431.

[38] Foody GM, Cutler ME, McMorrow J, Pelz D, Tangki H, Boyd DS, Douglas I. Mapping the biomass of Bornean tropical rain forest from remotely sensed data. Global Ecology & Biogeography. 2001;10:379–387.

[39] Zheng D, Rademacher J, Chen J, Crow T, Bresee M, Le Moine J, Ryu S. Estimating aboveground biomass using Landsat 7 ETM+ data across a managed landscape in northern Wisconsin, USA. Remote Sensing of Environment 2004; 93:402–411

[40] Lu D, Chen Q, Wang G, Liu L, Li G, Moran EA. A survey of remote sensing-based aboveground biomass estimation methods in forest ecosystems. International Journal of Digital Earth. 2016;9(1):63–105.

[41] Fraser RH, Li Z. Estimating fire-related parameters in boreal forest using SPOT VEGETATION. Remote Sensing of Environment. 2002;82:95–110.

[42] Fernández-Manso O, Fernández-Manso A, Quintano C. Estimation pf aboveground biomass in Mediterranean forests by statistical modelling of ASTER fraction images. International Journal of Applied Earth Observation and Geoinformation. 2014;31:45–56.

[43] Tompkins S, John FM, Carld MP, Donald WF. Optimization of Endmembers Mixture Analysis for Spectral. Remote Sensing of Environment. 1997; 59:472–489.

[44] Baccini A, Laporte N, Goetz MS, Dong H. A first map of tropical Africa's above-ground biomass derived from satellite imagery. Environmental Research Letters. 2008;3:045011.

[45] Zhang X, Kondragunta S. Estimating forest biomass in the USA using generalized allometric models and Modis land products. Geophysical Research Letters. 2006;33:L09402.

[46] Häme T, Salli A, Andersson K, Lohi A. A new methodology for the estimation of biomass of conifer-dominated boreal forest using NOAA AVHRR data. International Journal of Remote Sensing. 1997;18(15):3211–3243.

[47] Thenkabail PS, Stucky N, Griscom BW, Ashton MS, Diels J, Meer B, Enclona E. Biomass estimations and carbon stock calculations in the oil palm plantations of African derived savannas using IKONOS data. International Journal of Remote Sensing. 2004;25(23): 5447–5472.

[48] Palace M, Keller M, Asner GP, Hagen S, Braswell B. Amazon forest structure from IKONOS satellite data and the automated characterization of forest canopy properties. Biotropica. 2008;40:141–150.

[49] Leboeuf A, Beaudoin A, Fournier RA, Guindon L, Luther JE, Lambert MC. A shadow fraction method for mapping biomass of northern boreal black spruce forests using QuickBird imagery. Remote Sensing of Environment. 2007;110:488–500.

[50] Upgupta S, Singh S, Tiwari PS. Estimation of aboveground phytomass of plantations using digital photogrammetry and high resolution remote sensing data. Journal of the Indian Society of Remote Sensing. 2015;43(2):311–323.

[51] Sousa AMO, Gonçalves AC, Mesquita P, Marques da Silva JR. Biomass estimation with high resolution satellite images: a case study of *Quercus rotundifolia*. ISPRS Journal of Photogrametric and Remote Sensing. 2015;101:69–79.

[52] Zhu X, Liu D. Improving forest aboveground biomass estimation using seasonal Landsat NDVI time-series. ISPRS Journal of Photogrammetry and Remote Sensing. 2015;102:222–231.

[53] Quegan, S. Recent advances in understanding SAR imagery. In: Danson FM, Plummer SE, editors. Advances in Environmental Remote Sensing. Chichester: Wiley; 1995. p. 89–104.

[54] Imhoff ML. A theoretical analysis of the effect of forest structure on synthetic aperture radar backscatter and the remote sensing of biomass. IEEE Transactions on Geoscience and Remote Sensing. 1995;33:341–52.

[55] Kuplich TM, Freitas CC, Soares JV. 2000. The study of ERS-1SAR and Landsat TMsynergism for landuse classification. International of Remote Sensing. 2000;21(10):2101–2111.

[56] Castel T, Guerra F, Caraglio Y, Hollier F. Retrieval biomass of a large Venezuelan pine plantation using JERS-1 SAR data. Analysis of forest structure impact on radar signature. Remote Sensing of Environment. 2002;79:30–41.

[57] Sun W, Heidt V, Gong P, Xu G. Information fusion for rural land-use classification with high resolution satellite imagery. IEEE Transactions on Geoscience and Remote Sensing. 2003;3(1):3–5.

[58] Santos JR. Airborne Pband SAR applied to the aboveground biomass studies in the Brazilian tropical rainforest. Remote Sensing of Environment. 2003;87:482–493.

[59] Treuhaft RN, Law BE, Asner GP. Forest attributes from Radar interferometric structure and its fusion with optical remote sensing. BioScience. 2004;54:561–71.

[60] Garcia M, Riaño D, Chuvieco E, Danson FM. Estimating biomass carbon stocks for Mediterranean forest in Spain using height and intensity LiDAR data. Remote Sensing of Environment. 2010;114:816–830.

[61] Ni-Meister W, Lee S, Strahler AH, Woodcock CE, Schaaf C, Yao T, Ranson KJ, Sun G, Blair JB. Assessing general relationships between aboveground biomass and vegetation structure parameters for improved carbon estimate from LiDAR remote sensing. Journal of Geophysical Research. 2010;115:1–12.

[62] He Q, Erxue C, An R, Li Y. Above-ground biomass and biomass components estimation using LiDAR data in a coniferous forest. Forests. 2013;4:984–1002.

[63] Tian S, Tanase MA, Panciera R, Hacker J, Lowell K. Forest biomass estimation using radar and LiDAR synergies. In: Proceedings of the IEEE International Geoscience and Remote Sensing Symposium – IGARSS. 21–26 July 2013; Melbourne, Australia: IEEE; 2013. p. 2145–2148.

[64] Correia AV, Oliveira AC. Principais espécies florestais com interesse para Portugal: zonas de influência mediterrânica. [Main forest species with interest for Portugal: zones of Mediterranean influence]. Direcção-Geral das Florestas. Estudos e Informação; 1999. 318p.

[65] IFN5, 2010. Inventário Florestal Nacional. IFN5 2005–2006. Portugal Continental. [National Forest Inventory. IFN5 2005–2006. Continental Portugal]. Autoridade Florestal Nacional, Lisboa.

[66] Envi. Reference Guide—Exelis Visual Information Solutions. Boulder, Colorado: Exelis Visual Information Solutions [Internet]. 2009. Available from: http://www.exelis-vis.com/portals/0/pdfs/envi/envi_zoom_user_guide.pdf. Accessed 2012-11-27.

[67] Chavez Jr. PS. An improved dark-object subtraction technique for atmospheric scattering correction of multispectral data. Remote Sensing of Environment. 1988;24(3): 459–479.

[68] Rouse JW, Haas RH, Schell JA, Deering DW. 1973. Monitoring vegetation systems in the Great Plains with ERTS. In 3rd ERTS Symposium, NASA SP-351 I. 309–317.

[69] Definiens Imaging. eCognition Developer 8.0.1 Reference Book [Internet]. 2010. Available from: http://www.definiens.com. Accessed 2012-10-23.

[70] Sousa AMO, Mesquita PA, Gonçalves AC, Marques da Silva JR. 2010. Segmentação e classificação de tipologias florestais a partir de imagens Quickbird. [Segmentation and classification of forest typologies using Quickbird images]. Ambiência (special edition). 2010;6:57–66.

[71] Esri. ArcGIS Desktop: Release 10. Redlands, CA: Environmental Systems Research Institute [Internet]. 2010. Available from: http://www.esri.com. Accessed 2013-01-23.

[72] Avery TE, Burkhart HE. editors. Forest Measurements. 4th ed. New York: McGraw-Hill Inc.; 1994. 480p.

[73] R Development Core Team. R: A Language and Environment for Statistical Computing. R Foundation for Statistical Computing [Internet]. 2012. Available from: http://www.R-project.org. Accessed 2013-02-21.

[74] Sheather SJ. editor. A Modern Approach to Regression with R. New York: Springer Texts in Statistics; 2009. 393 p.

[75] Legendre P, Legendre L. editors. Numerical Ecology. 3rd ed. Amsterdam: Elsevier Science BV. 2012; 24. 1006 p.

[76] Pretzsch H. editor. Forest Dynamics, Growth and Yield: From Measurement to Model. Berlin Heidelberg: Springer-Verlag; 2009. 664 p.

[77] Burkhart HE, Tomé M. Modelling Forest Trees and Stands. Dordrecht: Springer Science +Business Media; 2012. 457p.

[78] Clutter JL, Fortson JC, Pienaar LV, Briester GH, Bailey RL. editors. Timber Management: A Quantitative Approach. New York: John Wiley & Sons, Inc; 1983. 333p.

[79] Myers RH. editor. Classical and Modern Regression with Applications. Chicago: Duxbury Press; 1986. 488 p.

[80] Paulo JA, Palma JHN, Gomes AA, Faias SP, Tomé J, Tomé M. Predicting site index from climate and soil variables for cork oak (*Quercus suber* L.) stands in Portugal. New Forests. 2015;46:293–307.

[81] Congalton RG, Oderwald RG, Mead RA. Assessing Landsat classification accuracy using discrete multivariate statistical techniques. Photogrammetric Engineering and Remote Sensing. 1983;49:1671–1678.

[82] Stehman SV. Estimating the kappa coefficient and its variance under stratified random sampling. Photogrammetric Engineering & Remote Sensing. 1996;62:401–407.

[83] Oliver CD, Larson BC editors. Forest Stand Dynamics. Update editions. New York: John Wiley & Sons, Inc; 1996. 544p.

[84] Schütz JP. Sylviculture 2. La gestion des forêts irrégulières et mélangées. [Silviculture 2. The management of uneven aged and mixed forests]. Collection Gérer L'environement, no. 13. Lausanne: Presses Polytechniques et Universitaires Romandes; 1997. 178 p. [in French].

[85] Ke Y, Quackenbush LJ. A review of methods for automatic individual tree-crown detection and delineation from passive remote sensing. International Journal of Remote Sensing. 2011;32(17):4725–4747.

[86] Chemura A, Duren I, Leeuwen LM. Determination of the age of soil palm from crown projection area detected from WorldView-2 multispectral remote sensing data: the case of Ejisu-Juaben district, Ghana. ISPRS Journal of Photogrammetry and Remote Sensing. 2014;100:118–127.

Section 2

Biomass Energy Valorization

Chapter 4

Fatal Accidents During Marine Transport of Wood Pellets Due to Off-Gassing: Experiences from Denmark

Frank Huess Hedlund and
Øssur Jarleivson Hilduberg

Additional information is available at the end of the chapter

http://dx.doi.org/10.5772/66334

Abstract

The atmosphere in unventilated wood pellet storage confinements, such as the cargo hold of marine vessels transporting pellets in solid bulk, can be severely oxygen deficient and contain deadly concentrations of harmful gasses, of which the most feared is the poisonous and odour-less carbon monoxide. The hazard has been known for over a decade and has been responsible for many accidents. We examine three fatal accidents on marine vessels in or near Danish waters and argue that they share strikingly similar aetiologies, if not repetitive patterns. It is generally recognized that accidents should be thoroughly investigated and lessons learned shared widely in order to minimize the number of times the same lessons have to be learned. The three Danish cases suggest that this learning process is deeply troubled for the solid biomass segment. The International Maritime Organization IMO/SOLAS has recently revised its guidance on entering enclosed spaces aboard ships in response to the ongoing problem of confined space incidents. We argue that the interpretation of the concept of an "enclosed space" is of utmost importance because accidents take place in rooms that are not considered enclosed by the crew.

Keywords: wood pellets, off-gassing, confined space, accident investigation

1. Introduction

This paper pays homage to what may be man's oldest accident prevention strategy: Learning from own—and better, other people's—past accidents and misfortune. In its modern form, the key characteristics of this strategy are that past mishaps shall be recorded and analysed to extract lessons learned which in turn shall be disseminated through broad feedback loops in order to prevent future similar, and not just identical, accidents.

Paraphrasing Petroski [1], the profession of accident prevention professionals is Janus-faced, looking always both backwards and forwards. The history of past accidents offers a repository of knowledge gained from painful experience. If heeded, the past thus provides caveats and lessons for future safe operations. If shunned, it will still haunt the future, always lurking in the shadows everyday work.

Having paid the price of an accident, we should use the opportunity to learn from it. The benefits of such learning are obvious—to avoid repetition, and to share the lessons learned in order to minimize the number of times the same lessons have to be learned. What is not so obvious, however, is how to make this seemingly simple and straightforward idea work in practice [2]. This challenge is reflected in common aphorisms, such as Santanyana's: Those who cannot remember the past are condemned to repeat it.

This paper examines three fatal accidents on marine vessels transporting sustainable, environmentally friendly and carbon-neutral wood pellets in bulk, in or near Danish waters, and argue that they share strikingly similar aetiologies, if not repetitive patterns—offering support to Sanatnyana's dismal adage. The paper examines some of the barriers that impede the capacity to learn from accidents on marine vessels in this particular segment.

2. Wood pellets in bulk

2.1. Solid biomass fuel in Danish thermal power plants

To reduce global carbon dioxide emissions, Denmark has converted many coal-fired thermal power plants to burn green and sustainable solid biomass. Solid biomass refers to non-fossilized carbon-based solid materials derived from plants, typically wood, but also straw, and certain energy crops grown primarily for the purpose of fuel utilization. Because Denmark's own woodlands are limited and covered by environmental protection policies, the domestic production of solid biomass is insignificant. Hence, essentially all biomass is imported, mostly from loggings in the Baltic countries and Russia, but also from North America and Africa [3].

Thermal power plants consume very large amounts of fuel; a single block can easily burn half a million metric ton per year. In order for the logistics chain to handle such quantities in an efficient manner, the irregularly sized and bulky biomass is processed into uniformly sized pellets, typically about 8 mm in diameter and 20–25-mm length.

In the production process, the wood is dried, milled to a fine powder and compacted in a pellet press that often employ die ring press and compression roller technology. The pellets are shipped in bulk in marine vessels to their final destination.

2.2. Wood pellet fire and explosion risks

The fire and explosion hazard profile of biomass pellets is much different from that of coal [3]. Pellets inevitably generate fine dust when handled in the logistics chain, e.g. in screw conveyors, band conveyors, cup elevators and when dumped into storage areas or hoppers. If made airborne and ignited, the wood dust is capable of producing severe dust explosions.

In addition, this dust is prone to ignition, e.g. from overheated electric motors or conveyor bearings, or from mechanical friction heat between conveyer belts and accumulated pellets, fines and/or dust. Small pieces of smouldering material are difficult to detect and they may migrate in the band conveyor systems and start smouldering fires in the storage areas which may develop into massive storage fires. Smouldering fires are furthermore notorious for their capacity to generate flammable pyrolysis gasses that travel and accumulate, leading to flash fires or explosion hazards elsewhere in the system [4].

2.3. Wood pellet chemical off-gassing risks

The milling process, the compaction and extreme shear forces in the pellet press, cause plant cell walls to rupture. This exposes cell cytoplasm and chemically reactive compounds such as resins, oils and fatty acids to air making them available to chemical oxidation and other reactions. As a result, freshly produced wood pellets are probably best thought of as a chemically active material.

The chemically active pellets may emit a range of gasses such as carbon monoxide, carbon dioxide, methane and volatile organic compounds (VOCs) that can accumulate in storage confinements and reach dangerous levels [5]. The release of the highly poisonous and odourless carbon monoxide is of particular concern. In addition, the chemical reactions may deplete the atmosphere in storage confinements of the oxygen required to sustain life. The topic has been investigated in several recent publications, e.g. [6, 7] but the exact nature of the chemical processes at play do still not appear to be well-understood.

Chemical oxidation may also explain why wood pellets can self-ignite. Heat loss is largely a surface-based phenomenon and because of the low surface-to-volume ratio of a large pile, any process that generates heat will slowly increase the temperature inside the pile. Pockets may form where the temperature of the contents can rise to the temperature necessary to produce spontaneous ignition. This produces an oxygen deficient smouldering fire deep inside the pile and the associated problems with flammable pyrolysis gasses that travel and accumulate, leading to flash fires or explosion hazards elsewhere in the system.

It is true that some biomass feedstocks species rich in resins are desirable in the pellet production process. The mechanical stability of pellets is an important property as this limits the amount of dust generated when they are handled in the logistics chain. Pellet stability is enhanced if the mill feed composition includes some biomass species rich in resins, e.g. pine wood, which can act as a natural binder system. Naturally occurring oils and resins can have other beneficial properties during production, for instance acting as lubricants in the pellet press. It will probably be difficult to reduce the problem with chemical reactivity of wood pellets through substitution of resin-rich species in the feedstock unless, of course, non-bio but safer additives are tolerated by pellet consumers.

What is clear though is that the massive increase in wood pellets consumption entails novel risks—wood pellets have a much different hazard profile than fossil fuel cargoes.

3. Case 1—Fatal accident on AMIRANTE, 2009

3.1. AMIRANTE

The ship AMIRANTE (IMO 7425334) was a general cargo ship enabling it to carry a variety of bulk and mixed cargos. It was built in 1978/1982 and had a gross tonnage of 4083 and a length of 82 m. The vessel was classed by DNV.

At the time of the accident, the vessel had a crew of eight, seven from the Russian Federation and one from Belarus. The flag State was St. Vincent & The Grenadines. AMIRANTE was mainly in a coastal trade in northern Europe and the Mediterranean.

3.2. Sources

Information on this accident is based on records from the Danish police [8] obtained through the Danish equivalent of a Freedom of Information Act unless otherwise noted.

3.3. The accident

On 15 July 2009 at 0300 (all times are local time), AMIRANTE left the port of Riga, Latvia, after having loaded a cargo of 2600 t of wood pellets in bulk. The port of destination was Copenhagen, Denmark, were the cargo was to be unloaded at the Amagerværket power plant owned by Wattenfall, which had recently converted one of its coal fired blocks from coal to solid biomass. The shipper was Granul Invest in Tallinn, Estonia and the operator was Alga Chartering in Copenhagen.

On the afternoon 15 July, during the passage of the Baltic Sea, four of the vessel's crew, three able seamen and the motorman, worked on the deck readying new mooring ropes on the forward part of the ship.

Shortly after 1800, the master was alerted by the ship's cook that one of the able seamen and the motorman had not shown up for dinner and their whereabouts were unknown. A search was initiated and at 1835, a crew member spotted the missing crew members lying lifeless at the bottom of the stairwell to the forepeak compartment. The crew member did not attempt to reach them fearing for his own safety.

The forepeak is a wedge-shaped stowage room at the bow of the ship, about 5 m deep, and a width ranging from 5 to 1.5 m. The room is reached from the deck through a stairwell climbing a narrow 18-m long fixed vertical steel ladder. The forepeak was used for stowing, e.g. paint containers, floor brushes, buckets, surplus wooden planks and mooring ropes.

The master rushed to the forepeak and climbed down the steel ladder. Halfway down he began to experience laboured breathing. Upon reaching the forepeak floor his breathing had become so laboured that he realized the atmosphere might be dangerous. The motorman was lying about 1 m from the ladder and the able seaman was lying just at the foot of the ladder, as if he had attempted to climb it, still wearing work gloves. Froth could be seen from the

http://dx.doi.org/10.5772/66334

corner of the mouth of both men. The master briefly tried to awaken them and then hurried up the ladder to the deck. The vessel was about 16 km north of Bornholm, a Danish island in the Baltic Sea about midway between Sweden and Poland. At 1845, the master radioed the Danish authorities requesting immediate medical assistance.

The crew worked frantically to extract the two men from the forepeak compartment. They switched the room's mechanical ventilation on and crew members climbed down the ladder to tie a rope around the chest of the victims. Hoisting the victims through the narrow stairwell proved difficult. When a doctor arrived by helicopter from the Danish Admiral Fleet, at 1945, one victim has been extracted and there was a work frenzy to hoist the other. The doctor could soon declare both men dead as they presented with livor mortis and beginning rigor mortis. The doctor estimated that death had occurred 2–3 h earlier.

3.4. The investigation

The vessel was redirected to the port of Rønne at Bornholm by Danish Admiral Fleet, where it arrived at 2300. The police was waiting at the quay and began a standard crime scene procedure, securing evidence and interviewing the crew. The forepeak was inspected by the fire and rescue services done full breathing protection at about 0100. At the time of the inspection, the compartment's mechanical ventilation had been switched off for about 40 min upon request. Oxygen levels were normal, CO levels were 100 ppm. The inspection team noticed a pair of glasses which had been placed on some stowed construction timber and an unopened packet of cigarettes lying on the floor.

A door in the forepeak compartment allowed access to the front cargo hold. The door had three closing hinges but only two were engaged. It was speculated that gasses might travel from the cargo hold to the forepeak through a crevice in the door opening where the third hinge was not engaged (**Figure 1**).

The deceased presented no signs of violence or trauma and there was no other evidence to suggest criminal actions, e.g. if they should have been pushed while standing in the door opening on the deck and then fallen the 18 m down the forepeak stairwell. There was briefly some speculation if red stains on the floor of the forepeak were bloodstains but they turned out to be red paint. Follow-up interviews with the master and other crew members were carried out in the afternoon to resolve some inconsistencies in reported times and observations, this time with the assistance of a certified interpreter. Only the master spoke some English, and he had acted as interpreter during the first interviews.

A sample of the wood pellets was sent for chemical analysis which found 150 ppm CO in the headspace of the specimen jar. The actual concentration is likely much higher, however, as 150 ppm simply appears to be the upper range of the measuring instrument. An autopsy confirmed death by CO poisoning, COHb blood saturation levels were measured at 52% (the motorman) and 60% (the able seaman). Tests for cyanide (blood) and alcohol (lacrimal fluid) were negative.

Figure 1. The door in the forepeak compartment, which led the front cargo hold. The door had three closing hinges but only two were engaged. Gasses travelled from the cargo hold to the forepeak through a crevice where the third hinge was not engaged. (AMIRANTE). Source: Police photo file #13 [8].

The crew was at pains to explain why the two seamen would go down the forepeak. They had no business down there whatsoever. It was suggested that they might want to check if there was sufficient room for stowage of the old mooring ropes.

The two deceased were experienced seamen. The motorman was born in 1981 in Russia, the able seaman in 1971 in Belarus. Records showed that both seamen had duly complied

http://dx.doi.org/10.5772/66334

with regulations concerning rest and work time. Both seamen had received the proper safety instruction for working in enclosed rooms in 2008 (July 8 and April 20, respectively) and they were not newcomers to the ship. Police records state that they should have known that full breathing protection was required before entering a compartment such as the forepeak. In the absence of evidence of the contrary, the police ruled the two deaths an unfortunate workplace accident and the crime scene investigation was closed.

3.5. Reflections

The direct cause of the accident is probably simple and understandable. Working new and bulky mooring ropes, often weighing several 100 kg, is a laborious and tedious task. According to the weather data from Rønne airport, the afternoon weather conditions were mostly cloudy, temperature 17–19°C, light winds about 3 m/s and occasional rain showers and light rains. While such ambient weather conditions definitely are tolerable to seasoned seamen, it is easy to imagine that the nearby forepeak might offer a convenient refuge for a clandestine cigarette break, in particular during a rain shower.

Important questions aimed at understanding *why* the accident happened were never raised, simply because they were out scope of a judicial police investigation. Were the crew aware of hazardous properties of the cargo? Was the forepeak considered to be a compartment with properties similar to those of enclosed cargo spaces? The starkest circumstantial evidence may be that the two seamen chose to enter the forepeak without switching on the mechanical ventilation. In all likelihood, they were clueless about the danger—otherwise, they would not have entered. Were the rest of the crew and in particular the master aware of the hazard? Although it is pure conjecture, the answer is probably negative. In the police interrogation reports, none of the crew mentioned the ability of wood pellets to generate carbon monoxide. The closest mention was the master stating that he suspected something was "wrong" with this particular cargo, because it had a "funny smell" unlike any of the 15 earlier occasions where AMIRANTE had carried wood pellets.

Was information about the hazardous nature of the cargo available to the crew? The question too was never raised and although the answer is uncertain it is probably negative. The manifest of cargo simply stated: "Wood Pellets in bulk—Cargo Status C—European Community Cargo" and the "Dangerous" field on the manifest form was left blank.

The criminal investigation was satisfied that the two seamen had received the proper safety instruction for working in enclosed rooms and concluded that they should have known that full breathing protection was required before entering a compartment such as the forepeak. The question was never raised, however, if the seamen, or the rest of crew for that matter, would consider the forepeak "an enclosed room". The police interrogation reports state that the practice on board the vessel was to lock open the door on the deck that provided access to the stairwell to the forepeak. The door was only closed during bad weather. The reason for this practice is unexplained. But in the absence of specific precautionary instructions or warnings, the crew may well have assumed that a room with a door locked open is not "enclosed".

3.6. Information sharing and learning

The local news media at Bornholm briefly covered the event, reporting that a coaster with two dead sailors had arrived at the port of Rønne. At the time of media reporting the cause had not yet been determined, and the media speculated that CO poisoning or oxygen deficiency in an enclosed room might be responsible for a workplace accident. The vessel's cargo was correctly identified as wood pellets, but there is nothing in the media coverage to suggest that there might be a causal link between the type of cargo and the presence of toxic gasses. For an accident prevention professional, the media coverage is worthless.

Because the police opened the case as a crime scene investigation, the case was technically registered as a criminal case. Criminal cases are kept confidential in order to protect the privacy of the individuals concerned. To date, very little information, if any, relevant to accident prevention professionals is available in open sources. The source material for this paper could only be obtained through persistent requests and the Danish equivalent of a Freedom of Information Act.

It is true that there is sporadic and passing mention of the AMIRANTE accident in the wood pellet literature, e.g. [9, 10] but descriptions are hazy and basic facts are corrupted. This is also true for the papers produced by authors in Denmark. For example [11] states that the seamen entered the cargo hold, which they did not, and [12] states that they died from asphyxia, which is also incorrect.

Because proper sharing of correct information about an accident is a basic precondition for learning, we conclude that learning processes were derailed at a very early stage. Therefore, it may be expected that repeat accidents will take place. As we will argue shortly, this is precisely what happened. Lamentably, the broader accident history reveals that the AMIRANTE accident was no first, but only one in a string of previous similar accidents with similar aetiologies.

4. Case 2—Fatal accident at "LADY IRINA", 2014

4.1. LADY IRINA

The ship LADY IRINA (IMO 9137038) was a general cargo ship with capacity to carry containers enabling it to carry a variety of bulk and general cargo. LADY IRINA was built in 1997 with a gross tonnage of 3323 and a length of 88 m. The vessel was classed by Lloyd's Register. At the time of the accident, the vessel had a crew of seven. The flag state was The Netherlands.

4.2. Sources

Information on this accident is based on a report from the Dutch Safety Board [13] unless otherwise noted.

4.3. The accident

On 5 July 2014, LADY IRINA departed from the Port of Arkhangelsk, Russia, with a cargo of wood pellets in bulk which was to be unloaded in Kolding, Denmark. Arkhangelsk is LADY

IRINA's permanent loading port for this type of cargo, and it has a dozen discharge ports in Europe, of which Kolding is one.

On the evening of 13 July 2014, the crew prepared for LADY IRINA's arrival in the Port of Kolding. At about 1900, the chief engineer met with the chief officer and decided to drain the anchor chain locker compartment before LADY IRINA expected arrival at about 2400 h. The anchor chain locker is located in the forward part of the ship. Access to the anchor chain locker is from the deck, entering the forecastle compartment, opening a door and climbing a vertical steel ladder in a stairwell to the bow thruster room, located directly beneath the forecastle room.

The forecastle compartment holds maintenance and safety equipment and other frequently used items such as tools, paint and brushes. The door to the forecastle room is therefore normally open throughout the day and only closed at the end of work, typically at about 1700, to secure the watertight integrity of the ship while the crew were off duty.

On 13 July, sometime after 1900, the chief engineer went to the engine room to start the pump draining the anchor chain locker and then went to the bow thruster room, probably to check if the draining operation was complete. The door to the forecastle had been closed since end of work Saturday, i.e. for at least 26 h.

At 2145, the chief officer went looking for the chief engineer. The door to the forecastle room was open and he found the chief engineer lifeless on the floor of the deck lower, on the floor of the bow thruster room. He examined the chief engineer and found that he was not breathing and had no heartbeat.

The chief officer rushed to the bridge to alert the master, who had the watch, and to summon assistance. The second mate, who happened to be on the bridge was ordered to go to the accident site with a stretcher. The master immediately decided to divert to the nearby port of Fredericia (Denmark), only 4 nautical miles away, and requested emergency assistance to be ready at the quay.

The chief officer went to the crew accommodation to alert the rest of the crew and directed them to the bow thruster room. Probably at about 2215, all crew members except the master were involved in the rescue operation in the bow thruster room. They first worked to resuscitate the chief engineer and then attempted to evacuate him through the stairwell on the stretcher, which proved unsuccessful as he was a heavily built man. The rescue involved six crew members: the first and second mates, the apprentice engineer, the cook and two seamen. They did not use the breathing apparatus that they had taken with them because they thought that the chief engineer had fallen from the ladder and that air quality was not a problem.

Around 2245, the vessel arrived in the Port of Frederica and the apprentice engineer and a seaman left the bow thruster room to assist with berthing the vessel. The bow thruster was required for berthing and had to be operated from the engine room. The second mate also left the room to collect a neck brace because the crew still thought that the chief engineer had fallen from the ladder, and an oxygen kit because the chief engineer was not breathing.

Upon returning to the forecastle room, a few minutes later, the second mate saw the chief officer unconscious, a crew member trying to carry him away and another crew member walking around as if drunk. The second mate climbed down ladder to the bow thruster room, placed a breathing apparatus over the mouth of the chief officer and urged the two other crew members to leave the room. He then left the bow thruster room and forecastle himself because he did not feel well. Before leaving the forecastle, he opened the medical oxygen kit and lowered it down the bow thruster room with a rope, in order to generate extra oxygen there.

The vessel berthed at 2250 and the local fire and rescue services evacuated the chief engineer and the chief officer from the bow thruster room. The other two crew members had managed to climb the ladder and get outside by themselves but were much weakened. The chief officer was given oxygen and regained consciousness on deck.

The chief officer and the two seamen were admitted to hospital and were diagnosed with carbon monoxide poisoning. The chief engineer had died from carbon monoxide poisoning. Danish news media [14] report that four, not three, seamen were admitted to hospital and two of them were treated in a hyperbaric oxygen therapy chamber. All crew appear to be of Russian nationality.

The local fire and rescue services ventilated the forecastle and bow thruster room with electrical ventilators for 1½ h and then took CO measurements. CO levels in the forecastle were measured at 80 ppm, bow thruster room at 20 ppm. After having the forecastle and bow thruster rooms closed for 36 h, the measurements were 690 and 555 ppm, respectively. The LADY IRINA is equipped with a fixed CO_2 system, which can inject carbon dioxide in the cargo hold for fire-fighting purposes. CO concentrations in the neighbouring carbon dioxide storage room, which had not been ventilated, were in excess of 2000 ppm, the upper limit of the measuring device.

4.4. The investigation

The investigation by the Dutch Safety Board [13] concluded that carbon monoxide generated by the wood pellets in the unventilated cargo hold could potentially have reached the bow area in several ways, e.g. via piping or ventilations ducts:

The CO_2 room, forecastle and bow thruster room are adjacent to each other and gasses might travel from one room to another through a number of connecting conduits. The shipping company contracted the classification society to investigate how carbon monoxide could have travelled from the cargo hold to the bow area. The investigation identified a minor gap in a section of the rim of an inspection cover on a ventilation duct that runs from the cargo hold through the forecastle as the most likely route (**Figure 2**).

The atmosphere in the bow area had been identified as potentially unsafe and warning signs on the forecastle access door stated: "DANGER. Low Oxygen Level". The bow area had no mechanical ventilation. In principle, according the rulebook, the bow area should probably be considered a confined space. At LADY IRINA, however, the daily work practice had evolved and adapted to the need to enter the room on a routinely and daily basis. The crew simply opened the door for 15–20 min to allow natural ventilation before entering and they would normally leave the door open.

Figure 2. The inspection hatch on a ventilation duct that ran from the cargo hold through the forecastle. A minor gap in a section of the rim (knife blade inserted) allowed gasses to travel from the cargo hold to the bow area. (LADY IRINA). Source: [13].

In response to the fatal accident, the shipping company acquired multi-gas meters and introduced a new work procedure mandating that the person making "First Entry" must carry an operational multi-gas meter. New warning placards with the instruction "ATTENTION—No first entry without use of multi-gas meter" were been put up on access doors.

The investigation report states that: "*The [Dutch Safety] Board is of the opinion that, in the light of this accident, it is good that the shipping company examines the working method that was common in practice. However, the new working method desires focus on definition and crew discipline during use. The working method should also be evaluated and improved, where necessary*" [13].

4.5. Reflections

The most striking feature of this accident is probably that the *entire crew*, except the master on the bridge, climbed down the ladder to the bow thruster room where dangerous levels of carbon monoxide were present and stayed there for more than 30 min, *completely oblivious of the danger*. It is easy to imagine that under just marginally different circumstances, the entire crew would have succumbed to the poisonous gasses down there, which in turn would have hampered the master's ability to berth the vessel and hence delayed the arrival of the rescue services. It can safely be stated that the death toll could have been higher.

The Dutch investigation observes that the relevant International Maritime Solid Bulk Cargoes (IMSBC) Code for wood pellets at the time of the accident stated that: "*all crew members must carry an oxygen and carbon dioxide meter and switch this on when they enter the cargo hold or adjacent confined spaces*". We observe that there is only superficial and passing mention of carbon monoxide in the Code. We also remark that an atmosphere with deadly levels of CO may present with normal readings for oxygen and CO_2.

Was information about the hazardous nature of the cargo available in a form that made the crew aware of the risk? The investigation does not clearly address this issue. The investigation simply states that the crew "*was informed of the risks of this cargo*" and that they "*knew that the hold should not be entered*". Furthermore that "*each new crew member had received safety instructions*" which comprised "*confined spaces and ventilation of holds in connection with possible lack of oxygen and other dangerous substances that emerge as a consequence of cargo. Also during the safety committee meetings attention was paid to entering confined spaces*".

We conclude that the answer probably is negative: information was not available to the crew. In any case, their actions demonstrate that they were clueless of breathing hazards of any kind, be it oxygen depletion, carbon dioxide, carbon monoxide, etc.

For a safety professional working with industrial accident prevention (onshore), it is unusual that rooms routinely entered on a daily basis are, first, identified and labelled as having a potentially unsafe atmosphere, and second, do not have mechanical ventilation. The absence of mechanical ventilation would appear to be a weakness in design, which ought to be eliminated in future designs and retrofitted in existing designs wherever possible. Mechanical ventilation is a *risk prevention* measure, whereas testing before entry is a *risk control* measure, merely to confirm that the preventive measures are successful. It is remarkable that a relatively recently built ship (1997) has no mechanical ventilation where dangerous gasses potentially may be present.

4.6. Information sharing and learning?

The accident has been investigated and the findings exhaustively communicated by the Dutch Safety Board [13]. The coverage in Danish new media correctly identifies the generation of CO from wood pellets as the culprit and has no factual errors [14]. A Maritime Merchant Periodical with international coverage [15] carried an article that summarized the findings of Dutch Safety Board's accident report, highlighting that other crew members could have been killed in the rescue attempt. The article is without factual errors. The information provided is sufficiently detailed to permit safety professionals elsewhere to prevent recurrence.

The basic preconditions for learning: broad sharing of relevant and factually correct information; are met. As we will argue later, this is a necessary, but not a sufficient precondition, for learning to take place. The information must also be delivered locally in a manner so its importance is recognized and it must be acted upon.

http://dx.doi.org/10.5772/66334

5. Case 3—Fatal accident at CORINA, 2015

5.1. CORINA

M/V CORINA (IMO 8908545) was a general cargo vessel with capacity to carry containers enabling it to carry a variety of bulk and general cargo, length 115 m, gross tonnage 5796. The vessel was built in 1990 and classed by Germanischer Lloyd. At the time of the accident, the vessel had a crew of eleven, all of Polish nationality. The flag state was Poland.

5.2. Sources

Information on this accident is based on a report from the Polish State Maritime Accident Investigation Commission [16], and records from the Danish police [17] and the Danish Working Environment Authority [18], both obtained through the Freedom of Information Act.

5.3. The accident

Between 11 April and 20 April 2015, CORINA loaded 6744 t of wood pellets in the Port of Arkhangelsk (Russia). On the 28 April, the vessel arrived at Hanstholm, Denmark, to unload 1921 t before proceeding to The Netherlands.

Company policy required the visible presence of a watchman near the access gangway at all times while at port and an able seaman had the watch from 1600 h. The cook expected him for dinner at about 1700 and eventually informed the chief officer that the able seaman had not shown up. The chief officer searched the ship twice and noticed that the door to the bosun store under forecastle deck was open. Inside, he noticed that the hatch cover to a lashing equipment room at a lower deck in the forepeak was open. He went down and found the seamen there, unconscious. He radioed for help and shortly after, at 1754, the master contacted the port in broken English and requested medical assistance to a heart attack victim.

Within about 15 min a bosun, an electrician and another seaman had come to the accident site to assist and taking turns with cardiac pulmonary resuscitation believing that the unconscious seaman, born in 1947 and overweight at 120 kg, had suffered a heart attack. At about 1810, the port officer arrived at the site with a defibrillator. The port officer soon did not feel well and asked the crew if the air was ok, which was confirmed. After some time one of the crew members, probably the chief officer, attempted to climb the ladder but fell down and received a blow to the head and had to sit down. The first ambulance arrived at 1813 and one of the paramedics soon arrived at the site and began administering oxygen to the able seaman. Six persons were now present in the room. After some time, the paramedic asked the crew if the air was ok, which was confirmed. Shortly after the port officer radioed, the master asking if the air was ok. Only then was the mechanical ventilation switched on. The paramedic ordered everybody out of the room because the air was not ok. Two crew members were unable to climb the ladder and had to be left behind with the unconscious seaman while the other four managed to move up the forepeak and reach the deck, some only with great difficulty.

Exact times are not available, but the on-scene commander arrived at 1845 at which four ambulances and three vehicles from the emergency services were at the site. A doctor arrived by helicopter at 1855. Fire fighters donned full respiratory protection extracted the two unconscious crew members who regained consciousness on the deck. At the entry door to the forepeak, CO levels were measured at 66 ppm. Further down 366 ppm were reported. It should be noted that these measurements were taken after mechanical ventilation had been switched on. The able seaman was difficult to extract through the narrow hatch because of his heavy build. He was declared dead at 1940. An autopsy confirmed CO poisoning with COHb measured at 60%. Four crew members and the port officer were admitted to hospital.

5.4. The investigation

CORINA was general cargo multipurpose vessel adapted to transport of containers. Containers must be lashed, and in order to ease the handling and movement of lashing gear, the vessel was designed with a 2100 mm × 1150 mm opening in the forward bulkhead of cargo hold no. 1. The opening allowed easy access to the neighbouring storage room, where chains, stacking cones and other lashing equipment for containers were kept (**Figure 3**). When carrying bulk cargo, the opening was closed with planks inserted into guide bars. This barrier was not gas tight and gasses from the cargo hold could therefore travel to the forecastle (**Figure 4**).

The crew was at pains to explain why the able seaman would enter the lashing equipment room. He had no business down there whatsoever and had not requested permission or informed anybody of his intention to go there. He was supposed to be on watch near the gangway.

Figure 3. An opening in the front cargo hold allowed easy access to the lashing equipment room. When carrying bulk cargo, the opening was closed with planks of wood inserted into guide bars (CORINA). Source: [16].

Figure 4. The plank barrier seen from the lashing equipment room. Wood pellets in the cargo hold are visible are visible between planks 10 and 11. The barrier is evidently not gastight. (CORINA). Source: Police photo file #23 [8].

In line with the International Maritime Solid Bulk Cargoes (IMSBC) Code, the shipper had declared the hazardous nature the cargo. The wood pellets were declared as a group B cargo and marked as Materials Hazardous only in Bulk (MHB), which covers materials that may possess chemical hazards when carried in bulk. The shipper had provided several safety recommendations, only allowing entry to the cargo hold after 2 h of natural ventilation and measurements of oxygen content and dangerous gasses, carbon monoxide included.

The Polish investigation concluded, however, that crew members likely were unaware of the risks associated with the cargo. Points of entry to hazardous or confined spaces were not marked with warning signs and potential access doors had not been locked to prevent inadvertent entry [16]. To the Danish police, the master explained that he knew that wood pellets could *"suck oxygen out of the air"* but also that he believed that gasses were present in the cargo hold only, although *"he was not a university professor"* [17].

5.5. Reflections

In this recent accident, the hazards associated with wood pellets and preventive measures are clearly stated in the cargo documents. The extent to which this information is fully grasped by the master can be questioned. Seemingly, air quality hazards only became apparent to him after the port officer radioed and asked if the air in the forepeak was ok. Only then did he arrange for the mechanical ventilation to be switched on. Information trickle down from the master to the rest of the crew is obviously doubtful. Again, the actions of the crew demonstrate that they were truly oblivious of the danger. And so were the Danish port officer, the paramedic and probably many more.

Language issues may also have played a role. In particular, the ability of seamen from Eastern Europe with limited proficiency in English to understand hazard codes and guidelines in English, even more so for Russian seamen who may have a preference for information printed in Cyrillic letters.

5.6. Information sharing and learning?

The accident was investigated and the findings exhaustively communicated by the Polish State Maritime Accident Investigation Commission [16]. The coverage in Danish media mainly paid attention to the irony that the would-be rescuers succumbed to gasses and themselves became victims but is otherwise riddled with misleading factual errors. For example, the accident is stated to have taken place in the cargo hold because of oxygen depletion, both facts are plain wrong. The source of the hazard, the wood pellets, is correctly identified, however.

6. Discussion

6.1. Accident history involving off-gassing from wood pellets

Lamentably, strikingly similar serious accidents related to marine transport of solid biomass continue to take place. Publicly available information on these incidents is often scant, however, and as the AMIRANTE case shows, at times incorrect or misleading. The list below (**Table 1**) has been compiled from various sources and does not claim to be exhaustive. But, it is evident that the release of hazardous gasses from wood pellets in bulk qualifies as an emerging risk for which proper control strategies have yet to be developed.

6.2. Repetitive patterns in accident aetiology

The three Danish cases have strikingly similar accident aetiologies:

- the accidents take place in compartments adjacent to the cargo hold
- the compartments are likely covered by IMO's technical definition of an "enclosed room" but the crew does perceive the rooms as being "enclosed"
- dangerous gasses from the cargo hold travel unexpectedly through minor crevices and small gaps unknown to the crew

- the crew appear oblivious of the hazardous properties of the cargo
- would-be rescuers subject themselves to grave danger
- victims in rooms only accessible using a vertical ladder can only be extracted with great difficulties, in particular if the victim is of heavy build

We are of the opinion that similar accident aetiologies apply to many of the other accidents in the list above though scant publicly available information (and time constraints) prevents a rigorous demonstration of this claim.

Year	Vessel name, location, accident	Fatal	Injured	Source
2002	VEAWER ARROW, Rotterdam (NL), entry into cargo hold with wood pellets	1		[10]
2005	EKEN, Gruvön (SE), entry into cargo hold with pulp logs	1		[10]
2005	SAGA FOREST CARRIER, Wilmington (US), working in open cargo hold with green lumber, oxygen depletion	1		[16, 19]
2006	SAGA SPRAY, Helsingborg (SE), entry into cargo hold with wood pellets	1	Several	[10]
2006	NOREN, Skelleftehamn (SE), entry into cargo hold with wood chips	1		[10]
2007	FEMBRIA, Timrå (SE), entry into cargo hold with timber	2		[10]
2009	AMIRANTE, Bornholm (DK), entry intro room adjacent to the cargo hold with wood pellets	2	0	[8]
2010	TPC WELLINGTON, New Zealand, entry into cargo hold with timber	2		[20]
2014	LADY IRINA, Fredericia (DK), entry intro room adjacent to the cargo hold with wood pellets	1	3	[13]
2015	CORINA, Hanstholm (DK), entry intro room adjacent to the cargo hold with wood pellets	1	5	[17, 16]

Table 1. Serious accidents related to marine transport of solid biomass known to the authors.

6.3. The IMO concept of a "confined room" and local sense-making

The Danish accidents illustrate how the explanatory framework for unsafe atmosphere accidents is based on a discourse that certain rooms are considered confined or enclosed. By implication, the hazard is perceived as being a property of the compartment and not a property of the atmosphere itself or the cargo.

This is particularly evident in the recently revised industry recommendations for entering enclosed spaced aboard ships issued by IMO [21]. The resolution's overall definition of an enclosed space is so broad that it can potentially apply to any space on the ship. However, the examples provided of enclosed spaces are characterized by being spaces that are not part of the normal workspace of the seaman, e.g. fuel tanks, ballast tanks, cargo pump rooms (used on e.g. chemical tankers), etc.

In the Danish cases, the accidents took place in spaces that were part of the normal working areas of the ship and were routinely accessed. The doors were kept open for practical purposes or convenience. If such rooms are labelled "enclosed" and warning signs posted, as in the LADY IRINA case, the crew may become desensitized and dulled to what is dangerous, because the warnings signs are inconsistent with daily routine and local sense-making.

6.4. Mental models of accident causation and stop rules

An accident investigation is a tremendously challenging task. The identification of accident causes is often subjective and guided by the tool box of the analyst rather than the nature of the accident. We subscribe to the ideas of Jens Rasmussen that there is a tendency to see what you expect to find [22], and his proposition [23] that this is a simple reflection of the nature of causal analysis and the fact that no objective "stop rule" exists to terminate the causal back tracking in search of a root cause.

In the AMIRANTE case, the stop rule was from what Jens Rasmussen has termed the "prosecutor's perspective"—the investigation was closed when it was not possible to find somebody to punish. In the LADY IRINA case, the stop rule was from the "therapist's perspective"—the search for causes stopped when a cure known to the analyst was identified (a new procedure and larger warning signs with more text).

The CORINA case was a hybrid, the search stopped when violations were identified (no warning signs and insufficient crew safety instructions) which conveniently also pointed to the cure: improve work procedures and safety training. The accident investigation outcomes above align well with the constructivist thinking of Donald Schön, that the moment we name the things to which we will attend we also frame the context in which we will attend to them [24].

We are of the opinion that it is important to be cognizant of such stop rules. In particular, we note that enhanced proceduralization of work may have limited impact on actual workplace safety. The use of administrative risk controls like procedures may from a managerial perspective be perceived as a time- and cost-effective tool to manage risk, but the actual impact on daily safety can be greatly limited when crew members face multiple and sometimes conflicting goals in their everyday work.

6.5. The coaster segment—a challenge to reach

Wood pellets are imported to Denmark on two types of ships: Small tonnage coastal ships (coasters) able to navigate in the shallow and narrow Danish waters calling smaller commercial ports, and larger bulk carriers used for transporting larger quantities to larger ports.

Contrary to coal, wood pellets do not tolerate moisture and cannot be stored in outdoor piles. Wood pellets must be handled under roof or in silos and storage capacity comes at a price. Former coal-fired power plants therefore often face storage capacity limitations. Wood pellets' energy content per unit volume is anywhere from half to one-third of that of coal, which furthermore aggravates storage capacity limitations [3]. The limitations call for frequent arrivals of relatively small cargo sizes, a market well-suited for the small coaster segment.

The coasters are usually general cargo ships, i.e. they are designed to carry many different kinds of cargo, e.g. goods on pallets, machinery and various bulk cargoes such as grain, coal and pellets. The high variability of the type of cargo sets limits to the degree of specialization and cargo domain safety knowledge that can be achieved in the small coaster segment.

In addition, coastal ships have, as most ships in international trade, crews of mixed nationality with varying background in training and experience, often from low-wage countries, and sometimes from Russia accustomed to the Cyrillic alphabet. We argue that cargo documents and safety recommendations in English may not be readily absorbed by this group and the work culture may have an oral rather than a written tradition.

6.6. Media shifting—a sub-optimality

Environmental interventions may be undertaken with insufficient attention paid to the environment of the workers. The concept of "media shifting" refers to a suboptimal situation which occurs if the "resolution" of a problem within the environmental domain gives rise to new, and unforeseen, problems within other domains, specifically the workplace safety domain [25].

Examples of insufficient attention to workplace safety are sometimes apparent in green reporting and environmental statements, if they omit or misrepresent sections covering occupational health and safety. For example, a study examined "Green Statements" of a biomass pellet facility certified to ISO 14001 and OHSAS 18001—a type of reporting which aims to demonstrate commitments towards sustainability performance. Fire station turnout statistics showed that fires at the facility were frequent. In the sustainability reporting, however, fire and explosion risks were only ranked 11 out of 14, long after more benign, if not trivial issues such as bad acoustics, psychosocial issues and indoor environment problems such as drafts, cold spots and radiant heat [3].

In recent years, critical studies that examine the general topic of balancing safety with sustainability have begun to emerge, see, e.g. [12, 26].

Denmark's climate change policy is very ambitious and comprises the conversion of coal-fired thermal power plants to solid biomass fuel in order to reduce carbon dioxide emissions [3]. This chapter points to evidence of media shifting of this climate change intervention that the transport of unprecedented quantities of wood pellets has created new workplace hazards in the marine transportation sector. We argue that the release of hazardous gasses from wood pellets in bulk present as an emerging risk for which proper control strategies have yet to be developed.

The domestic production of wood pellets in Denmark is limited (**Figure 5**), and more than 95% are imported and arrive by ship (**Table 2**). For the period 2001–2015, both years included, we have estimated the domestic production, imports and total consumption 2015 using simple linear inter-/extrapolation, all quantities in SI ton (t). For the same period, we have knowledge of three serious accidents on board marine vessel during the import voyage, which resulted in four fatalities and eight persons hospitalized after having been exposed to grave danger. For the domestic production, we have knowledge of one fatal accident and five persons hospitalized (at least) after having been exposed to grave danger due to fire and explosion, details cannot be disclosed, however, because of confidentiality agreements.

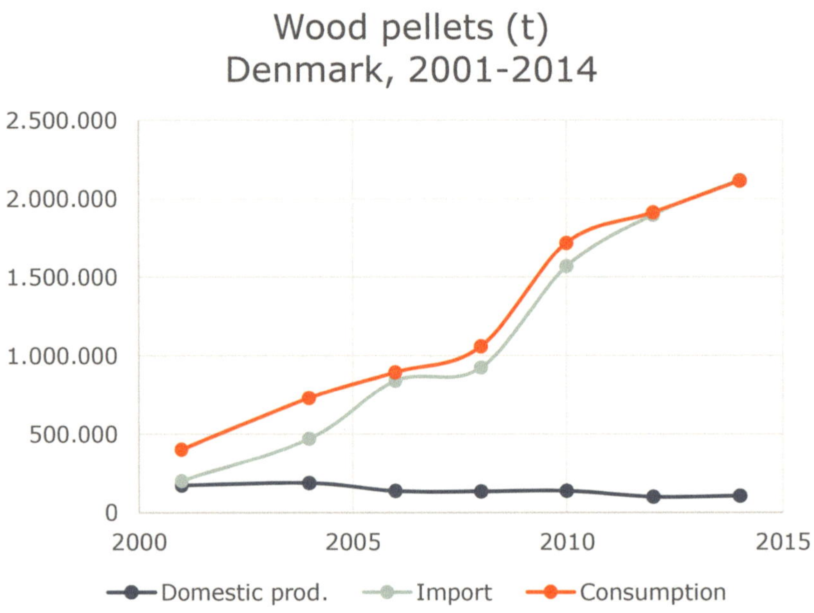

Figure 5. The consumption of wood pellets in Denmark has risen dramatically over the last 15 years.

Year	Domestic	Import	Consumption
2001	173,073	200,871	401,291
2004	187,458	470,588	731,134
2006	137,080	841,132	895,796
2008	134,280	925,401	1,059,519
2010	137,622	1,568,952	1,718,976
2012	99,930	1,898,143	1,914,048
2014	106,000	2,118,750	2,113,908
2001-2015 (est.)	2,100,650	17,452,647	19,191,795

Numbers do not sum due to stockpile changes and (small) exports (Data source: [27]). We estimate the total quantities (t) for 2001–2015.

Table 2. Danish domestic production, imports and consumption (t) of wood pellets for select years 2001–2014.

The numbers allow a crude estimation of the accident rate for imports and domestic production of wood pellets (**Table 3**). The numbers are not comparable because the import figure only comprises accidents (known to us) during the final leg of the marine voyage—excluding for example accidents during biomass harvest, land transport, wood pellet production in the country of origin and loading at the export port.

	Accident rate per Mt wood pellets	
	Fatal	**Grave danger**
Produced domestically	0.48	2.38
During marine voyage (import)	0.23	0.46

Table 3. The accident rate per million t wood pellets for accidents with fatal outcome and accidents where an individual is in grave danger, for wood pellets produced domestically and imports (final leg of marine voyage only).

7. Conclusion

Common sense tells us that having paid the hefty price of a serious accident, we should use every opportunity to learn from it and prevent recurrence. In the case of the fatal accident on AMIRANTE in 2009, such opportunities were wholly missed. The accident was investigated as a criminal case and the legal system is so designed that findings are kept confidential and out of reach of safety professionals and the general public. This approach has been criticized at an earlier occasion as being entirely misplaced in an accident prevention context [4]. It is a systemic deficiency by design—the bureaucratic judiciary system did not malfunction, it worked precisely as intended. The flag state, St. Vincent & The Grenadines, did not investigate. As a direct consequence, there was no learning.

Predictably, the accident recurred, at first not in Denmark, but elsewhere. In fact, it would turn out that similar accidents had happened before the AMIRANTE, but lessons learned had not been effectively shared and the crew at AMIRANTE were ignorant of the dangers of wood pellets in bulk.

After AMIRANTE, evidence slowly began to accumulate that wood pellets in bulk present chemical off-gassing hazards that qualify as an emerging risk. It is difficult not to mention the early work, foresight and leadership of the Canadian Wood Pellet association in shedding light on this novel issue [28].

By 2015, there is a comprehensive body of the literature describing the off-gassing hazards and the IMO has launched initiatives and revised the guidelines for management of entry into confined spaces. Unfortunately, very serious repeat accidents continue to recur. That implies that accidents can no longer be attributed to lack of knowledge but a failure to disseminate and use the knowledge available

The general cargo coaster segment appears difficult to reach. Because of the wide range of cargoes carried, there is little specialization and little accumulation of specific cargo safety knowledge.

We doubt if the recently revised IMO guidelines for management of entry into enclosed spaces will be effective. The guidelines are broad and their implementation will likely result in compartments—that are part of the normal working areas of the crew and routinely accessed—being labelled as potentially confined spaces. The crew may become desensitized and dulled to danger, because the hazard warnings are inconsistent with local sense-making. The accident history supports this prediction—confined space accidents routinely take place in rooms that are not considered confined spaces by the crew.

In addition, coasters generally have crews of mixed nationality, often from low-wage countries, and sometimes from countries accustomed to the Cyrillic alphabet. We are concerned that cargo documents and written safety recommendations in the English language may not be readily absorbed by this group and the work culture may have an oral rather than a written tradition. An entirely new approach must be sought. More research into this issue and into the existing legal and institutional barriers to learning is required.

The situation is special and delicate because the dramatic increase in the consumption of wood pellets arises out of an ambition to do good—to reduce carbon dioxide emissions and prevent climate change and global warming—often at steep costs to society [3]. It is morally problematic, if climate change environmental interventions are pursued so narrow-mindedly that sight is lost of the environment of the workers, and if workplace safety issues are starved of resources.

Utmost care should be taken to avoid media shifting—that the resolution of a problem within one domain, the environmental, creates a new problem in another, the workplace safety domain.

Acknowledgements

This article has been produced as voluntary work· Opinions expressed are those of the authors, not their employers or organizations. We are thankful to COWI for covering publication expenses.

Conflicts of interest

None of the contributing authors have conflicts of interest to declare.

Author details

Frank Huess Hedlund[1,2*] and Øssur Jarleivson Hilduberg[3]

*Address all correspondence to: fhuhe@dtu.dk

1 COWI, Kongens Lyngby, Denmark

2 Technical University of Denmark (DTU/Compute), Kongens Lyngby, Denmark

3 Danish Maritime Accident Investigation Board (DMAIB), Valby, Denmark

References

[1] H. Petroski, *Success through failure: the paradox of design*. Princeton University Press, Princeton, New Jersey, 2006

[2] F. H. Hedlund and H. B. Andersen, "Institutional support of learning from accidents: Some obstacles to getting a useful community-wide database in the EU," in *Society of Risk Analysis, SRA Europe, Annual Conference*, Ljubljana, Slovenia, 2006.

[3] F. H. Hedlund and J. Astad, "Solid biomass climate change interventions examined in a context of inherent safety, media shifting, and emerging risks.," *Human and Ecological Risk Assessment: An International Journal*, vol. 21, no. 5, pp. 1410–27, 2015.

[4] F. H. Hedlund, J. Astad and J. Nichols, "Inherent hazards, poor reporting and limited learning in the solid biomass energy sector: a case study of a wheel loader igniting wood dust, leading to fatal explosion at wood pellet manufacturer," *Biomass and Bioenergy*, vol. 66, pp. 450–459, 2014.

[5] G. Thek, I. Obernberger. *Pellet handbook: the production and thermal utilisation of biomass pellets*. Taylor and Francis, Florence, ISBN: 9781849775328

[6] L. Soto-Garcia, X. Huang, D. Thimmaiah, A. Rossner and P. K. Hopke, "Exposures to carbon monoxide from off-gassing of bulk stored wood pellets," *Energy & Fuels*, vol. 29, no. 1, pp. 218–26, 2014.

[7] F. Meier, I. Sedlmayer, W. Emhofer, E. Wopienka, C. Schmidl, W. Haslinger and H. Hofbauer, "Influence of oxygen availability on off-gassing rates of emissions from stored wood pellets," *Energy & Fuels*, vol. 30, no. 2, pp. 1006–12, 2016.

[8] "Aktindsigt – afsluttet straffesag vedr. arbejdsulykke på skibet Amirante. Journr. 2200-10170-00040-15 Bornholms Politi, Rønne. 2015 (Request for information through the Access to Public Administration Files Act)".

[9] S. Gauthier, H. Grass, M. Lory, T. Krämer, M. Thali and C. Bartsch, "Lethal carbon monoxide poisoning in wood pellet storerooms—two cases and a review of the literature," *Annals of Occupational Hygiene*, vol. 56, no. 7, pp. 755–63, 2012.

[10] U. Svedberg and A. Knutsson, "Hazards and health risk with storage and transport of wood pellets, wood shavings and timber in enclosed spaces. Kunskapsöversikt. Rapport 2011:2 (in Swedish)," Arbets-och miljömedicinska kliniken, Sundsvalls Sjukhus, Sundsvall, Sweden, 2011.

[11] W. Stelte, "Guideline: storage and handling of wood pellets. Resultat Kontrakt (RK) report," Danish technological institute, Energy and climate, Centre for renewable energy and transport, section for biomass, Taastrup, Denmark, 2012.

[12] B. K. Sovacool, R. Andersen, S. Sorensen, K. Sorensen, V. Tienda, A. Vainorius, O. M. Schirach and F. Bjørn-Thygesen, "Balancing safety with sustainability: assessing the risk

of accidents for modern low-carbon energy systems," *Journal of Cleaner Production*, vol. 112, pp. 3952–65, 2016.

[13] Carbon monoxide in bow thruster room," Dutch Safety Board (www.safetyboard.nl), The Hague, The Netherlands, 2015.

[14] "Mystery on ship at Fredericia: One dead and four admitted to hospital by invisible killer," BT, 29 4 2015. [Online]. Available: http://www.bt.dk/danmark/stor-mystik-paa-skib-vedfredericia-en-omkom-og-fire-blev-indlagt-af-usynlig-draeb [Accessed 29 4 2015]

[15] "Officer killed by cargo's fumes". *Telegraph*, vol. 48, no. 12), p. 12, Dec 2015 (Published by Nautilus International, London).

[16] "Very serious marine casualty. M/V Corina. Death of seaman and carbon monoxide poisoning of four other persons during port call in Hanstholm on 28 April 2015. Final report—WIM 13/1.5," Polish State Maritime Accident Investigation Commission, Warsaw, Poland, 2016.

[17] "Aktindsigt vedrørende ulykke på skibet Corina i Hanstholm, journr 4100-10170-00541-16, Midt- og Vestjyllands Politi, Holstebro, 2016. (Request for information through the Access to Public Administration Files Act).".

[18] "Aktindsigt – dødsulykke på skib, Corina, Hanstholm Havn. Sag 20160049016/2. Tilsynscenter Nord, Arbejdstilsynet, København, 2016. (Request for information through the Access to Public Administration Files Act)".

[19] S. Melin, "Research on off-gassing" and self-heating in wood pellets during bulk storage. Report prepared for Ethanol BC," Wood Pellet Association of Canada, 2011.

[20] J. Koppejan, A. Lönnermark, H. Persson, I. Larsson, P. Blomqvist, M. Arshadi, E. Valencia-Reyes, S. Melin, P. Howes, P. Wheeler and D. Baxter, "Health and safety aspects of solid biomass storage, transportation and feeding," *IEA Bioenergy*, 2013.

[21] "IMO Revised Recommendations for Entering Enclosed Spaces Aboard Ships. IMO Resolution A.1050(27).," The International Maritime Organization (IMO), 2012.

[22] J. Rasmussen, A. M. Pejtersen and L. P. Goodstein, *Cognitive Systems Engineering*, New York, USA: Wiley, 1994.

[23] J. Rasmussen, P. Nixon and F. Warner, "Human error and the problem of causality in analysis of accidents," *Philosophical Transactions of the Royal Society, Series B: Biological Sciences.*, vol. 327, no. 1241, pp. 449–62, 1990.

[24] D. A. Schön, The Reflective Practitioner: How Professionals Think in Action, Aldershot, England: Arena, 1995 edition.

[25] N. A. Ashford, "Industrial safety: the neglected issue in industrial ecology," *Journal of Cleaner Production, 5(1), 115–121.*, vol. 5, no. 1, pp. 115–21, 1997.

[26] S. S. Rivera, R. D. C. Olivares, P. A. Baziuk and J. E. N. McLeod, "Assessment of Biofuel Accident Risk: A Preliminary Study," in *Proceedings of the World Congress on Engineering*, London, U.K., July 1–3, 2015.

[27] M. T. Hansen, "The Danish market for wood pellets 2014 (Tasknummer: 115-25037)," FORCE Technology for Energistyrelsen, Kongens Lyngby, Danmark, 2016

[28] S. Melin, U. Svedberg and J. Samuelsson, "Emissions from wood pellets during ocean transportation (EWDOT). Research report.," Wood Pellet Association of Canada (lead), 2008.

Chapter 5

Biomass Valorization: Agricultural Waste in Environmental Protection, Phytomedicine and Biofuel Production

Inyinbor Adejumoke Abosede,
Oluyori Abimbola Peter and
Adelani-Akande Tabitha Adunola

Additional information is available at the end of the chapter

http://dx.doi.org/10.5772/66102

Abstract

Industrialization is a major promoter of any nation's economy; it is not without detrimental effects on our immediate environment. Human exposure to various pollutants discharged into the environment may lead to serious health challenges. In the same vein, discharge from the combustion of fossil fuel contributes a great deal to the environmental pollution. The resulting quest for a clean and sustainable environment has spurred myriads of research into advantageous utilization of waste biomass in industrial wastewater treatment and environmentally friendly/alternative energy. Similarly, diverse waste materials have been adopted as sources of important phytochemicals with different medicinal applications. This chapter focuses on the application of waste biomass in environmental remediation, curative medicine, and clean/alternative energy.

Keywords: waste, biomass, bioadsorption, phytochemicals, bioethanol

1. Introduction

Huge amounts of wastes generated during various agro-processes over time constitutes nuisance to the environment. Agro-wastes are cellulosic materials and depending on their sources, may possess high fixed carbon content, multifunctional group surface, and bioactive agents. These characteristics have been well explored and have led to their applications in environmental remediation, phytomedicine, and biofuel generation [1–3].

Biosorption, a sorption technique that utilizes biomass in sequestering various toxicants, is an attractive technique in wastewater treatment. It adds economic advantages to the operational simplicity and ability to remove very low concentration of toxicants offered by commercial activated carbon. Surface tailoring toward a specific pollutant is another unique advantage of biomass [4]. Multifunctional group surface and cellulosic backbone of biomass present them as suitable materials for pollutants uptake. Modification and surface treatment may enhance the sorption efficiency of raw biomass [5].

Similarly, waste biomasses have been explored for the beneficial phytochemicals, which they contain. These phytochemicals have been useful as antimicrobials, antioxidants, hepatoprotectives, and anticancer agents, which are generally preferred because of their minimal side effects on their consumers [1, 6]. Obtaining these useful phytochemicals from waste biomass is a very useful and timely development since threat to food security usually results whenever useful compounds of medicinal interest are identified in major food sources [7].

In the same vein, the use of agricultural waste in biofuel production has brought about provision of alternative energy and reduction in environmental pollution. The use of renewable resources such as cornstarch leads to competition with human food supply. In view of this, there is a drift in the use of agricultural waste for biofuel production [8]. Agro-wastes are lignocellulosic materials; their rich cellulose content qualifies them as feedstock in bioethanol production.

This chapter is aimed at explicitly depicting the potential of agricultural wastes in environmental remediation, curative medicine, and biofuel production with focus on research efforts, successes, and gaps.

1.1. Effective wastewater treatment: panacea to water pollution challenges

The advent of industrialization in developed economy brought to bear great environmental pollution challenges. The discharge of toxic wastewater and polluting gases increased exponentially with industrialization and economy [9]. The hydrosphere continuously receives wastewater with loads of toxicants. The effects of such discharge on the aquatic organism and aquatic ecosystem and by extension on human are enormous. Wastewater contains different categories of pollutants ranging from organics to inorganics. Organic pollutants such as dyes form a major part of wastewater discharged from dye-utilizing industries such as leather, textile, and paint industries.

Dyes are large organic molecules, which are nonbiodegradable and carcinogenic; they cause skin, eye, and respiratory tract irritations as well as the reduction of dissolved oxygen and the hindrance of sunlight penetration into water bodies [5]. Heavy metals, on the other hand, are inorganic pollutants with serious deleterious effects on plants and human [10, 11]. Effective treatment of wastewater containing the aforementioned toxicants before their discharge into the environment is thus very important.

Conventional methods of wastewater treatment exist, namely electrochemical treatment, ion exchange, precipitation, reverse osmosis, evaporation, solvent extraction, adsorption with activated charcoal, among others. Disadvantages associated with conventional methods of wastewater treatment may be a concern to the developed and the developing economy. For instance, toxic sludge generation in chemical precipitation subsequently results in soil and

http://dx.doi.org/10.5772/66102

water pollution. Inability of conventional methods to remove low concentration of pollutants also calls for concern. In addition, cost of installation, operation, and maintenance is a great concern to developing economies. Adsorption using activated charcoal is a way out of some of these earlier mentioned challenges. However, activated charcoal is very expensive.

Naturally occurring low-cost adsorbents/biosorbents, such as clay and agricultural wastes, particularly, crop residues, may effectively address the challenges associated with conventional methods of wastewater treatment. Due to the basic composition of crop residues, they have been widely investigated as a potential tool for effective wastewater treatment.

1.1.1. *Crop residues as a tool in environmental remediation: characteristics, uniqueness and versatility*

Crop residues are the part of crop production processes not used as food for human. They are not the primary products usually consumed as food. These may come in the form of stalks, straw, leaves, roots, peels, husks, cobs, shells, nuts, and even waste woods. The composition of these crop residues may vary depending on their sources [12]. However, they are characterized by similar features and basic composition.

1.1.1.1. *Polyols backbones*

The basic components of crop residues are cellulose, hemicellulose, lignin, and some extractives [13]. While cellulose is a crystalline α, β (1–4) linked D-glucose polymer with high molecular weight, hemicellulose is an amorphous, shorter chain polymer of various sugars. These two components (cellulose and hemicellulose) contain numerous hydroxyl groups and thus result in a characteristic polyols backbone in crop residues. The polyols surface presents suitable sites for the uptake of pollutants. The deprotonation of hydroxyl group leaves on the residue surface a number of electron-rich oxygen atoms. These electron-rich species could act as Lewis bases, thus attracting electron-deficient species such as metal ions as depicted in **Figure 1**.

1.1.1.2. *Multifunctional group surface*

Aside the basic hydroxyl groups present in these crop residues, their surfaces may also carry other functional groups such as carbonyl, carboxyl, methoxyl, sulfhydryl, ether, acetamido, or amino groups. These functional groups, however, may vary depending on the crop residue source, variety, age, and environmental and cultivation parameters [14–16]. Common to these functional groups is the possession of lone pairs with which they chelate (bite) metal ions. Metal and dye sequestering by different parts of the residue can occur via various processes among which are the following:

Complexation: A process in which metal ions form complexes with organic molecules. This organic species must have an atom or atoms having lone pair of electrons to donate.

Chelation: A process in which organic molecules containing more than one functional group with donor electron pairs simultaneously donate these to a metal atom resulting in the formation of a ring structure involving the metal atom. Due to the multiple binding sites, chelates are more stable than complexes.

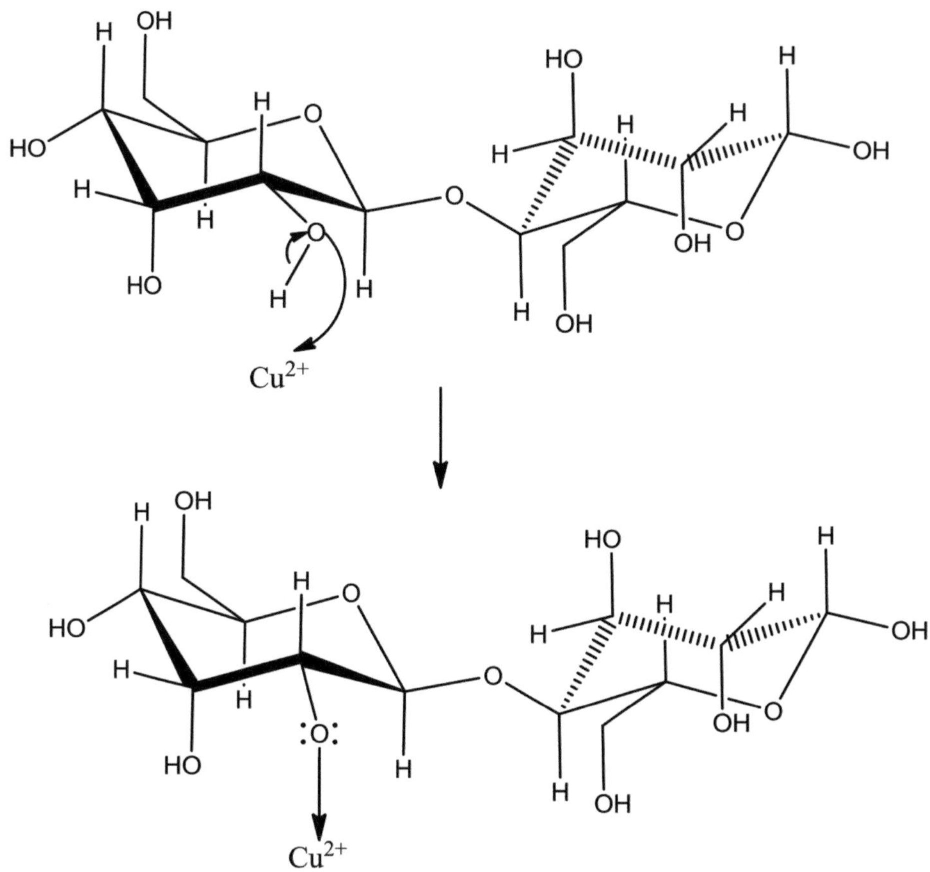

Figure 1. Possible dissociation in cellulosic chain and attraction between Lewis base and metalion. (Source: Authors).

Precipitation: Precipitation is a double decomposition reaction in which toxic metal ions are removed from solutions. This has been used severally in biosorption.

Reduction: Reduction is a process involving electron gain. Removal of toxic hexavalent chromium from aqueous solution by biosorption has been reported, and it is often associated with the simultaneous reduction of Cr (VI) to Cr (III).

Immobilization may be the result of more than one mechanism; for example, metal complexation may be followed by metal reduction or metal precipitation.

The high binding ability of the aforementioned functional groups with a wide range of pollutants via various mechanisms presents crop residues as excellent tools in environmental remediation.

1.1.1.3. Fibrous nature and possible porosity

The characteristics of crop residue are well understood and established by studying their surface morphology and porosity. While some crop residue depicts vivid threadlike surface fibers, pores scarcely appear in some, and some others may possess naturally wide-open pores and hollows. Fibrous character is synonymous with wood waste [17], and a range of water pollutants can be locked up within the crop residue fibers. Pores and hollows also serve as mediums for transporting various types of pollutants into the bioadsorbents. Mesopores (pores with size range between 2 and 50 μm) are more useful in the transport and trapping of large molecules such as dyes into the bioadsorbents [4]. Scanning electron microscopy studies (SEM) of two biomass (*Irvingia gabonensis* nut and *Raphia hookerie* epicarp) as investigated by Inyinbor and colleagues reveal the fibrous nature of *I. gabonensis* biomass, while several moderately large pores were observed in *R. hookerie* biomass (**Figure 2a** and **b**).

1.1.1.4. Carbon content and versatility

Crop residues are generally characterized by high carbon content. Hence, they have found great application in diverse fields. Their use in environmental remediation spans through utilizing raw bioadsorbents, modified biomass, as well as biomass-generated activated carbon for pollutant uptake. The high fixed-carbon content in raw biomass significantly aids pollutant uptake [18].

Pollutants may be chemisorbed via its interactions with surface functional group(s), locked in the fibers, or trapped within the biomass pores. Via one or a combination of these mechanisms, biomasses sequester pollutants from aqueous solutions. Although raw biomasses are effective in pollutant uptake, surface modification, as well as functionalization, increases the sorption capacity. Thermal and acid treatment increase sorbent porosity and subsequently its sorption capacity [4, 5]. Commonly used activating agents are KOH, NaOH, H_3PO_4, H_2SO_4, $ZnCl_2$, among others [19]. Highly porous-activated carbons have been prepared from various crop

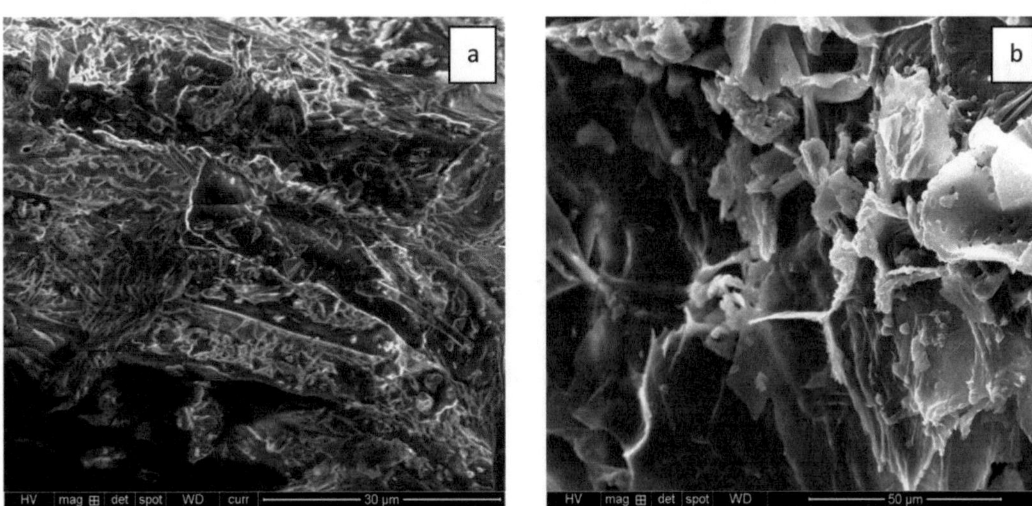

Figure 2. (a) Fibrous surface of *I. gabonensis* bioadsorbents (Source: Authors); (b) moderately large pores on the surface of *R. hookerie* bioadsorbent (Source: Authors).

residues using different activating agents. Prepared porous-activated carbons have been applied to the uptake of dyes and heavy metals and a few of previously reported work is summarized in **Table 1**.

Crop residue	Treatment	Surface area (m^2/g)	Pollutants	q_{max} (mg/g)	References
Corn cob	Thermal/H_3PO_4	600–700	Methylene blue	0.81–28.65	[20]
Palm stem	Thermal/K_2SO_4	—	Methylene blue	149	[21]
Rambutan peel	Microwave induced	971	Acid yellow	215.05	[22]
	KOH activation				
Olive mill	Thermal/KOH	1641	Cr (III)		[23]
Pineapple waste	$ZnCl_2$/Thermal	914.67	Methylene blue	288.34	[24]
Grape waste	$ZnCl_2$/Thermal	1455	Methylene blue	417	[25]
			Metanil yellow	386	
Silkworm cocoon	$(NH_4)_2HPO_4$/KOH	1153–2797	Congo red	512 g/kg	[26]
	Activation				
Durian seed	Thermal/KOH	980.62	Methyl red	384.62	[27]
Tomatoe waste	$ZnCl_2$/Thermal	1093	Methylene blue	400	[28]
			Metanil yellow	385	
Peanut shells	Microwave irradiation	370.10	Direct black	110.6	[29]
	Pyrolysis		Reactive red	284.5	
Potato peels	H_3PO_4/Thermal	1.04	Co^{2+}	373–480	[30]
Guava seed	Carbonization and physical activation	1000	Pb^{2+}	96	[31]
Tropical almond				112	

Table 1. Activated carbon preparation from different crop residue and their applications in dyes and heavy metal uptake.

1.1.2. Pretreatment methods and biosorbent efficiency

Surface treatment, modification, and/or functionalization enhance biosorption efficiency of bioadsorbents. The desired feature can be created in bioadsorbents via designed treatment and/or modification. Biomass treatment using concentrated acid may blow up the fibers subsequently leading to mesopore creation on biomass surface [5]. Mesopores are large enough to transport dyes into the bioadsorbents. The release of adventitious water from the biomass framework via thermal treatment may also result in generation of pores [4]. Texture and porosity of biomass can also be altered by microwave irradiation. Hollows, pores, and cavities serve as transport medium for the adsorbate into bioadsorbents [32].

The possibility of surface modification and/or functionalization gives additional advantage to crop residue. The surface of raw bioadsorbents may be tailored toward specific pollutants

http://dx.doi.org/10.5772/66102

via the introduction of specific functional group(s) that has(ve) affinity for the desired pollutant.

Surface modification may involve the grafting of functional group(s) onto the biomass frameworks, which have a high tendency to reduce sorbent porosity and subsequently its surface area [33, 34]. Interestingly, surface modified biomaterial usually generates adsorbents with far superior adsorption capacity. Sections 1.21 and 1.22 of this chapter introduced the possibility of pollutants uptake onto biomass via surface functional groups. Introduction of such functional groups with unique metal/dye binding characteristics enhances biomass sorption capacity. Metal ions can be complexed by thiol, amino, carboxylates, nitrogen, and oxygen-based moieties. Bonding via electrostatic interactions can also occur between metal and unprotonated carboxyl, oxygen, and sulphate groups. **Table 2** compares some reported literature works that reveal increase in sorption potential of modified biomass over their unmodified counterparts.

Crop residue	Pollutants	q_{max} (mg/g)	References
Citrus sinensis	Reactive yellow 42		[35]
Raw		36.36	
Immobolized		51.47	
Chemically treated		106.36	
Citrus sinensis	Reactive red 45		[35]
Raw		18.28	
Immobolized		39.06	
Chemically treated		31.45	
Irvingia gabonensis	Rhodamine B		[5]
Raw		212.77	
Acid treated		232.00	
Lagerstoemia speciosa bark	Cr (VI)		[36]
Raw		20.41	
HNO_3 modified		24.39	
Phytolacca americana L	Pb(II)		[37]
Raw		13.19	
HNO_3 modified		14.51	
Charred dika nut waste	Rhodamine B		[4]
Raw		52.90	
Chitosan coated		217.39	
Eucalyptus sheathiana bark	Zn (II)		[38]
Raw		128.21	
NaOH modified		250.00	

Table 2. Comparing the efficiency of raw and modified biomass.

1.1.3. Crop residue in dyes and heavy metal biosorption

Crop residues, based on their versatility earlier described, have been widely used in the uptake of diverse range of pollutants. Pollutants of interest here are dyes and heavy metals. This section, therefore, presents some recent works reported in literatures using crop residues in dye and heavy metal sorption.

The uptake of Cu (II) ions using almond shell has been reported. Binary metal mixture viz Cu (II) and Pb (II) was used in this study. The presence of lead was reported to affect copper uptake due to competition for sorption sites. The affinity of Pb (II) for almond shell was reported to be higher than that of Cu (II). While equilibrium sorption fitted best into the Sips model, maximum sorption was obtained to be 9.0 and 13.7 mg/g for Cu (II) and Pb (II), respectively [39]. Water chestnut peel was used in the uptake of Rhodamine B dye from aqueous solution with consideration of various sorption operational parameters. Heterogeneous sorption controlled by both liquid film and intraparticle diffusions was reported for the sorption mechanism. Adsorption data fitted best into the Freundlich isotherm model [40]. The use of *Trichoderma reesei* in the uptake of Co^{2+}, Cu^{2+}, Ni^{2+}, Pb^{2+}, and Zn^{2+} ions has been reported with consideration of various sorption operational parameters. While the uptake of the metal ions onto the bioadsorbents obeyed the Pseudo second kinetics models, biosorption process was affirmed to be spontaneous and feasible [41].

The uptake of Pb (II) from aqueous solution onto peanut shells has been reported. Decrease in bioadsorbents efficiency was observed with increased temperature. Equilibrium sorption data fitted best into the Langmuir isotherm model with maximum monolayer sorption capacity of 39 mg/g [42]. Batch sorption of Cd^{2+} onto Arundo donax reed leaves has been reported. Cd^{2+} removal was reported to be pH-dependent, and fast sorption kinetics was observed. The maximum monolayer sorption capacity was reported to be 27.90 mg/g. Equilibrium sorption data was best described by the Freundlich isotherm model, while the kinetic data best obeyed the pseudo second-order kinetics model [43]. The uptake of Cr (VI) using cork powder has been investigated and reported. Sorption parameters, such as contact time, pH, and concentration of Cr (VI), showed significant effects on pollutant uptake, and maximum uptake was observed between pH 2 and 3 [44]. The use of loquat (*Eriobotrya japonica*) leaves waste in Cd (II) uptake was investigated and reported. Uptake of Cd (II) was reported to be dependent on solution pH, initial cadmium ion concentrations, biosorbent dose, contact time, and temperature. Uptake of Cd (II) onto loquat leaves occurred via endothermic process; thus, bioadsorption was more favored at a high temperature. Equilibrium sorption data fitted well into both the Langmuir and Temkin isotherm models [45]. Removal of Malachite green using pea shells (*Pisum sativum L*) has also been studied and reported. Optimum operational parameter was established, and equilibrium sorption data was tested with some isotherm models. Freundlich isotherm model best described the sorption process, and pseudo second-order kinetics best explains the kinetics data [46]. Chir pine sawdust in the bioadsorption of Congo red and basic violet from aqueous has been reported. Equilibrium was attained between 45 and 60 min. While the equilibrium sorption data for the uptake of basic violet obeyed the Langmuir isotherm model, Freundlich isotherm model best described the uptake of Congo red onto the bioadsorbents. Maximum monolayer sorption capacity was 11.3 and 5.8 gkg for basic violet and Congo red, respectively. Chemisorption was reported for the uptake of Congo red onto Chir pine sawdust, while physical sorption ruled the uptake of basic violet onto Chir pine sawdust [47].

The use of water hyacinth leaves in the uptake of amaranth (acid red 27) anionic dye has also been reported. The effects of various operational parameters were reported, and the dye uptake depended greatly on these effects. Dye uptake was reported to increase with increased contact time and dye concentration. Sorption kinetic data was best explained by pseudo second-order kinetic model, while the equilibrium sorption data fitted best into the Langmuir isotherm model with 70 mg/g [48]. *Nauclea diderrichii* seed biomass with low surface area was used for the uptake of toxic Cr (III), and it was found to be effective. Fast kinetics was reported for this study, with data fitting best into the pseudo second-order kinetics. Equilibrium sorption data was reported to fit best into the Freundlich isotherm model, and the maximum monolayer sorption capacity was 483.81 mg/g [49].

The effectiveness of oat hull in the removal of malachite green was investigated and reported. Optimum sorption condition was established, and the pseudo second-order kinetic models best described sorption kinetic data. High temperature favored the uptake of malachite green, while the Freundlich isotherm model best described the sorption equilibrium data. The maximum monolayer sorption capacity was reported to be 83 mg/g [50]. The use of *R. hookerie* fruit epicarp in the removal of Rhodamine has been investigated and reported. The prepared bioadsorbents were found to be effective in the removal of the pollutants. Freundlich isotherm best described the equilibrium sorption data, and maximum monolayer sorption capacity was reported to be 666.67 mg/g [2].

1.1.4. Conventional methods versus biosorption

The preference for adsorption over other wastewater treatment techniques relates to its simplicity of operations and ability to remove pollutants of very low concentration. Biosorption, however, adopts the aforementioned characteristics as well as abundance, availability, and economic advantages. In order to attain zero pollution outflows by most industries, combinations of methods are employed. While adsorption may act effectively in the removal of residue concentration as well as protects resins, ion exchange resins are very efficient in the combat of highly metal-loaded wastewater [51]. The application of activated carbon, ion exchange resins, chelating resin, and bioadsorbents by batch adsorption in removal of metal ions from real and aqueous effluents is summarized in **Table 3**.

In conclusion, the uptake of dyes and heavy metals onto various crop residues has been widely investigated. Amazing results of their efficiency were recorded. Freundlich isotherms that suggest a multilayer sorption commonly describe the sorption process. With the understanding that sorption onto crop residue may either be via functional group or available pores or the combination of the two, multilayer sorption may be expected often in pollutant uptake using plant residue.

Optimum operational parameters that can give insight into further applications were well studied and reported. However, only a few reports exist for applications of biomass to real

Techniques	Pollutants	Percentage (%)	References
Ion exchange resin (mix)			[52]
	Cr	68.18–about 85%	
	Cu	82.09–87.25%	
	Fe	58.33–about 90%	
	Ni	50.00–75.58%	
Chelatin resin			[52]
	Cr	6.67–69.23%	
	Cu	54.48–85.19%	
	Fe	25.00–87.50%	
	Ni	11.11–39.54%	
Activated carbon			[52]
	Cr	46.67–about 85%	
	Cu	40.74–82.35%	
	Fe	58.33–83.33%	
	Ni	5.26–11.11%	
Ion exchange resin (INDION225H)			[51]
	Cu^{2+}	9.01–14.49%	
	50–100 mg/L		
Eucalyptus bark			[38]
	Zn^{2+}	33.75–40.12%	
	20–70 mg/L		
Olive stone			[53]
	Cu^{2+}	6.61–19.87%	
	10–300 mg/L		
Pine bark			[53]
	Cu^{2+}	35.25–76.00%	
	10–300 mg/L		
Peanut shell			[42]
	Pb^{2+}	65.73%	
	100 mg/L		

Table 3. Comparison of pollutant removal techniques and percentage removal.

industrial effluent. Result-oriented applications of crop residues to real industrial effluents present a solution to the lingering challenge of water pollution. Real industrial effluents are a composite of various ions and other materials such as starch and salt. Thus, pilot studies using binary/ternary effluents as well as dye-metal mixture will assist in understanding the behavior of crop residue in real industrial effluent applications.

1.2. Biomass valorization in phytomedicine

The use of plant materials as sources of beneficial nutritional component dates back to prehistoric time. Noah obtained alcohol from vine as recorded in the Bible; Hippocrates encouraged the consumption of medicinal food [54], and the list continues.

Centuries later, several plants have been identified as sources of phytochemicals with potent medicinal value. Initially, plant extracts were used directly, but in these modern times, bioactive extracts are fractionated, and pure compounds isolated therefrom are then investigated for their inherent pharmacological properties. This, of course, is preceded by the characterization of the isolates using spectroscopic tools such as 1D and 2D-Nuclear Magnetic Resonance, Infrared, and mass spectrometry.

Despite the advent of several synthetic drugs via improved scientific methods of drug development, natural products have continued to maintain relevance by consistently providing new compounds with interesting structural motifs, which serve as sources of new drug leads.

The consequence of this intense plant utilization is the generation of a huge amount of waste, which eventually constitutes environmental pollution among other problems. In line with waste valorization, which promotes environmental sustainability, several scientists have engaged in intense search into possible economically beneficial application of plant wastes.

In the same vein, the discovery of beneficial phytochemicals in waste plant materials would be a welcome development for scientists. This is because there would not be any threat to food security as is the case when an edible food source is found to supply uniquely valuable phytochemicals [55].

This section is, therefore, dedicated to the interesting medicinal application of phytochemicals obtained from waste biomasses.

1.2.1. Direct application of plant parts

Conventionally, the active components in different plant parts are first extracted with the aid of a suitable solvent, for example, water and ethanol. The extract solution is then concentrated to obtain the crude extract, which is then applied to proffer solution to different health challenges.

However, these same plant parts can be used directly for different purposes without conventional extraction. An example of such usage is the use of banana leaves, plantain leaves, and *Thaumatococcus daniellii* in food packaging. Particularly, these leaves have been employed in the packaging of bean pudding. Although the initial reason for employing these leaves as packaging materials was their surface area and their ability to withstand heat, it has also been observed that these leaves perform additional functions such as flavour enhancement and possible supply of medicinal compounds to the consumer [56]. Quantitative phytochemical analysis of the leaves of *T. daniellii* prior to and after usage revealed that some of these phytochemicals (saponins, flavonoids, and tanins) have leached either into the bean pudding or the water used for steaming [57].

It will be interesting to initiate a research that will reveal the exact position of these leached phytochemicals as the presence of saponins in food samples could constitute a natural preservative, which will prevent the growth of food spoilage organisms such as *Escherichia coli*. Saponins have been earlier implicated in antimicrobial activity [58]. Similarly, the presence of flavonoids in bean pudding would be advantageous since they are potent scavengers of free radicals in human systems.

Furthermore, agro-wastes can be utilized in biotreatment procedure for safe long-term storage of cereal products. An example is the inhibition of *Aspergillus parasiticus* and aflatoxin production during the storage of wheat, rice, and maize by powdered neem (*Azadiratcha indica*) and kika leaves (*Acia nilotica*). The agro-wastes (leaves) in their powdered forms were applied to the cereals at three concentrations 5, 10, and 20% out of which 20% neem gave the best overall result. Neem leave powder is, therefore, a safe and cheap natural additive for the improvement of the storage of selected cereal products [59]. Similarly, pomegranate peels powder at 1–3% concentration have been shown to provide high storage stability in prepared beef sausage samples, and the presence of antimicrobial/antioxidant phenolics explains this important activity [60].

1.2.2. Extraction of phytochemicals from waste biomass

In several homes, various plant parts such as lemon grass and tea have been boiled in water in order to obtain tonics, which are drunk afterward to achieve different objectives. This domestic procedure is known as extraction.

Extraction, in this context, is the process of separating active constituents or secondary metabolites of interest from waste plant matrix by the use of extractants (solvents). Different methods of extraction that have been utilized in the extraction of phytochemicals include the following:

1.2.2.1. Maceration and percolation

These are cold traditional methods of extraction in which water or organic solvents are made to permeate pulverized biomass with the aim of transferring the phytochemicals in the biomass into the solvent. The extract solution is afterward concentrated using an appropriate technique, for example, rotary evaporator to obtain the crude extract. While maceration emphasizes soaking the biomass in the choice solvent to achieve the extraction, percolation involves the permeation of the solvent through the biomass in the direction of gravity and filtration in order to obtain the extract solution. These traditional methods of extraction have been employed in the extraction of phytochemicals from sweet potato peels, leading to extracts and isolates with antioxidant and antifungal properties [1].

Although these methods may not be as fast as other methods discussed in this section, they do preserve phytochemicals which are thermolabile by nature, thus retaining the pharmacological activities of interest.

1.2.2.2. Hydrodistillation/assisted hydrodistillation

Hydrodistillation (HD) is a method, which is employed for the extraction of fragrances/flavors from plants. In principle, the vapors of the volatile constituents are conveyed by steam to a condenser, where they are returned to liquid state and form two layers: aqueous and oil-rich layers. The two methods via which hydrodistillation is executed are steam distillation and Clevenger distillation [61]. This important extraction method yielded the essential oils from tangerine peels, which were successfully adopted for fish preservation [62].

In recent times, the conventional HD has been assisted and improved with ohmic and microwave heating processes. Assisting hydrodistillation with ohmic heating reduced the extraction time by over 50%, that is, from 1 h to 24.75 min [63], while the GC-MS analysis conducted did not show any difference in the chemical constitution of the oil. Although a shorter extraction time was achieved using the microwave-assisted hydrodistillation [64], the ohmic-assisted HD was the greener method of extraction since its energy consumption is relatively low.

1.2.2.3. Sohxlet extraction

This is a hot method of extraction in which the soxhlet apparatus is employed to achieve the extraction process. The soxhlet apparatus is depicted in **Figure 3**.

This method was successfully utilized in our laboratory to extract constituents of *Ipomoea batatas* peel-waste [65]. Although this method has good speed and high yield as advantages, the possibility of thermal decomposition of the active principles remains a subject of huge concern and constitutes a limitation to this application.

1.2.2.4. Pressurized fluid extraction

One of the emerging methods that have been employed for extractions from waste biomass is pressurized fluid extraction (PFE). We will focus on two PFE methods in this section: supercritical extraction and subcritical extraction.

Supercritical extraction is an extraction method in which the conventional solvent has been replaced by a supercritical fluid: a supercritical fluid is a liquid substance at a temperature and

Figure 3. Three soxhlet apparatus on heating mantles during the hot extraction of the phytochemicals in water melon rind (Source: Authors).

pressure above its critical point—a point where distinct liquid and gas phases do not exist. A common example of this supercritical fluid is CO_2, which is sometimes modified by cosolvents such as ethanol and methanol [66]. This environment friendly method was successfully applied to citrus peels, a common by-product in citrus processing industries [66]. Fatty acid esters, phenols, coumarin derivatives, and terpene derivatives were among the useful compounds that were isolated in the process.

Although more expensive to set up, the advantage of this method is its environmental friendliness and extremely easy solvent removal.

The cost cut by improved percentage yield achieved by PFEs has not been compared with the cost of setting up pressurized fluid extraction apparatus. Despite this gap, the safety guaranteed by the absence of obnoxious chemicals continues to make PFE a preferred alternative.

In the same vein, subcritical water (pressurized hot water) has also been employed to achieve environment friendly extractions. Subcritical water is water under 0.2–21 MPa pressures and temperatures higher than its boiling point but lower than its critical temperature. Subcritical water was first reported in 1994 [67], and several other applications of subcritical water has been reported since then. Recently, this method was applied to several agro-waste biomasses in Canada. These waste biomasses are potato peel, barley hull, lentil husk, and flaxseed hull as contained in a recent review by Saldana and Valdivieso-Ramírez [68].

1.2.3. Effect of extraction method on bioactivity

The different extraction conditions associated with different extraction methods have considerable effect on the pharmacological potency of various phytochemicals. Such conditions include the following:

- Extraction pressure
- Extraction time
- Extraction solvent
- Extraction temperature

Trabelsi et al. studied the effect of high pressure (300 and 500 MPa) and time (3 and 10 min) on the antioxidant/antimicrobial activity of citrus peel extracts [66]. While the antimicrobial activity was not at all influenced by high pressure, the highest Total Phenolic Content (TPC) and antioxidant capacities were observed at 300 MPa and 3 min. Extraction at higher temperature and longer time probably drew out more "nonantioxidant" matter or decomposed some antioxidant constituents in the extracts, thus reducing the TPC per gram.

Furthermore, extraction solvent and extraction time have significant effect on the antimicrobial activity of bioactive extracts from watermelon seeds as shown by a recent study from the School of Science, Landmark University, Nigeria [58]. Cold methanol extract was more active against *Staphylococcus aureus* compared with the hot methanol extract. Thermal decomposition might have affected some of the phytochemicals that are responsible for the observed activity.

Furthermore, the cold chloroform extract that did not inhibit the growth of *S. aureus* at all was, however, active against *Pseudomonas aeruginosa*. This simply shows that the choice of a solvent in phytochemical extraction greatly influences the nature of compounds extracted and ultimately the biological activity of the extract. The effect of different extraction methods on percentage yield and bioactivity is vividly displayed in **Table 4**.

Generally, the introduction of pressure, higher temperature, and supercritical fluids led to the achievement of better yields and improved/preserved biological activity.

1.2.4. *Medicinal products from waste biomass*

Different kinds of plant biomass have been employed in the search for medicinal phytochemicals. Selected examples are shown in **Table 5**.

In line with the principles of green chemistry, waste biomass that otherwise would have been discarded have been exploited as sources of beneficial/medicinal phytochemicals by several research scientists.

1.2.4.1. *Antioxidants from waste biomass*

Antioxidants, as the name implies, are compounds that oppose or prevent oxidation. They are compounds that scavenge free radicals, which have been known to cause unwanted oxidation of important molecules in the body.

$$R + ArOH \rightarrow RH + ArO \quad (1)$$

The unwanted free radical (R) is reduced by the antioxidant compound (ArOH), resulting in the harmless RH and ArO which is a more stable radical.

Furthermore, a high consumption of antioxidants has been correlated with lower incidence of cancer [82]. During renovation of cultivations, the roots of *Rugosa rose* are usually treated as wastes. Olech et al. studied, among other parameters, the antioxidant potential of this postharvest residue with the aim of proposing new uses for it. Interestingly, these roots displayed significant antiradical activity compared with ascorbic acid. Hence, these roots could be explored as a source of antioxidant supplements in food and pharmaceutical industries [83].

In the same vein, tomato-processing by-products (TPP) have also been shown to contain antioxidants. Particularly, the presence of simple polyphenols in TPP suggests that it could be safely used as a functional ingredient in the production of antioxidant-rich foods [84].

Similarly, Oluyori et al. recently reported a successive extraction of sweet potato peels using ethyl acetate and methanol, respectively. While the extracts did not show significant antimicrobial activity against the selected microorganisms, the DPPH assay and the TPC determination revealed the methanolic extract as the extract with higher antioxidant capacity [6].

These and many other research efforts continue to prove that there is, indeed, a vast amount of wealth in agro wastes, which are usually discarded.

S/No	Extraction method	Process condition	Percentage of yield (%)	Remarks	Reference (s)
1	Room temperature water extraction	25 C; 3 h; 1:20; w/v	<30.35%	% yield of CPS by microwave-assisted extraction was the highest and the antitumor activity same extract was the best	[69]
	Hot water extraction	78 C; 3 h; 1:20; w/v	<30.35%		
	Microwave-assisted extraction	5 min; 280 W; 70°C; 1:20; w/v	30.35%		
	Ultrasound-assisted extraction	13 min; 250 W; 45 C	<30.35%		
	Cellulase-assisted extraction	55 C; 80 min; pH 5.0	32.50%		
2	Ultra-high pressure extraction (UPE)	60% ethanol; 400 MPa; 2 min; 30°C; 1:5 solid to liquid ratio	4.395% chlorogenic acid	UPE showed higher extraction yield, shorter extraction time, lower energy consumption, and higher purity of extracts	[70]
	Heat reflux extraction	60 C; 3 h	3.471% chlorogenic acid		
	Ultrasonic extraction	250 W; 30°C; 40 min; 42 KHz	3.601% chlorogenic acid		
	Soxhlet extraction	60 C; 2 h	3.793% chlorogenic acid		
3	Ultra-high hydrostatic pressure (UHP)	60% ethanol; 70:1 ml/g solvent: herb; 500 Mpa hydrostatic pressure; 3 min	26.12%	High efficiency at room temperature was observed for UHP; flavanoids and salidroside were used to monitor the extraction yield	[71]
	Reflux extraction	80 C; 120 min; 4:1 ml/g	13.83%		
	Soxhlet extraction	80°C; 240 min;	17.00%		
4	Supercritical fluid extraction (SFE)	250 bar and 40°C		SFE gave the best extraction yield and the most active extract	[72]
	Soxhlet extraction	Ethanol; 6 h	10.00%		
	Steam distillation	4 h	0.60%		
	Marceration	40 °C; 48 h; material: solvent 1:10	8.90%		

S/No	Extraction method	Process condition	Percentage of yield (%)	Remarks	Reference (s)
5	Soxhlet extraction	720 min; 200 ml solvent	9.58%	% yield of scopoletin was used to determine the extraction efficiency	[73]
	Reflux extraction	480 min; 100 ml solvent	10.95%		
	Ultrasound-assisted extraction	20 min; 60 W; 50 ml solvent	28.55%		
	Microwave-assisted extraction	1 min; 560 W; 50 month = l solvent	45.10%		
	Supercritical fluid extraction	60 min; 6500 psig;	7.99%		
6	Steam distillation	403–423 K; 24 h; atm pressure	2.32%	% yield of essential oil from Kabosu (*Citrus sphaerocarpa*) peel was used to measure the extraction efficiency	[74]
	Supercritical fluid extraction	313–352 K; 10–30 Mpa; 0.19–0.21 g/s CO_2 flow rate	1.55%		
	Cold press	100 ml distilled water; centrifuged at 10,000 G for 15 min/frozen at 253 K	0.12%		
7	Supercritical fluid extraction	600 bar/40°C	88.20 mg/100 g oil	Extraction of tocopherols	[75]
		600 bar/80°C	85.57 mg/100 g oil		
	n-hexane extraction	—	62.38 mg/100 g oil		

Table 4. Different extraction methods: process conditions and effects on percentage(%) of yield.

1.2.4.2. Antimicrobial compounds from waste biomass

The emergence of multidrug-resistant bacteria has continued to be a major concern across the globe. This is the reason for the increased attention given to the search for new antimicrobial agents.

Generally, many people after consumption discard the seeds of watermelon. However, several scientists have now shown that inherent in these seeds are phytochemicals with varying activities against *Staphylococcus* sp., *E. coli, Proteus* sp., *Klebsiella* sp., and *P. aeruginosa* [57]. Hence, these seeds could serve as renewable sources of new antimicrobial agents.

A similar study focused on the valorization of sweet potato peels. It was interesting to have isolated beneficial triterpenoids from sweet potato peels that we normally discard. Although the isolates (an example is shown in **Figure 4**) from the peels of *I. batatas* Lam exhibited low

S/No	Biomass type	Plant source	Medicinal phytochemical of interest	Application	References
1	Seed	Sour cherry seed kernel	Alpha-tocopherol, tocotrienols, tocopherol-like compounds	Antioxidants	[76]
2		Germinated fenugreek seed	Vitexin, Isovitexin	Antioxidants	[77]
3	Stem bark	*Mitrephora celebica*	Ent-tradyloban-19-oic acid and ent-kaur-16-en-19-oic acid	Antimicrobial	[78]
4	Leaves	Leaves of *X. laavigata*	Essential oils (γ-muurolene, δ-cadinene, α-copaene, germacrene-D, bicyclogermacrene and (E)-cariophyllene	Antitumor activity	[79]
		Leaves of *Clinacanthus nutaus lindau*	13(2)-hydroxy-(13(2)-R)-phaephytin b, 13(2)-hydroxy-(13(2)-R)-phaephytin a, 13(2)-hydroxy-(13(2)-S)-phaephytin a	Anti-herpes simplex activity, Anti-HSV-IF activity	[80]
5	Flowers	Elder flower: *Sambucus nigra*	Naringenin	Nutraceutical application (modulation of glucose metabolism)	[81]
6	Peels	Sweet potato peels	3-Friedelanol; Stigmasterol; Urs-13(18)-ene-3β-yl acetate	Antifungal	[1]

Table 5. Different plant biomass sources and the phytochemicals of interest.

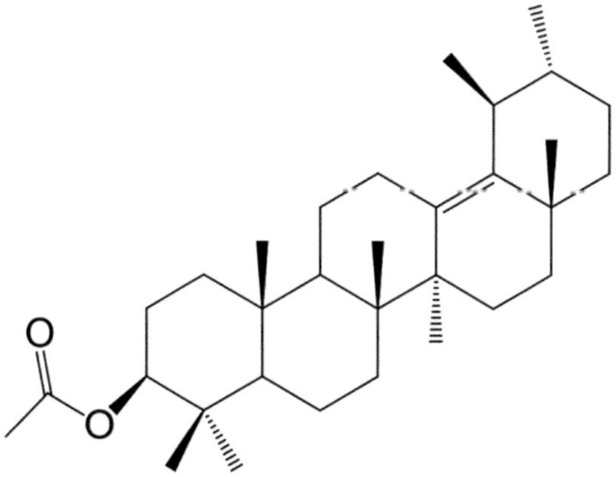

Figure 4. A natural antifungal compound from sweet potato peel: Urs-13(18)ene-3β-yl acetate (Source: Authors).

antibacterial activity, the same isolates were more active against two fungi: *Tricophyton metagrophytes* and *Sporothrix schenkii* (IC_{50}, 50 μg/ml) [1].

Similarly, extracts from drumstick peel recently exhibited significant activity against two human pathogenic bacteria: *S. aureus* and *E.coli*. This extract, which was nontoxic to red blood cells, also showed some photocatalytic property [85].

1.2.4.3. Anticancer compounds from waste biomass

Cancer is the second leading cause of death in the United States after heart disease. In fact, it has been said to be responsible for one out of every four deaths [86]. Although there are current cures for cancer, the high cost of treatment is a challenge to many cancer patients. In view of the increasing cost of cancer treatment, the presence of anticancer compounds in waste biomass is a blessing to many, especially, people in the developing world [87]. The roots of *R. rose,* which showed some antiradical activity as earlier mentioned, also exhibited significant anticancer potential. About 90% inhibition of cervical (HeLa) and breast (T47D) cancer cell lines was observed in the presence of the rugosa root extracts, showing that the roots that have hitherto been ignored are an important phytochemical source [83]. Similarly, Oluyori et al. recently carried out an anticancer guided fractionation/isolation of an alcoholic extract from sweet potato peels [88]. An anticancer fraction was obtained, and a glucocerebroside {N-((E)-3,4-dihydroxy-1-(((2R,3R,4S,5S,6R)-3,4,5-trihydroxy-6-(hydroxymethyl)tetrahydro-2H-pyran-2-yl)oxy)octadec-8-en-2-yl)-2-hydroxytetracosanamide} was isolated and characterized as the main anticancer compound. Our results suggest that the consumption of sweet potato peels may supply the body with phytochemicals that can prevent the onset or the development of cancer [89]. These findings among others project agro-wastes as potential sources of economically important compounds with multifaceted medicinal applications.

1.3. Harnessing agro-wastes for bioethanol production

Conversion of agricultural waste into fuel has become popular in the past few decades occasioned by the increasing demand for fossil fuel. Campbell and Laherrere [89] had predicted the reduction in availability of "cheap fuel" years ago. In a bid to forestall imminent fall in supply, many countries ventured into research seeking alternative energy sources. This led to the production of biofuel from various sources including cornstarch, sugar cane, and now from cellulosic biomass. Biofuels have several advantages over fossil fuel including being environment friendly, thus reducing pollution. There is also a reduction in the release of greenhouse gases and consequent global warming [90]. Biofuels are produced from renewable sources such as rind of various fruits, corn stover, "weeds" including switch grass, winter rye, and faba bean to mention a few [91]. This section elaborates the extent of research on bioethanol production from agricultural wastes, exposes possible challenges in bioethanol production from agro-wastes as well as exposes gaps from various previous research works efforts thus opening new areas of investigations to researchers.

1.3.1. A drift to biofuel as an alternative energy source

Biofuels are any kind of fuel made from living things or from the wastes they produce [92]. Biofuels can also be referred to as renewable energy produced from organic matter by converting its complex carbohydrates into energy. The carbon dioxide (CO_2) emitted *during biofuel production* is not larger than that consumed during photosynthesis. In order words, the process is CO_2 neutral and beneficial to the environment as opposed to combustion of fossil fuels that leads to increase in greenhouse gases, which causes ozone layer depletion. Biofuels include biogas, biodiesel, and bioethanol, but bioethanol has enjoyed vast investment and is popular as a transport fuel. It is presently the most widely used in the transport industry, and it is receiving attention at the international and national fronts [93]. Countries at the forefront of bioethanol production include the United States of America (USA) and Brazil. Blending ethanol with gasoline is advantageous because it brings about reduction in consumption of fossil fuel, and *in recent time, bioethanol has* provided a means of powering engines without using fossil fuels. Currently, there are vehicles that can run solely on bioethanol as well as in blends with gasoline, for example, E5, E10, and E5 implies that 5 parts of bioethanol is blended with 95 parts of gasoline. When gasoline is blended with bioethanol, it has a higher octane number, which implies that the engine has increased thermal efficiency and works more efficiently.

1.3.2. Second-generation bioethanol

For the production of bioethanol, sugars are fermented using yeast and other microorganisms. Commercial quantities of bioethanol have been produced from corn, sugar cane, sweet sorghum, and cassava [94]. Bioethanol produced in this manner is referred to as first-generation ethanol. In a bid to find feedstock that will not compete with (human) food and (animal) feed and also meet up with global demand for bioethanol production, second-generation bioethanol production was introduced [93]. This process exploits the use of lignocellulosic biomass such as agricultural wastes as feedstock because they are renewable, available in abundance, and relatively cheap. Utilization of such agricultural waste involves pretreatment of such waste, then hydrolysis of the pretreated material, and finally, fermentation of monomeric sugars released to yield bioethanol.

Lignocellulosic materials *currently in use* include agricultural wastes/residues; forest residues such as fallen branches, leaves, twigs, and sawdust, municipal solid wastes including paper and cardboard products as well as industrial wastes and energy crops. Energy crops include herbaceous crops such as switch grass, fast-growing hybrids of poplar and leucaena trees [95]. Among all these sources of lignocellulose, agricultural wastes/residues are the most abundant and provide higher quantities of cellulose. Agricultural wastes are parts of the plant that are left after the main crop has been obtained. Examples are corn stover, sugar cane baggase, spent sugar beets pulp, sweet sorghum straw, cotton seed hairs, oil palm frond, and banana stem. All these can be converted into bioethanol to provide energy to drive various engines. These materials are of great advantage because the crop is grown for dual purpose: first, as a source of food or feed and second, to provide lignocellulosic biomass, which will serve as feedstock for bioethanol production. This is of great advantage to the farmer who now obtains economic value from both the crop and its waste.

1.3.3. Pretreatment technologies

As stated earlier, agricultural wastes undergo pretreatment, and hydrolysis before the fermentation of sugars can be carried out. Pretreatment is a process of solubilization and separation of one or more components of a biomass. In this case, the lignocellulosic material breaks down thereby making cellulose more susceptible to enzymatic attack [95]. The cell wall of agricultural waste consists of cellulose, hemicelluloses, and lignin. Of these three, cellulose, which is the highest constituent, is responsible for the mechanical strength of the

http://dx.doi.org/10.5772/66102

cell wall. The glucose monomers of cellulose are connected by β-1,4 glycosidic bonds; hemicellulose is a branched heteropolymer of D-xylose, L-arabinose, D-mannose, D-glucose, D-galactose, and D-glucuronic acid, while lignin is composed of three major phenolic components, namely p-coumaryl alcohol, coniferyl alcohol, and sinapyl alcohol and other extractable components.

Various types of pretreatment ranging from physical, chemical to biological pretreatment have been described [96, 97]. Pretreatment is aimed at removing lignin and hemicellulose, reducing the crystallinity of cellulose, and increasing porosity of the lignocellulosic materials. Physical pretreatment could be a combination of mechanical methods whereby the waste materials are comminuted by chipping, grinding, and milling, leading to reduced cellulose crystallinity. Pyrolysis is another form of physical pretreatment that involves treating the agricultural waste at temperatures above 300°C. Another form of physical pretreatment of agricultural waste steam is explosion that involves heating the biomass to a desired temperature for a specified period of time [98]. Chemical pretreatment involves treating lignocellulosic material with chemicals, for example, ozonolysis necessitates using ozone to degrade lignin and hemicellulose. Acid [99] and alkaline pretreatments are other forms of chemical pretreatment using concentrated acids such as H_2SO_4 and bases, respectively, to treat lignocellulosic materials. Apart from the aforementioned chemical pretreatment methods, lignin degradation could be catalyzed by a process known as oxidative delignification in the presence of hydrogen peroxide. Combinations of physical, chemical, and biological pretreatment methods can be used on lignocellulosic agricultural waste [100, 101]. Examples of such methods are steam explosion (a physical pretreatment method) and ammonia fiber explosion (AFEX that is a chemical pretreatment method); ultrasonic (physical) pretreatment of rice hull followed by biological pretreatment with *Pleurotus ostreatus* [102].

Biological pretreatment is highly advocated as it can prevent the formation of inhibitory by-products that can adversely affect subsequent stages of hydrolysis and fermentation. Biological pretreatment is also said to be cost-effective, environment friendly compared with chemical methods. Brown rot, white rot, and soft rot fungi are currently used for biological pretreatment [103, 104].

Advantages of biological pretreatment include mild conditions to undertake the process and low energy requirements, ecofriendly, less hazardous to health, more economical, and more effective compared with other chemical and physiochemical methods of pretreatment [105]. White rot fungi has been recognized as efficient for lignin degradation, and their activity is said to enhance the subsequent enzymatic hydrolysis. These organisms degrade the lignin seal and thus give access to cellulases attack on cellulose, but this process is often slow, spanning from 28 days to about 120 days [104]. Apart from longer pretreatment period, biological pretreatment also requires a large space and carefully controlled growth conditions; sometimes the organisms used act on lignin as well as hemicellulose and cellulose [93, 103]. Agricultural wastes are abundant feedstock for bioethanol production; however, there is need to focus more on pretreatment preferably biological pretreatment in order to exploit the benefit of this feedstock for biofuel production.

1.3.3.1. Hydrolysis

After pretreatment, hydrolysis of cellulose is carried out; this involves the conversion of cellulose (and hemicelluloses) to monomeric sugars using chemicals or enzymes [97]. Cellulases are often used for enzymatic hydrolysis of the pretreated agricultural waste [103]. Enzymatic hydrolysis leads to the release of fermentable sugars from the pretreated waste. This process of enzymatic hydrolysis is dependent on some factors including pH of the reaction, temperature, substrates, time, and enzyme activity.

1.3.3.2. Fermentation

When hydrolysis and fermentation are carried out separately, it is called Single Hydrolysis and Fermentation (SHF), but when enzymatic hydrolysis and fermentation are carried out concurrently, the process is termed Simultaneous Saccharification and Fermentation (SSF). In another process known as simultaneous saccharification and cofermentation (SSCF), enzymatic hydrolysis of both glucose and hemicellulose is carried out followed by fermentation of the sugars (glucose and xylose) produced in the same vessel [104].

Consolidated bioprocessing, on the other hand, integrates cellulase production with hydrolysis and fermentation of lignocellulosic material for biofuel production. It has the advantage of potentially reduced production cost, higher energy input, and higher conversion efficiency. A major challenge has, however, been finding a suitable consortium of organism to drive this process. Organisms such as *Saccharomyces cerevisiae, Clostridium thermocellum,* and *C. phytofermentans* are, however, being genetically engineered to perform this purpose [3]. Agricultural residues such as winter rye straw, oilseed rape straw, and faba bean straw have been shown to yield 66, 70, and 52% (of theoretical calculation) ethanol after pretreatment by wet oxidation and fermentation by *S. cerevisiae* [106].

One of the challenges faced in the fermentation of monomeric sugars to ethanol is that ethanol produced in the medium is inhibitory to some organisms used for fermentation. In addition, certain products such as furfural and hydroxylmethylfurfural will limit or totally stop the activities of organisms used for fermentation. Apart from these, other factors such as high temperature, osmotic stress, and contamination by other organisms can adversely affect the process. Eventually, the ethanol produced by fermentation is made to undergo distillation [97].

Bioethanol yield of 11.11% was obtained when preheated rice husk was treated with *Trichophyton soudanense* that was fermented with palm wine yeast. But without fungal treatment, the heat-treated rice husk did not give up to 5% bioethanol yield [107]. Also, the highest cellulose and lignin yields recorded from rice husk in their study were 44.50 and 28.90% obtained from rice husk pretreated with *Trichophyton mentagrophyte* and a coculture of *T. soudanense* and *T. mentagrohyte*, respectively, while the highest hemicellulose value of 30.50% was obtained from preheated rice husk treated with *T. soudanense* .

Agro-waste such as coffee mucilage has also been reported to produce bioethanol. Unlike some agro-wastes that require pretreatment, it was established that coffee mucilage could provide up to 60 g/L of reducing sugar, which could be directly converted to bioethanol using *S. cerevisiae* Y2034 [108]. Pretreatment of wild cassava peels (*Manihot glaziovii*) with enzymes and a combination of enzyme with alkali followed by simultaneous saccharification and fermentation (SSF) was reported. Results revealed that the latter treatment led to higher ethanol production of about 95% theoretical yield [109].

Banana waste composed of approximately 14% of lignin, 14.8% of hemicellulose, and 13.2% of cellulose was reported to yield 7.45% v/v ethanol [95, 104]. Paulova et al. [110] also reported increased ethanol production as conversion of cellulose to glucose increased; in the same vein, as the quantities of furfural and hydroxymethyl-furfural reduced, ethanol yield increased. For example, 78% conversion of cellulose to glucose brought about 0.34 g/g yield of ethanol in the presence of 88 mg/l furfural and 6 mg/l hydromethyl-furfural. On the other hand, 64% cellulose to glucose conversion brought resulted in an ethanol yield of 0.04%, while 158 mg/g and 10 mg/g of furfural and hydromethyl-furfural were detected.

In conclusion, utilizing agricultural waste as energy source in the form of bioethanol is a viable way of converting waste to wealth. These wastes are available in abundance and renewable hence continuity in production of bioethanol is assured. Nevertheless, there is a need for further research on how to minimize production cost especially at the pretreatment stage. In the same vein, future research should focus on how to scale-up second-generation bioethanol production so that commercial production of bioethanol will not depend on food crops only.

Author details

Inyinbor Adejumoke Abosede[1]*, Oluyori Abimbola Peter[1] and Adelani-Akande Tabitha Adunola[2]

*Address all correspondence to: inyinbor.adejumoke@landmarkuniversity.edu.ng

1 Department of Physical Sciences, Landmark University, Omu Aran, Nigeria

2 Department of Biological Sciences, Landmark University, Omu Aran, Nigeria

References

[1] Oluyori A.P., Shaw A.K., Preeti R., Sammajay R., Atolani O., Olatunji G.A. and Fabiyi O. A. (2015) Natural antifungal compounds from the peels of *Ipomoea batatas Lam*; Natural Product Research; 30(18):2125–2129.

[2] Inyinbor, A.A., Adekola, F.A. and Olatunji, G.A.(2016) Kinetic and thermodynamic modeling of liquid phase adsorption of Rhodamine B dye onto *Raphia hookerie* fruit epicarp; Water Resources and Industry; 15:14–27.

[3] Joshi B., Bhatt M.R., Sharma D., Joshi S., Malla R., Sreerama L. (2011) Lignocellulosic ethanol production: current practices and recent developments; Biotechnology and Molecular Biology Review; 6(8):172–182.

[4] Inyinbor, A.A., Adekola, F.A. and Olatunji, G.A. (2016) Liquid phase adsorptions of Rhodamine B dye onto raw and chitosan supported mesoporous adsorbents: isotherms and kinetics studies; Applied Water Science, 1–11, http://link.springer.com/article/10.1007/s13201-016-0405-4

[5] Inyinbor, A.A., Adekola, F.A. and Olatunji, G.A. (2015) Adsorption of Rhodamine B dye from aqueous solution on *Irvingia gabonensis* biomass: kinetics and thermodynamics studies; South Africa Journal of Chemistry; 68:115–125.

[6] Oluyori A.P. and Olatunji G.A. (2016) Phytochemical analysis, antimicrobial and antioxidant activity of Peels' extracts from *Ipomoea batatas Lam*; Journal of International Research in Medical and Pharmaceutical Sciences; 6(4):157–164.

[7] Concannon, J.A. and DeMeo, T. (1997) Golden seal: facing a hidden crisis; Endangered Species Bulletin; 22(6):10–12.

[8] Imamoglu, E., and Sukan, F.V. (2014) The effects of single and combined cellulosic agrowaste substrates on bioethanol production; Fuel; 134:477–484.

[9] Chen, S. (2015) Environmental pollution emissions, regional productivity growth and ecological economic development in China; China Economic Review; 35:171–182.

[10] Anastopoulos, I. and Kyzas, G.Z. (2015) Progress in batch biosorption of heavy metals onto algae; Journal of Molecular Liquids; 209:77–86.

[11] Dada, A.O., Adekola, F.A. and Odebunmi, E.O. (2015) A novel zerovalent manganese for removal of copper ions: synthesis, characterization and adsorption studies; Applied Water Science; http://link.springer.com/article/10.1007/s13201-015-0360-5.

[12] Johnson, M.F.J., Barbour, N.W. and Weyers, S.L. (2007) Chemical composition of crop biomass impacts its decomposition; Soil Science Society of America Journal; 71:155–162.

[13] Patel, A., Sindhu, D.K., Arora, N., Singh, R.P., Pruthi, V. and Pruthi, P.A. (2015) Biodiesel production from non-edible lignocellulosic biomass of Cassia fistula L. fruit pulp using oleaginous yeast *Rhodosporidium kratochvilovae* HIMPA1; Bioresource Technology; 197:91–98.

[14] Bhatnagar, A., Sillanpaa, M. and Witek-Krowiak, A. (2015) Agricultural waste peels as versatile biomass for water purification—a review; Chemical Engineering Journal; 270:244–271.

[15] Jeong, C.Y., Dodla, S.K., Wang, J.J. (2016) Fundamental and molecular composition characteristics of biochars produced from sugarcane and rice crop residues and by-products; Chemosphere; 142:4–13.

[16] Sindhu, R., Gnansounou, E., Binod, P. and Pandey, A. (2016) Bioconversion of sugarcane crop residue for value added products: an overview, Renewable Energy; http://dx.doi.org/10.1016/j.renene.2016.02.057.

[17] Cretescu, I., Lupascu, T., Buciscanu, I., Balau-Mindru, T. and Soreanu, G. (2016) Low-cost sorbents for the removal of acid dyes from aqueous solutions; Process Safety and Environmental Protection; http://dx.doi.org/10.1016/j.psep.2016.05.016.

[18] Bello, O.S. Auta, M. and Ayodele, O.B. (2013) Ackee apple (*Blighiasapida*) seeds: a novel adsorbent for the removal of Congo red dye from aqueous solutions; Chemistry and Ecology; 21(1):58–71.

[19] Gao, Y., Yue, Q. and Gao, B. (2016) Insights into properties of activated carbons prepared from different raw precursors by pyrophosphoric acid activation; Journal of Environmental Sciences; 41:235–243.

[20] El-Sayed, G.O., Yehia, M.M., Asaad, A.A. (2014) Assessment of activated carbon prepared from corncob by chemical activation with phosphoric acid; Water Resources and Industry; 7–8:66–75.

[21] AlOthman, Z.A., Habila, M.A., Ali, R., Ghafar, A.A., Hassouna, M.S.E. (2014) Valorization of two waste streams into activated carbon and studying its adsorption kinetics, equilibrium isotherms and thermodynamics for methylene blue removal; Arabian Journal of Chemistry; 7:1148–1158.

[22] Njoku, V.O., Foo, K.Y., Asif, M. and Hameed, B.H. (2014) Preparation of activated carbons from rambutan (*Nephelium lappaceum*) peel by microwave-induced KOH activation for acid yellow 17 dye adsorption; Chemical Engineering Journal; 250:198–204.

[23] Bautista-Toledo, M.I., Rivera-Utrilla, J., Ocampo-Pẽrez, R., Carrasco-Marin, F. and Sănchez-Polo, M. (2014) Cooperative adsorption of bisphenol-A and chromium(III) ions from water on activated carbons prepared from olive-mill waste; Carbon; 7 (3):338–350.

[24] Mahamad, M.N., Zaini, M.A.A. and Zakaria, Z.A. (2015) Preparation and characterization of activated carbon from pineapple waste biomass for dye removal; International Biodeterioration & Biodegradation; 102:274–280.

[25] Sayğılı, H., Güzel, F. and Önal, Y. (2015) Conversion of grape industrial processing waste to activated carbon sorbent and its performance in cationic and anionic dyes adsorption; Journal of Cleaner Production; 93:84–93.

[26] Li, J., Ng, D.H.L., Song, P., Kong, C., Song, Y. and Yang, P. (2015) Preparation and characterization of high-surface area activated carbon fibers from silkworm cocoon waste for Congo red adsorption; Biomass and Bioenergy; 7(5):189–200.

[27] Ahmad, M.A., Ahmad, N. and Bello, O.S. (2015) Modified durian seed as adsorbent for the removal of methyl red dye from aqueous solutions; Applied Water Science; 5:407–423.

[28] Sayğılı, H. and Güzel, F. (2016) High surface area mesoporous activated carbon from tomato processing solid waste by zinc chloride activation: process optimization, characterization and dyes adsorption; Journal of Cleaner Production; 113:995–1004.

[29] Georgin, J., Dotto, G.L., Mazutti, M.A. and Foletto, E.L. (2016) Preparation of activated carbon from peanut shell by conventional pyrolysis and microwave irradiation-pyrolysis to remove organic dyes from aqueous solutions; Journal of Environmental Chemical Engineering; 4:266–275.

[30] Kyzas, G.Z., Deliyanni, E.A., and Matis, K.A. (2016) Activated carbons produced by pyrolysis of waste potato peels: cobalt ions removal by adsorption; Colloids and Surfaces A: Physicochemical and Engineering Aspects; 490:74–83.

[31] Largitte, L., Brudey, T., Tant, T., Dumesnil, P.C. and Lodewyckx, P. (2016) Comparison of the adsorption of lead by activated carbons from three lignocellulosic precursors; Microporous and Mesoporous Materials; 219:265–275.

[32] Inyinbor, A.A., Adekola, F.A. and Olatunji, G.A. (2016) Kinetics and isothermal modeling of liquid phase adsorption of rhodamine B onto urea modified *Raphia hookerie* epicarp; Applied Water Science (In Press), http://link.springer.com/article/10.1007/s13201-016-0471-7

[33] Xu, X., Gao, B., Tan, X., Yue, Q., Zhong, Q. and Li, Q. (2011) Characteristics of amine-crosslinked wheat straw and its adsorption mechanisms for phosphate and chromium (VI) removal from aqueous solution; Carbohydrate Polymers; 84:1054–1060.

[34] Kumar, R., Barakat, M.A., Daza, Y.A., Woodcock, H.L. and Kuhn, J.N. (2013) EDTA functionalized silica for the removal of Cu(II), Zn(II) and NI(II) from aqueous solution; Journal of Colloid and Interface Science; 408:200–205.

[35] Asgher, M, and Bhatti, H.N. (2010) Mechanistic and kinetic evaluation of biosorption of reactive azo dyes by free, immobilized and chemically treated Citrus sinensis waste biomass; Ecological Engineering; 36:1660–1665.

[36] Srivastava, S., Agrawal, S.B. and Mondal, M.K. (2015) Biosorption isotherms and kinetics on removal of Cr(VI) using nativeand chemically modified Lagerstroemia speciosa bark; Ecological Engineering; 85:56–66.

[37] Wang, G., Zhang, S., Yao, P., Chen, Y., Xu, X., Li, T., Gong, G. (2015) Removal of Pb(II) from aqueous solutions by *Phytolacca americana* L. biomass as a low cost biosorbent; http://dx.doi.org/10.1016/j.arabjc.2015.06.011

[38] Afroze, S., Sen, K.T. Ang, H.M. (2016) Adsorption removal of zinc (II) from aqueous phase by raw and base modified *Eucalyptus sheathiana* bark: kinetics, mechanism and equilibrium study; Process Safety and Environmental Protection; 102:336–352

[39] Ronda, A., Martìn-Lara, M.A., Dionisio, E., Blázquez, G. and Calero, M. (2013) Effect of lead in biosorption of copper by almond shell; Journal of the Taiwan Institute of Chemical Engineers; 44:466–473.

[40] Khan, T.A., Nazir, M. and Khan, E.A. (2013) Adsorptive removal of rhodamine B from textile wastewater using water chestnut (*Trapa natans* L.) peel: adsorption dynamics and kinetic studies; Toxicological and Environmental Chemistry; 95(6):919–931.

[41] Ghaedi, M., Hajati, S., Karimi, F., Barazesh, B. and Ghezelbash, G. (2013) Equilibrium, kinetic and isotherm of some metal ion biosorption; Journal of Industrial and Engineering Chemistry; 19:987–992.

[42] Taşar Ş., Kaya, F. and Özer, A. (2014) Biosorption of lead(II) ions from aqueous solution by peanut shells: equilibrium, thermodynamic and kinetic studies; Journal of Environmental Chemical Engineering; 2:1018–1026.

[43] Ammari, T.G. (2014) Utilization of a natural ecosystem bio-waste; leaves of Arundo donax reed, as a raw material of low-cost eco-biosorbent for cadmium removal from aqueous phase; Ecological Engineering; 71:466–473.

[44] Sfaksi, Z., Azzouz, N., Abdelwahab, A. (2014) Removal of Cr(VI) from water by cork waste; Arabian Journal of Chemistry; 7:37–42.

[45] Awwad A.M. and Salem, N.M. (2014) Kinetics and thermodynamics of Cd(II) biosorption onto loquat (*Eriobotrya japonica*) leaves; Journal of Saudi Chemical Society; 18:486–493.

[46] Khan, T.A., Rahman, R., Ali, I., Khan, E.A. and Mukhlif, A.A. (2014): Removal of malachite green from aqueous solution using waste pea shells as low-cost adsorbent; adsorption isotherms and dynamics; Toxicological and Environmental Chemistry; http://dx.doi.org/10.1080/02772248.2014.969268.

[47] Khan, T.A., Sharma, S., Khan, E.A. and Mukhlif, A.A. (2014) Removal of Congo red and basic violet 1 by chir pine (*Pinus roxburghii*) sawdust, a saw mill waste: batch and column studies; Toxicological and Environmental Chemistry; 96(4):555–568.

[48] Guerrero-Coronilla, I., Morales-Barrera, L. and Cristiani-Urbina, E. (2015) Kinetic, isotherm and thermodynamic studies of amaranth dye biosorption from aqueous solution onto water hyacinth leaves; Journal of Environmental Management; 152:99–108.

[49] Omorogie, M.O., Babalola, J.O., Unuabonah, E.I., Song, W. and Gong, J.R. (2016) Efficient chromium abstraction from aqueous solution using a low-cost biosorbent: *Nauclea diderrichii* seed biomass waste; Journal of Saudi Chemical Society; 20:49–57

[50] Banerjee, S., Sharma, G.C., Gautam, R.K., Chattopadhyaya, M.C., Upadhyay, S.N. and Sharma, Y.C. (2016) Removal of Malachite Green, a hazardous dye from aqueous solutions using *Avena sativa* (oat) hull as a potential adsorbent; Journal of Molecular Liquids; 213:162–172.

[51] Thakare, Y.N. and Jana, A.K. (2015) Performance of high density ion exchange resin (INDION225H) for removal of Cu(II) from waste water; Journal of Environmental Chemical Engineering; 3:1393–1398.

[52] Charles, J., Bradu, C., Morin-Crini, N., Sancey, B., Winterton, P., Torri, G., Badot, P. and Crini, G. (2016) Pollutant removal from industrial discharge water using individual and combined effects of adsorption and ion-exchange processes: chemical abatement; Journal of Saudi Chemical Society; 20:185–194.

[53] Blázquez, G., Martìn-Lara, M.A., Dionisio-Ruiz, E., Tenorio, G. and Calero, M. (2011) Evaluation and comparison of the biosorption process of copper ions onto olive stone and pine bark; Journal of Industrial and Engineering Chemistry; 17:824–833.

[54] Chishti, H. (1988). The Traditional Healer's Handbook. Vermont: Healing Arts Press. pp. 11. ISBN 0892814381.

[55] Concannon J.A. and DeMeo T.E. (1997). Golden seal: facing a hidden crisis; Endangered Species Bull; 22(6):10–12.

[56] Ojekale, A.B, Makinde, S.C. and Osileye, O. (2007). Phytochemistry and anti-microbial evaluation of *Thaumatococcus danielli,* Benn. (Benth.) leaves; Nigerian Food Journal; 25 (2):176–183.

[57] Dosumu, O.O. and Akinnuoye, G.A. 2014. Effect of steaming of beans pudding on the phytochemical composition of *Thaumatococcus Daniellii* Wrapper; NIFOJ; 32(1):110–116.

[58] Adunola A.T., Chidimma A.L., Olatunde D.S. and Peter O.A. (2015); Antibacterial activity of watermelon (*Citrullus lanatus*) seed against selected microorganisms; African Journal of Biotechnology; 14(14):1224–1229.

[59] Sultana, B., Naseer R., and Nigam P.. (2015) Utilization of agro-wastes to inhibit aflatoxins synthesis by *Aspergillus parasiticus*: a biotreatment of three cereals for safe long-term storage; Bioresource Technology; 197:443–450.

[60] El-Nashi H.B., Abdel Karim A.F., Abdel Rahman N.R. and Abd El-Razik M.M.. 2015 Quality characteristics of beef sausage containing pomegranate peels during refrigerated; Annals of Agricultural Sciences; 60(2):403–412.

[61] Gavahian, M., Farhoosh, R., Farahnaky, A., Javidnia, K. and Shahidi, F. (2015) Comparison of extraction parameters and extracted essential oils from Mentha piperita L. using hydrodistillation and steamdistillation; International Food Research Journal; 22(1):283–288.

[62] Qi H. and Kaijun, X. (2016) The effects of tangerine peel (*Citri reticulatae* pericarpium) essential oils as glazing layer on freshness preservation of bream (*Megalobrama amblycephala*) during superchilling storage; Original Research Article; Food Control; 69:339–345.

[63] Gavahian, M., Farahnaky, A., Javidnia, K. and Majzoobi, M. (2012) Comparison of ohmic-assisted hydrodistillation with traditional hydrodistillation for the extraction of essential oils from *Thymus vulgaris* L; Innovative Food Science and Emerging Technologies; 14:85–91.

[64] Gavahian, M., Farahnaky, A., Farhoosh, R., Javidnia, K. and Shahidi, F. (2015) Extraction of essential oils from *Mentha piperita* using advanced techniques: microwave versus ohmic assisted hydrodistillation; Original Research Article; Food and Bioproducts Processing; 94:50–58.

[65] Oluyori AP, Olatunji GA, and Atolani O. 2013. Alkyd resin from *Ipomoea batatas* Lam; Pakistan Journal of Scientific and Industrial Research; 56:57–58.

[66] Trabelsi, D., Aydi, A., Zibetti, G.W., Porta, D., Sconagmiglio, M., Cricchio, V., Langa, E., Abderrabba, M. and Mainar, A.M., Supercritical extraction from Citrus *Aurantium amara* peels using CO_2 with ethanol as co-solvent; The Journal of Supercritical Fluids; http://dx.doi.org/doi:10.1016/j.supflu.2016.07.003.

[67] Kubatov, A., Lagade, A.J.M., Miller, D.J., and Hawthorn, S.B. (2000) Selective extraction of oxygenates from savory and peppermint using subcritical water; Flavour and Fragrance Journal; 16:64–73.

[68] Saldana, M.D.A. and Valdivieso-Ramírez, C.S. (2015) Pressurized fluid systems: phytochemical production from biomass; Journal of Supercritical Fluids; 96:228–244.

[69] Zhu, Z., Dong, F., Liu, X., Qian, L., Yang, Y., Liu, F., Chen, L., Wang, T., Wang, Z. and Zhang, Y. (2016) Effects of extraction methods on the yield, chemical structure and anti-tumor activity of polysaccharides from *Cordyceps gunnii* mycelia; Carbohydrate Polymers; 140:461–471.

[70] Hu, W., Guo, T., Jiang, W., Dong, G., Chen, D., Yang, S. and Li, H. (2015) Effects of ultrahigh pressure extraction on yield and antioxidant activity of chlorogenic acid and cynaroside extracted from flower buds of *Lonicera japonica*; Chinese Journal of Natural Medicines; 13(6):445–453.

[71] Zhang, S., Bi, H. and Liu, C. (2007) Extraction of bio-active components from *Rhodiola sachalinensis* under ultrahigh hydrostatic pressure; Separation and Purification Technology; 57(2):277–282.

[72] Kotnik, P., Škerget, M. and Knez, Ž. (2006) Supercritical fluid extraction of chamomile flower heads: comparison with conventional extraction, kinetics and scale-up; The Journal of Supercritical Fluids; Selected Papers from the 3rd International Meeting on High Pressure Chemical Engineering, Erlangen, Germany, May 10–12, 2006; vol.43(2):192–198.

[73] Tatke, P. and Rajan, M. (2014) Comparison of conventional and novel extraction techniques for the extraction of scopoletin from *Convolvulus Pluricaulis*; Indian Journal of Pharmaceutical Education and Research; 48(1):27–31.

[74] Takuya, S., Masahiro, T., Hideo I., Teruaki, M., Yukihiro, K. Chiho, S., Yoshito, S., Munehiro, H., Armando, T., Mitsuru, S, Junshi S. and Motonobu G. (2013) Supercritical CO2 extraction of essential oil from Kabosu (*Citrus sphaerocarpa* Tanaka) peel; Flavour; 2:18. DOI: 10.1186/2044-7248-2-18.

[75] Abdalbasit A.M., Bertrand M. and Maznah I. (2011) Comparison of supercritical fluid and hexane extraction methods in extracting Kenaf (*Hibiscus cannabinus*) seed oil lipids; Journal of the American Oil Chemists' Society; 88(7):931–935.

[76] Bak, I., Lekli, I., Juhasz, B., Varga, E., Varga, B., Gesztelyi, R., Szendrei, L. and Tosaki, A. (2010) Isolation and analysis of bioactive constituents of sour cherry (*Prunus cerasus*) seed kernel: an emerging functional food; Journal of Medicinal Food; 13(4):905–10. DOI: 10.1089/jmf.2009.0188.

[77] Swati, K., Suchandra, C. Saroj, G. (2014) Bioactive constituents of germinated fenugreek seeds with strong antioxidant potential; Journal of Functional Foods; 6:270–279, DOI:10.1016/j.jff.2013.10.016.

[78] Zgoda-Pols, J.R., Freyer, A.J., Killmer, L.B. and Porter, J.R. (2002) Antimicrobial diterpenes from the stem bark of *Mitrephora celebica*; Fitoterapia; 73(5):434.

[79] Quintans, J.S., Soares, B.M., Ferraz, R.P., Oliveira, A.C., da Silva, T.B., Menezes, L.R., Sampaio, M.F., Prata, A.P., Moraes, M.O., Pessoa, C., Antoniolli, A.R., Costa, E.V. and Bezerra, D.P. (2013) Chemical constituents and anticancer effects of the essential oil from leaves of *Xylopia laevigata*; Planta Medica; 79(2):123–30.

[80] Sakdarat, S., Shuyprom, A., Pientong, C., Ekalaksananan, T. and Thongchai S. (2009) Bioactive constituents from the leaves of *Clinacanthus nutans Lindau*; Bioorganic and Medicinal Chemistry; 17(5):1857–60.

[81] Bhattacharya, S., Christensen, K.B., Olsen, L.C., Christensen, L.P., Grevsen, K., Færgeman, N.J., Kristiansen, K., Young, J.F. and Oksbjerg, N. (2013) Bioactive components from flowers of *Sambucus nigra* L. increase glucose uptake in primary porcine myotube cultures and reduce fat accumulation in *Caenorhabditis elegans*; Journal of Agricultural and Food Chemistry; 61(46):11033–40.

[82] Casquete, R., Castro, S.M., Martín, A., Ruiz-Moyano, S., Saraiva, J.A., Córdoba, G.M. and Teixeira, P. (2015) Evaluation of the effect of high pressure on total phenolic content, antioxidant and antimicrobial activity of citrus peels; Innovative Food Science & Emerging Technologies; 31:37–44.

[83] Olech, M., Nowak, R., Nowacka, N., Pecio, L., Oleszek, W., Los, R., Malm, A. and Rzymowska, J. (2015) Evaluation of rose roots, a post-harvest plantation residue as a source of phytochemicals with radical scavenging, cytotoxic, and antimicrobial activity; Industrial Crops and Products;69:129–136.

[84] Kalogeropoulos, N., Chiou, A., Pyriochou, V., Peristeraki, A. and Karathanos, V.T. (2012) Bioactive phytochemicals in industrial tomatoes and their processing by-products; LWT —Food Science and Technology; 49(2):213–216.

[85] Surendra T.V., Roopan S.M., Arasu M.V., Al-Dhabi, N.A., Sridharan M., (2016) Phenolic compounds in drumstick peel for the evaluation of antibacterial, hemolytic and photocatalytic activities; Journal of Photochemistry and Photobiology, B: Biology; 161:463–471.

[86] Cancer facts and figures 2015, American Cancer Society.

[87] Pinar, K., Jeffrey, S., Mustaqeem, A.S., Holly, V.H. and Nilay, D.S. (2011) Impact of new drugs and biologics on colorectal cancer treatment and costs. Impact of new drugs and biologics on colorectal cancer treatment and costs; Journal of Oncology Practice; 7(3):30–37.

[88] Oluyori, A.P., Shaw A.K., Olatunji G.A., Rastogi P., Sanjeev, M., Datta D., Arora, A., Sama A. and Reddy, P.S. Sweet potato peels and cancer prevention; Nutrition and Cancer; (In Press), 68(8):1330–1337.

[89] Campbell, C.J. and Laherrere, J.H. 1998. The end of cheap oil; Scientific America; 78–84.

[90] Vimmerstedt, LJ, Bush, BW, Hsu, DD, Inman, D and Peterson, S.O. 2015. Maturation of biomass-to-biofuels conversion technology pathways for rapid expansion of biofuels production: a system dynamics perspective; Biofuel, Bioproduct and Biorefining; 9 (2):158–176.

[91] Khan, A.M., Khaliq, S. and Sadiq R. 2015. Investigation of waste banana peels and radish leaves for their biofuels potential; Bulletin Chemical Society of Ethiopia; 29(2):239–245.

[92] Igwe, C., Agbaeze, E., Obike, I. and Sonde, U. (2012) Characterization and determination of ethanol fuel composite qualities of *Saccharum officinarum*, *Pennisetum purpureum* and *Costusafer*; Asian Journal of Plant Science and Research; 2(5):643–649.

[93] Sarkar, N., Ghosh, S., Bannerjee, S. and Aikat, K. (2012) Bioethanol production from agricultural wastes: an overview; Renewable Energy; 37:19–27.

[94] Ohmain, I. (2010) Emerging bio-ethanol projects in Nigeria: their opportunities and challenges; Energy Policy; 38:7161–7168.

[95] Gupta, A. and Verma, P. (2015) Sustainable bio-ethanol production from agro-residues: a review. Renewable and Sustainable Energy Reviews; 41:550–567.

[96] Sun, Y. and Cheng, J. (2002) Hydrolysis of lignocellulosic materials for ethanol production: a review; Bioresource Technology; 83:1–11.

[97] Hendrick A.T. and Zeeman G. (2009) Pretreatments to enhance the digestibility of lignocellulosic biomass; Bioresource Technology; 100:10–18.

[98] Ballesteros M., Oliva J.M., Negro M.J., Manzanares P. and Ballesteros I. (2004) Ethanol from lignocellulosic materials by a simultaneous saccharification and fermentation process (SFS) with *Kluyveromyces marxianus* CECT 10875; Process Biochemistry; 39:1843–1848.

[99] Lloyd T.A., Wyman C.E. (2005) Combined sugar yields for dilute sulfuric acid pretreatment of corn stover followed by enzymatic hydrolysis of the remaining solids; Bioresource Technology; 96:1967–1977.

[100] Ma F., Yang N., Xu C., Yu H., Wu J. and Zhang X. (2010) Combination of biological pretreatment with mild acid pretreatment for enzymatic hydrolysis and ethanol production from water hyacinth; Bioresource Technology; 101:9600–9604.

[101] Yu H., Zhang X., Song L., Ke J., Xu C., Du W. and Zhang J. (2010) Evaluation of white-rot fungi-assisted alkaline/oxidative pretreatment of corn straw undergoing enzymatic hydrolysis by cellulase; Journal of Bioscience and Bioengineering; 110(6):660–664.

[102] Yu J., Zhang J. He J., Liu Z. and Yu Z. (2009) Combinations of mild physical or chemical pretreatment with biological pretreatment for enzymatic hydrolysis of rice hull; Bioresourse Technology; 100:903–908.

[103] Yu H., Guo G., Zhang X., Yan K. and Xu C. (2009) The effect of biological pretreatment with the selective white-rot fungus *Echinodontium taxodii* on enzymatic hydrolysis of softwoods and hardwoods; Bioresoure Technology; 100:5170–5175.

[104] Anwar, Z., Gulfraz, M. and Irshad, M. (2014) Agro-industrial lignocellulosic biomass a key to unlock the future bio-energy: a brief review; Journal of Radiation Research and Applied Sciences; 7:163–173.

[105] Saritha M., Arora A., Lata. (2012) Biological pretreatment of lignocellulosic substrates for enhanced delignification and enzymatic digestibility; Indian Journal Microbiology; 52(2):122–130, https://www.ncbi.nlm.nih.gov/pmc/articles/PMC3386436/.

[106] Petersonn A., Thomsen M.H., Hauggaard-Nielsen H. and Thomsen A. 2007. Potential bioethanol and biogas production using lignocellulosic biomass from winter rye, oilseed rape and faba bean; Biomass and Bioenergy; 31:812–819.

[107] Ezeonu, C.S., Onwurah, I.N.E., Ubani, C.S., Ejikeme, C.M. and Ogodo, A.C. (2016) Achievements in the life sciences *Trichophyton Soudanense* and *Trichophyton*; Achievements in the Life Sciences; 10:72–79.

[108] Yadira, P.B., Sergio, S., Fernando, S.E.L., Sebastian, P.J. and Eapen, D. (2014) Bioethanol production from coffee mucilage; Energy Procedia; 57:950–956.

[109] Moshi, A.P., Temu, S.G., Nges,I.A., Malmo, G., Hosea, K.M.M., Elisante, E. and Mattiasson, B. (2015), Combined production of bioethanol and biogas from peels of wild Cassava *Manihot glaziovii*; Chemical Engineering Journal; 279:297–306.

[110] Paulova, L., Patakova, P., Rychtera, M. and Melzoch, K. (2014) High solid fed-batch SSF with delayed inoculation for improved production of bioethanol from wheat straw; Fuel; 122:294–300.

Chapter 6

Modeling Biomass Substrates for Syngas Generation by Using CFD Approaches

Nuno Couto and Valter Silva

Additional information is available at the end of the chapter

http://dx.doi.org/10.5772/65857

Abstract

Recent reports from top universities state that in spite of having great national importance, there are dozens of fields of study that are suffering due to a lack of funding. Perhaps the greatest tool available to assist researchers with this regard is numerical simulation. This tool allows cutting costs, decreasing the necessary design cycle and allows an enormous amount of physical insight on the process itself). Numerical model's ability to correctly predict a complex system was tested in this chapter by drawing from a previously developed computational fluid dynamics model for biomass gasification. Numerical results were compared with both experimental results (pilot scale plant) and available literature. Results from common Portuguese biomass substrates were found to be within a satisfactory margin of error of 20%. Influence of all major operational conditions was then investigated and the model was once again able to predict all the expected trends. All the relevant process products were also analyzed. Finally, the numerical model was coupled with an optimization model. Maximum efficiency value was found at 900°C with a SBR of 1.5 for MSW and 1 for forest residues. Results showed that numerical models could have a preponderant impact on biomass gasification field.

Keywords: numerical simulation, biomass, municipal solid wastes, gasification, optimization

1. Introduction

In the last few years, biomass has become an important source of energy and it most often refers to any organic matter derived from plant-based materials. Biomass includes, among others, forestry and agricultural residues, organic waste, energy crops, sewage sludge, and woody plants [1, 2].

Among all the thermochemical conversion processes, gasification has been emerging as a very promising technology due to being environmental friendly, having high efficiency and flexible enough to use different substrates with a wide range of applications for heat, electricity, chemicals, and transportation purposes [3]. Gasification can be defined as the conversion of a solid waste to synthetic gas or fuel by the partial oxidation of the feedstock under stoichiometric combustion conditions. The synthetic gas is generally called "producer gas" or syngas and contains mainly carbon monoxide and hydrogen. However, some undesired products can be also found such as tar, alkali metals, chloride and sulfide, among others [3]. Gasification presents several advantages over waste combustion, namely and among others [3]: (a) effective response to increasingly environmental restrictive regulations; (b) syngas can be used not only in highly efficient internally fired cycles but also in producing valuable products as chemicals and fuels; and (c) flexible use under different operating conditions and reactors.

However, some drawbacks need further investigation. Besides high ranges of operating and capital costs than those of conventional combustion-based plants, the syngas generated from biomass and especially from municipal solid wastes (MSWs) gasification is unstable due to changes in the feedstock properties. Indeed, the high heterogeneous nature of MSW implies significant variations in syngas yield and quality [4]. The negative impact from the heterogeneous nature of MSW can be reduced by implementing strategies as MSW preprocessing, where undesired MSW components are removed before sending them to the gasifier. Further improvements can also be found by blending MSW with other feedstock with more favorable characteristics [5]. Portugal has a major potential considering biomass resources, only forestry and pruning residues have potential to produce 13,800 GWh, about 13% of the total primary energy demand in Portugal [6]. Also, operating the MSW gasifier under certain operating conditions will allow reliable operation with stable and improved syngas generation.

Because experimental runs conducted on industrial gasification plants or even on pilot scale gasification plants are very expensive, predictable models able to simulate the syngas composition and other responses of interest are required. Accurate predictions by gasification models require the simulation of different kinetic and hydrodynamic phenomena and we should not forget that gasification always involves complex chemistry. Concerning this degree of complexity, and also, the high investment costs involved to perform the experimental studies, the use of tools involving numerical simulation, such as computer fluid dynamics (CFD), exhibit utmost importance [7–9].

There are different designs to build a gasifier but one with special relevance is the fluidized bed configuration [10]. Basically, in this kind of reactor, both fuel and inert bed material act as a fluid. To obtain this behavior a gas is forced to pass through the solid particles of the reactor. The great advantages of operating such a system are its ability to handle with different feedstock due to the easy control of temperature, to deal with fine grained materials and the fact that generates an intense mixing enhancing the transference of mass and heat. Finally, the fluidized bed gasifiers are flexible enough to use additives with the goal to remove pollutants and increase tar conversion [11].

The fluidized bed gasification is essentially a multiphase problem. There are two main strategies that are commonly implemented to handle this kind of gasification: Euler-Lagrange and Euler-Euler approaches [12]. In the Euler-Lagrange strategy, the dispersed phase is characterized by following a wide number of particles through information given from the continuum phase. Both phases exchange mass, momentum, and energy. One major assumption relies on the fact that volume fraction of the dispersed phase is lower than about 10%, even the mass fraction could be higher.

Snider et al. [13] implemented a 3-D model based on the previous approach to simulate the behavior of a coal gasifier. The model also included energy transport equations and homogeneous and heterogeneous chemistry described by reduced chemistry. The model was able to identify zones with stagnant particles and then provide important data for reactor design purposes.

The Euler-Euler approach considers the different phases as interpenetrating continua, meaning that both volume phases depend on each other regarding time and space. Each phase obeys the conservation laws and constitutive relations from empirical data are required to close the equations system. Regarding the Euler-Euler approach, there are three multiphase models: (a) the volume of fluid; (b) the mixture model and; (c) the Eulerian model, which is indicated to model the discussed gasification system in this chapter.

Xue et al. [14] followed the Eulerian-Eulerian approach to predict the syngas generation from wood gasification using air as a fluidization agent. The developed model was able to provide detailed information on several processes, such as char elutriation, species profiles along the reactor and gas composition at reactor outlet.

This work aims at presenting an advanced modeling strategy within the framework of the CFD ANSYS Fluent program combined with suitable experimental data input and user defined the code to predict a gasification process under several operating conditions and robust enough to be applied both to biomass and MSW. Model development is detailed highlighting the set of complex choices that have to be taken and corresponding implications and consequences. Numerical results for both biomass and MSW were validated against experimental results, and then, the effect of using different operating conditions is deeply discussed as well as the main syngas quality indices. The combined use of the developed numerical model and the design of experiments (DoE) procedure is also discussed for optimization purposes. Finally, a reflection section is included to highlight the efficiency and effectiveness of using numerical methods to describe physical problems.

2. Experimental setup

The numerical model (presented in Section 3) was developed using experimental data collected in the gasification plant from the School of Technology and Management (ESTG) of the Polytechnic Institute of Portalegre (IPP). The plant is based on fluidized bed technology, with a 200 kW reactor that produces an average syngas flow rate close to 100 Nm^3/h. **Figure 1** depicts a diagram of the biomass gasification unit used in the experiments.

The system comprises two biomass silos connected to a worm screw, which forms the feeding system. The gasifier itself is just slightly above 4 m high and half a meter wide with a biomass intake capacity of 100 kg/h. The synthetic gas is cooled by two heat exchangers, with the first

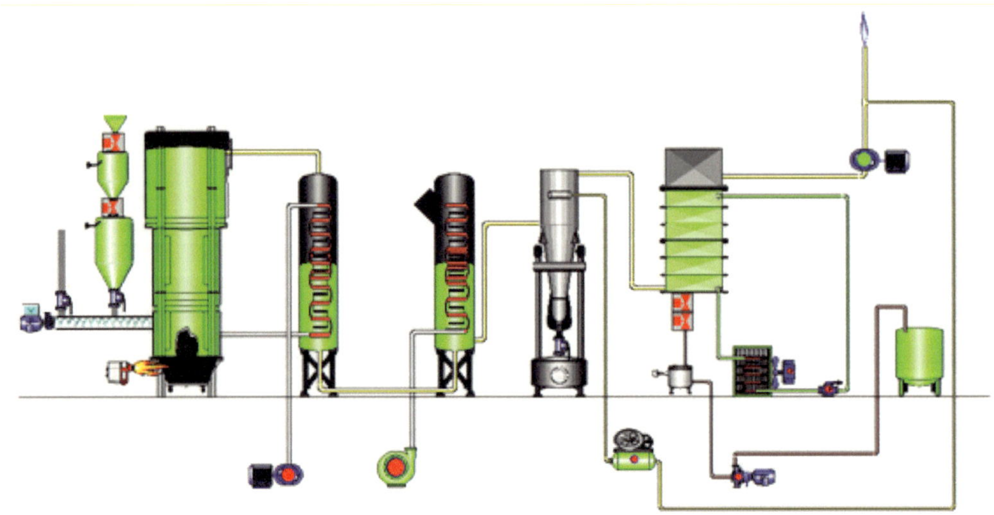

Figure 1. Schematics of the gasification plant.

one being also responsible for preheating the gas. The ash and char are removed to the bag filter, a system with seven filters cleaned with shots taken from synthetic gas through a compressor downstream of the vacuum pump. The purpose of the third heat exchanger is to remove tars from the system by condensation, which are then forwarded to a deposit. The last component of the system is a vacuum pump, which guarantees that the entire system control.

Once the predetermined temperature and biomass input flow are secured, the system stabilizes for about 2 h and then the test starts. The test lasts 2 h each, including syngas collections (usually taking two different samples) and recording temperatures, flows, ash, and tars. The schematics as well as an extensive description of the gasification plant can be found elsewhere [15, 16].

The described system was first used to analyze the gasification process of Portuguese biomass substrates and their potential for fossil fuel replacement [15, 17]. Recently, to counter the improper disposal of municipal waste in landfills, the focus shifted from Portuguese biomass gasification to Portuguese municipal solid waste (referred from now on as PMSW for simplicity) gasification.

The characterization and analysis of Portuguese MSW were carried out using data from the Oporto metropolitan area obtained from LIPOR, entity responsible for the management, treatment, and recovery of solid waste municipal produced in the city. From the pretreatment of MSW conducted by LIPOR, usually via shredding and dehydration, a refuse-derived fuel (RDF) containing only cellulosic and plastics is obtained [18] (chemical composition of the MSW is presented in **Table 1**).

However, since at that moment, the reactor could not handle LIPOR's wastes, the model had to be validated using data collected from the literature. To properly assess the potential of PMSW,

Category	% Weight	Chemical formula
Cellulosic material	85.42	*
Polyethylene	10.99	$(C_2H_4)_n$
Polyethylene terephthalate	2.02	$(C_{10}H_8O)_n$
Polypropylene	0.81	$(C_3H_6)_n$
Polystyrene	0.76	$(C_8H_8)_n$

Table 1. Chemical composition of the MSW in Oporto in 2014.

Substrate properties	Forest residues	PMSW
Elementary analysis (dry ash free)		
N (%)	2.4 ± 0.3	1.4 ± 0.2
C (%)	43 ± 4.1	48 ± 4.4
H (%)	5 ± 0.6	6.3 ± 1.4
O (%)	49.6 ± 5.2	43.6 ± 3.6
Humidity (%)	11.3 ± 1.7	17.6 ± 2.3
Density (Kg/m^3)	650 ± 70	247 ± 23
Lower heating value (MJ/Kg biomass)	21.2 ± 1.8	14.4 ± 1.1
Mean particle size (mm)	5 ± 2	20 ± 10
Proximal analysis (%)		
Ash	0.2 ± 0.1	14.9 ± 1.2
Volatile matter	79.8 ± 3.1	76.62 ± 2.9
Fixed carbon	20 ± 2.5	8.46 ± 1.5

Table 2. Ultimate and proximate analyses of forest residues and PMSW.

a previously studied Portuguese biomass substrate will be used as benchmark. Forest residues [17] (in pellet form) were selected for this purpose, since it revealed relevant energetic as well as economic benefits.

Prior to the actual gasification process, biomass analysis was carried out in the Laboratory of Chemistry of the High School of Technology and Management located in Portalegre, Portugal, since biomass characteristics can provide valuable information on how the gasification process will occur. This kind of analysis also provides crucial data to treat the implemented numerical model. The instruments used in the performed analysis are: thermal gravimetric analysis—data for proximal analysis; elemental analysis—determination of biomass composition with respect to the percentage of C, H, N, and O; humidity—sample moisture content assessment; calorific value—appraisal of energy contained in biomass. Data regarding proximate and ultimate analysis for the referred substrates are presented in **Table 2**.

It should be noted that regarding the experimental gasification runs, as well as the analyses for the studied substrates, every run was performed twice in order to avoid measurements. When deviation was higher than 5%, extra runs were performed to ensure reproducibility below 5%, which is typical for this kind of system.

3. Mathematical model

Experimental studies conducted in pilot scale or industrial reactors like the one presented in the previous section are fairly absent from the available literature. The reason is due to the difficulty in regulating operating parameters but primarily due to the high cost of a gasification plant, which can reach tens of millions of Euros depending on the generated power [19].

Mathematical models, with the ability to theoretically simulate any physical condition, allow studying the gasification process without resorting to major investments and/or the need for long-waiting periods (with all the bureaucratic and logistical problems associated). However, due to the extreme complexity of the gasification process, largely due to the chemical and physical interactions that occur throughout, the ability of numerical models to correctly predict experimental data collected from pilot scale or industrial reactors is usually very limited. In fact, the lack of reliable models to describe the gasification process in the open literature was the main motivation for our team to develop a new CFD model able to predict the syngas generation from biomass gasification in a pilot scale gasifier [20].

3.1. Gas-solid interaction

When modeling bubbling fluidized-bed reactors, like the one described, the two-phase flow theory of fluidization is usually applied for the description of the process hydrodynamics. Because of this, correctly modeling the interaction between gas and solid phases is crucial since they exchange heat by convection, mass over the heterogeneous chemical reactions, and momentum due to the drag between gas and solid phase. The main equations governing both phases are depicted in **Table 3**.

3.1.1. Granular Eulerian model

According to Goldschmidt et al. [21], two-phase flows can be modeled using two different approaches: the Lagrangian-Eulerian and the Eulerian-Eulerian models. With considerable similarities, the fundamental difference between them lies in the way the particles are treated. The former describes the solid phase at the particle level while the latter treats the particles as a continuum. In industrial applications, typically composed of millions of particles, following individual particles become excessively time consuming and for this reason the Lagrangian approach tends to be less used [22].

The Eulerian approach not only requires lower computational resources and calculation times, but also allows a detailed analysis of the disperse phase flow field, which is convenient for engineering design applications. For this reason, the granular Eulerian model was applied to our model.

Gas phase	Solid phase
Energy:	
$\frac{\partial(\alpha_g \rho_g h_g)}{\partial t} + \nabla \cdot (\alpha_g \rho_g \vec{u}_g h_g) = -\alpha_g \frac{\partial(p_g)}{\partial t}$ $+\overline{\overline{\tau}}_g : \nabla(\vec{u}_g) - \nabla \vec{q}_g + S_g + \sum_{g=1}^{n} (\vec{Q}_{sg} + \dot{m}_{sg} h_{sg})$	$\frac{\partial(\alpha_s \rho_s h_s)}{\partial t} + \nabla \cdot (\alpha_s \rho_s \vec{u}_s h_s) = -\alpha_s \frac{\partial(p_s)}{\partial t}$ $+\overline{\overline{\tau}}_s : \nabla(\vec{u}_s) - \nabla \vec{q}_s + S_{ps} + \sum_{s=1}^{n} (\vec{Q}_{sg} + \dot{m}_{sg} h_{sg})$
Mass:	
$\frac{\partial(\alpha_g \rho_g)}{\partial t} + \nabla \cdot (\alpha_g \rho_g \vec{u}_g) = -M_C \sum \gamma_C R_C$	$\frac{\partial(\alpha_s \rho_s)}{\partial t} + \nabla \cdot (\alpha_s \rho_s \vec{u}_s) = M_C \sum \gamma_C R_C$
Momentum:	
$\frac{\partial(\alpha_g \rho_g \vec{u}_g)}{\partial t} + \nabla \cdot (\alpha_g \rho_g \vec{u}_g \vec{u}_g) = -\alpha_g \nabla p_g + \alpha \rho_g g$ $+\beta(u_g - u_g) + \nabla \cdot \alpha_g \overline{\overline{\tau}}_g + S_{sg} U_S$	$\frac{\partial(\alpha_s \rho_s \vec{u}_s)}{\partial t} + \nabla \cdot (\alpha_s \rho_s \vec{u}_s \vec{u}_s) = -\alpha_s \nabla p_s + \alpha \rho_s g$ $+\beta(u_s - u_s) + \nabla \cdot \alpha_s \overline{\overline{\tau}}_s + S_{sg} U_S$

Table 3. Governing equations for gas and solid phases.

The granular Eulerian model is described by the following conservation equation for granular temperature:

$$\frac{3}{2}\left[\left(\frac{\partial(\rho_s \alpha_s \Theta_s)}{\partial t}\right) + \nabla \cdot (\rho_s \alpha_s \vec{v}_s \Theta_s)\right] = (-p_s \bar{I} + \overline{\tau}_s) : \nabla(\vec{v}_s) + \nabla \cdot \left(k_{\Theta_a} \nabla(\Theta_s)\right) - \gamma_{\Theta a} + \phi_{ls} \quad (1)$$

This expression is obtained from the kinetic theory of gases. The term $(-p_s \bar{I} + \overline{\tau}_s) : \nabla(\vec{v}_s)$ describes the generation of energy by the solid stress tensor, ϕ_{ls} stands for the energy exchange between the fluid and solid phase, γ_{Θ_a} stands for the collisional dissipation of energy and $k_{\Theta_a} \nabla(\Theta_s)$ stands for the diffusion energy, in which k_{Θ_a} is the diffusion coefficient.

3.2. Turbulent model

In turbulent flows, like the one being studied, transported quantities like momentum, energy, and species concentration tend to fluctuate. Modeling fluctuations can be to computationally expensive, which is why instantaneous governing equations are usually replaced with their time-averaged, ensemble-averaged, or otherwise manipulated to remove the small time scales.

The standard *k*-*ε* model was used to simulate the turbulent flow due to its suitability for a wide range of wall-bounded and free-shear flows. The model is the simplest turbulence two-equation model in which the solution of two separate transport equation allows the turbulent velocity and length scales, which are to be independently determined. Turbulence kinetic energy (*k*) and dissipation rate (*ε*) are, respectively, given by:

$$\frac{\partial}{\partial t}(\rho k) + \frac{\partial}{\partial x_i}(\rho k u_i) = \frac{\partial}{\partial x_j}\left[\left(\mu + \frac{\mu_t}{\sigma_k}\right)\right] G_k + G_b - \rho\varepsilon - Y_M + S_k \tag{2}$$

$$\frac{\partial}{\partial t}(\rho\varepsilon) + \frac{\partial}{\partial x_i}(\rho\varepsilon u_i) = \frac{\partial}{\partial x_j}\left[\left(\mu + \frac{\mu_t}{\sigma_\varepsilon}\right)\right] + C_{1\varepsilon}\frac{\varepsilon}{k}(G_\varepsilon + C_{3\varepsilon}G_b) - C_{2\varepsilon}\rho\frac{\varepsilon^2}{k} + S_\varepsilon \tag{3}$$

To determine the turbulence kinetic energy as well as the dissipation rate, the following constants were assumed: G_k = 1.0 and G_ε = 1.3 stand for the turbulent Prandtl numbers for k and ε, respectively, $C_{1\varepsilon} = 1.44$, $C_{2\varepsilon} = 1.92$, and $C_{3\varepsilon} = 0$ are default constants commonly used in fluent and S_k and S_ε are user-defined source terms.

3.3. Chemical reaction model

In this study, two different chemical reaction models were used in the simulation: the finite-rate/Eddy-dissipation model was used to describe homogeneous reactions while the kinetic/diffusion surface reaction model was employed for heterogeneous ones. The main distinction between these two models is associated with how the carbon species are treated. The homogeneous gas reaction assumes the carbon species gasified straightaway, and that the carbon is treated as a gas, while heterogeneous particle-gas reaction treats carbon as solid particles and they go through finite-rate reaction via a typical reaction at particle surface. **Table 4** presents the main reactions as well as the corresponding reaction rates for these two models.

Above reactions can slightly change if different gasifying agents other than air are considered. Previously published models using steam [23] and carbon dioxide [24] address those exact changes.

Reactions	Reaction rate
Homogeneous reactions:	
$CO + 0.5O_2 \rightarrow CO_2$	$r_1 = 1.0\times10^{15}\exp\left(\frac{-16,000}{T}\right)C_{CO}C_{O_2}^{0.5}$
$CO + H_2O \rightarrow CO_2 + H_2$	$r_2 = 5.159\times10^{15}\exp\left(\frac{-3430}{T}\right)T^{-1.5}C_{O_2}C_{H_2}^{1.5}$
$CO + 3H_2 \leftrightarrow CH_4 + H_2O$	$r_3 = 3.552\times10^{14}\exp\left(\frac{-15,700}{T}\right)T^{-1}C_{O_2}C_{CH_4}$
$H_2 + 0.5O_2 \rightarrow H_2O$	$r_4 = 2780\exp\left(\frac{-1510}{T}\right)\left[C_{CO}C_{H_2O} - \frac{C_{CO_2}C_{H_2}}{0.0265\exp\left(\frac{3968}{T}\right)}\right]$
$CH_4 + 2O_2 \rightarrow CO_2 + 2H_2O$	$r_5 = 3.0\times10^{5}\exp\left(\frac{-15,042}{T}\right)C_{H_2O}C_{CH_4}$
Heterogeneous reactions:	
$C + 0.5O_2 \rightarrow CO$	$r_6 = 596T_P\exp\left(\frac{-1800}{T}\right)$
$C + CO_2 \rightarrow 2CO$	$r_7 = 2082.7\exp\left(\frac{-18,036}{T}\right)$
$C + H_2O \rightarrow CO + H_2$	$r_8 = 63.3\exp\left(\frac{-14,051}{T}\right)$

Table 4. Chemical reaction model.

3.4. Model expansion for MSW

In the above subsection, we purposely did not address the mathematical treatment for the devolatilization phenomenon. The reason why being that the devolatilization section was only properly included when the model was expanded to handle the heterogeneity of MSW. Until that point, a single rate model, developed by Badzioch and Hawksley [25], was assumed which computed reliable devolatilization rates in a simple way.

To cope with the heterogeneity of said substrate a pyrolysis model with secondary tar generation was implemented. MSW is mainly composed of cellulosic and plastic components, and while cellulosic material can be divided in cellulose, hemicellulose, and lignin [26], plastics comprise polyethylene, polystyrene, and polypropylene, among others. To distinguish the several components that comprise the MSW, the pyrolysis reactions of cellulosic and plastic groups are considered individually and following an Arrhenius kinetic expression, as shown in **Table 5**.

3.5. Numerical procedure

Fluent, a finite volume method-based CFD solver, was employed in this work to solve the stated problem. Regarding the geometry modeling there are some simplifications that one can make in order to make the computation less expensive. Since the described reactor type is cylindrical, one can use a 2D axisymmetric problem setup.

Mesh was built using GAMBIT software and a total number of 83,000 quadrilateral cells of uniform grid spacing were generated. The mesh density in a finite element model is an important topic because of its relationship to accuracy and cost. In this particular case, the chosen cell size was about 12 times larger than the average particle size which was shown to be able to effectively capture the hydrodynamics in fluidized bed gasifier [27].

In such a complex model, it is sometimes difficult to define a good initial condition. For this reason, the process was first simulated considering only flow and nonreacting heat transfer (also known as "cold flow") and after reaching conversion reactive multiphase flow was added.

Substrate and air inlet were defined as "velocity inlets". Each velocity-inlet surface was identified by mass fractions, temperature, and a velocity magnitude. The flow direction was kept normal to

Reactions	Reaction rate
Cellulose $\rightarrow \alpha_1$volatiles $+ \alpha_2 TAR + \alpha_3$char	$r_9 = A_i\exp\left(\frac{-E_i}{T_s}\right)(1-a_i)^n$
Hemicellulose $\rightarrow \alpha_4$volatiles $+ \alpha_5 TAR + \alpha_6$char	$r_{10} = A_i\exp\left(\frac{-E_i}{T_s}\right)(1-a_i)^n$
Lignin $\rightarrow \alpha_7$volatiles $+ \alpha_8 TAR + \alpha_9$char	$r_{11} = A_i\exp\left(\frac{-E_i}{T_s}\right)(1-a_i)^n$
Plastics $\rightarrow \alpha_{10}$volatiles $+ \alpha_{11} TAR + \alpha_{12}$char	$r_{12} = \left[\sum_{i=1}^{n} A_i\exp\left(\frac{-E_i}{RT}\right)\right]\rho_v$
Primary$TAR \rightarrow$ volatiles + SecondaryTAR	$r_{13} = 9.55\times10^4\exp\left(\frac{-1.12\times10^4}{T_g}\right)\rho_{\mathrm{TAR1}}$

Table 5. Devolatilization model.

the surface. Turbulence of the inlet surfaces was detailed by turbulence intensity and hydraulic diameter. Outlet was set as the pressure outlet, specified by a gauge pressure value of zero.

For the pressure-velocity coupling, the widely used SIMPLE algorithm was enabled. For all simulations presented in this paper, a first-order upwind scheme was used for all equations. The standard scheme was used for interpolation methods of pressure. This means that the solution approximation in each finite volume was presumed to be linear, leading to less computational expenses. In order to properly justify using a first-order scheme, it was necessary to make sure that the grid used in this work had adequate resolution to accurately capture the physics occurring within the domain. In other words, the results needed to be independent of the grid resolution. Nonetheless, for better accuracy, this was later changed to the second-order upwind scheme.

Similarly, the underrelaxation factors were initially set at 0.5 (except for turbulent viscosity, pressure, and body forces, which were kept the same as default), and were then gradually conveyed to their default values with convergence. Convergence criteria were set that normalized residuals for all equations must fall under 10^{-6}.

4. Results and discussion

4.1. Model validation for forest residues

The main goal of researchers when it comes to numerical simulation is to increase the system complexity and accuracy while minimizing, if not complete eliminating, physical testing. There are two key systematic processes for confirming numerical results: the verification and the validation processes. The former tries to answer "did I solve the model right?" while the latter asks "did I solve the right model?" and this is where one checks against experimental data.

In the gasification process, this step becomes crucial since one is dealing with an extremely complex multiphase model where gas and solid phases exchange heat, momentum, and mass. To make matters worse, the hydrodynamic phenomena on a laboratory scale fluidized bed are not the same as on large scales [28]. To overcome this problem, the numerical results were compared with gasification tests performed in a pilot scale gasification plant, installed in the Industrial Park of Portalegre, Portugal. A wide range of operational conditions were tested for several Portuguese biomass substrates. **Table 6** shows three of these tested operating conditions for one particular substrate used to validate the numerical model and respective results.

The presented results show that the developed numerical model has the ability to predict the obtained synthetic gas composition within a satisfactory margin of error of 20%, commonly found in similar studies [29]. The highest deviation was observed for CH_4, which was expected since smaller fractions tend to produce higher relative errors. Furthermore, all light hydrocarbons and tar can lump into CH_4, which can explain the disagreement sometimes found in Ref. [15]. Very similar margins of error were found for different biomass substrates analyzed in the same pilot scale.

Nevertheless, some differences can be observed due to some simplifying assumptions followed by our model:

http://dx.doi.org/10.5772/65857

Experimental conditions		Forest residues		
Run		1	2	3
Temperature (°C)		815	815	790
Admission biomass (Kg/h)		63	74	63
Air flow rate (Nm^3/h)		94	98	98
Syngas fraction (dry and inert basis)				
H_2	Experimental	8.2	8.4	7.6
	Numerical	7.5	7.7	6.8
CO	Experimental	18.6	18	17.9
	Numerical	20.9	20.6	20.1
CH_4	Experimental	4.6	4.4	4.4
	Numerical	3.9	3.7	3.7
CO_2	Experimental	16.7	17.1	17.1
	Numerical	15.9	16.5	16.2

Table 6. Operating conditions for validation proposes and respective results.

- A more detailed devolatilization approach should be attempted.
- In addition, tar decomposition and chemical reactions for ash and some light hydrocarbons were not included.
- Ideal gas principles apply for the gases.
- Syngas is only formed by H_2, CO, CO_2, and C_nH_m and it is at chemical equilibrium.
- Heat losses from the components are neglected.

Some of these simplifications were corrected when the model was expended to deal with the heterogeneity of MSW as will be shown in the next subsection.

4.2. Model validation for PMSW

As stated in the mathematical model, to cope with the heterogeneity of MSW, the devolatilization model had to be restructured. Ideally, one would like to validate the new upgraded model with the experimental setup used earlier. However, due to unfortunate logistical and bureaucratic setbacks this was not possible.

To work around the problem, it was decided to validate the model using data collected from the literature [30]. **Table 7** shows the operating conditions used to validate the numerical model.

Figure 2 shows the comparison of the composition of obtained gas, estimated by the model, with that measured in the experiments.

Comparison between **Figure 2** and **Table 6** shows that the advancement in the model allowed a more complex system to perform in a similar manner and in some cases to predict syngas

Run	1	2	3	4	5	6	7	8	9
Temperature (°C)	493	705	602	507	687	593	691	593	507
MSW admission (Kg/h)	2.3	3	3	3	4	4	6	6	6
ER	0.5	0.4	0.4	0.4	0.3	0.3	0.2	0.2	0.2
Preheated air (°C)	290	352	296	281	352	307	352	308	279

Table 7. Operating conditions for the experimental gasification runs [30].

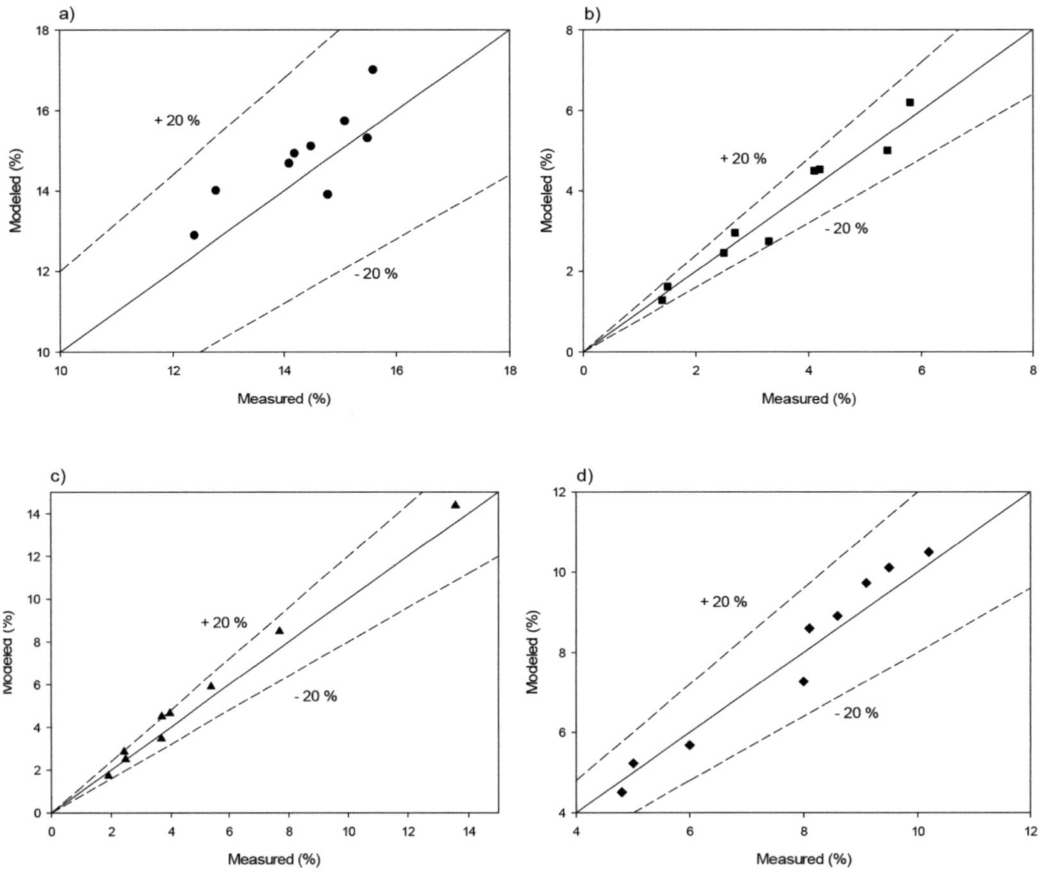

Figure 2. Comparison of the modeled and measured composition for (a) CO_2, (b) H_2, (c) C_nH_m, and (d) CO.

composition slightly better. This was due to a more realistic devolatilization model and the inclusion of light hydrocarbons.

4.3. Assessment of operational conditions

When scrutinizing a system (whatever it may be) it is imperative to devote a significant effort determining the influence of the main parameters on the system's output. In fact, the aim of the present subsection is to analyze the influence of several gasifier operating conditions including equivalence ratio (ER), steam-to-biomass ratio (SBR), carbon dioxide-to-biomass (CO_2/B) ratio, and reactor temperature on the obtained gas composition.

4.3.1. Equivalent ratio

Perhaps the most common, and the most analyzed, parameter regarding the gasification process is the air flow rate. Air is the cheapest gasifying agent but it also produces the poorest gas because it is highly diluted in N_2 [31]. To solve this problem, pure oxygen can be used to obtain a much higher quality gas (since no nitrogen will be present or at least a much lower fraction). However, implementation is only possible through high investment causing uncertainty when it comes to large facilities [32].

Furthermore, the O_2content in an air mixture can sometimes be misleading due to nitrogen dilution and the effects of small variations on substrate admission in the syngas composition going unnoticed, hence our decision to study ER. ER is one of the most significant parameters, which have effect on the gasification process including syngas composition. In accordance with previous gasification studies [33], the equivalence ratio can be defined as:

$$\mathrm{ER} = \frac{\text{oxygen mass/dry MSW mass}}{\text{stoichiometric oxygen/MSW ratio}} \tag{4}$$

The ratio was maintained between 0.15 and 0.35 since all of the experiments conducted to validate the model must be in this range and also because the ER values most suitable for gasification found in the current literature range between 0.2 and 0.4 [34].

The model predictions for the described reactor about the influence of ER on syngas molar fraction are shown in **Figure 3**.

A quick analysis shows that increasing ER suppresses the formation of combustible gases (H_2, CO, and C_nH_m) while promoting the formation of CO_2 contents. Since increasing the ER leads to more inert gas to enter the reactor, the obtained gas will be more diluted in nitrogen resulting in a poorer gas. A decrease in H_2 and CO can also be explained by a decrease in the residence time, considering that as air flow rate increases, it is no longer sufficient for CO and H_2 formation reactions to occur. Although to a smaller degree, ER negatively affects the C_nH_m content by enhancing steam reactions at higher temperatures leading to methane decomposition [35]. Finally, carbon dioxide fraction is expected to increase since combustion reactions (that consume CO and H_2 to produce CO_2) will be promoted. Results are consistent with the current literature [36, 37].

Despite presenting similar trends, studied substrates exhibit significant differences in the relative syngas molar fraction. One may explain this difference by recurring to the fuel's chemical composition. As demonstrated by Silva et al. [15], higher biomass calorific values result in higher calorific syngas production. This relationship between the biomass calorific content and the syngas lower heating values (LHV) can be explained considering, first, that the biomass calorific value is related to the amount of carbon (C) and hydrogen (H) present in the biomass, and second, a larger amount of these two elements allows production of larger quantities of hydrogen and carbon monoxide, the major contributors for the calorific value of

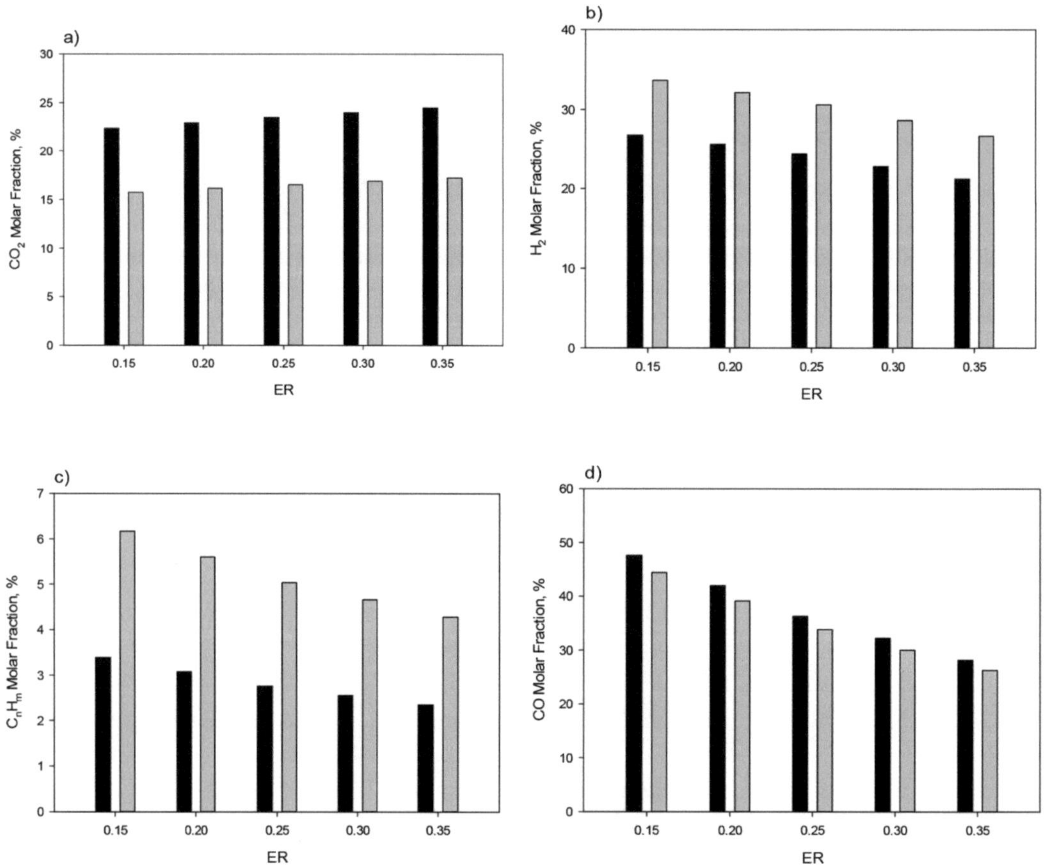

Figure 3. Comparison between PMSW (black columns) and forest residues (gray columns) as a function of ER for (a) CO_2, (b) H_2, (c) C_nH_m, and (d) CO (operating conditions: fuel feed rate = 25 kg/h; gasification temperature = 750°C).

the syngas. Nevertheless, since there are other biomass properties that can greatly influence the gasification process one must not take these conclusions as absolutes.

4.3.2. Steam-to-biomass ratio

Among oxidizing agents, steam gasification has received special attention since it produces a fuel gas with medium lower heating values of 12–18 MJ/Nm^3 [38], which is considerably higher than those from air gasification, while being less costly than oxygen gasification.

Steam-to-biomass ratio (SBR) is used in order to emphasize the effects of small variations on biomass admission, which often go unnoticed [39]. The SBR can be defined as the steam mass flow rate divided by the fuel mass flow rate (dry basis):

$$\mathrm{SBR} = \frac{\text{Steam mass flow rate}}{\text{Biomass substrate mass flow rate}} \tag{5}$$

The SBR was varied over a range of values from 0 to 2 by holding the other variables constant. The range was selected based on previous findings from our research team using the same facilities (**Figure 4**).

An increase in SBR leads to an increase in H_2 and CO_2 and a decrease in CO and C_nH_m. SBR will mostly favor char and tar steam reforming as well as the water-gas shift reaction, which in turn will lead to an increase in CO_2 and H_2 content at the expense of CO and C_nH_m. In fact, according to Hernández et al. [40], for steam gasification, the water-gas shift reaction will dominate over the Boudouard one and CO will be consumed to produce CO_2 and H_2. These results are consistent with the current literature [41].

An increase in the CH_4 content relates to the decrease in oxidation of volatile matter, which is not balanced out by the consumption of CH_4 in the reforming reactions. These reactions have lower rates than oxidation ones but are most favored by low temperatures. However, at higher steam levels, the steam reforming can in fact shift CH_4 consumption will also be affected.

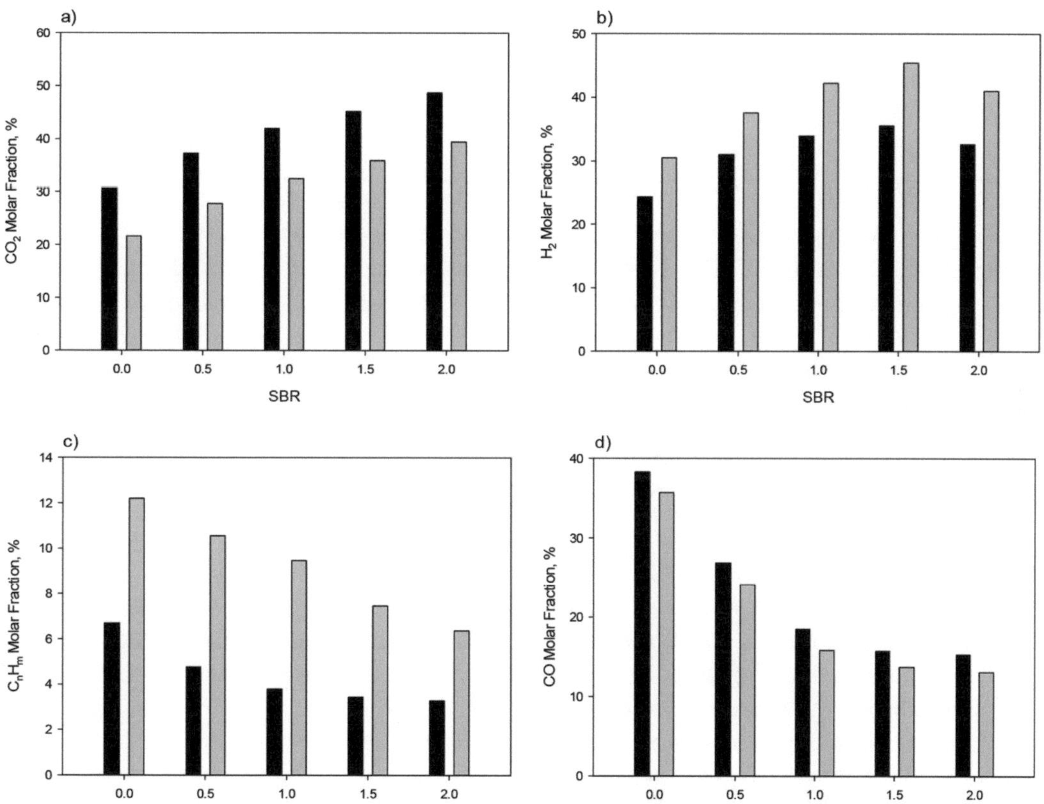

Figure 4. Comparison between PMSW (black columns) and forest residues (gray columns) as a function of SBR for (a) CO_2, (b) H_2, (c) C_nH_m, and (d) CO (operating conditions: fuel feed rate = 25 kg/h; gasification temperature = 750°C).

Excessive steam intake will lead to a significant decrease in gasification temperature, which in turn will have a negative effect on endothermic reactions, impairing product generation, which explains the decrease in H_2 after SBR=1.5, and producing insufficient heat to promote steam reforming and primary water-gas reactions. Furthermore, excessive steam could shift the steam reforming and water-gas reactions backward, consuming CO and H_2 to produce CO_2 and H_2O [42].

4.3.3. Carbon dioxide-to-biomass ratio

Even though steam gasification presents several advantages over air gasification, there are still some associated setbacks, such as the consumption of H_2O, which is an increasingly scarce resource.

Although poorly studied, carbon dioxide as a gasifier agent has been showing promising results, for starters, it consumes an unwanted end product (CO_2) of various industrial processes [43]. Furthermore, it also enhances both char gasification and pyrolysis and has the ability to act as a catalyst to enhance the thermal cracking of volatiles leading to tar mitigation.

The effect of carbon dioxide as a gasifying agent was studied using carbon dioxide-to-biomass ratio. The ratio was varied over a range of values from 0 to 1 by maintaining the other variables constant. To the best of our knowledge, regarding MSW, this ratio has only been investigated by our team [24], although some work has been published using biomass substrates by other researchers [44–46].

Results from **Figure 5** show that an increase in CO_2/B promotes the formation of CO and CO_2 while suppressing H_2 and C_nH_m. This can be explained with the fact that a higher CO_2 content mainly promotes Boudouard and reverse water-gas shift reactions, which leads to increase in CO fraction while H_2 decreases. On the other hand, C_nH_m molar fraction slightly decreases due to being consumed via CH_4reforming to produce CO and H_2 [47]. CO_2 content increases since a considerable fraction of the gasifying agent leaves the reactor unreacted.

Despite no results in the literature were found for MSW, those available for biomass are in agreement with the obtained results [44, 45, 48].

4.3.4. Gasification temperature

Gasification temperature is one of the most influential factors affecting the product gas composition and respective properties. The main reactions of the gasification are endothermic and thus strengthened by increasing temperature. Since the steam reforming, Boudouard, water-gas, and water-gas shift reactions occur simultaneously, the contents and ratios of considered species in the product gas are affected by temperature and partial pressures of reactants. Therefore, the reactor temperature significantly influences the syngas compositions. **Figure 6** show the influence of reactor temperature on final syngas composition for both substrates.

According to the Le Chatelier's principle, an increase in gasification temperature will favor products in endothermic reactions. Furthermore, promotion of endothermic reactions will lead to an increase in CO and H_2 content formation while decreasing CO_2 and C_nH_m [49].

http://dx.doi.org/10.5772/65857

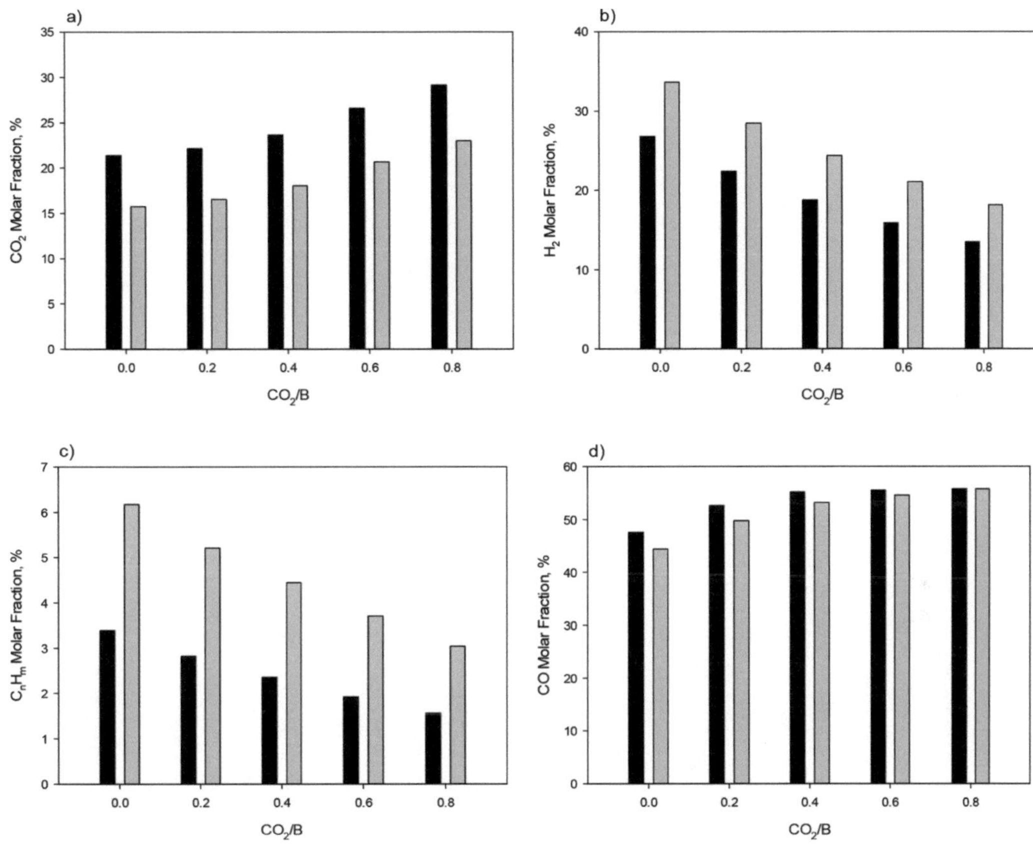

Figure 5. Comparison between PMSW (black columns) and forest residues (gray columns) as a function of CO_2/B for (a) CO_2, (b) H_2, (c) C_nH_m, and (d) CO (operating conditions: fuel feed rate = 25 kg/h; gasification temperature = 750°C).

Careful analysis shows that above 750°C hydrogen production becomes less pronounced while carbon monoxide further increases. In fact, in this range, the standard Gibbs free energy of the Boudouard reaction (responsible for CO production) becomes less than that of the water-gas reaction (main responsible for hydrogen production at lower temperatures), meaning that the former dominates over the latter as temperature increases. This is in agreement with the current literature [50].

Comparison between substrates was intentionally not discussed in the two previous subsections. The reason behind it is simply that all operating conditions share the same trends when it comes to different substrates, the underlining key is the biomass properties themselves.

4.4. Syngas quality indices

Syngas quality indices such as CH_4/H_2 and H_2/CO not only give a good indication of process efficiency, but also give its most suitable application. For instances, syngas with high CH_4/H_2 ratios tends to be used in domestic purposes while high H_2/CO ratios tend to be preferred in the chemical industry [51].

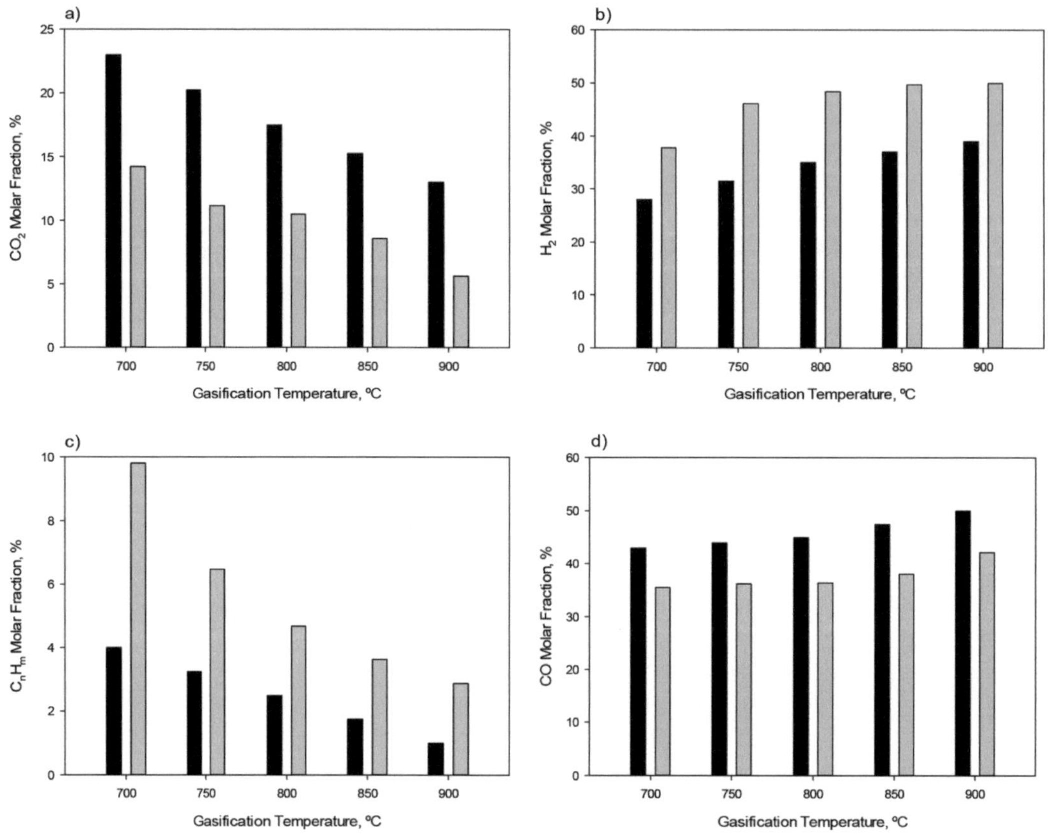

Figure 6. Comparison between PMSW (black columns) and forest residues (gray columns) as a function of gasification temperature for (a) CO_2, (b) H_2, (c) C_nH_m, and (d) CO. Dry and N_2-free basis (operating conditions: ER – 0.25; SBR – 1; CDMR – 0.4; MSW admission – 25 kg/h).

Besides syngas quality indices, there are other gasification products that can help determine process quality, namely carbon conversion (CC), cold gas efficiency (CGE), and tar content.

4.4.1. Methane-to-hydrogen ratio

Figure 7 displays the effect of gasification temperature on the syngas CH_4/H_2 ratio using the operating conditions from **Figure 6**.

Figure shows that the CH_4/H_2 ratio suffers a steep decrease with an increase in temperature. As can be seen from **Figure 6** (and corresponding explanation), an increase in gasification temperature will promote the formation of H_2. Meanwhile, due to the strengthening of the endothermic steam-methane reactions a decrease in CH_4 fraction is also expected.

At higher temperatures, usually above 800°C, one may notice that ratio decrease becomes less pronounced. This can be explained with the fact that, simply because within the studied

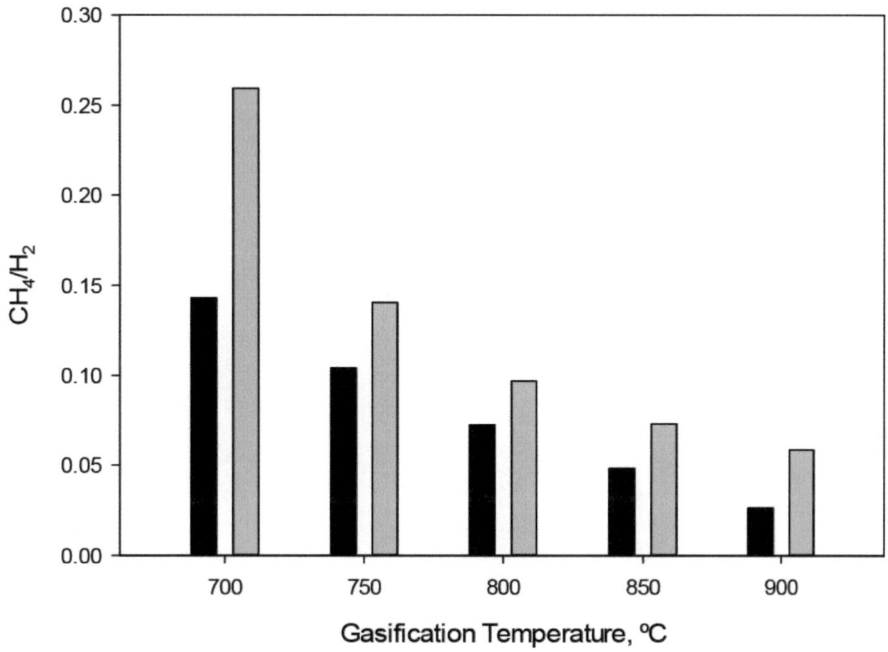

Figure 7. Syngas CH_4/H_2 ratio as a function of the gasification temperature for PMSW (black columns) and forest residues (gray columns) (dry and N_2-free basis).

temperature range, a decrease in CH_4 is somewhat constant while at higher temperature hydrogen production will also be less pronounced [33].

Forest residues presented a higher CH_4/H_2 ratio, especially for lower temperatures. Pinpointing exactly why this is can be extremely challenging due to a number of biomass properties that influence the gasification process. One can argue that since forest residues have a higher carbon and hydrogen content (according to our data, if we consider the proximate analysis as received from PMSW, C content is just 33.66 and H 4.42%) which will lead to more combustible gases. Furthermore, an increase in the CH_4 content is related to higher levels of volatile matter, which also might explain why forest residues present the higher ratio.

4.4.2. Hydrogen-to-carbon monoxide ratio

As estimated by Butterman and Castaldi [43], the H_2/CO ratio has an obvious impact on the ideal application for a given substrate. Higher H_2/CO ratios permit for the operation of solid oxide fuel cells (SOFC) [52] while medium H_2/CO ratios are suitable for FT synthesis of liquid fuels. Mid-to-lower ratios are mainly appropriate for catalyst-based FT synthesis while very low ratios are particularly appropriate for the production of a specific biomass-derived liquid chemical [53].

Figure 8 displays the effect of gasification temperature on the syngas H_2/CO ratio using the operating conditions from **Figure 6**.

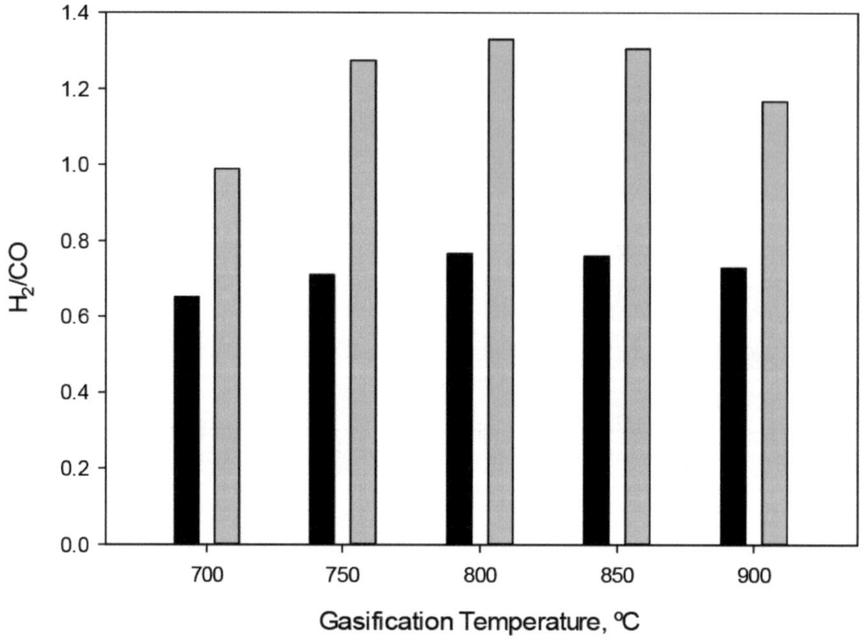

Figure 8. Syngas H_2/CO ratio as a function of the gasification temperature for PMSW (black columns) and forest residues (gray columns) (dry and N_2-free basis).

Conversely to CH_4/H_2, the H_2/CO ratio presents two distinct trends: from 700 to 800°C, gasification temperature has a positive influence on the H_2/CO ratio whereas an increase beyond 800°C shows the opposite trend. This can be explained with the fact that at low temperatures H_2 production is enhanced by primary water-gas reaction as well as steam-methane reforming reactions while at higher temperatures primary water-gas as well as Boudouard reaction will favor CO production.

Perhaps most importantly, in the range between 750 and 800°C, the standard Gibbs free energy of the Boudouard reaction (responsible for CO production) becomes less than that of the water-gas reaction (main responsible for hydrogen production at lower temperatures), indicating that the former dominates over the latter as temperature increases, leading to a decrease in the H_2/CO ratio, which is in agreement with the literature [54].

To make it useful for the chemical industry to synthesize products such as methanol and virgin naphtha, H_2/CO ratios higher than 1.70 should be presented, which neither of these two substrates were able to achieve, although forest residues were able to reach close to 1.4. The H_2/CO ratio would be increased by an injection of water [51].

4.4.3. Carbon conversion

Carbon conversion can be defined by the fraction of carbon from the substrate converted into carbon in syngas composition, and can give an good indication of the amount of unconverted materials, thus providing a measure of chemical efficiency of the process. It can be expressed as:

$$\text{Carbon Conversion} = \frac{12 \times M}{X_c \times m} \tag{6}$$

where M stands for the total molar flow rate of carbon in syngas composition, X_c stands for the carbon fraction in the MSW, and m stands for the MSW flow rate into the gasifier.

Figure 9 displays the effect of gasification temperature on carbon conversion using the operating conditions from **Figure 6**.

Figure shows that gasification temperature has a positive effect on carbon conversion. Higher temperatures will favor tar reforming leading to an increase in gas yield and carbon conversion [55]. Furthermore, an increase in temperature enhances steam reforming reactions, which in turn promote carbon conversion [41].

Although operational conditions were kept constant to ensure uniform residence time, substrates with different size particles lead to different residence times [47, 52]. In addition, the increasing residence time promotes gasification and carbon conversion reactions, leading to a higher gas yield [53]. Since forest residues have a smaller particle it leads to higher carbon conversion values.

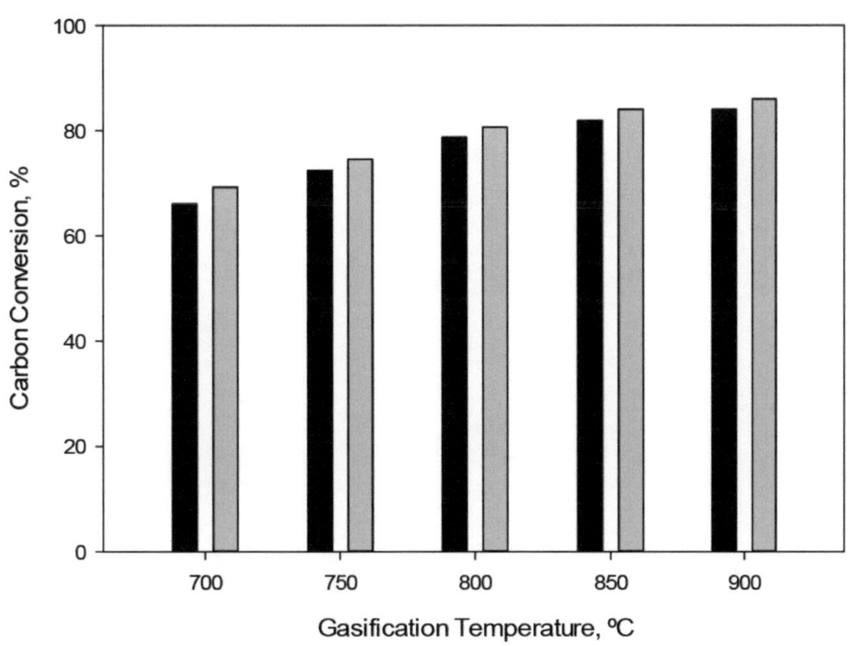

Figure 9. Carbon conversion as a function of the gasification temperature for PMSW (black columns) and forest residues (gray columns) (dry and N_2-free basis).

4.4.4. Cold gas efficiency

CGE can be defined as the percentage of the heating value of MSW converted into the heating value of the product gas. It can be computed as follows:

$$CGE = \frac{\text{Gas yield} \times \text{HHV of product gas}}{\text{HHV of fuel} + \text{Heat addition}} \tag{7}$$

Figure 10 displays the effect of gasification temperature on CGE using the operating conditions from **Figure 6**.

Similarly to CC, gasification temperature also has a positive influence on CGE. This is expected since the main gasification reactions are endothermic and thus strengthened by increasing temperature as well as gas yield, the two main factors responsible for CGE increase. Results are in agreement with the current literature [56, 57].

As expected, the calculated CGE for MSW is significantly lower due to a combination of low gas yield and poor syngas LHV. Efficiency values for both forest residues and MSW are within range from what is commonly found in the current literature [58].

4.4.5. Tar content

One of the major problems to deal with during biomass gasification is tar formation. At reduced temperature, tar condenses, blocking and fouling process equipment (such as engines

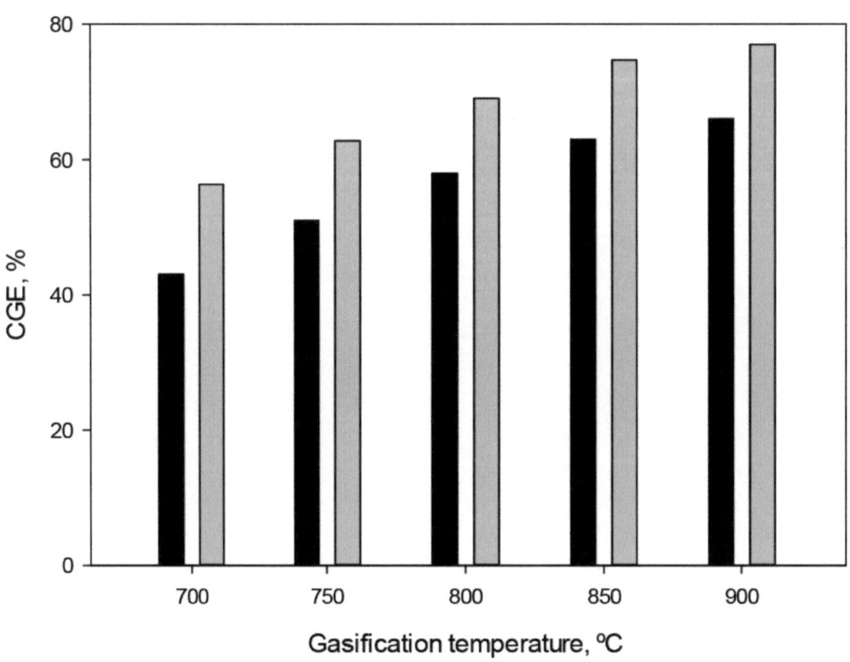

Figure 10. CGE as a function of the gasification temperature for PMSW (black columns) and forest residues (gray columns) (dry and N_2-free basis).

and turbines). According to Devi et al. [59], tar should be prevented or eliminated in the gasifier by manipulating factors such as operating conditions, addition of active bed materials, and possible reactor design modifications.

Figure 11 displays the effect of gasification temperature on the tar content using the operating conditions from **Figure 6**.

Figure shows that gasification temperature strongly influences the tar content. This can be explained with the fact that a high temperature favors destruction and reforming of tar, leading to a decrease in tar content [60].

As previously mentioned, a higher volatile content leads to an increase in residence time that, in turn, favors gasification reactions [61]. According to Aljbour and Kawamoto [62], an increase in residence time can lead to a reduction in the tar content. Since forest residues have a slightly higher volatile content, it comes with no surprise that it also consist a slightly lower tar content.

4.5. Process optimization

As shown all throughout the paper, gasification parameters have a strong influence on the overall process quality. Trying to have a better understanding of the underpinning mechanisms requires performing several gasification runs studying different operating conditions and parameters. Without a systematic approach such as the design of experiments this can become extremely time-consuming as well as expensive. Combining optimization methodologies such as the DoE with numerical models avoids expensive and time-consuming

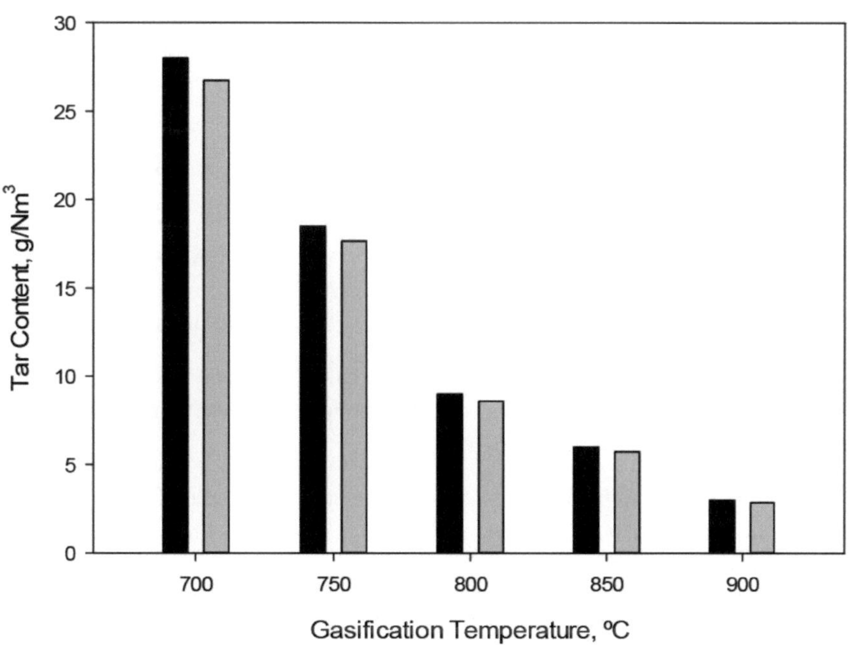

Figure 11. Tar content as a function of the gasification temperature for PMSW (black columns) and forest residues (gray columns) (dry and N_2-free basis).

experiments while obtaining the optimal operational condition set according to the desired response [63].

To test this hypothesis, an exergy efficiency optimization model was built using data obtained from the numerical model. The optimization model aimed to establish where the maximum exergy efficiency is according to the substrate and operating conditions chosen.

Particularly, in order to promote a more hydrogen-rich gas, the optimization model focused on an exergy flow rate of the produced hydrogen instead of syngas.

The empirical model was built minimizing the sum of the residues square to give the parameters of a second-order response model.

$$Y = B_0 + \sum_{i=1}^{3} B_i \times X_i + \sum_{i=1}^{3} B_{i,i} \times X_i^2 + \sum_{j=2}^{3} \sum_{i<j}^{3} B_{i,j} \times X_i \times X_j \quad (8)$$

where Y is the response, the X_i terms are the main factors ($-1 \leq X_i \leq 1$), temperature (1), steam-to-biomass ratio (2), and biomass type (3) and the B_i terms are the equation coefficients related to the main factors. The B_0 term is the interception coefficient, the $B_{i,i}$ terms are the quadratic effects (give the curvature to the response surface), and the $B_{i,j}$ terms symbolize the cross-interactions between factors. The present design does not consider the use of replicates because the results are obtained by computer simulations. In this case, a test on lack of fit and analysis concerning pure error are not provided [64]. Despite these circumstances much of the standard statistical analyses remain relevant, including measurements of model-fit such as PRESS (predicted residual sum of squares).

ANOVA analysis with high values for PRESS and "R-squared predicted" (not shown) also reinforced that the exergy efficiency response is well described by the empirical model in the design space.

Figures 12 and **13** show the hydrogen exergy efficiency as a function of the temperature and SBR for MSW and forest residues, respectively.

Similarly to what was shown in previous subsections, both substrates present very similar trends. In general, hydrogen efficiency increases with SBR since adding steam increases its chemical energy and exergy content. However, adding steam also demands additional exergy.

Regarding influence of gasification temperature one can clearly see that it has a positive effect on hydrogen efficiency all throughout the studied range. This can be explained by an increase in gas yield and enthalpy of gas component [56, 65].

The interpolating polynomial indicated in Eq. (8) provides the maximum values for hydrogen efficiency for both studied substrates. The maximum efficiency value was found at 900°C with a SBR of 1.5 for MSW and 1 for forest residues, with respective values of 50.6 and 50.2%.

Surprisingly, forest residues and MSW presented virtually the same maximum hydrogen efficiency. This can be explained with forest residues having the higher hydrogen molar composition and the higher gas yield, while MSW presenting the lower substrate exergy value,

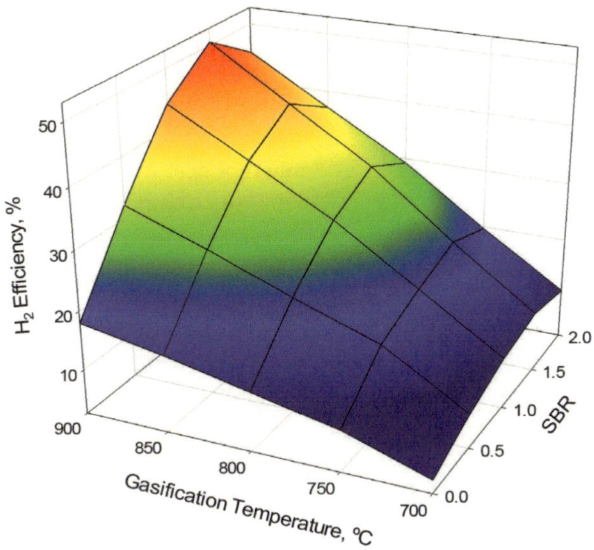

Figure 12. Hydrogen exergy efficiency as a function of the temperature and SBR for MSW.

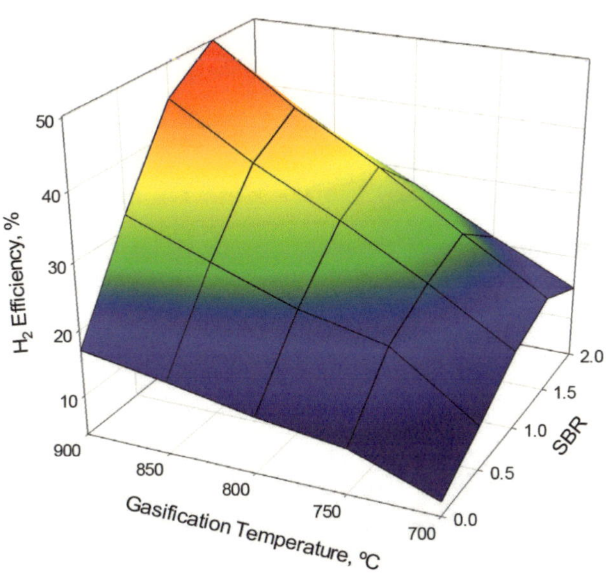

Figure 13. Hydrogen exergy efficiency as a function of temperature and SBR for forest residues.

leading to the latter having a slightly higher maximum hydrogen exergy value. Perhaps more important is the fact that forest residues have a much higher C_nH_m fraction (which is much more calorific than H_2 or CO), leading to a substantially higher CGE. When considering hydrogen efficiency, we are only considering hydrogen fraction and gas yield, so higher hydrocarbons do not influence the results. Available literature supports the conclusions made in this study [56, 65].

Even though results from MSW were not on par with those from forest residues in regard to CGE, exciting results were seen in both tar content and hydrogen efficiencies.

4.6. Role of numerical simulation in research assistance

Most real-life problems are simply not solved by analytical methods, especially if they involve interaction between physical processes or when comparison is made with real-experimental data. The use of numerical simulation is becoming increasingly imperative with the increase in computing power available by allowing researchers to get results not achievable by other means.

Most simulators of physical phenomena nowadays are using numerical methods rather than closed-form solutions. Examples would be solvers for Maxwell's equations (electromagnetic phenomena), Navier-Stokes equations (fluid mechanics), Schrödinger's equation (quantum mechanics), Fourier equations (heat conduction), and so on. Even "simple" simulators like the SPICE circuit simulator use numerical methods for nearly all but the simplest resistor circuits.

Numerical solutions open new possibilities for analysis but solutions can sometimes be difficult to interpret. Furthermore, simulators such as fluent are not 100% accurate or certain to give the user a physically real answer, one still has to know the basic physics to even setup most such simulators and one needs to know something about numerical analysis to know how and why such methods typically fail to simulate if you want identify or debug simulation failures.

The key determining factor is the level of complexity the researcher is willing to take (in other words what computational cost is the researcher willing to take). In the presented work, there were several choices that were necessary to make in order to ensure work feasibility:

- What kind of geometry simplifications could one make without jeopardizing physical model?
- What would the optimal grid density be? A very courser mesh could lead to very poor results, but a very fine mesh could mean dozens or even hundreds of hours extra just to improve results by a few percentage points.
- Is it better to use faster approaches like PDF flamelet models even though their fundamental assumption of infinitely fast chemistry renders them inapplicable for modeling low-Btu, low-exit velocity flaring?
- Using a radiation model can definitely improve the simulation results but requires a considerable amount of additional computational time, is it worth?
- When using a chemistry set, some authors opt to use detailed mechanisms (sometimes containing hundreds of reactions and dozens of species), while others prefer to use reduced or even simplified versions.

These are just a small fraction of all choices necessary to make when dealing with such a complex system as the one described in Section 3. Depending on system complexity, even with all the necessary adjustments to ensure faster convergence, some simulations will require optimal planning and patience. Some known examples by ANSYS:

- NETL fluidization challenge problem, able to simulate 16 s of flow time per day on 12 processors with a complete particle size distribution (PSD);
- Flow of solids in a riser, approximately 1 s of flow time simulated in a day in a single processor.

However, all this effort is rewarded when researchers are able to apply their results in creating credible model descriptions of new systems, which can be used in new designs or to improve the parameters on working systems and decision making for its modernization.

Nevertheless, perhaps the greatest benefit of using numerical models is the impact that these can bring not only for industrial purposes but also in important areas such as financial, social, and environmental areas.

5. Conclusions

Ideally, researchers would always want to test their hypothesis using the latest equipment in the most advanced laboratories with a large team of experts available to assist them. However, only an insignificant fraction has access to such conditions, and even those who do still have to go through tedious bureaucratic and logistical setbacks. Numerical simulation methods are used to study the behavior of systems whose mathematical models are too complex to provide analytical solutions in a time and cost-efficient way.

In this chapter, one of the most challenging energetic systems to predict was analyzed, the gasification process. In fact, gasification is considered to be one of the most difficult processes to model due to the chemical and physical interactions that occur throughout. A previously developed numerical model was used and its results validated using data collected from the literature and from a semiindustrial gasifier. Forest residues (one of the most abundant substrates in Portugal) and municipal solid wastes (one of the greatest challenges facing modern society) were analyzed. The model able not only to satisfactory the experimental results but also to correctly predict the trends of all studied parameters.

Finally, the numerical model was coupled with an optimization model designed to predict the optimal operation conditions for obtaining a more hydrogen-rich gas. This further states the ability of numerical simulations not only to assist analyzing the key trends of a given process but also how they can be used to predict the optimal operating point for a given optimal response.

Nomenclature

A, B	calibration constants
A_i	preexponential factor
$C_{1\varepsilon}$, $C_{2\varepsilon}$, $C_{3\varepsilon}$	constants
C_p	specific heat capacity
D_0	diffusion rate coefficient

E_i	activation energy
G_k	generation of turbulence kinetic energy due to the mean velocity gradients
G_b	generation of turbulence kinetic energy due to buoyancy
h_q	specific enthalpy of phase
h_{pq}	heat transfer coefficient between the fluid phase and the solid phase
k	thermal conductivity
Nu	Nusselt number
$\dot{m}$	biomass flow entering into the gasifier
M	total mole flow of carbon in the syngas components
M_i	molecular weight of each the species
M_c	molecular weight
$M_{w,i}$	molecular weight of i component
p	gas pressure
Pr	Prandtl number
p_s	particle phase pressure due to particle collisions
$\vec{Q}_{pq}$	heat transfer between p^{th} and q^{th} phases
$\vec{q}_q$	heat flux
q^{th}	specific enthalpy
R	universal gas constant
R_i	net generation rate of specie i due to homogeneous reaction
Re	Reynolds number
R_c	reaction rate
S_i	source term of the species i production from the solid heterogeneous reaction
S_k	user-defined source terms
S_q	source term due to chemical reactions
$S\varepsilon$	user-defined source terms
T	temperature
t_s	particle phase stress tensor
U	mean velocity
v	instantaneous velocity
X_C	carbon fraction in the biomass (obtained from the ultimate analysis)
Y	mass fraction
Y_M	contribution of the fluctuating dilatation in compressible turbulence to the overall dissipation rate

Other symbols

α	volume fraction
β	gas-solid interphase drag coefficient
γ_c	stoichiometric coefficient
γ_{Θ_a}	collisional dissipation of energy
ε	dissipation rate
ρ	density

φ_{ls}	energy exchange between gas and solid phases
k_{Θ_a}	diffusion coefficient
$k_{\Theta_a}\nabla(\Theta_s)$	diffusion energy
$(-p_s\bar{I} + \bar{\tau}_s) : \nabla(\vec{v}_s)$	generation of energy by the solid stress tensor.
τ	tensor stress
μ	viscosity

Subscripts

g	gas phase
s	solid phase
i	component

Acknowledgements

We would like to express our gratitude to the Portuguese Foundation for Science and Technology (FCT) for the support to the grant SFRH/BD/86068/2012 and the project PTDC/EMS-ENE/6553/2014 as well as IF/01772/2014. We also would like to thank Dr. Paulo Brito for providing experimental data needed to complete this paper.

Author details

Nuno Couto and Valter Silva*

*Address all correspondence to: vsilva@inegi.up.pt

INEGI-FEUP, Faculty of Engineering, University of Porto, Porto, Portugal

References

[1] V. Silva, E. Monteiro, A. Rouboa, An analysis on the opportunities, technology and potential of biomass residues for energy production in Portugal, in: A. Méndez-Vilas (Ed.) Materials and Processes for Energy, Formatex Research Center, Spain 2013, pp. 190–201.

[2] W.C. Tuckenger, Renewable energy technologies, in: U.W. UNDP (Ed.) World Energy Assessment of the United Nations, USA 2000, pp. 219–272.

[3] U. Arena, Process and technological aspects of municipal solid waste gasification. A review, Waste Management, 32 (2012) 625–639.

[4] G. del Alamo, A. Hart, A. Grimshaw, P. Lundstrøm, Characterization of syngas produced from MSW gasification at commercial-scale ENERGOS Plants, Waste Management, 32 (2012) 1835–1842.

[5] M. Lapuerta, J.J. Hernández, A. Pazo, J. López, Gasification and co-gasification of biomass wastes: Effect of the biomass origin and the gasifier operating conditions, Fuel Processing Technology, 89 (2008) 828–837.

[6] S. Ferreira, N.A. Moreira, E. Monteiro, Bioenergy overview for Portugal, Biomass and Bioenergy, 33 (2009) 1567–1576.

[7] D. Baruah, D.C. Baruah, Modeling of biomass gasification: A review, Renewable and Sustainable Energy Reviews, 39 (2014) 806–815.

[8] F. Tabet, I. Gökalp, Review on CFD based models for co-firing coal and biomass, Renewable and Sustainable Energy Reviews, 51 (2015) 1101–1114.

[9] S. Kraft, M. Kuba, F. Kirnbauer, K. Bosch, H. Hofbauer, Optimization of a 50 MW bubbling fluidized bed biomass combustion chamber by means of computational particle fluid dynamics, Biomass and Bioenergy, 89 (2016) 31–39.

[10] M. Siedlecki, W. De Jong, A.H.M. Verkooijen, Fluidized bed gasification as a mature and reliable technology for the production of bio-syngas and applied in the production of liquid transportation fuels: A review, Energies, 4 (2011) 389.

[11] G. Mirmoshtaghi, H. Li, E. Thorin, E. Dahlquist, Evaluation of different biomass gasification modeling approaches for fluidized bed gasifiers, Biomass and Bioenergy, 91 (2016) 69–82.

[12] R.I. Singh, A. Brink, M. Hupa, CFD modeling to study fluidized bed combustion and gasification, Applied Thermal Engineering, 52 (2013) 585–614.

[13] D.M. Snider, S.M. Clark, P.J. O'Rourke, Eulerian–Lagrangian method for three-dimensional thermal reacting flow with application to coal gasifiers, Chemical Engineering Science, 66 (2011) 1285–1295.

[14] Q. Xue, R.O. Fox, Multi-fluid CFD modeling of biomass gasification in polydisperse fluidized-bed gasifiers, Powder Technology, 254 (2014) 187–198.

[15] V. Silva, E. Monteiro, N. Couto, P. Brito, A. Rouboa, Analysis of syngas quality from Portuguese biomasses: An experimental and numerical study, Energy & Fuels, 28 (2014) 5766–5777.

[16] N. Couto, V. Silva, E. Monteiro, S. Teixeira, R. Chacartegui, K. Bouziane, P.S.D. Brito, A. Rouboa, Numerical and experimental analysis of municipal solid wastes gasification process, Applied Thermal Engineering, 78 (2015) 185–195.

[17] N. Couto, V. Silva, E. Monteiro, P. Brito, A. Rouboa, Using an Eulerian-granular 2-D multiphase CFD model to simulate oxygen air enriched gasification of agroindustrial residues, Renewable Energy, 77 (2015) 174–181.

[18] S. Teixeira, E. Monteiro, V. Silva, A. Rouboa, Prospective application of municipal solid wastes for energy production in Portugal, Energy Policy, 71 (2014) 159–168.

[19] Z.A.B.Z. Alauddin, P. Lahijani, M. Mohammadi, A.R. Mohamed, Gasification of lignocellulosic biomass in fluidized beds for renewable energy development: A review, Renewable and Sustainable Energy Reviews, 14 (2010) 2852–2862.

[20] N. Couto, V. Silva, E. Monteiro, P.S.D. Brito, A. Rouboa, Experimental and numerical analysis of coffee husks biomass gasification in a fluidized bed reactor, Energy Procedia, 36 (2013) 591–595.

[21] M.J.V. Goldschmidt, J.A.M. Kuipers, W.P.M. van Swaaij, Hydrodynamic modelling of dense gas-fluidised beds using the kinetic theory of granular flow: Effect of coefficient of restitution on bed dynamics, Chemical Engineering Science, 56 (2001) 571–578.

[22] P. Cornejo, O. Farías, Mathematical modeling of coal gasification in a fluidized bed reactor using a Eulerian granular description. International Journal of Chemical Reactor Engineering 9 (1) 2011.

[23] N. Couto, V. Silva, A. Rouboa, Assessment on steam gasification of municipal solid waste against biomass substrates, Energy Conversion Management 124 (15) 2016.

[24] N. Couto, V. Silva, A. Rouboa, Municipal solid waste gasification in semi-industrial conditions using air-CO2 mixtures, Energy, 104 (2016) 42–52.

[25] S. Badzioch, P.G.W. Hawksley, Kinetics of thermal decomposition of pulverized coal particles, Industrial & Engineering Chemistry Process Design and Development, 9 (1970) 521–530.

[26] O. Onel, A.M. Niziolek, M.M.F. Hasan, C.A. Floudas, Municipal solid waste to liquid transportation fuels – Part I: Mathematical modeling of a municipal solid waste gasifier, Computers & Chemical Engineering, 71 (2014) 636–647.

[27] S.J. Gelderbloom, D. Gidaspow, R.W. Lyczkowski, CFD simulations of bubbling/collapsing fluidized beds for three Geldart Groups, AIChE Journal, 49 (2003) 844–858.

[28] N. Couto, V.B. Silva, C. Bispo, A. Rouboa, From laboratorial to pilot fluidized bed reactors: Analysis of the scale-up phenomenon, Energy Conversion and Management, 119 (2016) 177–186.

[29] V. Silva, A. Rouboa, Combining a 2-D multiphase CFD model with a response surface methodology to optimize the gasification of Portuguese biomasses, Energy Conversion and Management, 99 (2015) 28–40.

[30] G. Xiao, B.-S. Jin, Z.-P. Zhong, Y. Chi, M.-j. Ni, K.-F. Cen, R. Xiao, Y.-J. Huang, H. Huang, Experimental study on MSW gasification and melting technology, Journal of Environmental Sciences, 19 (2007) 1398–1403.

[31] M. Campoy, A. Gómez-Barea, F.B. Vidal, P. Ollero, Air–steam gasification of biomass in a fluidised bed: Process optimisation by enriched air, Fuel Processing Technology, 90 (2009) 677–685.

[32] K. Maniatis, Progress in biomass gasification: An overview, in: Progress in Thermochemical Biomass Conversion, Blackwell Science Ltd., United Kingdom 2008, pp. 1–31.

[33] N.D. Couto, V.B. Silva, E. Monteiro, A. Rouboa, Assessment of municipal solid wastes gasification in a semi-industrial gasifier using syngas quality indices, part 1, Energy, 93 (2015) 864–873.

[34] L. Wang, C.L. Weller, D.D. Jones, M.A. Hanna, Contemporary issues in thermal gasification of biomass and its application to electricity and fuel production, Biomass and Bioenergy, 32 (2008) 573–581.

[35] Z. Wang, T. He, J. Qin, J. Wu, J. Li, Z. Zi, G. Liu, J. Wu, L. Sun, Gasification of biomass with oxygen-enriched air in a pilot scale two-stage gasifier, Fuel, 150 (2015) 386–393.

[36] S.S. Thanapal, K. Annamalai, J.M. Sweeten, G. Gordillo, Fixed bed gasification of dairy biomass with enriched air mixture, Applied Energy, 97 (2012) 525–531.

[37] N. Couto, E. Monteiro, V. Silva, A. Rouboa, Hydrogen-rich gas from gasification of Portuguese municipal solid wastes, International Journal of Hydrogen Energy, 41 (2016) 10619–10630.

[38] P. McKendry, Energy production from biomass (part 3): Gasification technologies, Bioresource Technology, 83 (2002) 55–63.

[39] Q. Miao, J. Zhu, S. Barghi, C. Wu, X. Yin, Z. Zhou, Modeling biomass gasification in circulating fluidized beds: Model sensitivity analysis, International Journal of Energy and Power Engineering, 2 (2013) 57.

[40] J.J. Hernández, G. Aranda, J. Barba, J.M. Mendoza, Effect of steam content in the air–steam flow on biomass entrained flow gasification, Fuel Processing Technology, 99 (2012) 43–55.

[41] J. Wang, G. Cheng, Y. You, B. Xiao, S. Liu, P. He, D. Guo, X. Guo, G. Zhang, Hydrogen-rich gas production by steam gasification of municipal solid waste (MSW) using NiO supported on modified dolomite, International Journal of Hydrogen Energy, 37 (2012) 6503–6510.

[42] C. Franco, F. Pinto, I. Gulyurtlu, I. Cabrita, The study of reactions influencing the biomass steam gasification process, Fuel, 82 (2003) 835–842.

[43] H.C. Butterman, M.J. Castaldi, CO2 as a carbon neutral fuel source via enhanced biomass gasification, Environmental Science and Technology, 43 (2009) 9030–9037.

[44] T. Renganathan, M.V. Yadav, S. Pushpavanam, R.K. Voolapalli, Y.S. Cho, CO_2 utilization for gasification of carbonaceous feedstocks: A thermodynamic analysis, Chemical Engineering Science, 83 (2012) 159–170.

[45] L. Garcia, M.L. Salvador, J. Arauzo, R. Bilbao, CO_2 as a gasifying agent for gas production from pine sawdust at low temperatures using a Ni/Al coprecipitated catalyst, Fuel Processing Technology, 69 (2001) 157–174.

[46] A.M. Parvez, I.M. Mujtaba, T. Wu, Energy, exergy and environmental analyses of conventional, steam and CO_2-enhanced rice straw gasification, Energy, 94 (2016) 579–588.

[47] O. Corigliano, P. Fragiacomo, Technical analysis of hydrogen-rich stream generation through CO_2 reforming of biogas by using numerical modeling, Fuel, 158 (2015) 538–548.

[48] Y. Cheng, Z. Thow, C.-H. Wang, Biomass gasification with CO_2 in a fluidized bed, Powder Technology, 296 (2016) 87–101.

[49] C. Loha, P.K. Chatterjee, H. Chattopadhyay, Performance of fluidized bed steam gasification of biomass – Modeling and experiment, Energy Conversion and Management, 52 (2011) 1583–1588.

[50] T. Song, J. Wu, L. Shen, J. Xiao, Experimental investigation on hydrogen production from biomass gasification in interconnected fluidized beds, Biomass and Bioenergy, 36 (2012) 258–267.

[51] P. De Filippis, C. Borgianni, M. Paolucci, F. Pochetti, Prediction of syngas quality for two-stage gasification of selected waste feedstocks, Waste Management, 24 (2004) 633–639.

[52] M. Marco, M. Giovanni, P. Lorenzo, R. Rosario, A quasi-3D computer model of a planar SO fuel cell stack, in: 3rd International Energy Conversion Engineering Conference, American Institute of Aeronautics and Astronautics, San Francisco, CA 2005.

[53] X. Peng, B. Toseland, A. Wang, G. Parris, Progress in development of LPDME process: Kinetics and catalysts, in: Coal Liquefaction & Solid Fuels Contractors Review Conference, Pittsburgh, Pennsylvania, 1997.

[54] J. Xiao, L.H. Shen, X. Deng, Z.M. Wang, X.L. Zhong, Study on characteristics of pressurized biomass gasification, Zhongguo Dianji Gongcheng Xuebao/Proceedings of the Chinese Society of Electrical Engineering, 29 (2009) 103–108.

[55] S. Luo, B. Xiao, Z. Hu, S. Liu, X. Guo, M. He, Hydrogen-rich gas from catalytic steam gasification of biomass in a fixed bed reactor: Influence of temperature and steam on gasification performance, International Journal of Hydrogen Energy, 34 (2009) 2191–2194.

[56] Y. Zhang, B. Li, H. Li, B. Zhang, Exergy analysis of biomass utilization via steam gasification and partial oxidation, Thermochimica Acta, 538 (2012) 21–28.

[57] Y. Wu, W. Yang, W. Blasiak, Energy and exergy analysis of high temperature agent gasification of biomass, Energies, 7 (2014) 2107.

[58] K.J. Ptasinski, M.J. Prins, A. Pierik, Exergetic evaluation of biomass gasification, Energy, 32 (2007) 568–574.

[59] L. Devi, K.J. Ptasinski, F.J.J.G. Janssen, A review of the primary measures for tar elimination in biomass gasification processes, Biomass and Bioenergy, 24 (2003) 125–140.

[60] X. Zhang, S. Deng, J. Wu, W. Jiang, A sustainability analysis of a municipal sewage treatment ecosystem based on energy, Ecological Engineering, 36 (2010) 685–696.

[61] F. Pinto, R.N. André, C. Carolino, M. Miranda, P. Abelha, D. Direito, N. Perdikaris, I. Boukis, Gasification improvement of a poor quality solid recovered fuel (SRF). Effect of using natural minerals and biomass wastes blends, part B, Fuel, 117(2014) 1034–1044.

[62] S.H. Aljbour, K. Kawamoto, Bench-scale gasification of cedar wood – Part II: Effect of operational conditions on contaminant release, Chemosphere, 90 (2013) 1501–1507.

[63] V. Silva, A. Rouboa, Optimizing the gasification operating conditions of forest residues by coupling a two-stage equilibrium model with a response surface methodology, Fuel Processing Technology, 122 (2014) 163–169.

[64] M.J. Anderson, P.J. Whitcomb, RSM Simplified: Optimizing Processes Using Response Surface Methods for Design of Experiments, Productivity Press, New York, 2005.

[65] A. Abuadala, I. Dincer, Efficiency evaluation of dry hydrogen production from biomass gasification, Thermochimica Acta, 507–508 (2010) 127–134.

Chapter 7

Sustainability of the Biowaste Utilization for Energy Production

Thorsten Ahrens, Silvia Drescher-Hartung and Olga Anne

Additional information is available at the end of the chapter

http://dx.doi.org/10.5772/65825

Abstract

This article presents strategies for the development and practical modelling of biogas processes under economy market conditions. Herewith, anaerobic digestion results out of practical tests in different scales (lab to pilot) and different substrate mixtures were taken into account. Two lab-scale reactors on the one hand and pilot-scale examinations of chosen substrate mixtures on the other hand led to workable conclusions such as mixture suitability for biogas production, gas amounts and technical demands for full-scale implementation under market economy conditions. A comparison of both laboratory and pilot system performance with a full-scale biogas system has been done; herewith, the suitability of the corresponding practical process upscaling simulation has been proven. On the basis of the results, calculations regarding the necessary full-scale fermenter sizes and the required substrate amounts as well as the disposable (reusable) fermentation residues were made. The conclusion of outputs on biogas technology particularly with regard to the demand-driven production of electricity (500 + 250 kW flexible) as a special request for energy from renewable sources is given. As a further result, a general outlook and estimation for the economical implementation on a common Baltic Sea region country basis have been developed.

Keywords: biowaste, biogas, pilot scale, full scale, circular economy, waste-to-energy, sustainability, process simulation

1. Introduction

The European Union (EU) calls on all parties to the efficient use of energy and accelerates the integration of energy recovery technologies from wastes that emit less carbon dioxide [1].

Biogas production by anaerobic digestion is a well-known method of energy recovery from biomass. Nowadays, such technologies are widely implemented in accordance with the EU policies for green and sustainable energy supply, for example, in all Baltic Sea region countries, but in different percentages. As an example, in Lithuania, mostly sewage sludge was used for biogas production till year 2012. Half of the Lithuanian energy demand in 2009 was covered by fossil fuels, 16% by renewable sources [2]. As different publications of FNR (Fachagentur Nachwachsende Rohstoffe, Germany) show that the anaerobic digestion is well investigated, but due to the technology development, there is need of research, especially in terms of market economy conditions, because funding for such processes will end in the near future [3]. Funding strategies should be taken into account in order to allow the development and pilot implementation of new technologies. In any case, new technologies need to position themselves in market economy, which means that a long-lasting funding (e.g. for more than 20 years) should not be taken into account. In terms of biogas processes, long-lasting funding took place and market demands were not considered in most cases, corresponding solutions for the future technology development shall be addressed in this article. For example, in order to identify substrate potentials (especially from waste materials) for biogas production in connection with suitable operational strategies for different parts of the Baltic Sea region, two EU financed BSR programme research projects were performed—regional mobilizing of sustainable waste-to-energy production (REMOWE) and ABOWE (implementing advanced strategies for biological utilization of waste) [4, 5]. The technological and economic efficiency analyses of the biogas development were fulfilled within above-mentioned projects. Therefore—besides to operational data—costs (investment and operating costs) and expected revenues from the products were considered, because efficiency analyses constitute the basis for the process and evaluation of decision-making process [6]. A standard method for the execution of company evaluations is the calculation of discounted cash flow via assessing the value of future cash flows via discounting to the valuation date and deduction of the initial net investment [7]. The energy demand, which is presented by load profiles in the article, shows the difficulty of the energy management from different renewable energy sources like wind and solar power. From the authors point of view, an adapted feeding and demand-driven operation of biogas plant provides a possibility to solve this problem.

2. Renewable energy production: initial situation

The production of renewable energy currently plays an important role in the European Union and will gain importance in the future. Renewable energy sources are wind power, solar power, hydroelectric power, tidal power, geothermal energy, biomass and renewable parts of waste. Between 2003 and 2013, the production of renewable energy within the European Union increased by 84.4%. Thereby, the most important source was biomass and renewable waste. The renewable electric energy production had a share of 25.4% of the gross electricity consumption in the EU. Here the main growth could be registered in three renewable energy sources, which are wind turbines, solar power and biomass [8]. In Germany, the production of renewable energies had a share of 30% of the gross electricity production; in Lithuania, a rate of 16% is reached. Until 2025 the share, for example, in Germany should be increased up to 40–45% and until 2035 up to 55–60% [9].

2.1. Load profiles vs. renewable energy from wind and solar power

The term load profile relating to electricity describes a detailed listing of the yearly energy consumption or demand [10]. That means the load profile provides a pattern of the electricity usage of one or more and also different customers. Herewith, the current collection will be projected at 15-min intervals. Unless the energy consumption is not being measured directly, there are standard load profiles available, which will be used for forecasting calculations [11]. Based on measured data or standard load profiles, the yearly electricity demand of a certain region or customer group can be determined. An example of an annual load profile is being shown in **Figure 1**. The figure constitutes a simplified version. The curve was straightened and therefore shows no 15-min-pattern.

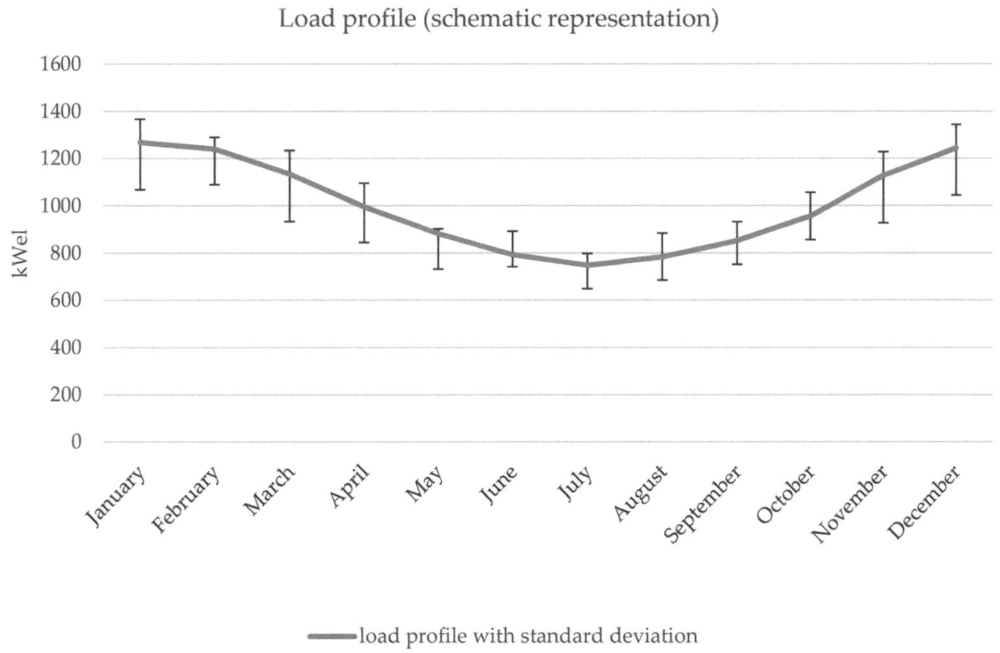

Figure 1. Load profile electricity demand.

As shown in **Figure 1**, the electricity demand is not evenly during the year. In current energy management concepts, the fluctuating demand is being compensated by the quick ramp-up of natural gas power plants [12]. The production of electricity from renewable energy sources had a high growth in between the last year. The relative share of electricity in the total quantity of electricity which was produced from biomass (incl. renewable waste) rose to 17.8%, the share of wind power to 26.5% and the share of solar power to 9.6% [8]. Anyhow wind and solar power are fluctuating producing electricity systems. Energy storage (for wind and solar power) is one of the possibilities to solve this problem [13]. Because of the increasing rate of fluctuating energies, it becomes more important to modify the electricity system, respectively, to find possibilities for energy storage or adapted energy production. **Figure 2** shows in an exemplary presentation how the covered amount of electricity produced by wind and solar power could exemplary look like. Data about the share of these techniques on the total energy demand are herewith being considered in percentage rates.

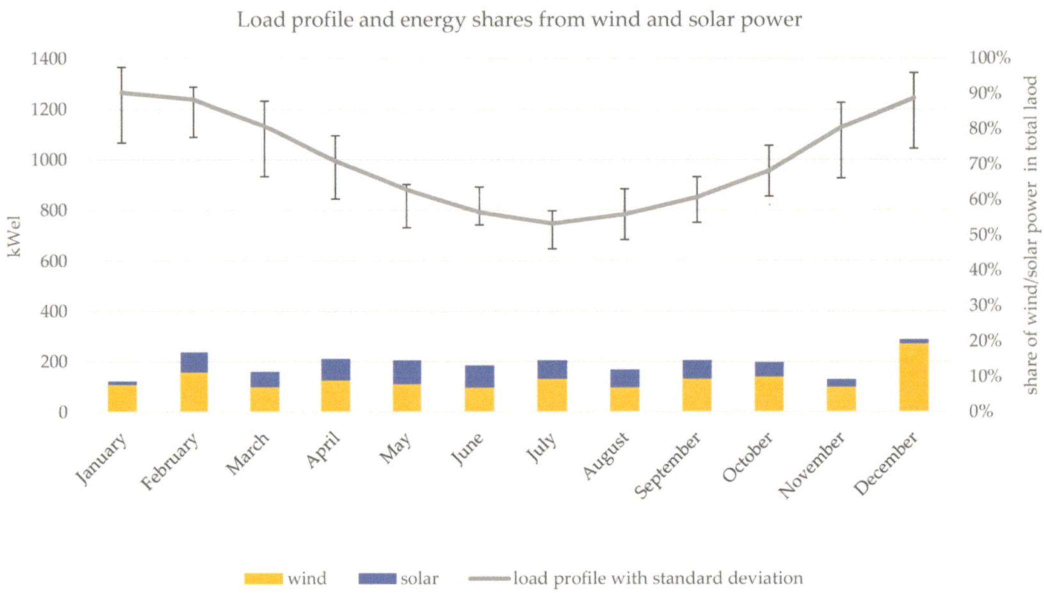

Figure 2. Load profile and energy shares from wind and solar power, data from Refs. [11, 14], adapted.

The annual load profile shows the variations in the electricity demand, which are not consistent. The problem of the electricity from renewable sources becomes obvious as it has been described above. Energy production from wind and solar power cannot be adapted, and it is fluctuating. The operation of biogas plants and production of biogas could be adapted with suitable operational methods, but nowadays the energy production is mostly even at a 24/7 rate. In order to solve the problem of adapted electricity production from renewable sources, there are essentially two possibilities available: (a) to store it or (b) to operate biogas plants adapted to the energy demand. Option (b) means that the production of electricity from biogas provides more features than wind or solar power. The produced biogas or also the produced electricity could be stored for later use. Moreover, a biogas plant provides the option to be operated energy demand oriented. Here the production of electricity depends on the feeding quantity, and therefore, an adapted feeding would lead to a demand-adapted production of electricity.

3. Substrate demand for adapted biogas scenarios

For production of biogas, many different substrates are used [3]. Primarily, manure or sewage sludge was used, but the kind of substrates used for biogas production changed in the last years. Many organic materials, for example, by-products in industrial or agricultural processes or also organic fractions from wastes like residual waste or biowaste from households, can be used. Many data concerning new substrates were collected during European-funded projects REMOWE and ABOWE at Ostfalia University [15]. Because some substrates are arising fluctuating and also with changing biogas yields, a targeted substrate management for an adapted operation of biogas plants is required.

3.1. Optimal operation of biogas plants adapted to the annual load profile

Based on load profiles (graph of the variation in the electricity demand of a region/city/single electricity consumer or other), the operation of a biogas plant can be planned and adapted, so that the electricity will be produced demand-based. Load profiles are provided by the power supplier or can be determined by use of standard load profiles (see Chapter 2.1). The substrate amounts, which are necessary for the calculation (see Chapter 7), will be determined on the basis of regional structural data or known arising substrates. As a conclusion, the exact quantity of demanded energy can be determined. Consequently, the biogas plant can be operated according to a well-structured substrate management.

3.2. Substrate management adapted to the annual load profile

Ostfalia University performed extensive studies concerning the energy demand and the adapted operation of biogas plants within several non-published student thesis works, including the consideration of the substrate demand. Herewith, the load profiles of different energy demands were used for the determination of regional-specific load profiles. The results of an exemplary adapted operation are shown in **Figure 3**.

In **Figure 3**, the exemplary adapted feeding of a biogas plant with biowaste and corn is demonstrated. The biowaste, which arises regularly but not evenly and furthermore with different biogas potentials (dependent on the time of year), has to be fed constantly because it cannot be stored for a longer period.

Therefore, energy crops like ensiled corn, which can be stored over a longer period of time, are being used for the additional feeding. The amount of corn, which has to be fed to reach the amount of the energy demand (based on an exemplary load profile), is being calculated in reference to laboratory substrate assessment results in Chapter 4 of the article.

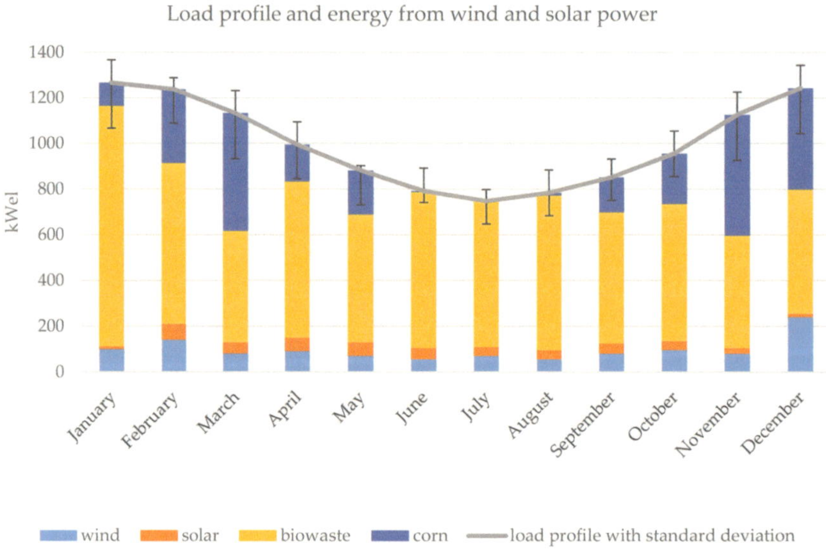

Figure 3. Load profile vs. energy potentials from wind, solar and biomass, data from Refs. [11, 14] and own calculations.

4. Biogas plant operation: from laboratory tests to large-scale plant operation

In order to be able to estimate the gas potentials of pure substrates as well as from substrate mixtures, practical data assessment is required. The following explanations are about samples being exemplary chosen by their availability in the Šilutė region of the country of Lithuania.

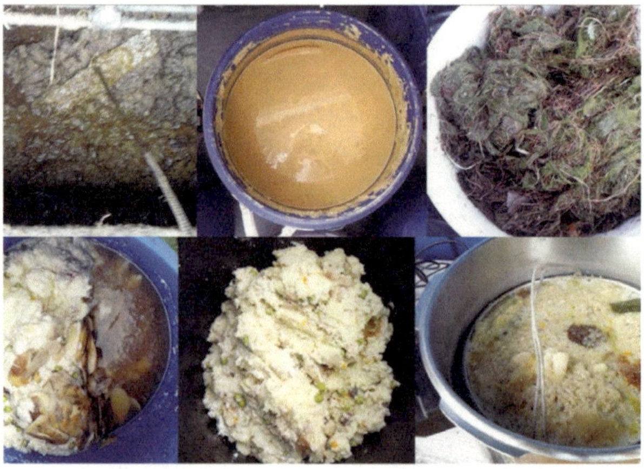

Figure 4. Substrates used in Lithuania. Cow manure (top left), distillery leftovers (top middle), algae (top right), food waste (original, mashed, sanitized) (from bottom left to right) [16].

The Šilutė region is widely settled by farmers mostly cow farms; furthermore, there is a spirit distillery factory which produces considerable amounts of bioethanol process leftovers and residues of food (food waste) from catering business. Other sources for biomass are macro- and microalgae. These have been collected at different locations at the Curonian Lagoon near Klaipeda. The cow manure was delivered by a local farmer, and the distillery leftovers were provided by a local spirit distillery factory. The food waste has been collected in Lithuanian kindergartens.

The substrates are shown in **Figure 4**, and the physical properties are listed in **Table 1**.

Substrate	Methane potential (m_N^3 CH_4/Mg FM)	Methane content (%)	Dry matter content (DM) (%)	Organic dry matter content (oDM) (%)
Food waste	85	57	23.5	22.7
Distillery leftovers	40.5	57	12.3	11.5
Algae, fresh	14.4	55	22.8	14.1
Cow manure	19.3	57	24	10.7

Table 1. Physical properties of the substrates used for the scenario in Lithuania.

4.1. Laboratory tests: batch and continuous

For the determination of anaerobic fermentation data of different substrates, two basic steps have to be done. These are batch fermentation tests to determine the maximum biogas potential and continuous tests to investigate the long-term fermentation behaviour of the tested substrate or substrate mixture. The batch tests are performed in 5-l flasks over a period of 35 days. The vessels for the continuous tests have a volume of 15 l and are being operated constantly over a longer period of time. In **Figures 5** and **6**, batch and continuous fermentation reactors in laboratory scale can be seen.

Figure 5. Batch tests in heating cabinet.

Figure 6. Continuous lab fermenter.

4.2. Pilot-scale digester

The pilot plant used during the research is a 550-l plug flow dry digester that was designed for long-term continuous operation to estimate the biogas potential of various substrates. The digester is situated inside a 20-ft container that is equipped with all necessary laboratorial equipment for daily diagnostic of anaerobic digestion conditions, and thus, this pilot plant is totally mobile (**Figures 7** and **8**).

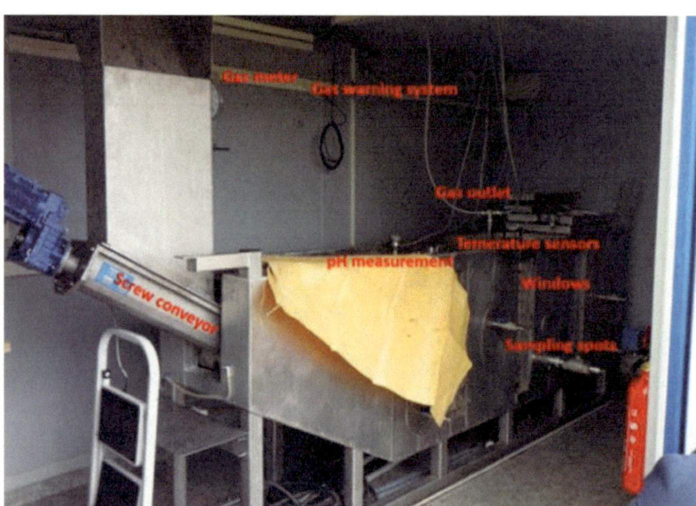

Figure 7. Plug flow dry digestion fermenter with components [16].

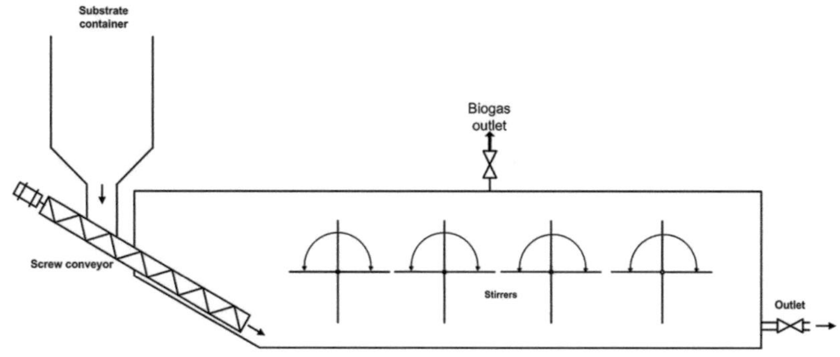

Figure 8. Schematic figure of the pilot plant [17].

Due to its size, real-size material (as for full-scale plants) can be used. This allows the process simulation of full-scale biogas plants.

5. Experimental proof of concept for fermentation strategies

The information obtained from plant operation is compared with the results of a parallel laboratory analysis of the substrates used during the testing period regarding materials and methods. **Figure 9** represents the methane yields of the continuous tests and the expected yields calculated from the batch tests. The results show that these tests correlate to each other. Therefore, the comparability of the continuous tests—relating to the benchmarking of fermentation performances—with the estimated results from previous batch tests is proven [16].

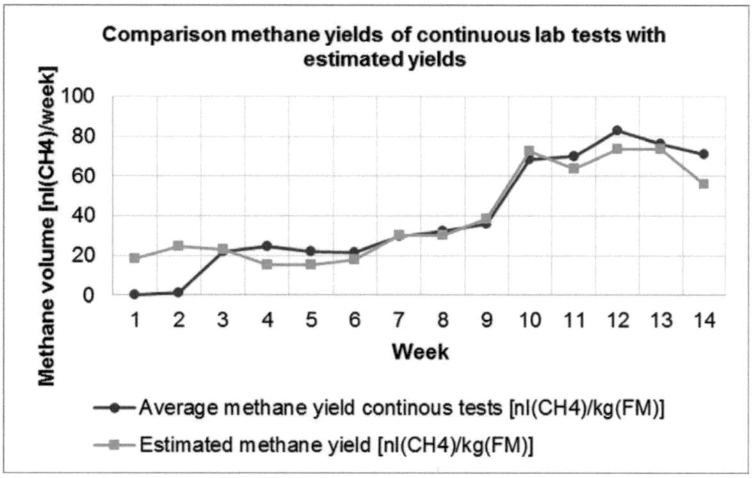

Figure 9. Comparison of methane yields of continuous lab tests with estimated yields calculated from batch tests (substrates: cow manure, food waste, algae, distillery leftovers) [16].

As well as the continuous lab tests, also the measured methane yields from the pilot plant tests correlate to estimated methane yields. The results are shown in **Figure 10**. Accordingly, the comparability of the pilot plant with estimated results from previous performed batch and continuous tests is proven. Because of previous overfeeding, the amount of methane produced in the first 4 weeks is high, but the almost parallel development of the two curves is clearly visible and obvious [16].

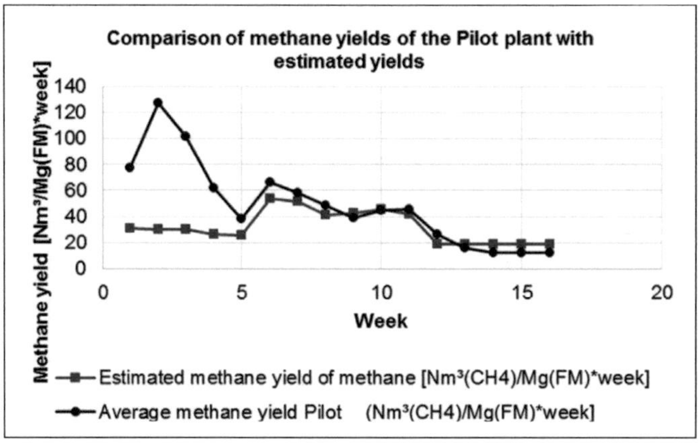

Figure 10. Comparison of methane yields of pilot plant with estimated yields calculated from batch tests (substrates: cow manure, food waste, algae, distillery leftovers).

Finally, the methane yields of the pilot plant were compared with the yields from a German full-scale biogas plant under comparable substrate and operation conditions. As **Figure 11** shows, the comparability of both plants was proven as well. The results of the full-scale plant were thankfully provided by the plant operator (input engineers) and adapted by Bansen [18].

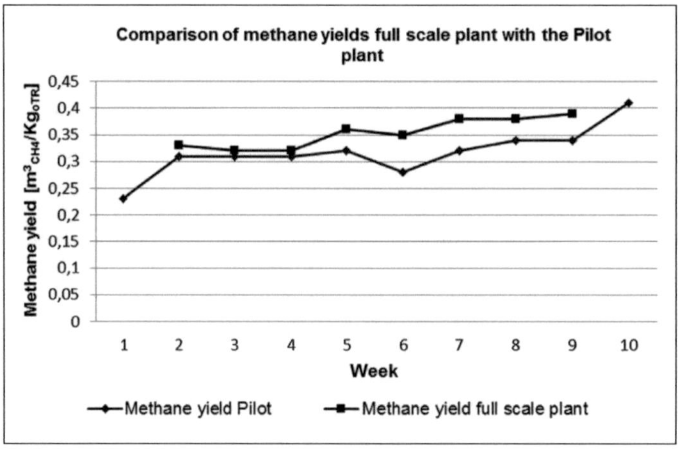

Figure 11. Comparison of methane yields of full-scale plant vs. pilot plant.

http://dx.doi.org/10.5772/65825

The experimental conditions for the lab-scale as well as pilot-scale tests are shown in **Table 2**.

	Temperature (°C)	Retention time (days)
Continuous lab test	42*	56*
Pilot-scale test	42*/55**	50–333*/40**
Full-scale plant	55	40

*For Lithuanian tests;
**for comparison of pilot-scale with full-scale plant.

Table 2. Experimental conditions of lab-scale, pilot-scale and full-scale tests.

A successful process simulation of the fermentation in pilot scale vs. the full scale has been also obtained by Cavinato [19] biogas production tests in 2010 and by Scano et al. [20] where fruit and vegetable wastes were used. Besides applicable experimental results, an economic analysis showed the possible profitability of the planned biogas project. It is important to emphasize that an improvement in the biogas yield was assessed by the operation of the pilot-scale fermenter, which means that the operation of plant engineering in pilot scale is absolutely suitable for technical and economic simulation and development of upscaling scenarios.

6. Financial implementation of biogas technology under market economy conditions

The implementation of biogas technology without any public funding provides the basis for the following considerations concerning the profitability. Implementing biogas technology requires the assessment of many aspects concerning the economic efficiency. The main financial and economic aspects of biogas plant exploitation include investment costs, plant operating cost and possible revenues achieved by the production and use of biogas, utilization of wastes and use of digestate as fertilizer. The resulting operative cash flow (OCF) calculation is based on the difference between annual revenues and annual expenses as well as investments. The discounted cash flow (DCF), that is, the calculation of future cash flows' present value, was a result of the annual OCF multiplied by a discount factor.

The result for the investor is just the amount of money, which could be received after a period of time. Thereby, the time value of money has to be taken into account [21]. The possible cost factors of biogas plants depend on the different sizes, and noticeable biogas plant characteristics concerning the size of the plant and the substrates, which will be used as input materials, are specified.

6.1. Cost factors

For building the biogas plant, investment costs as well as operational plant costs (both in extracts) have to be considered:

- Investment costs: engineering, permission of the authority (application costs), connection to the public grid, land costs, vehicles, offices and functional units such as substrate delivery and pre-treatment, digester, gas storage, biogas treatment, CHP unit, pumps, piping, digestate storing and others.
- Operational expenses: substrate costs, costs for analysis, process energy, consumables, costs for maintenance and repair, labour costs, land costs and others [3].

In particular, the specific investment costs vary, depending on the size of the biogas plant [22].

For the calculation of average investment costs of biogas plants, key values were determined by referring to investment costs of different German agricultural biogas plants and publications during project ABOWE [4, 16].

The operation of biogas plants with substrates, which require special treatment like municipal solid waste or biowaste, causes additional costs for pre-treatment and other necessary modifications. The use of household biowaste especially demands special treatment such as the sanitation of the material [16]. The hygienisation of biowaste used for anaerobic treatment is regulated by EU hygiene regulation (VO 1774/2002/EG) [23] and country-specific by, amongst others, the German Biowaste Ordinance [24]. Accordingly, biowaste could be sanitized by heating it up to 70°C for a time period of 1 h. Consequently, it has to be taken into account that investment costs for biogas plants with plug flow fermenter and the use of biowaste as substrate are about one-third higher than for the use of renewables [25]. For biowaste or MSW, the garage fermenter system makes biogas production with less pre-treatment of the substrate possible, which potentially reduces the operational as well as the investment costs.

In any case, the specific investment costs tend to decrease with larger plant capacities. When assessing the different types of the investment expenses, it becomes obvious that part of the costs for planning and construction of the biogas plant is personnel expenses. These costs should be considered separately, because, as it was pointed out in project ABOWE, there are considerable variations in different countries [4].

Another point is that some parts of the biogas plant have to be replaced regularly, because certain components such as pumps, stirrers and also the CHP unit have a short operational lifespan. The estimated lifetime of pumps is 4 years and that of CHP units about 6 years [26].

One of the most important aspects when implementing biogas technology is to ensure reliable availability of substrates. Biogas plants demand a continuous supply with substrates. Furthermore, the utilization of the produced energy or the conditioned biogas itself as well as the resulting heat or the electric energy generated by the CHP unit need to be guaranteed.

In addition, it has to be taken into consideration that substrates with a high-energy potential should be used, so that costs and expenses for transport are kept low [4, 16].

Considering further economic aspects, the biogas plant (pumps, stirrer and others) itself demands electrical energy amounting to 5–20% of the total of electrical energy produced by CHP technology [25]. The heat demand of the biogas plant (for the heating of the fermenter), which is about 5–25% [25] (according to FNR even at 28% [22]) of the produced heat, can be covered by the heat of the CHP unit. Here it might be economical to sell all of the energy produced and buy the energy needed from the national energy supplier (depending on the feed-in tariff of the electricity generated) [27].

6.2. Country-specific aspects

For the economic and financial implementation, the different cost items vary according to the country in which the biogas plant shall be built.

The calculation done in the project is based on specific data, which vary according to the relevant countries. Therefore, some prior general considerations are necessary:

- Investment costs: it has to be considered which parts of the plant can most cost-effectively be manufactured in the country in which the plant shall be operated.
- Operational costs: most specific costs depending on the relevant countries; the personnel costs in particular are varying strongly.
- Revenues: the feed-in tariffs for electricity and heat are country-specific, also the price for the sale of digestates [4, 16, 17].

General data were collected with respect to the investment and operational costs of biogas plants in EU developing counties and furthermore with data of plant operators as well as plant construction companies.

Based on all above-mentioned aspects, results of the pilot plant operation in certain EU developing and developed countries, lab tests at Ostfalia University and a numerous collection of data (own investigations, literature sources, data from biogas plant operators), many key values were determined and used for the calculation of the operative cash flows over a period of 20 years. The cumulative discounted cash flow then gives an impression of the profitability of a planned biogas project. Spoken fundamentally, a detailed calculation and estimate of the operative cash flows are only possible by defining real system models. Based on these data (referring to commercial offers) and on the investigation of the economic aspects of a planned biogas plant, detailed cash flow calculations are possible.

7. Full simulation of a practical application: implementation of a biogas plant from energy demand to profitability

For illustration of above-constituted considerations concerning the implementation of biogas technology with no public funding, necessary substrate management and load profile adapted operation a scenario were developed. Substrates mentioned in Chapter 4 were used for a complete implementation scenario of biogas technology in Lithuania.

Herewith, food waste, distillery leftovers and algae serve as basic substrates for the production of biogas during the year. Cow manure is available as additional substrate and will be fed as compensation substrate to get an adapted feeding for the production of the needed electricity.

According to the previously described strategy, the following amounts (in **Table 3**) of substrates and methane potentials were used for further calculations. For the scenario, the size of the biogas plant was chosen with 500 kWel.

Substrate	Methane potential (Nm^3 CH_4/Mg FM)	Quantity used (Mg/a)
Food waste	85	8000
Distillery leftovers	40	5000
Algae, fresh	14	2000
Cow manure	19	10,000

Table 3. Substrates used for an implementation scenario (Ostfalia lab).

Table 3 shows the biogas yields, which arose from batch tests of food waste, distillery leftovers and algae (see Chapter 4.1). Starting from the assumption that the waste is arising constantly with a constant biogas yield (algae occur from April/May to September/October), the electricity production would be evenly because of a constantly daily feeding amount. The load profile, which constitutes the electricity demand, would therefore not be covered (sum of electricity from food waste, distillery leftovers and algae). Consequently, cow manure will be used as additional feeding material, see **Figure 12**.

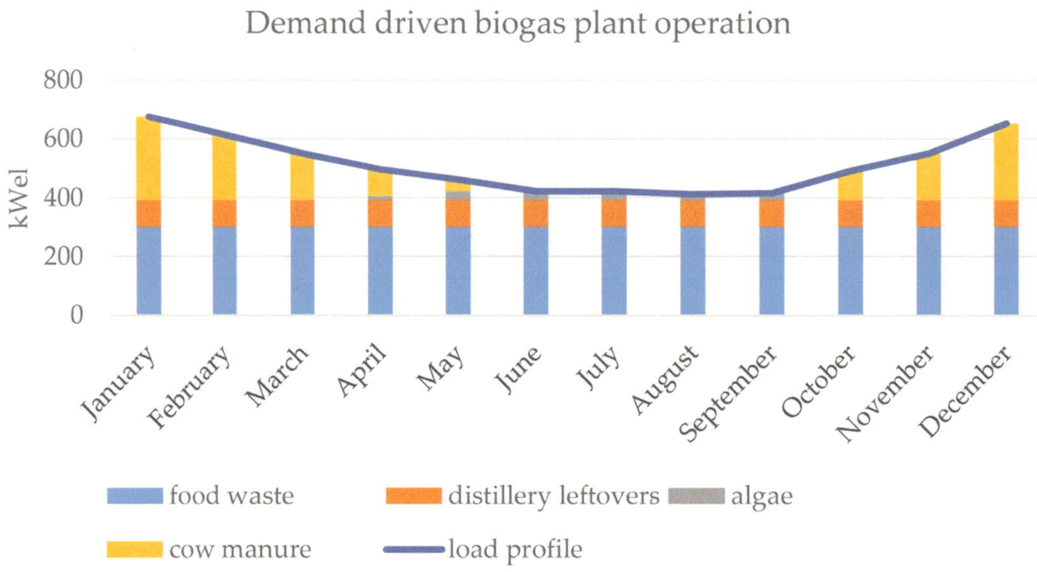

Figure 12. Power from methane production (scenario).

Based on the results of the performed batch tests and the developed scenario, the substrate amounts and mixtures were tested in continuous fermenter tests. The long-term fermentation behaviour is therefore proven, before tests with these substrates will be proceeded in the pilot-scale fermenter. The results of this test procedure lead to the planning of a demand-driven supplied biogas plant by meanings of an adapted production of electricity, no demand for long-time storing capacities and strict avoiding of energy undersupply.

The corresponding calculations were done in reference to basic data, (especially concerning the substrates and feed-in tariffs and other incentives). These data concerning substrate amounts and tariffs are stated by the prospective customer or need to be determined. Therefore, all data are exchangeable and can easily be adapted to other scenarios besides the exemplarily chosen Lithuanian scenario. Assumptions made for the following calculations are summarized in **Table 4**.

	Tariffs (€/Mg)	
	Costs	**Income**
Substrates:	8.66	20
Food waste	–	–
Distillery leftovers		25–50
Algae		
Cow manure		
Digestate		5
Electricity	Public grid: 0.126 (€/kWh_{el}) or less	Feed-in: 0.09 (€/kWh_{el})
Heat		0.04–0.06 (€/kWh_{th}) (by season)
Produced energy	3,929,083 kW_{el}/a and 3,929,083 kW_{th}/a; efficiency rate $\eta_L = 40\%$	
Digestate	22,646.55 Mg/a	
Biogas plant: plug flow fermenter system, investment costs plus 30% for waste hygienisation (assumption based on Ref. [25]).		
CHP unit, 1 × 500 kWel, 1 × 250 kWel: production of electricity and heat, each re-invested after 72-month operating time.		
Substrates: food waste, distillery leftovers, fresh algae, cow manure.		
Sale of electricity and heat (minus own demand).		

Table 4. Tariffs and further background assumptions.

The operational strategy resulting from **Figure 12** requires a flexible operation of the biogas plant. Thus, the biogas plant should be operated with two CHP units, one with an installed power of 500 kWel and another with 250 kWel. The 500 kWel system is in continuous operation and the other is operated in modulation and on-demand. In any case, a suitable dimensioned gas storage facility needs to be provided.

With respect to this substrate scenario, an exemplary financial implementation of the biogas system is developed. The results of the corresponding economic calculations are shown in

Figure 13 as a cumulative discounted cash flow curve progression starting with the initial investment costs and cumulating the operative cash flows over a period of 20 years.

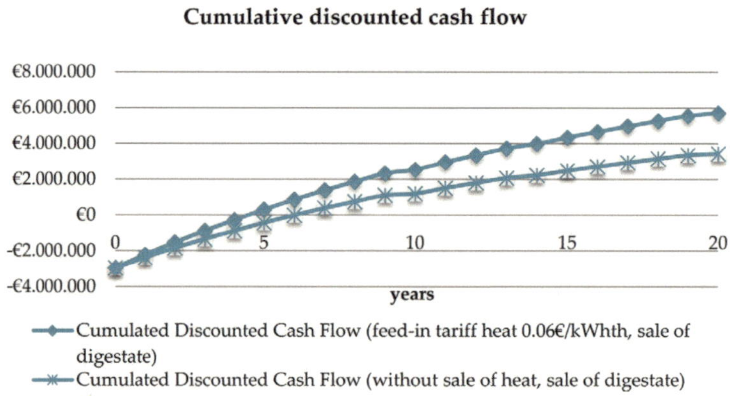

Figure 13. Cumulative discounted cash flow—comparison of two exemplary sale scenarios.

Assuming that the biogas plant operator gets gate fees from the acceptance of waste and all of the produced electricity and heat (minus own heat demand) will be sold, the biogas plant operation reaches the zero line after about 4.5 years. The second curve results in the supposition that the heat could not be sold. Here the zero line would be reached after about 6 years. This example of the variation of just one parameter shows that it is really easy for the user to create scenarios, to vary parameters, to find the most influencing parameters and to get first information about the profitability of a planned project.

The chosen scenario as well as the calculated model biogas plant constitutes a theoretical exemplary situation. The corresponding assumptions are not being meant for a generalized use, because they depend on scenario-specific aspects and were just chosen for the exemplary developed scenario. Nevertheless, all economic influencing variables can easily be changed and adapted to other scenarios and regions, and the resulting impacts are illustrated in the same way as described above. This calculation model provides a good impression of the profitability of a planned biogas implementation project. Therefore, the essential feature of the presented model is a suitable opportunity for easy and rapid profitability calculation for the usual users, respectively, investors such as farmers or rural area authorities.

8. Perspectives towards sustainability and impact to global environmental problems

From the sustainability and circular economy point of view [28], the idea of the biogas production from biowaste is in line with industrial symbiosis mechanism that is essential for achieving social-economic and environmental benefit [29]. Except economic and social aspects that have close relationship and depend on profitability and subsequent improvement in the

http://dx.doi.org/10.5772/65825

quality of life, the environmental aspects should be discussed as well. LCA is often used as a tool to evaluate a process, service or product impact to the global environmental problems. The method of LCA is described in ISO standards [30] and starts from the finding of the study system boundaries. The magnitude of the environmental impact of the biomass anaerobic digestion and CHP unit depends on numerous factors, such as fermentation conditions (mesophilic, thermophilic), kinds of substrates (cow manure, bioethanol residue, algae, food wastes), biogas yield, energy demand for the own process of digestion and sanitation, transportation, digestate storage and method of utilization and climate condition. The scientific results related to influence of the renewable energy production and transportation processes to global problem, especially greenhouse effect, sometimes differ a lot [31–33]. Nevertheless, there are some common features, which allow determining the main tendency of the above-mentioned processes and products contribution towards sustained environment.

Global warming potential. The main impact of the GHG emissions is expected from biogas combustion at CHP and methane emissions from the storage of the digestate. Anaerobic biomass digestion itself does not pollute environmental air and does not use oxygen/air for biogas production.

Ozone depletion potential. During biogas burning process at the energy production plant, some amount of CFC/HCFC can get in to the air. In addition, the manufacturing of the fermenter and energy plant could cause CFC/HCFC release in the atmosphere.

Photochemical oxidation potential. This kind of impact relates to the secondary methane emission source—emission from the digestate storage.

Acidification potential. Environmental acidification potential expressed as sulphur dioxide equivalent is the biggest for hydrogen sulphide and ammonia and mostly connected with ammonia emissions due to the digestate storage.

Eutrophication potential. Nitrogen and phosphorus compounds contribute to an increased biomass production in aquatic environment, and these results in additional oxygen consumption for biomass decomposition. Release of ammonia from the digestate storage can increase the eutrophication potential.

Abiotic depletion potential. This kind of depletion can be decreased due to the usage of digestates instead of traditional fertilizers. Anaerobic digestion demand of additional energy consumption increases abiotic depletion, but use of traditional fossil fuel for traditional energy production plant does not reduce abiotic depletion as well. The recycling of the fermenter and CHP plant constructions after usage prevents damage for the environmental and resources.

The environmental impact assessment does not take the digestate spreading into consideration because spreading of fertilizers exists in any case, but the method of the spreading can have significant impact to soil acidification and aquatic body eutrophication. The evaluation of environmental aspects of the study shows that biomass anaerobic digestion and CHP system for energy production have savings in global warming and abiotic depletion potentials. Scientific publications [34–36] regarding biomass anaerobic digestion and energy production support present study environmental results and recommendations.

9. Conclusions

9.1. Technical conclusions

Digestion technology adapted for a Lithuanian scenario with lab-scale examination in batch and continuous tests was realized. Four different kind of substrates were applied: cattle manure, bioethanol distillery waste, food waste and Curonian Lagoon algae.

The similarity of the fermentation process in lab-scale, pilot-scale and full-scale plant was shown.

The technological adequateness of biogas technology for substituting fossil energy resources with respect to fluctuating energy demand and fluctuating supply of renewable energy has been demonstrated, and the possibility of demand-driven operation of a biogas plant has been proven.

9.2. Financial-economic conclusions

The financial-economic aspects as the most important part of the chain of waste-to-energy advanced strategy have been confirmed.

The feasibility study based on certain regional data was detected as the most realistic for biogas development benefit for local society and environment.

A detailed perspective into the strategy of digestion for biogas production is given. Starting with energy demand, load profiles and substrate demand, a detailed substrate management has been performed. Based on this, the profitability and thereby financial implementation without any public funding could be considered. Herewith, a special focus is given in the potential implementers' and investors' demands, which shall be enabled to take reliable decisions on technologies for implementing advanced digestion technology as full-scale plants. With the corresponding developed strategies (experimental as well as economical), new possibilities of producing renewable energy can be implemented and fostered in the Baltic Sea region as well as in other regions of the EU.

Author details

Thorsten Ahrens[1*], Silvia Drescher-Hartung[1] and Olga Anne[2]

*Address all correspondence to: th.ahrens@ostfalia.de

1 Institute for Biotechnology and Environmental Engineering, Ostfalia University of Applied Sciences, Wolfenbüttel, Germany

2 Department of Natural Science, Klaipeda University, Klaipeda, Lithuania

References

[1] Huopana, T. et al: A regional model for sustainable biogas production. Case study:North Savo, Finland, REMOWE project, Savonia University of Applied Sciences (publisher), Kuopio. 2012. ISBN 978-952-203-169-3

[2] Luostarinen, S. (ed.): Energy potential of manure in the Baltic Sea region: biogas potential & incentives and barriers for implementation. Knowledge Report: BALTIC-MANURE WP6 Energy Potentials of Manure. 2013. Available from: http://www.baltic-manure.eu/download/Reports/bm_energy_potentials_web.pdf [accessed: June 2016]

[3] Fachagentur Nachwachsende Rohstoffe e.V. (FNR): Guide to Biogas-from Production to Use. 232 pp. Available from: http://mediathek.fnr.de/media/downloadable/files/samples/g/u/guide_biogas_engl_2012.pdf [accessed: June 2016]

[4] ABOWE project reports, 2014. Available from www.abowe.eu

[5] REMOWE project reports, 2012. Available from www.abowe.eu

[6] Ullrich, H.: economic planning and execution of process plants. Vulkan Verlag, Essen. 1996. 248 pp.

[7] Schacht, U., Fackler, M.: Practical handbook company evaluation. 2. Auflage. Springer Gabler, Wiesbaden. 2009. ISBN 978-3-8349-0633-5

[8] Eurostat, Statistics explained: renewable energy statistics. Available from: http://ec.europa.eu/eurostat/statistics-explained/index.php/Renewable_energy_statistics [accessed: June 2016]

[9] Bundesministerium für Wirtschaft und Energie (BMWi): The Renewable Energy Act, 2014. The key facts on the reform of the EEG. 2014. Available from: https://www.erneuerbare-energien.de/EE/Redaktion/DE/Downloads/Hintergrundinformationen/eeg-2014-infobroschuere-bf.pdf?__blob=publicationFile&v=9 [accessed: June 2016]

[10] Elexon: Load profiles and their use in electricity settlement. 2013. Available from: https://www.elexon.co.uk/wp-content/uploads/2013/11/load_profiles_v2.0_cgi.pdf [accessed: June 2016]

[11] BDEW (Bundesverband der Energie- und Wasserwirtschaft): Standardlastprofile Strom. Available from: https://www.bdew.de/internet.nsf/id/DE_Standartlastprofile [accessed: June 2016]

[12] DVGW/VDE 1. Munich Energy Days: electricity and gas: joint solutions to the pathway to renewable resources in energy supply. Energie, wasser-praxis. Interview. 2013. Available from: [http://www.muenchener-energietage.de/fileadmin/dvgw/angebote/berufsbildung/pdf/1301engelber.pdf [accessed: June 2016]

[13] Ferreira, H. L., Garde, R., Fulli, G., Kling, W., Lopes, J.P.: Characterisation of electrical energy storage technologies. Energy, 2013; 53: 288–297. Available from: http://ac.els-

cdn.com/S0360544213001515/1-s2.0-S0360544213001515-main.pdf?_tid=e34b1ade-2f02-11e6-a511-00000aab0f6c&acdnat=1465560193_bce0f75dddfbb0d2d2572de061785876 [accessed: June 2016]

[14] Burger, B.: Electricity production from solar and wind energy in 2011. Fraunhofer-Institute for Solar Energy Systems ISE. Available from: https://www.ise.fraunhofer.de/de/downloads/pdf-files/aktuelles/vortragsfolien-stromproduktion-aus-solar-und-windenergie-im-jahr-2011.pdf [accessed: June 2016]

[15] Behrendt, A., Drescher-Hartung, S., Ahrens, Th.: Optimization of biogas processes: European experiences. In: E. Dahlquist: Technologies for Converting Biomass to Useful Energy. Sweden: CRC Press; 2013, pp. 293–313. ISBN: 978-0-415-62088-8

[16] ABOWE project reports, 2013. Available from www.abowe.eu

[17] Drescher-Hartung, S., Anne, O., Ahrens, T.: Availability and characteristics of different waste substrates for biogas quality and potential in the Baltic Sea Region. In Book of Proceedings, Bioenergy 2013, Conference and Exhibition, Jyväskylä, Finland. pp. 221–225. Available from: http://www.iat.eu/files/book_of_proceedings_bioenergy2013.pdf [accessed: August 2016]

[18] Bansen, B.: Vergleich der Biogasanlage Hoheneggelsen mit einer Versuchsanlage im Pilotmaßstab zur Entwicklung von prozessspezifischen Kennwerten. Ostfalia University of Applied Sciences Braunschweig/Wolfenbüttel and INPUT engineers. 2010. Sehnde, Germany; not published.

[19] Cavinato, C., Fatone, F., Bolzonella, D., Pavan, P.: Thermophilic anaerobic co-digestion of cattle manure with agro-wastes and energy crops: comparison of pilot and full-scale experiences. Bioresource Technology, 2010; 101: 545–550. ISSN: 0960-8524

[20] Scano, E.A., Asquer, C., Pistis, A., Ortu, L., Demontis, V., Cocco, D.: Biogas from anaerobic digestion of fruit and vegetable wastes; experimental results on pilot-scale and preliminary performance evaluation of a full-scale power plant. Energy Conversion and Management, 2014; 77: 22–30.

[21] Investopedia: Discounted cash flow (DCF). Available from: http://www.investopedia.com/terms/d/dcf.asp [accessed: April 2016]

[22] Fachagentur Nachwachsende Rohstoffe e.V. (FNRa): Faustzahlen Biogas. Available from: http://biogas.fnr.de/daten-und-fakten/ [accessed June 2016]

[23] EU Hygiene Regulation: Regulation (EC) No 1774/2002 of the European Parliament and of the Council of 3 October 2002 laying down health rules concerning animal by-products not intended for human consumption. Available from: http://eur-lex.europa.eu/legal-content/EN/TXT/?uri=celex%3A32002R17742002 [accessed: June 2016]

[24] Federal Ministry of Justice and Consumer Protection: Regulation on the Recycling of biowaste on agricultural, forestry and horticultural used soil (Biowaste Ordinance-

BioAbfV). 1998. Available from: https://www.gesetze-im-internet.de/bioabfv/ [accessed: June 2016]

[25] HEI Hornbachner Energie Innovation. Available from: http://www.biogas-netzeinspeisung.at/technische-planung/ [accessed: May 2016]

[26] Personal information: Armin Weiss, Biokraft, Germany

[27] Al Seadi, T., Rutz, D., Prassl, H., Köttner, M., Finsterwalder, T., Volk, R., Janssen, S.: Biogas handbook. Denmark: University of Southern Denmark Esbjerg [web document]. 2008. Available from: http://www.lemvigbiogas.com/BiogasHandbook.pdf [accessed: June 2016]

[28] European Commission (2015) Closing the loop. an action plan for the European Circular Economy. Available online at: http://ec.europa.eu/environment/circular-economy/index_en.htm

[29] Ghisellini, P., Cialani, C., Ulgiati, S.: A review on circular economy: the expected transition to a balanced interplay of environmental and economic systems. Journal of Cleaner Production. 2015, xxx (online), 1-22. doi: 10.1016/j.jclepro.2015.09.007

[30] Life-Cycle-Based-Sustainability-Standards-Guidelines. Available from https://www.pre-sustainability.com/download/Life-Cycle-Based-Sustainability-Standards-Guidelines.pdf [accessed: June 2016]

[31] Fusi, A., Bacenetti, J., Fiala, M., Azapagic, A.: Life cycle environmental impacts of electricity from biogas produced by anaerobic digestion. Front Bioeng Biotechnol. 2016; 4: 26.

[32] Olsson, L.: Sociotechnical system studies of the reduction of greenhouse gas emissions from energy and transport systems. Dissertations. 2015; 54. doi: 10.3384/diss.diva-116685 (ISSN 0345-7524; 1656, Linköping University Electronic Press)

[33] Holm-Nielsen, J. B., Al Seadi, T., Oleskowicz-Popiel, P.: The future of anaerobic digestion and biogas utilization. Bioresource Technology. 2009; 100: 5478–5484. doi: 10.1016/j.biortech.2008.12.046

[34] Whiting, A., Azapagic, A. Life cycle environmental impacts of generating electricity and heat from biogas produced by anaerobic digestion. Energy. 2014; 70: 181–193. doi: 10.1016/j.energy.2014.03.103 [accessed June 2016]

[35] Poeschl, M., Ward, S., Owende, P.: Environmental impacts of biogas deployment—part I: life cycle inventory for evaluation of production process emissions to air. Journal of Cleaner Production. 2012; 24: 168–183. doi: 10.1016/j.jclepro.2011.10.039

[36] Al Seadi T., Lukehurst C.: Quality management of digestate from biogas plants used as fertiliser. IEA Bioenergy; 2012, May, 1-40. Available from www.iea-biogas.net/files/datenredaktion/download/publi-task37/digestate_quality_web_new.pdf [accessed June 2016] http://www.iea-biogas.net/files/daten-redaktion/download/publi-task37/digestate_quality_web_new.pdf

Chapter 8

Effects of Fertilizers on Biomass, Sugar Content and Ethanol Production of Sweet Sorghum

Tran Dang Xuan, Nguyen Thi Phuong and Tran Dang Khanh

Additional information is available at the end of the chapter

http://dx.doi.org/10.5772/66814

Abstract

Sweet sorghum (*Sorghum bicolor*) is a promising alternative crop for bioethanol production in developing countries. However, to extend the cultivative area of this crop, it needs to develop an appropriate growing protocol for farmers. This chapter describes the examination of different doses of fertilizers combined with manure and micronutrients, in various applied times, on biomass, sugar content and ethanol production of sweet sorghum. It was observed that the application of 90 N + 90 P_2O_5 + 60 K_2O provided maximum stem yield and optimum contents of sugar and ethanol yield, however nontreatment of any among P, P_2O_5 and K_2O caused significant reduction of biomass and ethanol production. Higher fertilization >90 N may provide greater productivity of this crop but it may cause lodging and economic deficit for farmers in developing countries. It was also found that the applied times of fertilization should be at 3–4 to 7–8 leaf stage. In contrast, when the fertilization was as close to the flowering stage caused remarkable reduction of stem yield and ethanol production. The supplementation of $(NH4)_2MO_7O_2.4H_2O$ at 5 kg/ha provided an increase of 10–12 tons/ha of stem yield and a remarkable enrichment of ethanol production. Findings of this study are useful for farmers and agricultural extensionists to promote biomass and ethanol productivity of this crop for bioethanol production. This research also highlights a greater possibility of exploiting sweet sorghum cultivation in infertile and hilly, abandoned areas for ethanol production.

Keywords: sweet sorghum, fertilizer, stem yield, biomass, ethanol yield, growing protocol, biofuel crops

INTECH
open science | open minds

1. Introduction

It has been estimated that there are 700 million light duty vehicles, automobiles, light trucks, SUVs and minivans on roadways around the world. These numbers may increase to 1.3 billion by 2030 and to over 2 billion vehicles by 2050, with most come from developing countries [1]. This growth will affect the stability of ecosystems and global climate as well as global oil reserves [2]. The use of biofuels can contribute to the mitigation of greenhouse gas emissions, provide a clean and sustainable energy source and may promote the agricultural income for rural poor in developing countries. Currently, biofuels are predominantly produced from biomass resources, because they are the renewable resource that could be sustainably developed, resulting in no net release of carbon dioxide and very low sulfur content and economic potential as fossil fuel prices can be increased in the future [2]. The principal sources of biofuels are liquid or gaseous fuels made from plant matter and residues, including agricultural crops, municipal wastes and agricultural and forestry by-products.

Bioethanol has a high octane number (108), broader flammability limits, higher flame speeds and higher heats of vaporization. It helps for a higher compression ratio and shorter burn time, which lead to efficiency advantage over gasoline in an IC engine [3]. Bioethanol can be directly used or it can be blended with gasoline in different ratio from 3 to 25%, depending on countries. The most common that has been widely used is E10 (10% bioethanol blended with 90% gasoline), however in Brazil, bioethanol 100% (E100) or bioethanol 25% (E25) are used [4].

Recently, the production of bioethanol has focused on crops, such as maize, sugarcane and soybean. It includes three major groups: sucrose-containing feedstocks (sugarcane, sugar beet, sweet sorghum and fruits), starchy materials (corn, milo, wheat, rice, potatoes, cassava, sweet potatoes and barley) and lignocellulosic biomass (wood, straw and grasses) [5]. Major countries produced bioethanol include Brazil (sugarcane 100%), the USA (corn 98%, sorghum 2%), China (corn 70%, wheat 30%), EU (wheat 48%, sugar beet 29%) and Canada (corn 70%, wheat 30%) [2]. Although the production of bioethanol can provide environmental and economic benefits, but food security in developing countries has been a key factor to the instability, whereas a large amount of bioethanol is produced from crops in developed countries, that has been the crucial social controversy for bioethanol production.

Sweet sorghum (*Sorghum bicolor*) is a C_4 crop and is considered as one of the most important food and fodder crops in arid and semiarid regions of the world [6]. This crop occupies approximately 45 million ha in Africa and India, accounts for >80% of the global cultivated area. Similar to grain sorghum, sweet sorghum has rapid growth and wide adaptability [7], good drought resistance [8], salinity tolerance [9] and high yield of biomass [10]. Sweet sorghum has played an important role in promoting the development of agricultural production, livestock husbandry [11], biofuel [12, 13], refining sugar and paper making [10]. Carbohydrates in sweet sorghum can be nonstructural as sugars and starch, or structural such as cellulose, hemicellulose and pectic substances [14]. The major sugars in sweet sorghum are the monosaccharides, glucose and fructose; the disaccharides, sucrose and maltose; and the trisaccharide, raffinose. Depending on the kind of sugar in the stalk, there are saccharin and syrup types of sweet sorghum [14]. Sugars in sweet sorghum stalk are mostly sucrose and invert sugars are glucose, fructose, maltose and xylose [15]. It was reported that other sugars such as mannose, galactose and arabinose were not found in juice of the crop; therefore, the production of bioethanol from sweet sorghum is appropriate, as the carbohydrates present are easily converted to bioethanol. The bagasse and grain of the crop are also possible to produce bioethanol, but they are not economically efficacy [16].

Although sweet sorghum has been suggested as a potential crop for bioethanol production in developing countries [6, 17, 18], but it is still acknowledged as a new crop in many countries. To earn permission for cultivation in a large area for bioethanol production, the cultivation protocol for this crop should be critically examined. In our previous report, we have selected a cultivar 4a from 66 sweet sorghum cultivars through 3 years of cultivation in Vietnam and evaluated the effects of sowing times, densities and soils to biomass and bioethanol production [19]. In this chapter, we describe the application of different fertilizers, including the combinations, doses and time of fertilization and examined their effects on the productivity of biomass, sugar content and ethanol production of sweet sorghum.

2. Materials and methods

The cultivar 4a (India origin) was selected from a previous experiment as the most potential of high biomass and sugar content [19]. Seeds of this cultivars were tested and their germination rate was >95%.

The experiments were conducted in three different locations in northern Vietnam including Phu Tho, Hoa Binh and Bac Giang provinces, in continuous cropping seasons from 2010 to 2013. In Vietnam, all major ethanol factories are currently using cassava as a material for production, with exception of Phu Tho factory where both cassava and sugarcane can be produced for ethanol production. Therefore, Phu Tho province is the first priority to be chosen as a place of sweet sorghum production. Bac Giang and Hoa Binh are selected because their close distance to the ethanol factory in Phu Tho province (less than 100 km). In addition, the three provinces have a large area of hilly and abandoned areas which can be exploited for cultivation of sweet sorghum.

2.1. Applied doses

Different ratios of N, P_2O_5 and K_2O from 30, 60, 90 kg/ha with 2 tons/ha of manure were combined. The commercial fertilizers were purchased from Ha Bac nitrogenous fertilizer & chemicals company limited (Bac Giang city, Vietnam). The details of fertilizer ratios are presented in **Table 1**. Each treatment was conducted in thrice in a plot of 30 m^2 and repeated in three provinces Hoa Binh, Phu Tho and Bac Giang provinces, Vietnam. Effects of different-treated doses on stem yield, sugar content and ethanol production of the cultivar 4a were measured.

Doses	Fertilizer combinations
1	0 N + 60 P_2O_5 + 60 K_2O + 2 tons/ha manure
2	30 N + 60 P_2O_5 + 60 K_2O + 2 tons/ha manure
3	60 N + 60 P_2O_5 + 60 K_2O + 2 tons/ha manure
4	90 N + 60 P_2O_5 + 60 K_2O + 2 tons/ha manure
5	90 N + 0 P_2O_5 + 60 K_2O + 2 tons/ha manure
6	90 N + 30 P_2O_5 + 60 K_2O + 2 tons/ha manure
7	90 N + 90 P_2O_5 + 60 K_2O + 2 tons/ha manure
8	90 N + 60 P_2O_5 + 0 K_2O + 2 tons/ha manure
9	90 N + 60 P_2O_5 + 30 K_2O + 2 tons/ha manure
10	90 N + 60 P_2O_5 +90 K_2O + 2 tons/ha manure

Table 1. Applied doses of fertilizers.

2.2. Chemical composition of soils in locations selected for the experiment

The criteria for selecting locations in each Hoa Binh, Phu Tho and Bac Giang provinces are where there are abandoned and difficult for cultivating food crops because of their low nutrient soils, to examine the possibility of exploiting the cultivation of sweet sorghum for ethanol production. There were nine villages were selected (total 27 villages) and soils in the selected area in each villages were taken and analyzed for pH, contents of percentages of N, P, K, OC (organic carbon), CEC (cation exchanged capacity) (cmol$^{(+)}$/kg), Zn (mg/kg), Fe (mg/kg) and Al (mg/kg) (**Table 2**). The analysis of these criteria was conducted by conventional methods.

Locations	pH	N (%)	P (%)	K (%)	OC (%)	CEC (cmol$^{(+)}$/kg)	Zn (mg/kg)	Fe (mg/kg)	Al (mg/kg)
Hoa Binh	6.4	0.234	0.059	1.392	0.286	8.06	265.48	58116.93	85879.64
Phu Tho	5.1	0.219	0.062	0.561	0.174	5.48	89.84	93138.45	75458.45
Bac Giang	5.6	0.027	0.224	0.135	0.962	5.46	64.01	66112.50	78512.50

Table 2. Chemical compositions of soils for sweet sorghum cultivation in different locations.

2.3. Applied time of fertilization

This experiment was conducted in a plot of 30 m^2 in thrice, in the three locations in Hoa Binh, Phu Tho and Bac Giang provinces in summer-autumn cropping season 2010 and repeated in 2011. The dose of fertilizers per ha included 90 N +60 P_2O_5 +60 K_2O + 2 tons manure. Applied times of fertilization are presented in **Table 3**.

2.4. Improvement of biomass and ethanol production by micronutrient fertilizers

This experiment was conducted to examine the role of micronutrient fertilizers by an additional application of $(NH4)_2MO_7O_2.4H_2O$ including 54% Mo (5kg/ha) on biomass and ethanol production of the sweet sorghum 4a. The based fertilization (BF) was 90 N + 60 P_2O_5 + 60 K_2O + 2 tons/ha manure. Four trials were conducted: (i) BF, (ii) BF + $(NH_4)_2MO_7O_2.4H_2O$, (iii) BF + 60 kg/ha P_2O_5 and (iv) BF + 60 kg K_2O. This experiment was also conducted in the designated areas in the three provinces in the cropping season of summer-autumn 2010 and repeated in 2011.

Trials	Applied time
1	1st fertilization: plants at 3–4 leaf period: 1/2 total N + 1/2 total % K_2O 2nd fertilization: plants at 7–8 leaf period: 1/2 total N + 1/2 total % K_2O
2	1st fertilization: plants at 3–4 leaf period: 1/3 total N + 1/3 total % K_2O 2nd fertilization: plants at 7–8 leaf period: 1/3 total N + 1/3 total % K_2O, trimming seedlings at 20 cm space per plant 3rd fertilization: after the 2nd fertilization 20 d, provided the remained fertilizers
3	1st fertilization: plants at 3–4 leaf period: 1/3 total N 2nd fertilization: plants at 7–8 leaf period: 1/4 total N + 1/2 total K_2O 3rd fertilization: before flowering period: 1/2 total N + 1/2 total % K_2O

Table 3. Times of fertilization application.

2.5. Determination of biomass, sugar content (Brix %) and ethanol production

Sweet sorghum at flowering stages in different trials was harvested in random. The biomass weight, sugar content (Brix %) and ethanol production were determined by a method described in Ref. [19].

3. Statistical analysis

All trials of the study were conducted in a completely randomized design with three replications. Means and differences between the treatments were determined by using a two-way analysis of variance (ANOVA) with values of least significant difference (LSD) at $P \leq 5\%$ level from IRRISTAT program version 5.0.

4. Results

4.1. Effects of different applied doses of fertilizers on stem yield, sugar content and ethanol production of sweet sorghum

There were 10 different applied doses were used in the designated locations in Hoa Binh, Phu Tho and Bac Giang provinces. The effects of these different fertilization doses on stem yield, sugar content and ethanol production of the cultivar 4a are presented in **Table 4**.

It is observed that the stem yield of sweet sorghum varied between 44.7 and 77.3 tons/ha, depending on applied doses of fertilizers and locations (**Table 1**). The most effective dose was 90 N + 90 P_2O5 + 60 K_2O, followed by 90 N + 60 P_2O5 + 60 K_2O. It was found that nonapplication of any N, K_2O and P_2O_5 caused severe marked reduction of stem yield, up to 20 tons/ha and the amount of N played a more significant role on the yield of this crop, than either K_2O or $P2O_5$ application (**Table 1**). Accordingly, the ethanol yield was the highest in the doses of 90 N + 90 P_2O5 + 60 K_2O and 90 N + 60 P_2O5 + 60 K_2O. Nontreatment of N, K_2O and P_2O_5 also resulted in a critical decrease of ethanol yield and the lowest ethanol yield was the nontreatment of nitrogen fertilizer (2963.6 l/ha). Differently, the applied doses did not strongly influence on sugar contents of the cultivar 4a. The dose 90 N + 90 P_2O5 + 60 K_2O also recorded the highest sugar quantity, but the treatment of 90 N + 60 P_2O5 + 60 K_2O showed a significantly lower sugar content (12.0–13.3%). In addition, the dose 90 N + 60 P_2O5 + 90 K_2O did not provide the highest stem yield (53.2–59.5 tons/ha), but it has a remarkable high sugar content and ethanol yield than other doses. Regarding to the location, in general, Bac Giang province has a significantly higher sugar content and ethanol yield than Hoa Binh and Phu Tho provinces (**Table 4**).

Doses	Evaluated categories								
	Stem yield (tons/ha)			Sugar content (Brix %)			Ethanol yield (l/ha)		
	Hoa Binh	Phu Tho	Bac Giang	Hoa Binh	Phu Tho	Bac Giang	Hoa Binh	Phu Tho	Bac Giang
90 N + 60 P_2O_5 + 60 K_2O	64.6	66.0	66.9	12.0	12.0	13.3	4283.0	4375.8	4916.0
0 N + 60 P_2O_5 + 60 K_2O	44.7	45.0	47.7	12.0	12.0	12.7	2963.6	2983.5	3347.0
30 N + 60 P_2O_5 + 60 K_2O	53.2	53.2	55.6	13.0	13.0	13.0	3821.1	3842.6	3993.5
60 N + 60 P_2O_5 + 60 K_2O	58.0	58.0	61.0	12.0	12.0	13.3	3845.4	3845.4	4482.4
90 N + 0 P_2O_5 + 60 K_2O	46.0	46.0	54.3	12.0	12.0	10.3	3049.8	3049.8	3090.1
90 N + 30 P_2O_5 + 60 K_2O	48.7	48.5	66.4	12.0	13.0	11.7	3228.8	3483.5	4292.3
90 N + 90 P_2O_5 + 60 K_2O	68.6	69.0	77.3	14.0	14.0	13.7	5306.2	5337.2	5851.0
90 N + 60 P_2O_5 + 0 K_2O	49.2	53.0	50.5	12.0	12.0	12.0	3262.0	3513.9	3315.0
90 N + 60 P_2O_5 + 30 K_2O	52.6	52.0	54.4	13.3	13.3	10.0	3778.0	3734.9	3005.6
90 N + 60 P_2O_5 + 90 K_2O	53.2	53.5	59.5	14.0	14.0	13.3	4408.9	4138.2	4372.2
LSD (5%)	8.76			0.15			1.1		
CV (%)	9.7			3.3			5.1		

Table 4. Effects of applied doses of fertilizers on sugar content, sugar yield and ethanol production of the sweet sorghum variety 4a.

http://dx.doi.org/10.5772/66814

4.2. Effects of applied times of fertilizers on stem yield, sugar content and ethanol production of sweet sorghum

There were three trials of fertilizer application in different times and fertilizer doses (**Table 4**), applied in the three locations in Hoa Binh, Phu Tho and Bac Giang provinces (**Table 5**). It indicated that times of fertilizer application at 3–4 to 7–8 leaf stages provided the highest stem yield and ethanol yield in every location and their values were markedly different as compared among the 1st, 2nd and 3rd applied times (**Table 5**). However, when the amount of fertilizers were divided into three times of application, of them, the third fertilization was either after 20 d of the second application (trial 2) and before flowering, the yields of stems and ethanol production were significantly reduced, as compared with other treatments (**Table 5**). The stem yield was markedly decreased >20 tons/ha, whereas the amount of ethanol production was significantly reduced > tons/ha. For location, Phu Tho province obtained higher yields of stems and ethanol production than Hoa Binh and Bac Giang provinces. For the sugar content (Brix %), different times of fertilizer application did not influence on the sugar content of sweet sorghum, as no significant difference among treatments were observed (**Table 5**).

Locations	Trials	Stem yield (tons/ha)	Sugar content (Brix %)	Ethanol yield (l/ha)
Hoa Binh	1	68.6	14.0	5306.2
	2	62.7	14.0	4849.9
	3	47.5	14.0	3674.1
Phu Tho	1	72.5	14.0	5607.9
	2	65.5	14.0	5066.4
	3	49.5	14.0	3828.8
Bac Giang	1	69.5	14.0	5375.8
	2	61.5	14.0	4757.0
	3	46.6	13.0	3347.0
LSD (5%)		0.4	1.4	60.0
CV (%)		0.7	5.8	0.7

Note: Trials 1–3 are different applied times as indicated in **Table 3**.

Table 5. Effects of different times of fertilizer application on stem yield, sugar content and ethanol production of sweet sorghum.

4.3. Effects of micronutrient fertilizers on stem yield, sugar content and ethanol production of sweet sorghum

In this experiment, the control was the based fertilization (BF) that included 90 N + 60 P_2O_5 + 60 K_2O + 2 tons/ha manure. Other treatments consisted of BF + $(NH4)_2MO_7O_2.4H_2O$ (Mo), BF + 60 K_2O and BF + 60 P_2O_5. The effects of these treatments on stem yield, sugar content and ethanol productivity are presented in **Table 6**.

Treatments	Stem yield (tons/ha)			Sugar content (Brix %)			Ethanol yield (l/ha)		
	Hoa Binh	Phu Tho	Bac Giang	Hoa Binh	Phu Tho	Bac Giang	Hoa Binh	Phu Tho	Bac Giang
Based fertilization (BF)	64.4	66.0	62.4	14.8	15.0	14.8	5266.0	5469.8	5012.5
BF + Mo	74.3	79.0	75.9	14.9	15.0	14.9	6116.6	6547.1	6248.3
BF + P_2O_5	70.4	75.4	71.5	15.5	15.0	15.5	6028.9	6248.8	6123.1
BF + K_2O	68.5	69.5	70.5	15.5	15.0	14.9	5866.2	5759.8	5073.7
LSD (5%)	0.9			0.8			138.7		
CV (%)	0.7			3.1			11.3		

Table 6. Effects of micronutrient fertilizers on stem yield, sugar content and ethanol production of sweet sorghum.

It was found that by the application of based fertilization, the stem yield of sweet sorghum was 62.4–66.0 tons/ha. However, the supplemented application of Mo significantly increased the stem yield 10–12 tons/ha and it was higher than either the supplementation of P_2O_5 or K_2O (**Table 6**). For the sugar content, there was no remarkable difference among treatments (14.8–15.5 Brix %). Following the promoted amount of stem yield, the ethanol productivity was the maximum with Mo application, followed by P_2O_5 and K_2O supplementation. They were all significantly higher than the application of unique-based fertilizers (**Table 6**).

5. Discussion

Although sweet sorghum is well adaptable to many soils and climates and has a strong resistance to pests and diseases, an effective fertilization protocol is crucial to farmers and agricultural extensionists to promote the biomass and ethanol productivity of this crop. Trials of this study were replicated in the three locations in three different provinces, where are the most suitable for ethanol production in the North of Vietnam [19]. Among different doses of fertilizers, this study found that the application of 90 N + 90 P_2O_5 + 60 K_2O was the best for maximum biomass and ethanol productivity of the crop. Their values were significantly higher than those of other treatments (**Table 4**). Nontreatment of N, P_2O_5 and K_2O was examined and the lack of fertilization of any among these nutrients caused severe markedly reduction of stem yield, sugar content and ethanol production; however, N was the most important for biomass and ethanol productivity (**Table 4**).

It has been described that sweet sorghum can become a profitable crop for farmers in developing countries, however its agronomic practices should be optimized to improve the productivity. Among them, nitrogen is the most important nutrient element as this crop responds well to N fertilization [20]. Turgut et al. [21] investigated the optimum N input for maximizing sweet sorghum yield in a clay loam soil. Reddy et al. [22] also studied the effects of N input on sweet sorghum growth. Uchino et al. [20] reported that the amount of N from 90 to 120 kg/ha was the optimum for ethanol productivity, but the amount of N >150 kg/ha may cause lodging. In this study, the application of N >90 kg/ha was not studied as it may cause economic troublesome for farmers in developing countries such as Vietnam.

This research also verified the application of K_2O and P_2O_5 in combination with nitrogen and manure (**Table 4**) to reduce the dependency on nitrogen application. The use of manure, K_2O and P_2O_5 with reduced quantity of nitrogen fertilizers are efficacy for soil health and maintain the biomass and ethanol productivity of sweet sorghum. The time of application was also examined to optimize the time of application and doses of fertilizers. The best time for fertilization was found to be on 3–4 to 7–8 leaf stage (**Table 5**). Findings of this study indicate that fertilization should not be conducted after 7–8 leaf stage, as the application of fertilizers at 20 d after the second application (7–8 leaf stage) to flowering stage caused significant reduction of stem yield and ethanol production, although the sugar content (Brix %) was not influenced (**Table 5**). This also helps to save the labor costs as the too much fertilization may attribute to heavy fieldwork and economic deficit in sweet sorghum cultivation.

In this study, for micronutrient fertilization, 5 kg/ha of $(NH4)_2MO_7O_2.4H_2O$ (Mo) was used as supplement to the base fertilizer application of 90 N + 60 P_2O_5 + 60 K_2O + 2 tons/ha manure. The supplements of 60 kg/ha of P_2O_5 and K_2O were also conducted to compare with that of the Mo supplement. Interestingly, the use of Mo was the most effective in enhancing the biomass and ethanol production and it was higher than that of either P_2O_5 or K_2O supplements (**Table 6**). The use of Mo was found to be more effective than the adding of P_2O_5 and K_2O. In addition, the application of Mo at 5 kg/ha provided less heavy fieldwork than the supplementation of 60 kg of either P_2O_5 or K_2O. The application of micronutrients such as $ZnSO_4$ and sodium borate was effective on the stalk yield of the sweet sorghum ICSV 93046 variety, which is widely cultivated in India and Philippines [23]. The supplements of micronutrients of Zn and Fe significantly promote the stalk yield of sweet sorghum [24]. It is therefore proposed that the application of micronutrient fertilizers is effective to improve the productivity and profitability of sweet sorghum than the supplement of unique either P_2O_5 or K_2O.

The influences of fertilization including times of application and applied doses on energy crops have been extensively studied. It has been reported that the application of N fertilizer significantly improves the production of bioenergy crops, such as switchgrass (*Panicum virgatum* L.) and corn (*Zea mays* L.) [25]. Individual applications of N, P, or K fertilizers markedly improved cassava (*Manihot esculenta* Crantz) yield. The best yield performance in this research to increase the yield of cassava (62.8 t/ha) was the use of 400, 200 and 400 kg/ha of N, P, K fertilizers, respectively, although the use of P fertilizer incorporation with high levels of N and K fertilizers was not effective to improve cassava yields [26]. In addition, Olugbemi and Ababyomi [27] examined different treatment of nitrogen fertilizer on the growth and ethanol yield of four sweet sorghum varieties and found that the application of 120 kg N/ha was the best for ethanol production. There are also several research studies on optimizing the nitrogen requirements of sweet sorghum for various environments and soil types [20–21]. The applied doses on sugarcane were important on the yield and ethanol production in Brazil [28]. Appropriate doses of fertilizers are useful to increase the yield and ethanol production of biofuel crops, but the overuse of fertilizers may cause critical environmental pollution [29]; therefore, the examination of applied doses and times of application as conducted in this research is indispensable for the cultivation of biofuel crops.

6. Conclusions

Sweet sorghum is a promising alternative crop for bioethanol production in developing countries to reduce the social criticisms on the use of food crops for biofuel production. Findings of this study, together with a previous work, examined the sowing times, densities and soil chemicals [19] to determine the cultivar 4a which was continuously examined in this research for different doses of fertilizers and times of application. Achievements of the two research studies are useful to establish a protocol of sweet sorghum cultivation for farmers and agricultural extensionists in developing countries, as the fertilization of different doses and times was examined in different locations with low nutrient contents that were difficult to cultivate other crops. This study also highlighted the higher possibility of using sweet sorghum in infertile soil in hilly areas than other bioenergy crops. The combination of manure and micronutrients with a reduced amount of inorganic fertilizers is crucial to establish a sustainable agriculture in sweet sorghum production.

Acknowledgements

The author express sincere thanks to Ministry of Trade, Vietnam, for funding to this research. Thanks are also due to Phung Thi Tuyen, La Hoang Anh, Truong Ngoc Minh and Do Tan Khang for their assistance to the manuscript's preparation.

Author details

Tran Dang Xuan[1,*], Nguyen Thi Phuong[2] and Tran Dang Khanh[3]

*Address all correspondence to: tdxuan@hiroshima-u.ac.jp

1 Graduate School for International Development and Cooperation (IDEC), Hiroshima University, Higashi Hiroshima, Japan

2 Vietnam Academy of Agricultural Science (VASS), Hanoi, Vietnam

3 Agricultural Genetics Institute (AGI), Hanoi, Vietnam

References

[1] Hansen G. Driving technology in the motor vehicle industry. In: Proceedings of the IPCC expert meeting on the industrial technology development, transfer and diffusion. Tokyo, September 21–23. 2004.

[2] Balat M, Balat H. Recent trends in global production and utilization of bio-ethanol fuel. Applied Energy 2009; 86: 2273–2282. DOI: 10.1016/j.apenergy.2009.03.015

[3] Balat M. Global bio-fuel processing and production trends. Energy Exploration & Exploitation. 2007; 25 (3): 195–218. DOI: 10.100/15435070802107322

[4] Oliveria MED, Vaughan BE, Rykiel Jr. EJ. Ethanol as fuel: energy, carbon dioxide balance and ecological footprint. Bioscience 2005; 55(7): 593–602. DOI: 10.1525/bio.2013.63.4.6

[5] Balat M. Global trends on the processing of bio-fuels. International Journal of Green Energy 2008; 5(3): 212–238. DOI: 10.1080/15435070802107322

[6] Basavaraj G, Rao PP, Basu K, Reddy CR, Kumar AA, Rao PS, Reddy BVS. Assessing viability of bio-ethanol production from sweet sorghum in India. Energy Policy 2013; 56: 501–508. DOI: 10.1016/j.enpol.2013.01.012

[7] Reddy BVS, Ramesh S, Reddy PS, Ramaiah B, Salimath PM, Kachapur R. 2005. Sweet sorghum - a potential alternate raw material for bio-ethanol and bio-energy. International Sorghum Millets Newsletter 2005; 46, 79–86.

[8] Tesso TT, Claflin LE, Tuinstra MR. Analysis of stalk rot resistance and genetic diversity among drought tolerant sorghum genotypes. Crop Science 2005; 45(2): 645–652. DOI: 10.2135/cropsci2005.0645

[9] Almodares A, Hadi MR, Ahmadpour H. Sorghum stem yield a soluble carbohydrate under phonological stages and salinity levels. African Journal of Biotechnology 2008; 7(2), 4051–4055.

[10] Almodares A, Hadi M.R. Production of bioethanol from sweet sorghum: a review. African Journal of Agricultural Research 2009; 4(9), 772–780.

[11] Fazaeli H, Golmohhammdi HA, Almodares A, Mosharraf S, Shaei A. Comparing the performance of sorghum silage with maize silage in feedlot calves. Pakistan Journal of Biological Sciences 2006; 9, 2450–2455. DOI: 10.3923/pjbs.2006.2450.2455

[12] Nahvi I, Almodares A, Motocalemi M. 1994. Comparing ethanol production from sweet sorghum juice and sugar beet molasses. In: Proceeding of the 1st Iranian Congress on Chemical Engineering (Eds. I Nahvi, A Almodares and M. Motocalemi). Tehran, Iran.

[13] Nahvi I, Almodares A, Rasuli A. 1994. The effect of environmental factors on ethanol production from sweet sorghum juice. In: Proceeding of the 1st Iranian Congress on Chemical Engineering (Eds. I Nahvi, A Almodares and M. Motocalemi). Tehran, Iran.

[14] Anglani C. Sorghum carbohydrates – a review. Plant Foods for Human Nutrition 1998; 52(1): 77–83. [PMID: 9839837]

[15] Almodares A, Taheri R, Adeli S. Stalk yield and carbohydrate composition of sweet sorghum [(*Sorghum bicolor* (L.) Moench) cultivars and lines and different growth stages. Journal of Malaysian Applied Biology Journal. 2008, 37(1):31–36.

[16] Jacques K, Lyons TP, Kelsal DR. The Alcohol Textbook. Nottingham University Press: United Kingdom 1999. 3rd ed. p. 388.

[17] Liu SY, Lin CY. Development and perspective of promising energy plants for bioethanol production in Taiwan. Renewable Energy. 2009; 34: 1902–1907. DOI: 10.1016/j.renene.2008.12.018

[18] De Vries SC, Vandeven GWJ, Ittersum MKV, Giller KE. The production-ecological sustainability of cassava, sugarcane and sweet sorghum cultivation for bioethanol in Mozambique. GCB Bioenergy 2012; 4 (1): 20–35. DOI: 10.1111/j.1757-1707.2011.01103.x

[19] Xuan TD, Phuong NT, Khang DT, Khanh TD. Influence of sowing times, densities and soils to biomass and ethanol yield of sweet sorghum. Sustainability 2015; 7(9): 11657–11678. DOI: 10.3390/su70911657

[20] Uchino H, Watanabe T, Ramu K, Sahrawat KL, Marimuthu S, Wani SP, Ito O. Effects of nitrogen application on sweet sorghum (*Sorghum bicolor* (L.) Moench) in the semi-arid tropical zone of India. The Japan Agricultural Research Quarterly 2013; 47(1): 65–73. DOI: 10.6090/jarq.47.65

[21] Turgut I, Bilgili U, Duman A, Acikgoz E. Production of sweet sorghum (*Sorghum bicolo*r L. Moench) increases with increased plant densities and nitrogen fertilizer levels. Acta Agriculturae Scandinavica, Section B: Soil & Plant Science 2005; 55 (3): 236–240. DOI: http://dx.doi.org/10.1080/09064710510029051

[22] Reddy PS, Reddy BVS, Kumar AA, Rao S. Standardization of nitrogen fertilizer rate for sugar yield optimization in sweet sorghum. Journal SAT Agricultural Research 2008; 6, 1–4.

[23] Rao S, Kumar CG, Malapaka J, Kamal A, Reddy BVS. Effect of micronutrient treatments in main and ratoon crops of sweet sorghum cultivar ICSV 93046 under tropical conditions. Sugar Tech 2012; 14: 370. DOI: 10.1007/s12355-012-0172-y

[24] Ismail S, Arbad BK, Syed JJ. Impact of integrated nutrient management practices on yield, juice quality and nutrient uptake in sweet sorghum [*Sorghum biocolor* (L.) Moench] grown on vertisol. Asian Journal of BioScience 2009; 4(1): 100–102.

[25] Stewart CE, Follett RF, Pruessner EG, Varvel GE, Vogel KP, Mitchell RB. N fertilizer and harvest impacts on bioenergy crop contribution to SOC. GCB Bioenergy 2016; 8: 1201–1211. DOI: 10.1111/gcbb.12326

[26] Wilson H, Ovid A. Influence of fertilizers on cassava production under rainfed conditions. Journal of Plant Nutrition 1994; 17 (7): 1127–1135. DOI: 10.1080/01904169409364793.

[27] Olugbemi O, Ababyomi YA. Effects of nitrogen application on growth and ethanol yield of sweet sorghum (*Sorghum bicolor* L. Moench) varieties. Advances in Agriculture 2016; Article ID 8329754, 7 pages. DOI: 10.1155/2016/8329754

[28] Martinelli LA, Filoso S. Expansion of sugarcane ethanol production in Brazil: environmental and social challenges. Ecological Applications 2008; 18 (4) 885–898. [PMID: 18536250]

[29] Elobeid A, Carriquiry M, Dumortier J, Rosas F, Mulik K, Fabiosa JF, Hayes DJ, Babcock BA. Biofuel expansion, fertilizer use and GHG emission: unintended consequences of mitigation policies. Economics Research International 2013; Article ID 708604, 12 pages. DOI: 10.1155/2013/708604

Chapter 9

Biomass as Raw Material for Production of High-Value Products

Sibel Irmak

Additional information is available at the end of the chapter

http://dx.doi.org/10.5772/65507

Abstract

Industrial production of a wide range of value-added products heavily relies on fossil resources. Lignocellulosic biomass materials are receiving increased attention as a renewable, economical, and abundant alternative to fossil resources for the production of various value-added products. Biomass feedstocks utilized for these productions include energy crops, agricultural biomass residues, forest biomass, and food-based biomass wastes. Various conversion technologies are used for production value-added products from biomass. Efficiencies of conversion technologies highly depend on the types of biomass used as raw materials that differ in contents and compositions of cellulose, hemicellulose, and lignin structures in biomass. In some conversion technologies, such as chemical, biochemical, and hydrothermal conversion techniques, biomass materials must be first broken down into smaller molecular weight components (e.g., oligosaccharides and monosaccharides) in order to be efficiently converted into target products. In this matter, pretreatment and hydrolysis play critical roles on the yield of the product(s). The chapter describes lignocellulosic materials that are used for production of top value-added products and conversion technologies to produce products in high yields. Future developments in the conversion of lignocellulosic biomass into value-added products are directly correlated to improvements of conversion technologies and selection the right types of biomass in the process.

Keywords: biomass types, value-added products, biofuels, bioproducts, conversion methods

1. Introduction

Being a nonedible portion of the plant, lignocellulosic biomass materials are attractively growing the attention as sustainable and renewable energy sources. Biomass materials can be used for producing a wide range of value-added products, including biofuels (ethanol, hydrogen, etc.), bioproducts products (sugar and sugar alcohols, etc.), and industrially important chemicals (e.g., solvents) [1]. Conversion can be performed using a variety of methods, including chemical, biochemical, and thermochemical processes. Each method offers several advantages or disadvantages for high yielding of a certain product.

Biomass can be derived from forestry wastes such as residues of the trees and shrubs, energy crops like sorghum, miscanthus, kenaf, switchgrass, corn, sugarcane, and any agricultural residues such as corn stovers, wheat straw, etc. The diversity in the chemical composition of biomass (cellulose, hemicellulose, and lignin constituents) can affect the conversion technologies employed for production of high-value products.

2. Top value-added products from biomass

2.1. Biofuels

Ethanol is a renewable fuel made from corn, sugarcane, sweet sorghum, and other variety of carbohydrate containing sources. In the United States, the ethanol fuel industry is largely based on corn. The production consists of the fermentation of sugars to produce a dilute alcohol solution, which is followed by distillation and dehydration to produce fuel-grade ethanol. Ethanol is blended with gasoline in various amounts for use in vehicles (E10, E15, and E85). It was reported that the use of 10% ethanol (E10) blends decreases greenhouse gas emissions by 12–19% compared to petroleum-based fuels [2].

Butanol is also a biomass-based renewable fuel that can be produced by alcoholic fermentation of biomass feedstocks [3]. Butanol is commonly produced using fossil fuels, but it can also be produced from biomass. Butanol can blend with gasoline better than ethanol since it has longer hydrocarbon chain thus, has lower polarity. Butanol has an energy density similar to that of gasoline [4]. However, because of limitations and difficulties in its production it is not considered a viable biofuels like ethanol. Biobutanol is produced by ABE (acetone-butanol-ethanol) fermentation process using various substrates. ABE fermentation suffers from limitations and difficulties including low butanol titer, low butanol yield, high substrate cost (grain and molasses), end-product inhibition, and high product recovery cost by distillation [5].

Methanol is another alcoholic fuel that has a high octane rating, easily distributable, and has low volatility. Methanol has chemical and physical fuel properties similar to ethanol. Methanol is mostly produced by the catalytic conversion of syngas from fossil sources but it can be also produced from lignocellulosic biomass materials [6]. Methanol can be used directly as fuel or fuel in fuel cells or as a feedstock for the gasoline additive methyl tertiary butyl ether.

Biodiesel is another most widely used liquid biofuel; however, its production does not rely on lignocellulosic fraction of biomass. Biodiesel is produced from vegetable oil or animal fat with an alcohol and a catalyst through transesterification reaction.

Producer gas is the mixture of gases produced by the gasification of biomass materials. It is composed of CO, H_2, CO_2, and low molecular weight hydrocarbons such as CH_4. Producer gas

http://dx.doi.org/10.5772/65507

can be utilized as a fuel gas for heat or for electricity generation [7]. *Syngas* (synthesis gas) is a mixture of CO and H_2. Syngas has the ability to replace natural gas as a more thermally efficient liquid fuel. Electricity can be generated from the power provided by the combustion of syngas. It can also be used as a fuel source or as an intermediate for the production of other chemicals [8, 9].

Hydrogen is considered as a promising gas fuel. Hydrogen has the highest specific energy content of all conventional fuels. Its main use as fuel is in fuel cells. Fuel cells are being considered an attractive option for power generation, because of their high efficiency with no pollution [10]. Its only waste or byproduct is pure water, while hydrocarbon fuels produce massive amounts of carbon dioxide, a greenhouse gas. The development of less expensive and convenient methods for hydrogen production is challenging issue that limits its use. Among other various conversion methods, aqueous-phase reforming (APR) technique is a promising method for high-yielding hydrogen gas production [11, 12].

2.2. Bioproducts and industrially important chemicals

Nonedible lignocellulosic biomass materials are attracting increasing attention as renewable, economical, and abundant resources to reduce dependency on petroleum resources and minimize energy and material feedstock costs. In addition to energy and fuels, biomass can be used to create valuable carbon-based chemicals and materials, known as bioproducts. These products are sugars and sugar alcohols, glycerine, furfurals, cellulose fiber and derivatives, carbonaceous materials, resins, bioplastics, etc.

Activated carbons have been used for many applications including wastewater treatment (as adsorbent and filter) [13], catalyst [14], catalyst support [15], storage material [16], etc., and can be prepared from lignocellulosic biomass. *Mesoporous carbons* play important roles as catalysts supports as well as adsorbents, membranes, supercapacitors, chemical sensors, etc. [17, 18]. Mesoporous carbons can be prepared from lignin and enhance its mesoporosity by physical and chemical activation methods.

Furfural is a natural precursor to furan-based chemicals. It has been considered an important building block for the production of nonpetroleum-derived chemicals, a new generation of bioplastics, and potential biofuels or fuel additives. Furfural and its derivatives have been used to make jet and diesel fuel range alkanes, to serve as gasoline blend stock, and to develop a new generation of biofuels and bioplastics [19, 20].

Sugar alcohols are important products in the food industry. For instance, xylitol is a pentose sugar alcohol used as a sugar substitute in the food industry because of its low caloric and anticarcinogenic properties [21]. In addition, xylitol is a building block for a variety of commodity chemicals.

Addition to sugar alcohols, following C5 and C6 sugar-derived *platform chemicals* can be transformed into new families of useful molecules: 1,4-diacids (succinic acid, fumaric acid, and malic acid), 2,5-furan dicarboxylic acid, 3-hydroxy propionic acid, aspartic acid, glucaric acid, glutamic acid, itaconic acid, levulinic acid, 3-hydroxybutyrolactone, and glycerol [22].

Biomass feedstocks are converted into plenty of *other intermediate platforms, building blocks, secondary chemicals, and products* that are used in industry, transportation, textiles, food supply, environment, housing, etc. For instance, 1,3-butadiene is the building block for the production of rubbers, which are used in the production of tires for light vehicles. Ethyl lactate is a biodegradable solvent produced by the esterification of ethanol and lactic acid, which are produced from biomass. It is used in industrial applications to replace volatile organic petroleum-derived compounds. Lactic acid, which is mostly produced by microbial fermentation of carbohydrates, is used in many applications, including in food, pharmaceuticals, polymers, etc. Succinic acid, which is a dicarboxylic acid, is precursor for the synthesis of high-value products such as commodity chemicals, polymers, surfactants, and solvents.

Lignocellulosic biomass-based *polymers* have been considered as good candidates for sustainable development as well as ecofriendly environment. A wide range of polymers can be prepared from biomass-derivatives by different reaction routes [22]. For instance, C5 and C6 sugars and their derivatives can be either incorporated in polymer backbone or be used as pendant groups to prepare glycopolymers that mimics structural and functional responsibilities of glycoproteins [23].

3. Parameters affecting the product yield

3.1. Pretreatments

Despite their potential, the complex and rigid structures of biomass materials limit their use in many applications. Biomass materials must first be broken down into components with smaller molecular weights (e.g., oligosaccharides and monosaccharides) in order to be efficiently converted into a range of products. The goal of pretreatment is to make the cellulose accessible to breakdown process (hydrolysis) for conversion to fuels or value-added products. Various pretreatment techniques change the physical and chemical structure of lignocellulosic biomass and improve hydrolysis rates. Pretreatment makes biomass accessible to deconstruction by altering structural features of biomass such as removing lignin and reducing cellulose crystallinity, thereby increasing porosity. Successful production of biofuels and other bioproducts from lignocellulosic biomass depends on the pretreatment and deconstruction methods applied as well as the physical and chemical properties of the biomass. An efficient pretreatment method followed by solubilization in aqueous media without using toxic and hazardous chemicals is necessary to obtain reduced molecular weight of carbohydrates from biomass to produce various biofuels and bioproducts other value-added products.

In *ammonia fiber explosion* (AFEX) pretreatment method, liquid ammonia is mixed with biomass under moderate pressure (0.7–2.8 MPa) and temperature (70–200°C) before rapidly releasing the pressure. The temperature of the reaction, residence time, and ammonia concentration are main parameters that affect the dissolution [24]. The sudden release of pressure in the system explosively ruptures the biomass fibers and eventually breaks down the cellulose, hemicellulose, and lignin polymers into smaller components, and increases the size and the number of micropores in the cell wall.

http://dx.doi.org/10.5772/65507

Steam explosion is one of the most common and widely employed physicochemical pretreatments for lignocellulosic biomass. Biomass is treated with high-pressure (0.69–4.83 MPa) saturated steam at high temperature (160–260°C) for a short time (several seconds to a few minutes) and then the pressure is rapidly released. This treatment destroys the fibril structure and makes the materials undergo an explosive decompression [25]. After the treatment, hemicellulose is degraded, part of lignin is solubilized, and cellulose binding is reduced. Partial degradation of hemicellulose and formation of toxic compounds are the main drawbacks of steam explosion process. Toxic and/or inhibitory compounds can affect and reduce the performance of fermentations steps [26]. Substances that may act as inhibitors of microorganisms include phenolic compounds and other aromatics, aliphatic acids, furan aldehydes, inorganic ions, and bioalcohols or other fermentation products [27].

Alkali pretreatment involves the use of bases, such as sodium, potassium, calcium, and ammonium hydroxides, for pretreatment of biomass. Sodium hydroxide is the most common base used in this application. Alkali pretreatment can be carried out at ambient conditions, but pretreatment times take hours or days rather than minutes or seconds [28]. Alkali pretreatment disrupts the lignin structure, causes partial decrystallization of cellulose, removes acetyl, and the various uronic acid substitutions on hemicellulose and increases the accessibility of enzymes to cellulose and hemicellulose [29]. A neutralizing step to remove lignin and inhibitors (salts, phenolic acids, furfural, and aldehydes) is required before enzymatic hydrolysis. Compared with acid processes, alkaline processes cause less sugar degradation, and many of the caustic salts can be recovered and/or regenerated [28].

Acid pretreatment is based on mixing biomass with concentrated or diluted acids at temperatures between 130°C and 210°C. The use of concentrated acid can count as a treatment method rather than pretreatment as explained in the hydrolysis section in the detail. The most commonly used acid in the acid pretreatment is dilute sulfuric acid (H_2SO_4). Other acids such as hydrochloric acid (HCl), phosphoric acid (H_3PO_4), and nitric acid (HNO_3) can also be used. As an alternative to these inorganic acids, organic acids such as maleic acid and fumaric acid can also be used for dilute acid pretreatment [30]. The application time could take from a few minutes to hours depending on the type of acid, concentration, and temperature used in the process. This pretreatment method solubilizes hemicellulose fraction and releases monomeric sugars and soluble oligomers from the cell wall matrix into the hydrolysate. After removing hemicellulose, increase in porosity improves enzymatic digestibility. However, the sugars released from hemicellulose can be further degraded to furfural and hydroxymethyl furfural, which takes role as inhibitors in fermentation process.

Other common pretreatment methods used for biomass are *oxidative delignifications*. These methods are (a) lignin biodegradation catalyzed by the peroxidase enzyme in the presence of H_2O_2, (b) lignin degradation caused by ozone, which attacks and cleaves aromatic ring structures, while hemicellulose and cellulose are hardly decomposed (*ozonolysis*) [25], (c) the internal lignin and hemicellulose bonds in presence of an organic are broken by using methanol, ethanol, acetone, ethylene glycol, triethylene glycol, or tetrahydrofurfuryl alcohol or aqueous organic solvent mixture in presence of inorganic acid catalysts such as HCl and H_2SO_4 or organic acid catalysts such as oxalic, acetylsalicylic, and salicylic acids (*organosolv process*) [28], and (d) biomass is treated with oxygen or air as an oxidizer in combination with water at elevated temperature (e.g., 195°C) and pressure (e.g., 1.2 MPa) (*wet oxidation*). Wet oxidation mainly solubilizes hemicellulose and removes lignin from biomass structure. Lignin is decomposed into carbon dioxide, water, and carboxylic acids. In this method, the formation of strong inhibitors such as furfural and 5-hydroxymethylfurfural is very low.

There are also some methods that use *ultrasound, microwave, and radiofrequency radiations* as pretreatments methods. These pretreatments are used in conjunction with other methods. Ultrasound produces sonochemical and mechanoacoustic effects, which affect the chemical and physical composition of biomass. The mechanoacoustic alters the surface structure of the biomass while the sonochemical production of oxidizing radicals is effective on biomass components [31]. It was reported that sonication of biomass-water mixture under 20 kHz ± 50 Hz ultrasound for 8 min did not change the hydrolysis that was performed in subcritical water after sonication. Although hydrolyzed solid biomass fraction remained the same, the molar mass of polysaccharides fractions in the hydrolysates notably decreased because of degradation. The amount of monosaccharides in sonicated hydrolysates was high that confirming the positive effect of sonication on degradation of polysaccharides. For instance, hydrolysis of kenaf biomass at 250°C subcritical water treatment resulted in 10% of more xylose release in the sonicated hydrolysate compared to nonsonicated one [32].

On the other hand, microwave treatment has enhanced the surface disruption and the breaking of lignin structures in switchgrass and improved enzymatic saccharification 53% more compared to conventional heating [33]. Microwave pretreatment has positive effect on solubilization of switchgrass in subcritical water. Hydrolysis percentages and total organic carbons released into solution are higher in microwave-treated biomass samples. When microwave pretreatment is applied at higher temperature, solubilization significantly increases (**Table 1**).

Biomass[a]	Hydrolysis[b] (%)	TOC (mg L^{-1})
Untreated	49.01 ± 0.9	1132 ± 13
MW—120°C	57.09 ± 0.6	1541 ± 14
MW—150°C	62.91 ± 0.7	1679 ± 17

[a]MW: microwave-treated biomass.
[b]Determined by weighing microwave-treated biomass samples before and after solubilization in subcritical water treatment.

Table 1. Solubilization of microwave pretreated samples in subcritical water.

Microwave pretreatment was used for solubilization of lignocellulosic biomass in combination with acid and alkali treatments followed by enzymatic hydrolysis [33–35]. Chimentão et al. [36] investigated hydrolysis of dilute acid-pretreated cellulose in a conventional oven and under microwave heating. Although the method was called "mild hydrothermal conditions," the hydrolysis process was accelerated using acids (sulfuric and oxalic acids).

http://dx.doi.org/10.5772/65507

Radio frequency (RF) heating is another promising dielectric heating technology, which is used as an initial breakdown of the lignocellulosic matrix. Dielectric heating transforms electromagnetic energy into heat that is effective on breakdown of biomass structure. The electromagnetic field could generate nonthermal effects, which can also accelerate the destruction of the crystallinity structure [37]. RF heating prevents uncontrolled heating and overheating that protects the product from degradation. RF has large penetration depth (10–30 m) and higher energy efficiency than microwave [38–40]. Efficiency of pretreatment highly depends on temperature, frequency, and type of product/biomass (water content, chemical composition, etc.). Radio frequency-assisted dielectric heating was usually combined with alkaline pretreatment for destruction of biomass materials [41, 42].

*Biological pretreatment*s are attractive approaches for delignification of lignocellulosic biomass. Advantages of these processes are mild operation conditions, environment friendly, and low energy requirement [43]. The formations of toxic materials (furfural, hydroxymethylfurfural, etc.) are low. However, these pretreatments are time consuming [44]. The microbial treatment includes fungi, such as white-rot fungi, brown-rot fungi, and soft-rot fungi, actinomycetes, and bacteria to degrade recalcitrant polymeric structures in biomass [45]. Brown rots are mainly effective on polysaccharides with minimal lignin degradation, whereas white and soft rots attack carbohydrates and lignin. Most of the brown-rot fungi degrade cellulose and hemicellulose more rapidly than lignin in lignocellulosic biomass. White-rot fungi are the most effective for biological pretreatment of lignocellulosic materials since they are able to efficiently degrade all components of plant cell walls, both carbohydrates and lignin. Enzymes such as peroxidases and laccase take roles on lignin degradation by white-rot fungi. Bacteria and actinomycetes are not as efficient as white- and brown-rot fungi in pretreatment of biomass. Lignocellulosic biomass materials such as wheat, corn and rice straws, corn stover, corn stalks, beech wood chips, pine wood chips, and switchgrass were successfully pretreated by these methods using white-rot fungi, brown-rot fungi, and soft-rot fungi [46–48].

Many applications have combined one pretreatment with other one for effective breakdown of biomass structure.

3.2. Hydrolysis of biomass materials

The major hydrolysis processes typically used for the solubilization of biomass require either use of toxic, corrosive, and hazardous chemicals (e.g., acid and alkali treatments) or longer retention times (e.g., enzymatic hydrolysis), which collectively make the process environmentally unsafe and/or expensive. Mineral acids are commonly used to dissolve hemicelluloses, whereas lignin is typically dissolved by alkaline or organosolv pretreatments [45, 49]. Recovery of the chemical catalyst is often crucial to the success of these processes [24]. On the other hand, generally harsh conditions (e.g., high temperatures and high acid concentrations) are needed to release glucose from biomass complex structures. Pyrolysis and other side reactions at higher temperatures become very important, and the amount of undesirable byproducts (tars) increases as the temperature is increased above 220°C [50].

Concentrated acid hydrolysis has been applied to breakdown lignocellulosic efficiently. The hydrolysis reaction for cellulose conversion into sugars is principally the degradation of chemical bonds in cellulose, involving the hydrolytic cleavage of beta-1,4-glycosidic bond, which is catalyzed by H_3O^+ ions of an acid. The reaction rate depends on the H_3O^+ ion concentration, the reaction temperature, and the chemical environment of the glycosidic bond [51]. The acid hydrolysis process usually employs sulfuric acid and hydrochloric acid at concentrations of 1–10% using a moderate temperature (in the range of 100–150°C) [52]. A two-step sulfuric acid hydrolysis is a widely used technique for releasing sugars from biomass. Biomass is first treated with concentrated sulfuric acid at a low temperature and then hydrolyzed with diluted sulfuric acid at an elevated temperature. Concentrated acid recrystallizes cellulose to less crystallized oligosaccharides followed by less concentrated and higher reaction temperature for converting recrystallized oligosaccharides to monosaccharides. Advantages of concentrated acid hydrolysis process are higher conversion from polysaccharides to monosaccharides with minimum formation of reaction byproducts with careful control of reaction conditions. The use of concentrated acid for biomass hydrolysis has also several drawbacks such as energy consumption, equipment corrosion, handling of nonsafe chemicals, an added necessary step of acid neutralization, the formation of byproducts that create an inhibitory effect in the fermentation [53, 25], and other negative environmental impacts. On the other hand, *hydrothermal treatments* (subcritical and supercritical water) are alternative ways to hydrolyze lignocellulosic biomass in an environmentally friendly manner by only operating temperature and pressure conditions. The main shortcoming of these applications is their very high investment cost. **Figure 1** shows morphological changes occur in biomass structure (kenaf) after subcritical water hydrolysis treatment [32].

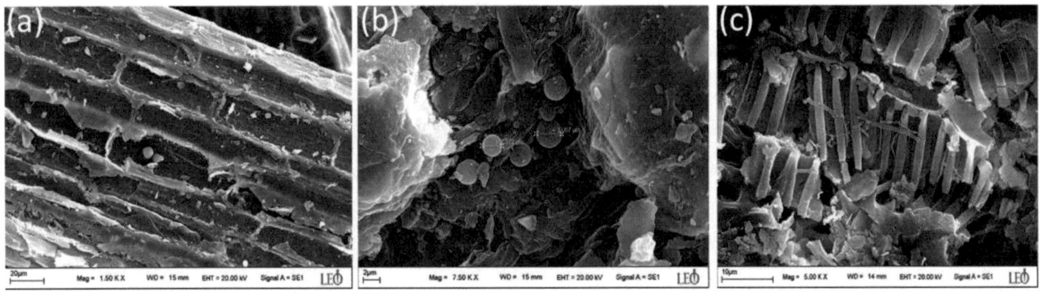

Figure 1. SEM images of untreated (a), 200°C (b) and 250°C (c) subcritical water treated kenaf samples.

Before treatment biomass exhibits rigid and highly ordered fibrils. The cell walls are visible creating a "brickwork-like" appearance to surface (a). The compacted outer layer is partially removed at 200°C and a range of discrete droplet morphologies that contain lignin is appeared on the cell. The 250°C-subcritical water treatment reduces and degrades lignocellulosic structure, leaving highly degraded solids. The maximum solubilization yield of wheat straw and kenaf biomass were found to be 70–75% in subcritical water medium (under 250°C and 27.58 MPa carbon dioxide pressure conditions) [11, 32]. However, the hydrolysates obtained in this process contained high molecular weight polysaccharides that were difficult to utilize for producing value-added products such as gas biofuel hydrogen.

http://dx.doi.org/10.5772/65507

3.3. Chemical composition of biomass materials

Chemical composition of biomass and structures of biopolymers (cellulose, hemicellulose, and lignin) are two important factors affecting the yield of the biofuels, bioproducts, and chemicals produced from biomass. Composition of lignin, cellulose, and hemicellulose in biomass materials significantly differ among biomass species. For instance, some biomass materials such as hardwoods contain more cellulose in their structures, whereas others such as straws and leaves have more of hemicelluloses. It is known that lignin content of herbaceous plants such as grasses is very low compared to softwoods, which are known to have highest amount of lignin in their structures [54]. On the other hand, polymerization degree and(or) structures of biopolymers can also considerably varies among biomass species. For instance, the chemical structure of lignin is based on syringyl (S), guaiacyl (G), and p-hydroxyphenyl (H) units. Softwood lignins are mainly composed of residues derived from guaiacyl units (lignin type G), whereas hardwood lignins contain both syringyl and guaiacyl units with minor amount of p-hydroxyphenyl (lignin type GS). Lignins from grasses are composed of the three basic precursors (lignin type HGS) [55, 56]. Hemicellulose fractions of softwoods mainly have D-mannose-derived structures such as galactoglucomannans, whereas hemicelluloses in hardwoods have D-xylose-derived structures such as arabinoglucuronoxylan. Xylan is a polypentose hemicellulose structure in biomass materials that displays a wide range of compositions, molecular sizes, and structures depending upon its source [57, 58]. This diversity among biomass materials can significantly affect the yield of value-added products directly produced from biomass as raw materials. On the other hand, this diversity can also affect solubilization efficiency of the biomass materials and, therefore, contents and compositions of the biomass components in the hydrolysates. The differences in the hydrolysates will considerably change the yield of the target compounds and byproducts produced from these biomass hydrolysates. For instance, molecular weight distribution of carbohydrates in the hydrolysates can significantly affect the method employed for biofuel or useful chemicals production. The more degraded organics containing hydrolysates are preferable for the production of certain various value-added products from biomass. For example, high-yielding hydrogen gas production from biomass hydrolysates requires reduced molecular weight oxygenated compounds containing biomass hydrolysates as feeds in aqueous-phase reforming gasification process [59].

3.4. Conversion methods

3.4.1. Chemical

Sugars released from biomass can be hydrogenated to C5-6 polyols (*sugar alcohols*) by using a chemical reducing agent such as sodium borohydride ($NaBH_4$). Biomass hydrolysates can also be utilized for direct production of derivatives, such as *furfural, hydroxymethyl furfural*, and/or *levulinic acid*. It is also possible to produce C2-3 *glycols* from biomass hydrolysates by hydrogenolysis. Sugar containing hydrolysates are further upgraded via oxidation or halogenation reactions [60]. Sugars obtained from biomass hydrolysis can be converted to N-heterocyclic components, pyrones, and aromatics that can be further converted into a *variety of chemical intermediates*.

Furfural is a valuable compound for a variety of chemical applications and it serves as a precursor for the synthesis of many fine chemicals and biofuels. It is produced industrially by acid-catalyzed hydrolysis and dehydration of pentoses (mainly xylose) in lignocellulosic feedstocks (sugarcane bagasse, corn cobs, sunflower stalk, etc.) at temperatures ranging from 153 to 240°C [61]. During the initial stage, the hemicellulose is hydrolyzed to xylans, which generate pentose carbohydrates to be further converted into furfural. Commercially, furfural is produced using sulfuric acid as a homogeneous catalyst. Significant quantities of steam are used in the process in order to strip the furfural and to avoid its further degradation.

Xylitol is currently produced through chemical reduction of xylose derived from birch wood chips and sugarcane bagasse hemicellulose hydrolysate. The chemical process adapted for xylitol production from xylan-rich biomass demands high production costs in terms of temperature and pressure input, as well as the formation of byproducts that require expensive separation and purification steps [62].

3.4.2. *Thermochemical*

Thermochemical conversion processes are combustion, pyrolysis, gasification, and liquefaction.

Combustion of biomass is the least complex conversion method to transform biomass into energy. Combustion process takes places at 800–1000°C. Complete combustion involves the production of heat as a result of the oxidation of carbon and hydrogen-rich biomass to CO_2 and H_2O. The biomass with high moisture content is not suitable for this process, predrying may be necessary in some cases. The high-pressure steam formed in the process can be utilized as hot air, hot water, steam, or electricity to produce hot gases. Combustion systems for electricity and heat production are similar to most fossil-fuel fired power plants. Combustion technologies can produce high NOx emissions.

Pyrolysis is thermal degradation of lignocellulosic biomass in the absence of oxygen around 500°C. The products are combination of solid (biochar) and gaseous fractions and liquid biooil (biocrude). These products can be used directly or after processing as fuel. Biochar and biooil can also be utilized for production of chemicals and value-added products [63]. Pyrolysis is a part of gasification and combustion, which consists of a thermal degradation of the initial solid biomass into gases and liquids without an oxidizing agent. Thermal decomposition of organic components in biomass starts at 350–550°C and goes up to 700–800°C in the absence of air/oxygen [64]. Pyrolysis can be fast or slow process depending on time taken for processing the feed into pyrolysis products. In slow pyrolysis, primarily biochar is produced at lower temperatures. At higher temperatures biooils are produced through fast pyrolysis. In slow pyrolysis, biomass is typically heated at slow heating rates (up to 10–20°C min^{-1} with sufficient time allowed for repolymerization reactions to maximize the solid yields while fast pyrolysis uses much higher heating rates ˃10–200°C s^{-1}), higher processing temperature and short vapor residence times (less than 2 s) that produce 50–70 wt% biooil yield (dry biomass basis). Fast pyrolysis usually requires a finely ground biomass feed (<1 mm); however, slow hydrolysis accepts a wide range of particle sizes (5–50 mm) [65].

Gasification is the conversion of biomass into a combustible gas mixture at 700–1600°C. The produced gas is rich in CO, H_2, CH_4, and CO_2. This combustible gas mixture can be used for different purposes after cleaning. The clean gas can be used directly as an engine fuel or upgraded to liquid fuels or converted into chemical feedstocks by different methods including biological fermentation or catalytic upgrading through the Fischer-Tropsch process [7, 19].

Lignocellulosic biomass can also be converted into liquid and gaseous fuels by *hydrothermal gasification processes*. Hydrothermal gasification processes are applied in either gaseous phase (steam reforming), in supercritical water, or in liquid phase (aqueous phase reforming). Steam reforming needs the lignocellulosic material to be dried prior to gasification. The temperatures applied in this conversion are also high, i.e., 800°C or above and considerable tar and char are formed. The thermodynamic critical points for water are 374.3°C and 221.2 bar for temperature and pressure, respectively. Effective gasification of lignocellulosics in supercritical water needs to go quite higher values than the critical points, e.g., above 600°C for complete gasification [66]. APR gasification process is a promising process since conversion reactions take place at moderate temperatures (225–265°C, and at pressures of 27–54 bar) in presence of a reforming catalyst (such as precious metal, Pt). The water-gas shift reaction ($CO + H_2O \leftrightarrow H_2 + CO_2$) is thermodynamically favored at processing temperatures. Biomass hydrolysates, which are composed of biomass-derived mixed oxygenated hydrocarbons, are successfully converted into gaseous products that are rich in hydrogen [12, 59, 67].

Hydrothermal liquefaction of biomass is the thermochemical conversion of biomass into liquid fuels at 280–370°C and 10–25 MPa in aqueous medium [68]. In this process, water is an important reactant and catalyst. It is still in liquid state and has a range of promising properties at the liquefaction conditions [68]. In this process, lignocellulosic biomass is break down into soluble low molecular weight products that undergo further condensation, cyclization, and polymerization reactions forming biooil that is soluble in organic solvents. These biooils are most commonly used as a replacement for petroleum fuels but they can be also served as feedstock for production of a wide range of value-added products. For instance, lignin-derived phenolics can be converted into aromatic chemicals by hydrogenation [69], and carbohydrates are utilized for catalytic production of hydrogen or consumed in fermentation [70].

Various value-added products can be produced or synthesis from biomass or biomass-derived compounds by thermochemical methods. For instance, sugars can be reduced into sugar alcohols by thermochemical reduction. In industrial reaction, sorbitol production by thermochemical reduction of glucose is mainly carried out discontinuously in stirred tank reactors at 100–180°C under 5–15 MPa of H_2 pressure in the presence of a catalyst, usually raney-type nickel, or ruthenium catalysts. **Table 2** shows thermochemical sugar alcohol production from glucose, simple biomass model compound, at various processing temperatures under 2.0 MPa H_2 pressure in the presence of carbon supported ruthenium catalyst. The reaction time was 60 min [71].

The results showed that 80°C hydrogenation temperature is not enough for complete reduction of glucose solution. Further increase of the process temperature causes complete reduction of all glucose; however, the contents and compositions of reduced products (sugar alcohols) significantly change as temperature increased. The main reduction product of glucose is sorbitol and its highest yield is at 100°C at the processing conditions described above.

Temperature (°C)	Sugar alcohols (% concentration)				
	Glucose	Mannitol	Sorbitol	Xylitol	Other products [b]
80	8.8	0.9	90.3	n.d.	n.d.
100	n.d.[a]	2.2	97.3	n.d.	n.d.
120	n.d.	5.9	88.6	n.d.	5.5
140	n.d.	16.5	59.7	2.7	21.1
160	n.d.	21.8	41.0	3.6	33.6
180	n.d.	23.4	20.7	4.1	51.8
200	n.d.	19.7	12.1	2.1	66.1

[a]n.d., not detected.
[b]Total of other reduced products (glycerol, erythritol, etc.).

Table 2. Thermochemical sugar alcohol production from reduction of glucose at various temperatures.

Other types of value-added products that can be produced by a thermochemical method are carbonaceous materials. Nonsolubilized residue containing mostly lignin and lignin destruction products can be utilized for production of carbon-based materials such as activated carbon and mesoporous carbons, etc. [72].

3.4.3. *Biochemical*

Biochemical conversion can also be used for breaking down biomass into sugars that can then be converted into biofuels (gaseous or liquid fuels) and bioproducts through the use of microorganisms and enzymes. This process is usually used for treating high moisture containing organic wastes. Biochemical processes are relatively slow processes that require more time for conversion of biomass into useful compounds [73]. The most popular biochemical technologies are *anaerobic digestion* and *fermentation* [74]. Biomass from various sources is biochemically degraded by *anaerobic digestion* in highly controlled, oxygen-free condition circumstances to produce biogas containing mostly methane and carbon dioxide. Biogas can be burned directly for heat or steam or converted to electricity. It can also be upgraded to biomethane or reformed into hydrogen fuel. In anaerobic digestion, bacteria are used to hydrolyze carbohydrates into sugars that are digestible by other bacteria. Methane and carbon dioxide are produced from those digestible components.

Fermentation is used commercially on a large scale to produce ethanol from carbohydrate containing materials. Structural carbohydrates in biomass are broken down into sugars by

http://dx.doi.org/10.5772/65507

using enzymes. Then, released sugars are transformed into alcohols, organic acids, or hydrocarbons by microorganisms in fermentation. The intermediate sugars can also be utilized to obtain other value-added chemicals. Conversions in this process take place at 25–70°C under atmospheric pressure conditions.

Not only biofuels but also a wide range of value-added products such as food-based sugar alcohols (e.g., xylitol) can be produced by combined processes such as thermochemical process combined with a chemical or biochemical method [62].

4. Comparison of various types of biomass materials for high-yielding value-added products

4.1. Energy crops

Energy crops are specifically grown for its fuel value or to produce bioenergy. These plants usually require low cost and low maintenance to grow and they are utilized to make biofuels or directly exploited for their energy contents. Energy crops can be food crops (corn, sugarcane, sugar beet, sweet sorghum, etc.) or nonfood crops (poplar trees, switchgrass, miscanthus, kenaf, etc.). A major focus among them is nonfood energy crops.

Perennial grasses (C4 species) such as *switchgrass* and *miscanthus* are promising potential lignocellulosic biomass sources for biofuel production. They require less fertilizer, water, and energy for production. Since these grasses regrow from their roots, they do not need to be replanted each year. Switchgrass and miscanthus are being developed as bioenergy crops because they have high yield potential and desirable agronomic traits. Switchgrass is a native prairie plant that grows about 10 feet tall. Annual dry matter yields are 15–30 t ha^{-1}. It exhibits adaptation to relatively wet and fertile soils. Switchgrass is very tolerant of poor soils, flooding, and drought. It greatly reduces soil erosion. Giant miscanthus is a perennial warm-season (C4) sterile hybrid grass. It is native to subtropical and tropical regions of Africa and southern Asia, which has also been used as an ornamental plant in many countries. Giant miscanthus can grow best in the soils that are well drained and have a pH between 5.5 and 7.5 under medium to high fertility [75]. Giant miscanthus has produced more than double the biomass of upland switchgrass per acre. Yield estimates from 10–15 tons per acre. Harvestable stems of miscanthus are usually more than 9 feet long [76].

Kenaf (*Hibiscus cannabinus* L.), is a warm season annual herbaceous crop that belongs to Malvaceae family and grows best in the tropics and subtropics where the mean daily temperatures during the growing season exceed 20°C. Kenaf is a good candidate for bioenergy because it grows fast without special care and produces large amount of biomass. Under good conditions, kenaf will grow to a height of 5–6 m in 6–8 months and produce up to 30 t ha^{-1} y^{-1} of dry stem material [77].

Corn is one of the most important crops in the world, and the United States is the largest corn producer in the world. The total production of corn in the United States for the year 2014–2015 was reported to be 14.215 billion bushels [78]. A major use of corn in the United States is ethanol production (30.5%). Other uses beyond animal feed include the production of high fructose corn syrup (3.6%), sweeteners (2.1%), starch (1.8%), cereals (1.5%), beverage alcohol (1.0%), etc. [78]. The amount of cropland devoted to growing corn has considerably increased in recent years and now, the United States has excess corn production. Ethanol production capacity from corn has reached about 15.25 billion gallons in 2014/15, exceeding projected capacity demand of 15.1 billion gallons per year [79]. Most ethanol in the United States is made from corn because it is a relatively low-cost source of starch that can easily be converted into sugar, fermented, and distilled into ethanol. Today, more than 10% of vehicle fuel comes from ethanol. Producing ethanol increases the cost of corn, which leads to *higher* prices for other commodities such as dairy, meat, and livestock feed that farmers rely on. Since ethanol use has increased incredibly since 2000, the demand and the price of corn have increased. This has forced the farmers to use more pristine lands to plant corn. This land conversion and the associated use of agricultural fertilizers and pesticides have caused significant increases in greenhouse gas emissions. Even more emissions were released when ethanol consumed as fuels. Because ethanol has less energy content than gasoline, vehicles have to consume more fuel and thus produce more emissions.

Sugarcane (*Saccharum* spp.) is an efficient tropical/subtropical grass with nitrogen-fixing symbionts, thus needing little fertilizer. Brazil is the largest producer (31 million tons year^{-1}) and produces 25% of the world's sugarcane [80]. Commercial average of sugar cane yield (Australia, Colombia, and South Africa) and total dry matter (biomass) production are reported to be 84 and 39 t ha^{-1} yr^{-1}, respectively [81]. A cycle of sugarcane planting and harvesting is about 12 months. The average sugarcane yield from four cycles is about 60 t ha $^{-1}$ yr^{-1} [82]. The sugarcane industry's waste product, bagasse, can also be used to produce biofuels. Bagasse is the fibrous residue left over when sugarcane is squeezed for its juice. It was reported that energy balance of sugarcane-based ethanol is seven times greater than that of corn-based ethanol. Energy balance defined as the difference between the energy consumed to produce ethanol from crop and the energy released when it is consumed as biofuel. Only certain percent of corn starch can be converted into ethanol that required an initial breakdown process to release sugars from starch first. This step significantly increases the operation costs compared to sugar-based ethanol [83].

Poplar trees are already known to be good candidates for bioenergy because of their fast growth rates (up to 4 m annually and matures in about 6 years) and large biomass production. Poplar can grow on 2–3 year rotations, growing back after it is harvested, with seven or more cycles possible before replanting. After harvest, the tree resprouts from the same root stalk, a very efficient way to produce biomass. Hybrid poplars have been planted for their potential as an energy crop due to their high yield rates and adaptability to many growing conditions. Hybrid poplar growth and biomass yield could vary significantly depending on climatic conditions, soil characteristics, and species genotype [84, 85]. Poplar plantations have many environmentally desirable applications, including use as buffer strips to decrease erosion and nitrate in run-off from highly erodible fields, treatment, and removal of toxic materials from landfills and other soil contaminations, and an excellent sink of atmospheric CO_2 [86]. Poplar woods do not need to be stored like other bioenergy crops such as grasses; it can be harvested throughout the year. Poplar requires less water and other inputs than crops such as corn. The

http://dx.doi.org/10.5772/65507

higher pest resistance, increased site adaptability, and development of poplar genotypes with improved yield make poplar trees important biomass materials for biofuel production.

Sweet sorghum (*Sorghum bicolor* L. Moench) is a widely adapted sugar crop that can be produced at less cost than corn [87]. It is a highly productive and versatile crop that can be cultivated in temperate climates. Sweet sorghum can be grown in many regions of the United States. Under favorable growing conditions, sorghum can produce high biomass yields with low rates of nitrogen fertilizer. Individual stalks can be over 10 feet tall. Two fractions of sweet sorghum can be used for biofuel production. The stalk and seed are used directly for biomass energy, and their high sugar content allows them to be fermented to make ethanol. Rather than producing starch, sweet sorghum carbohydrates are stored in the stalk as sugar, with sugar concentrations of 8–20% [88]. As the plant matures, carbohydrates are translocated from the stalks to the head to make starch in the seeds. Conversion of these sugar components into ethanol requires less energy than starch for same reason discussed for sugarcane. Sweet sorghum has the potential to produce up to 6000 L ha^{-1} of ethanol equivalent to corn grain yields of approximately 20 Mg ha^{-1} [89]. The main drawback in sweet sorghum is; since sugars in the sweet sorghum are not stable and rapidly degrade it is required to squeeze the juice out immediately for conversion. This reduces flexibility in harvesting and can increase transportation costs [90].

4.2. Agricultural biomass residues

Agricultural residues from well-established production chains are important sources of biomass that can provide a substantial amount of biomass for production of a wide range of value-added products. Since these residues are a natural byproduct of the food crop, they can be used as promising low-cost feedstocks without increasing the amount of land used for agriculture. Agricultural biomass includes *bagasse, straw, stem, stalk, leaves, husk, shell, peel, pulp, stubble*, etc. is produced annually worldwide and is not utilized to any significant extent.

Cereal straw is a typical example of an agriculture byproduct. Rice straw (includes stems, leaf blades, leaf sheaths, and the remains of the panicle after threshing) and wheat straw (includes nongrain portion of the wheat plant; stems, leaves, and chaff) are two of the most abundant cereal-based lignocellulosic wastes in the world. Wheat is the second most important grain that is cultivated in the United States, following only corn. The United States is a major wheat-producing country ranked third in production volume of wheat in the world. Total wheat production in United States was about 55 million tons in 2015. A good wheat crop yields between 1 and 1.2 tons of straw per acre on a dry matter basis. The annual global rice straw production is 731 million tons and Asia alone produces 667.6 million tons [1]. One ton of rice paddy produces approximately 290 kg rice straw. Rice husk is one of the major byproducts from the rice milling process and constitutes about 20% of paddy by weight. One ton of rice paddy produces roughly 220 kg rice husk [2].

High-yielding corn production activities generate large amounts of corn stover that is roughly 80% of all agricultural residues produced in the United States [91]. It has been estimated that corn in the United States produces 1.7 times more residue than other cereals. Corn stover contains 27.5% stalk, 8.2% cob, and 7.0% husk [92].

Biomass	Advantages and disadvantages
Switchgrass and miscanthus	Nonedible Require less fertilizer, water, and energy for production Do not need to be replanted each year Miscanthus produces more biomass than switchgrass
Kenaf	Nonedible Grows fast in the tropics and subtropics Does not need special care for production produces large amount of biomass
Corn	Edible Mostly used for ethanol production because of relatively low-cost source of starch Large amount of cropland devoted to corn Excess corn production in the United States
Sugarcane	Edible Grows in tropics and subtropics Needs little fertilizer Sugarcane juice is directly used in conversion techniques (no pretreatment needed) Waste product, bagasse, is an important feedstock for value-added products
Sweet sorghum	Needs little fertilizer Sugars in the juice can be directly utilized for value-added products (no pretreatment needed) Required to squeeze the juice out immediately (not stable) Limited flexibility in harvesting and transportation costs
Poplar trees	Nonedible Grow fast Produce large amount of biomass Can be harvested throughout the year Have many environmentally desirable applications such as reducing erosion
Cereal straw	Nonedible Natural byproduct of the food crops rice, wheat, corn, etc. Produce large amounts of biomass
Corn stover	Nonedible Natural byproduct of the food crop, corn Produce large amounts of biomass (more than cereals)
Forest biomass	Nonedible Largest source of lignocellulosic biomass High costs of harvesting and transportation Widely used in combustion process but not in gasification, pyrolysis, and fermentation
Food biomass wastes	Edible/nonedible Have low values Released in large amounts from food industries

Table 3. Comparison of biomass materials for production of value-added products.

4.3. Forest biomass

Forests are the largest source of lignocellulosic biomass that can be substitute for fossil fuels in the production of energy and other value-added products. Since it is a nonfood type of biomass, it is a promising feedstock for these conversions [93]. Forest biomass includes material left on logging sites (*trunks, crowns, and branches*), *unused wood* from forests and *wood manufacturing, and processing residues*. The wastes from wood industries are sawdust, off-cuts, trims, and shavings. Forest biomass utilization has some challenges that restrict fully production of value-added products from forests. The main issue is high costs of harvesting and transportation of these biomass materials. Second, forest biomass has been widely used in combustion process, however, other potential conversion technologies such as gasification, pyrolysis, and fermentation are not applied to forest biomass [94].

4.4. Food biomass wastes

The food industry produces a large number of residues and byproducts that can be used as biomass energy sources.

Bagasse is a promising biomass material that is produced during sugarcane processing. After sugarcane is milled for juice extraction, bagasse is obtained as a residue, which is about 25% of the total weight and contains 60–80% carbohydrates [95].

Pomace is also important food waste-based biomass candidate from wine industry that remains after the grapes are pressed and consists of skins, seeds, and a small percentage of stems. It is the most abundant wine-making waste representing about the 20% (w/w) of grapes used for the production of wine. It is rich in carbohydrates [96].

Nuts hulls, peanuts shells, corn stover, rice hull/husk, and other grain biomass residues can also be considered food biomass wastes. On the other hand, *corn bran*, a byproduct of the corn milling process, can also be counted as lignocellulosic biomass. A large starch producer in the United States releases approximately 9×10^5 tons of corn bran per year [97]. However, this byproduct currently has low value and is often used for animal feed alone or combined with corn germ cake or meal.

Table 3 summarizes advantages and disadvantages of biomass materials for utilization as raw materials for production of value-added products.

Author details

Sibel Irmak

Address all correspondence to: sibel.irmak@unl.edu

Biological Systems Engineering, Industrial Agricultural Products Center, University of Nebraska-Lincoln, Nebraska, USA

References

[1] Clark J, Deswarte F. Introduction to Chemicals from Biomass. 2nd ed. Chichester: Wiley; 2015. pp. 114–281. DOI: 10.1002/9781118714478

[2] Saini JK, Saini R, Tewari L. Lignocellulosic agriculture wastes as biomass feedstocks for second-generation bioethanol production: concepts and recent developments. 3 Biotech. 2015;5:337–353. DOI: 10.1007/s13205-014-0246-5

[3] Hansen AC, Kyritsis DC, Lee CF. Characteristics of biofuels and renewable fuel standards. In: Vertes AA, Blaschek HP, Yukawa H, Qureshi N, editors. Biomass to biofuels-strategies for global industries. New York: John Wiley; 2009. DOI: 10.1002/9780470750025.ch1

[4] Jin C, Yao M, Liu H, Lee CF, Ji J. Progress in the production and application of n-butanol as a biofuel. Renew Sustain Energy Rev. 2011;15:4080–4106. DOI: 10.1016/j.rser.2011.06.001

[5] Sabra W, Groeger C, Sharma PN, Zeng AP. Improved *n*-butanol production by a non-acetone producing *Clostridium pasteurianum* DSMZ 525 in mixed substrate fermentation. Appl Microbiol Biotechnol. 2014;98:4267–4276. DOI: 10.1007/s00253-014-5588-8

[6] Yin X, Leung DYC. Characteristics of the synthesis of methanol using biomass-derived syngas. Energ Fuels. 2005;19:305–310. DOI: 10.1021/ef0498622

[7] Wang L, Weller CL, Jones DD, Hanna MA. Contemporary issues in thermal gasification of biomass and its application to electricity and fuel production. Biomass Bioenergy. 2008;32:573–581. DOI: 10.1016/j.biombioe.2007.12.007

[8] De María R, Díaz I, Rodríguez M, Sáiz A. Industrial methanol from syngas: kinetic study and process simulation. Int J Chem Reac Eng. 2013;11:469–477. DOI: 10.1515/ijcre-2013-0061

[9] Khandan N, Kazemeini M, Aghaziarati M. Direct production of dimethyl ether from synthesis gas utilizing bifunctional catalysts. Appl Petrochem Res. 2012;1:21–27. DOI: 10.1007/s13203-011-0002-2

[10] Dresselhaus MS, Thomas IL. Alternative energy technologies. Nature. 2001;414:332–337. DOI: 10.1038/35104599

[11] Meryemoglu B, Hesenov A, Irmak S, Atanur OM, Erbatur O. Aqueous-phase reforming of biomass using various types of supported precious metal and raney-nickel catalysts for hydrogen production. Int J Hydrogen Energy. 2010;35:12580–12587. DOI: 10.1016/j.ijhydene.2010.08.046

[12] Cortright RD, Davda RR, Dumesic JA. Hydrogen from catalytic reforming of biomass-derived hydrocarbons in liquid water. Nature. 2002;418:964–967. DOI: 10.1038/nature01009

[13] Baccar R, Bouzid J, Feki M, Montiel A. Preparation of activated carbon from Tunisian olive-waste cakes and its application for adsorption of heavy metal ions. J Hazard Mater. 2009;162:1522–1529. DOI: 10.1016/j.jhazmat.2008.06.041

[14] Zhou F, Lu C, Yao Y, Sun L, Gong F, Li D, Pei K, Lu W, Chen W. Activated carbon fibers as an effective metal-free catalyst for peracetic acid activation: implications for the removal of organic pollutants. Chem Eng J. 2015;281:953–960. DOI: 10.1016/j.cej.2015.07.034

[15] Tsyntsarski B, Stoycheva I, Tsoncheva T, Genova I, Dimitrov M, Petrova B, Paneva D, Cherkezova-Zheleva Z, Budinova T, Kolev H, Gomis-Berenguer A, Ania CO, Mitov I, Petrov N. Activated carbons from waste biomass and low rank coals as catalyst supports for hydrogen production by methanol decomposition. Fuel Process Technol. 2015;137:139–147. DOI: 10.1016/j.fuproc.2015.04.016

[16] Ramesh T, Rajalakshmi N, Dhathathreyan KS. Activated carbons derived from tamarind seeds for hydrogen storage. J Energ Storage. 2015;4:89–95.

[17] Saha D, Deng S. Adsorption equilibrium and kinetics of CO_2, CH_4, N_2O, and NH_3 on ordered mesoporous carbon. J Colloid Interface Sci. 2010;345:402–409. DOI: 10.1016/j.jcis.2010.01.076

[18] Lee J, Yoon S, Hyeon T, Oh SM, Bum Kim K. Synthesis of a new mesoporous carbon and its application to electrochemical double-layer capacitors. Chem Commun. 1999;21:2177–2178. DOI: 10.1039/A906872D

[19] Huber GW, Iborra S, Corma A. Synthesis of transportation fuels from biomass: chemistry, catalysts, and engineering. Chem Rev. 2006;106:4044–4098. DOI: 10.1021/cr068360d

[20] Xing R, Subrahmanyam AV, Olcay H, Qi W, van Walsum GP, Pendse H, Huber GW. Production of jet-fuel-range alkanes from hemicellulose-derived aqueous solutions. Green Chem. 2010;12:1933–1946. DOI: 10.1039/C0GC00263A

[21] Ko BS, Kim J, Kim JH. Production of xylitol from D-xylose by a xylitol dehydrogenase gene-disrupted mutant of *Candida tropicalis*. Appl Environ Microbiol. 2006;72:4207–4213. DOI: 10.1128/AEM.02699-05

[22] Werpy TA, Petersen G. Top value added chemicals from biomass. Volume I. Results of screening for potential candidates from sugars and synthesis gas. N. R. E, Laboratory Report DOE/GO-102004–1992. U.S. Department of Energy (DOE), 2004.

[23] Godula K, Bertozzi CR. Synthesis of glycopolymers for microarray applications via ligation of reducing sugars to a poly(acryloyl hydrazide) scaffold. J Am Chem Soc. 2010;132:9963–9965.

[24] Mosier N, Wyman C, Dale BE, Elander R, Lee YY, Holtzapple M, Ladisch M. Features of promising technologies for pretreatment of lignocellulosic biomass. Bioresour Technol. 2005;96:673–686. DOI: 10.1016/j.biortech.2004.06.025

[25] Sun Y, Cheng J. Hydrolysis of lignocellulosic materials for ethanol production: a review. Bioresour Technol. 2002;83:1–11. DOI: 10.1016/S0960-8524(01)00212-7

[26] Olivia JM, Sáez F, Ballesteros I, Gónzalez A, Negro MJ, Manzanares P, Ballesteros M. Effect of lignocellulosic degradation compounds from steam explosion pretreatment on ethanol fermentation by thermotolerant yeast Kluyverinyces narxianus. Appl Microbiol Biotechnol. 2003;105:141–153. DOI: 10.1385/ABAB:105:1-3:141

[27] Jönsson LJ, Alriksson B, Nilvebrant NO. Bioconversion of lignocellulose: inhibitors and detoxification. Biotechnol Biofuels. 2013;6:16. DOI: 10.1186/1754-6834-6-16

[28] Kumar P, Barrett DM, Delwiche MJ, Stroeve P. Methods for pretreatment of lignocellulosic biomass for efficient hydrolysis and biofuel production. Ind Eng Chem Res. 2009:48:3713–3729. DOI: 10.1021/ie801542g

[29] Chang VS, Holtzapple MT. Fundamental factors affecting biomass enzymatic reactivity. Appl Biochem Biotechnol. 2000;84–86:5–37. DOI: 10.1385/ABAB:84-86:1-9:5

[30] Kootstra AMJ, Beeftink HH, Scott EL, Sanders JPM. Comparison of dilute mineral and organic acid pre-treatment for enzymatic hydrolysis of wheat straw. Biochem Eng J. 2009;46:126–131. DOI: 10.1016/j.bej.2009.04.020

[31] Bussemaker MJ, Zhang D. Effect of ultrasound on lignocellulosic biomass as a pretreatment for biorefinery and biofuel applications. Ind Eng Chem Res. 2013;52:3563–3580. DOI: 10.1021/ie3022785

[32] Ozturk I, Irmak S, Hesenov A, Erbatur O. Hydrolysis of kenaf (Hibiscus cannabinus L.) stems by catalytical thermal treatment in subcritical water. Biomass Bioenerg. 2010;34:1578–1585. DOI: 10.1016/j.biombioe.2010.06.005

[33] Hu Z, Wen Z. Enhancing enzymatic digestibility of switchgrass by microwave-assisted alkali pretreatment. Biochem. Eng J. 2008;38:369–378. DOI: 10.1016/j.bej.2007.08.001

[34] Zhu S, Wu Y, Yu Z, Zhang X, Wang C, Yu F, Jin S, Zhao Y, Tu S, Xue Y. Simultaneous saccharification and fermentation of microwave/alkali pre-treated rice straw to ethanol. Biosyst Eng. 2005;92:229–235. DOI: 10.1016/j.biosystemseng.2005.06.012

[35] Marx S, Ndaba B, Chiyanzu I, Schabort C. Fuel ethanol production from sweet sorghum bagasse using microwave irradiation. Biomass Bioenerg. 2014;65:145–150. DOI: 10.1016/j.biombioe.2013.11.019

[36] Chimentão RJ, Lorente E, Gispert-Guirado F, Medina F, López F. Hydrolysis of dilute acid-pretreated cellulose under mild hydrothermal conditions. Carbohydr Polym. 2014;111:116–124. DOI: 10.1016/j.carbpol.2014.04.001

[37] Hu Z, Wang Y, Wen Z. Alkali (NaOH) pretreatment of switchgrass by radio frequency-based dielectric heating. Appl Biochem Biotechnol. 2008;148:71–81. DOI: 10.1007/s12010-007-8083-1

[38] Balakrishnan PA, Vedaraman N, Sunder VJ, Muralidharan C, Saminathan G. Radio frequency heating - a prospective leather drying system for future. Drying Technol. 2004;22:1969–1982. DOI: 10.1081/DRT-200032738

[39] Izadifar M, Baik OD, Mittal GS. Radio frequency-assisted extraction of podophyllotoxin: prototyping of packed bed extraction reactors and experimental observations. Chem Eng Process. 2009;48:1437–1444. DOI: 10.1016/j.cep.2009.07.011

[40] Piyasena P, Dussault C, Koutchma T, Ramaswamy HS, Awuah GB. Radio frequency heating of foods: principles, applications and related properties - a review. Crit Rev Food Sci Nutr. 2003;43:587–606. DOI: 10.1080/10408690390251129

[41] Wang X, Taylor S, Wang Y. Improvement of radio frequency (RF) heating-assisted alkaline pretreatment on four categories of lignocellulosic biomass. Bioprocess Biosys Eng. 2016; 39:1539–1551. DOI: 10.1007/s00449-016-1629-2

[42] Iroba KL, Tabi LG, Dumonceaux T, Baik OD. Effect of alkaline pretreatment on chemical composition of lignocellulosic biomass using radio frequency heating. Biosyst Eng. 2013;116:385–398. DOI: 10.1016/j.biosystemseng.2013.09.004

[43] Okano K, Kitagaw M, Sasaki Y, Watanabe T. Conversion of Japanese red cedar (*Cryptomeria japonica*) into a feed for ruminants by white-rot basidiomycetes. Animal Feed Sci Technol. 2005;120:235–243. DOI: 10.1016/j.anifeedsci.2005.02.023

[44] Cardona CA, Sanchez OJ. Fuel ethanol production: process design trends and integration opportunities. Bioresour Technol. 2007;98:2415–2457. DOI: 10.1016/j.biortech.2007.01.002

[45] Galbe M, Zacchi G. Pretreatment of lignocellulosic materials for efficient bioethanol production. Adv Biochem Eng Biotechnol. 2007;108:41–65. DOI: 10.1007/10_2007_070

[46] Xu C, Ma F, Zhang X, Chen S. Biological pretreatment of corn stover by Irpex lacteus for enzymatic hydrolysis. J Agric Food Chem. 2010;58:10893–10898. DOI: 10.1021/jf1021187

[47] Canam T, Town JR, Tsang A, McAllister TA, Dumonceaux TJ. Biological pretreatment with a cellobiose dehydrogenase-deficient strain of *Trametes versicolor* enhances the biofuel potential of canola straw. Bioresour Technol. 2011;102:10020–10027. DOI: 10.1016/j.biortech.2011.08.045

[48] Hatakka A. Biodegradation of lignin. In: Hofrichter M, Steinbuchel A, editors. Biopolymers: Biology, Chemistry, Biotechnology, Applications, Vol. 1 Lignin, Humic Substances and Coal. Weinheim: Wiley-VCH; 2001. PP. 129–180. DOI: 10.1002/1521-3773(20020603)41:11<1963::AID-ANIE1963>3.0.CO;2-F

[49] Alvira P, Tomás-Pejó E, Ballesteros M, Negro M. Pretreatment technologies for an efficient bioethanol production process based on enzymatic hydrolysis: a review. Bioresour Technol. 2010;101:4851–4861. DOI: 10.1016/j.biortech.2009.11.093

[50] Brennan AH, Hoagland W, Schell DJ. High temperature acid-hydrolysis of biomass using an engineering scale plug flow reactor: results of low solids testing. Biotechnol Bioeng Symp. 1986;17:53.

[51] Klemm D, Philipp B, Heinze T, Heinze U, Wagenknecht W. General considerations on structure and reactivity of cellulose. Comprehensive Cellulose Chemistry: Fundamentals and Analytical Methods, Vol. 1. Verlag GmbH & Co. KGaA: Wiley-VCH; 1998. pp. 83–86. DOI: 10.1002/3527601929.ch2a

[52] Wingren A, Galbe M, Zacchi G. Techno-economic evaluation of producing ethanol from softwood: comparison of SSF and SHF and identification of bottlenecks. Biotechnol Prog. 2003;19:1109–1117. DOI: 10.1021/bp0340180

[53] Galbe M, Zacchi, G. A review of the production of ethanol from softwood. Appl Microbiol Biotechnol. 2002;59:618–628. DOI: 10.1007/s00253-002-1058-9

[54] Jorgensen H, Kristensen JB, Felby C. Enzymatic conversion of lignocellulose into fermentable sugars: challenges and opportunities. Biofuels Bioprod Biorefin. 2007;1:119–134. DOI: 10.1002/bbb.4

[55] Capanema EA, Balakshin MY, Kadla JF. Quantitative characterization of a hardwood milled wood lignin by NMR spectroscopy. J Agric Food Chem. 2005;53:9639–9649. DOI: 10.1021/jf0515330

[56] Lewis NG, Yamamoto E. Lignin: occurrence, biogenesis and biodegradation. Ann Rev PlantPhysiolPlantMolBiol.1990;41:455–496.DOI:10.1146/annurev.pp.41.060190.002323

[57] Ebringerova A. Hemicelluloses as biopolymeric raw materials. Papier Proceedings Paper: Cellulose-chemists roundtable discussion on environmental considerations during production, processing and use of dissolving pulps. Baden, Germany, 1992). 1992;46:726–733.

[58] Puls J. Chemistry and biochemistry of hemicelluloses. Relationship between hemicellulose structure and enzymes required for hydrolysis. Macromol Symposia. 1997;120:183–196. DOI: 10.1002/masy.19971200119

[59] Irmak S, Öztürk İ. Hydrogen rich gas production by thermocatalytic decomposition of kenaf biomass. Int J Hydrogen Energy. 2010;35:5312–5317. DOI: 10.1016/j.ijhydene. 2010.03.081

[60] de Wit D, Maat L, Kieboom APG. Carbohydrates as industrial raw materials. Ind Crops Products. 1993;2:1–12. DOI: 10.1016/0926-6690(93)90004-S

[61] Agirrezabal-Telleria I, Gandarias I, Arias PL. Production of furfural from pentosan-rich biomass: analysis of process parameters during simultaneous furfural stripping. Bioresour Technol. 2013;143:258–264. DOI: 10.1016/j.biortech.2013.05.082

[62] Rafiqul ISM, Sakinah AMM. Processes for the production of xylitol-a review. Food Rev Int. 2013;29:127–156. DOI: 10.1080/87559129.2012.714434

[63] Yin R, Liu R, Mei JY, Fei W, Sun X. Characterization of bio-oil and bio-char obtained from sweet sorghum bagasse fast pyrolysis with fractional condensers. Fuel. 2013;112:96–104. DOI: 10.1016/j.fuel.2013.04.090

[64] Fisher T, Hajaligol M, Waymack B, Kellogg D. Pyrolysis behaviour and kinetics of biomass derived materials. J Appl Pyrolysis. 2002;62:331–349. DOI: 10.1016/S0165-2370(01)00129-2

[65] Chhiti Y, Kemiha M. Thermal conversion of biomass, pyrolysis and gasification: a review. Int J Eng Sci. 2013;2:75–85. ISSN: 2319 – 1813 ISBN: 2319-1805

[66] Xu X, Matsumura Y, Stenberg J, Antal MJ Jr. Carbon-catalyzed gasification of organic feedstocks in supercritical water. Ind Eng Chem Res. 1996;35:2522–2530. DOI: 10.1021/ie950672b

[67] Irmak S, Kurtulus M, Hasanoğlu (Hesenov) A, Erbatur O. Gasification efficiencies of cellulose, hemicellulose and lignin fractions of biomass in aqueous media by using Pt on activated carbon catalyst. Biomass Bioenerg. 2013;49:102–108. DOI: 10.1016/j.biombioe.2012.12.016

[68] Behrendt F, Neubauer Y, Oevermann M, Wilmes B, Zobel N. Direct liquefaction of biomass - review. Chem Eng Technol. 2008;31:667–677. DOI: 10.1002/ceat.200800077

[69] Bu Q, Lei H, Zacher AH, Wang L, Ren S, Liang J, Wei Y, Liu Y, Tang J, Zhang Q, et al. A review of catalytic hydrodeoxygenation of lignin-derived phenols from biomass pyrolysis. Bioresour Technol. 2012;124:470–477. DOI: 10.1016/j.biortech.2012.08.089

[70] Wei Z, Zeng G, Huang F, Kosa M, Sun Q, Meng X, Huang D, Ragauskas AJ. Microbial lipid production by oleaginous Rhodococci cultured in lignocellulosic autohydrolysates. Appl Microbiol Biotechnol. 2015;99:7369–7377. DOI: 10.1007/s00253-015-6752-5

[71] Irmak S, Meryemoglu B, Hasanoglu H, Erbatur O. Does reduced or non-reduced biomass feed produce more gas in aqueous-phase reforming process? Fuel. 2015;139:160–163. DOI: 10.1016/j.fuel.2014.08.028

[72] Meryemoglu B, Irmak S, Hasanoglu A. Production of activated carbon materials from kenaf biomass to be used as catalyst support in aqueous-phase reforming process. Fuel Proc Technol. 2016;151:59–63. DOI: 10.1016/j.fuproc.2016.05.040

[73] Balan V, da Costa Sousa L, Chundawat SP, Vismeh R, Jones AD, Dale BE. Mushroom spent straw: a potential substrate for an ethanol-based biorefinery. J Ind Microbiol Biotechnol. 2008;35:293–301. DOI: 10.1007/s10295-007-0294-5

[74] McKendry P. Energy production from biomass (part 2): conversion technologies. Bioresour Technol. 2002;83:47–54. DOI: 10.1016/S0960-8524(01)00119-5

[75] Allen C, Kaiser J, Cordsiemon R. Fact Sheet for Planting and Managing Giant Miscanthus in Missouri for the Biomass Crop Assistance Program (BCAP); 2011. USDA-Natural Resources Conservation Service, Plant Materials Center. Elsberry, MO 63343.

[76] Pyter R, Voigt T, Heaton E, Dohleman F, Long S. Growing Giant Miscanthus in Illinois. Growers Guide, University of Illinois; 2007. Available from: http://miscanthus.illinois.edu/wp-content/uploads/growersguide.pdf

[77] Wood I. Kenaf: The Forgotten Fiber Crop. The Australian New Crops Newsletter 10; 2003. Available from: http://www.newcrops.uq.edu.au/newslett/ncn10212.htm

[78] USDA, FAS. Grain: World Markets and Trade, Jan. 12; 2015. Marketing Year Oct. 1, 2014-Sept. 30, 2015. Available from: http://www.worldofcorn.com/pdf/WOC-2015.pdf

[79] US Weekly Corn Update: Ethanol Demand Weighs on Corn Futures, See It Market; 2015. Available from: http://www.seeitmarket.com/us-weekly-corn-update-ethanol-demand-weighs-on-corn-futures-14780/

[80] UNICA. Statistics of Sugarcane Sector-2009. Available from: http://www.unica.com.br/dadosCotacao/estatistica/

[81] Waclawovsky AJ, Sato PM, Lembke CG, Moore PH, Souza, GM. Sugarcane for bioenergy production: an assessment of yield and regulation of sucrose content. Plant Biotechnol J. 2010;8:263–276. DOI: 10.1111/j.1467-7652.2009.00491.x

[82] Silalertruksa T, Gheewala SH. Environmental sustainability assessment of bio-ethanol production in Thailand. Energy. 2009;34:1933–1946. DOI: 10.1016/j.energy.2009.08.002

[83] Mejean A, Hope C. Modeling the costs of energy crops: a case study of US corn and Brazilian sugarcane. Energy Pol. 2010;38:547–561. DOI: 10.1016/j.enpol.2009.10.006

[84] Laureysens I, Bogaert J, Blust R, Ceulemans R. Biomass production of 17 poplar clones in a short-rotation coppice culture on a waste disposal site and its relation to soil characteristics. Ecol Manage. 2004;187:295–309. DOI: 10.1016/j.foreco.2003.07.005

[85] Kopp RF, Abrahamson LP, White EH, Volk TA, Nowak CA, Fillhart RC. Willow biomass production during ten successive annual harvests. Biomass Bioenergy. 2001;20:1–7. DOI: 10.1016/S0961-9534(00)00063-5

[86] Stanton B, Eaton J, Johnson J, Rice D, Schuette B, Moser B. Hybrid poplar in the pacific northwest: the effects of market-driven management. J Forest. 2002;100:28–33.

[87] Smith GA, Buxton DR. Temperate zone sweet sorghum ethanol production potential. Bioresour Tech. 1993;43:71–75. DOI: 10.1016/0960-8524(93)90086-Q

[88] Rains GC, Cundiff JS, Vaughan DH. Development of a whole-stalk sweet sorghum harvester. Trans ASAE. 1990;33:56–62. DOI: 10.13031/2013.31294

[89] Hunter EL, Anderson IC. Sweet sorghum. In: Janick J, editor. Horticultural reviews 21. New York, NY: John Wiley and Sons; 1997. pp. 73–104. DOI: 10.1002/9780470650660

[90] Stevens G. Sweet Sorghum for Biofuel Production; 2014. Available from: http://articles.extension.org/pages/26634/sweet-sorghum-for-biofuel-production

[91] Kadam KL, McMillan JD. Availability of corn stover as a sustainable feedstock for bioethanol production. Bioresour Technol. 2003;88:17–25. DOI: 10.1016/S0960-8524(02)00269-9

[92] Pordesimo L, Edens WC, Sokhansanj S. Distribution of above ground biomass in corn stover. Biomass Bioenerg. 2004;26:337–343. DOI: 10.1016/S0961-9534(03)00124-7

[93] Bach QV, Chen WH, Chu YS, Skreiberg Ø. Predictions of biochar yield and elemental composition during torrefaction of forest residues. Bioresour Technol. 2016;215:239–246. DOI: 10.1016/j.biortech.2016.04.009

[94] Guo Z, Sun C, Grebner DL. Utilization of forest derived biomass for energy production in the U.S.A.: status, challenges, and public policies. Int Forest Rev. 2007;9:748–758. DOI: 10.1505/ifor.9.3.748

[95] Betancur GJV, Pereira N Jr. Sugar cane bagasse as feedstock for second generation ethanol production. Part I: diluted acid pretreatment optimization. Electron J Biotechnol. 2010;13:1–9. ISSN: 0717-3458

[96] Kingston D, Novelli GF, Cerrutti P, Recupero MN, Blasco M, Galvagno MA. Use of grape pomaces to produce biomass of a *Komagataella pastoris* strain expressing a bovine chymosin activity. Food Sci Nutr. 2014;2:734–743. DOI: 10.1002/fsn3.128

[97] Agger J, Viksø-Nielsen A, Meyer AS. Enzymatic xylose release from pretreated corn bran arabinoxylan: differential effects of deacetylation and deferuloylation on insoluble and soluble substrate fractions. J Agric Food Chem. 2010;58:6141–6148. DOI: 10.1021/jf100633f

Chapter 10

Catalytic Biomass Valorization

Aiguo G. Wang, Danielle Austin and Hua Song

Additional information is available at the end of the chapter

http://dx.doi.org/10.5772/65826

Abstract

Biomass is a significant non-conventional energy reserve, which has been considered as a promising alternative over other renewable sources such as solar, wind or hydroelectric storage due to its comparatively ample availability. A variety of biomass types can be converted into useful products via bioenergy technologies. The deep understanding and knowledge of these processes are necessary for optimization and advancement in a cost-effective way. A comprehensive comparison and discussion is conducted with respect to biochemical and thermochemical conversion technology such as microbic digestion and fermentation, pyrolysis, liquefaction and gasification. Pyrolysis is the process of converting biomass into bio oil, charcoal and gaseous factions by heating anaerobically to above 500°C. Liquefaction is a low temperature (LT) and high-pressure thermochemical process to produce marketable liquid over suitable catalysts under hydrogen or reductive environment. Gasification is the conversion of biomass into preferred combustible gas mixture (syngas) via the partial oxidation at high temperature, typically in the range of 800–900°C. The product gas is more versatile and can be burned in gas turbine for electricity production or synthesis of high-value chemicals. The parametric impact, mechanism, development status and future direction have been summarized for each of these technologies with the aim to pave the way for optimization of future investigation.

Keywords: biomass, pyrolysis, gasification, liquefaction

1. Introduction

The declining reserves and fluctuating prices of fossil fuels necessitate a switch from conventional to renewable power sources, for instance, solar, wind, biomass and hydroelectric generation [1]. Biomass is considered the fourth largest source of energy in the world, currently supplying more than 10% of primary energy. It also provides a potential source of valuable chemicals, such as reducing sugars, furfural, ethanol and other products by using biochemical or chemical and thermochemical procedures [2, 3]. Due to its widespread availability, renewable nature and neutral relation to global warming, the potential of biomass to meet the need for the world energy has been widely recognized. Furthermore, biomass utilization has an advantage over other renewable sources on account of its topographical independence and easy storage and transportation.

Biomass is the term for all biologically produced matter, including land and water-based vegetation, as well as organic wastes, which can be used directly or indirectly by converting it into liquid or gaseous fuel. As the necessity for a renewable and sustainable energy becomes more significant, research interest in biomass conversion to fuel or high value-added chemicals has increased dramatically. The distinction between the energy carriers originated from biomass can be made based on their ability to provide heat, electricity and engine fuels. The famous *Van Krevlen* diagram is usually used for comparing biomass and fossil fuels in terms of their O/C and H/C ratios. The lower the respective ratios, the greater the energy content of the material. More details about this diagram can be obtained from the literature [4].

Currently, there are a number of different approaches used for biomass conversion, such as direct combustion for heat or electricity, thermochemical conversion to bio-oil, chemicals and syngas, and biochemical transformation of biogas or bioethanol. Combustion can be widely used on various scales to obtain heat or electricity from biomass; however, the energy efficiency of this conversion is low. Co-combustion of biomass in coal fire plants is an attractive option to improve the conversion efficiency. Biochemical processing involves anaerobic digestion, the decomposition of biomass via bacterial activity without oxygen to yield biogas and alcoholic fermentation, which is used commercially on a large scale to produce ethanol from sugar and starch crops. In this chapter, more attention has been concentrated on the thermochemical conversion of biomass, including gasification, pyrolysis and liquefaction, due to the diversity of products and high conversion efficiency in the process. The differences of the three process conditions are shown in **Table 1**.

An in-depth survey is carried out regarding the mechanism, parametric impacts, products, techno-economics, development status and future direction for each of the three thermochemical technologies. There are a wealth of articles regarding biomass conversion, and this chapter provides a comprehensive overview with a direct comparison of the current technologies highlighting the benefits and drawbacks of each with emphasis on catalytic biomass conversion.

Process	Temperature, K	Pressure, MPa	Drying	Catalyst
Pyrolysis	650–800	0.1–0.5	Unnecessary	Yes
Liquefaction	525–600	5–20	Necessary	No
Gasification	873–1273	1–2	Necessary	Yes

Table 1. Comparison of thermochemical process conditions [5].

2. Biomass valorization

2.1. Pyrolysis

2.1.1. Mechanism

Pyrolysis is the thermal decomposition of biomass in the absence of oxygen, as illustrated in **Figure 1**. Slow pyrolysis occurs when there is a slow heating rate over a longer residence time (5–30 min). The extended residence time and slow heating rate facilitate the increased production of char products and have been used throughout history for the production of charcoal. Fast pyrolysis occurs when there is a much faster heating rate with much shorter residence times. Increased heating rates and faster residence time result in greater liquid and gas products, making fast pyrolysis more desirable for the purposes of producing bio-oil and valuable chemicals [7]. Typically, decomposition occurs in just a few seconds at atmospheric pressure and relatively low temperatures between 300 and 600°C [8, 9]. The quick reaction time and mild operating conditions make pyrolysis an attractive option for producing bio-oil and valuable chemicals from the cheap feedstock of biomass.

A mixture of a complex feedstock and multiphase products make it difficult to study the exact mechanism of pyrolysis, but there is a general understanding. Initially upon heating, water is removed, resulting in a dry fuel that undergoes primary pyrolysis. Decomposition occurs upon the transfer of heat from the surrounding area to the biomass particles resulting in solid char, bio-oil, gas and water, constituting the primary stage of pyrolysis [8]. An increased

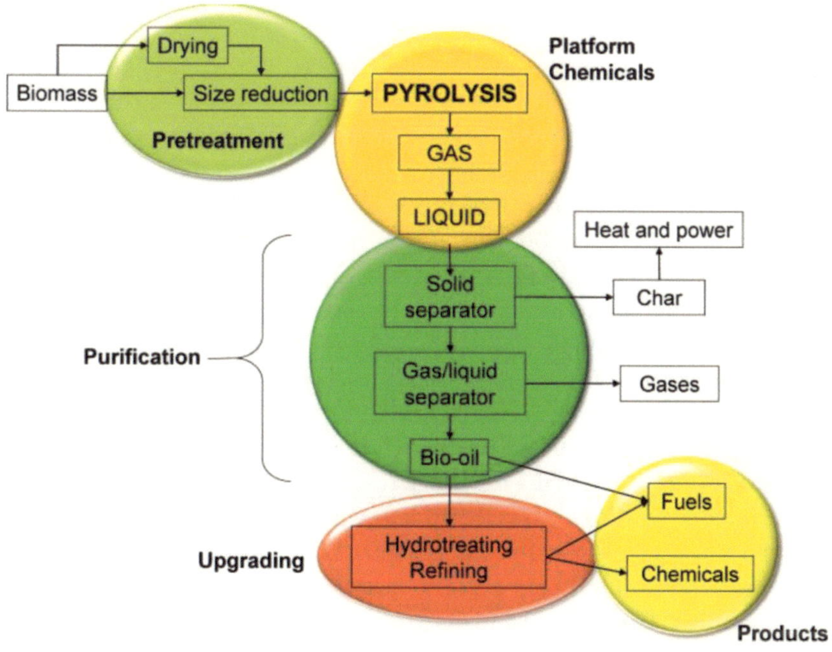

Figure 1. Process schematic of biomass pyrolysis and upgrading [6].

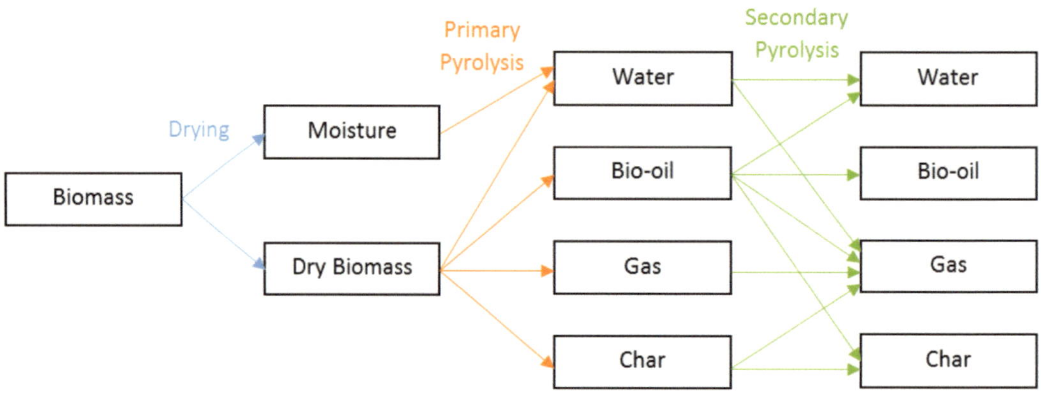

Figure 2. Primary and secondary pyrolysis including arrows indicating reaction pathway [8].

surface area results in better heat transfer, which causes a faster and more complete reaction [9, 10]. For this reason, biomass is typically mechanically ground into a fine powder prior to the reaction. During the secondary stage, further decomposition occurs in addition to various other reactions resulting in an increase in gaseous products. A summary of these processes can be seen in **Figure 2**. Fast pyrolysis can have a biomass to bio-oil conversion of 70–80%. Liquid yield can be maximized by a fast residence time, preventing the secondary decomposition of products. Due to the importance of residence time and heat transfer in pyrolysis, much research has gone into reactor design which is reviewed in reference [9].

2.1.2. Products

Pyrolysis results in a wide range of products that can be divided into the categories of solid char, liquids (water and bio-oil) and gases. Char is the term used for all of the solid products including organic matter (high carbon content) and ash [8]. Water is produced as a product of pyrolysis as well as during the initial drying stage via evaporation [8, 9]. The bio-oil component is a brown, polar liquid comprised of a mix of oxygenated compounds, which differs with feedstock and reaction conditions [9, 10]. At moderate temperatures, gas products are mainly comprised of CO, CO_2 and CH_4, with H_2 and C_XH_Y gases also being produced at higher temperatures [8, 10]. The amount of solid, liquid and gas produced varies greatly depending on the biomass composition as well as the reaction temperature. The temperature has a greater effect on the amount of product in each phase, while the feedstock composition has a greater effect on the bio-oil elemental composition. A higher C/O ratio in the feedstock results in a higher liquid yield and lower gas yield, while the char yield remains relatively unaffected. An increase in hydrogen in the feedstock results in more water formation, more bio-oil and less carbon dioxide and C_xH_y gases [8]. At low temperatures (<350°C) and slow residence times, char is the favoured product [8–10]. Longer residence times allow for the reaction of gas and liquid products to form char [7]. The best yields of liquid product are achieved at moderate temperatures (400—550°C) and fast residence times. These conditions allow for primary pyrolysis to occur, but the fast residence time allows for products to be collected before secondary reactions can take place [8–10]. At very high temperatures (HT) (>800°C), a higher yield of gas products is observed due to the presence of secondary reactions, which further break down the liquid and char components into gases.

The liquid product composition generally mimics the biomass carbon, hydrogen and oxygen composition. Due to the high oxygen content in biomass, the liquid product of pyrolysis is made up of a large portion of oxygen-containing compounds. The oxygen-containing compounds including alcohols and acids make the bio-oil reactive and acidic causing issues for storage and transportation through established pipelines [11–14]. As a result, bio-oil must be upgraded before it can be transported or used as a fuel. An increase in temperature leads to a slight decrease in oxygen content and a slight increase of carbon and hydrogen content in the bio-oil. This results in a small increase of aromatic compounds, which marginally improves the oil quality [8].

2.1.3. Developmental status

The majority of research on pyrolysis involves optimizing the upgrading process to acquire a higher quality product. Hydrodeoxygenation and catalytic cracking are the most popular processes currently being utilized for upgradation [12]. Hydrodeoxygenation involves removing oxygen from the liquid product in the form of water through a hydrogen supply [9, 15]. This process is expensive due to the high operating pressures (up to 20 MPa) and the high cost of hydrogen [9, 15]. Catalytic cracking is a fairly new technology, which uses a catalyst to promote a range of reactions including dehydration, decarboxylation, aromatization, alkylation and polymerization [12]. Zeolite catalysts have found to be particularly effective at aromatization and have been found to remove oxygen in the form of carbon dioxide [9, 14]. The addition of metals such as silver or nickel can increase selectivity and allow for optimized products [12]. Catalytic cracking is significantly cheaper than hydrodeoxygenation due to the relatively mild operating conditions and absence of hydrogen; however, the coking of catalysts is a significant concern and due to the lack of hydrogen supply, the product suffers from a low H/C ratio [12, 16].

At the current time, the combination of pyrolysis and catalytic cracking in one process has been attracting more attention [10]. One research group is combining pyrolysis with catalytic cracking under a methane environment, in a novel process called methanolysis [16, 17]. This process is operated at atmospheric pressure and moderate temperatures on a zeolite catalyst with a supply of methane. Methanolysis is economically attractive due to the relatively low operating conditions and the wide availability and low cost of natural gas of which methane is the major component. The aim of this research is to develop a catalyst that will allow pyrolysis to occur in a methane environment, while at the same time upgrading the liquid product. Promising results were found during feasibility studies, where it was shown that not only was the bio-oil produced of greater quality, but the methane was incorporated into the product [16, 17]. **Table 2** compares pyrolysis in inert gas, hydrogen and methane environments with and without catalysts for a sawdust feedstock. An increased oil yield accompanied by a decrease in oxygen content and an increase in H/C ratio can be observed in a methane environment using a catalyst. It is also clear that the metal used in the catalyst can affect both oil yield and quality. These results are very promising for creating a high quality bio-oil in an economical and environmentally friendly way.

Trials	Oil* yield (wt %)	Water formed (mg/g)	Oil quality		
			H/C ratio	O content (wt %)	O/C atomic ratio
Biomass, inert	5.47	97.0	1.62	5.25	0.226
Biomass, inert, Ag/ZSM-5	4.07	135.6	1.29	0.18	0.009
Biomass, 30% H_2	4.17	73.4	1.46	3.41	0.145
Biomass, 30% H_2, Ag/ZSM-5	3.42	100.2	1.45	0.45	0.024
Biomass, 30% CH_4	4.68	119.0	1.38	0.22	0.009
Biomass, 30% CH_4, Ag/ZSM-5	4.85	128.3	1.76	0.07	0.003
Biomass, 30% CH_4, Ag/P-Ce-ZSM-5	6.89	110.9	2.26	7.35	0.356
30% CH_4, Ag/ZSM-5	0	0	–	–	–

*Moisture-free liquid collections with boiling point less than 150°C.

Table 2. Results of the pyrolysis of sawdust under different environments [16].

2.1.4. Future direction

The next steps of pyrolysis research are optimizing product yield and quality, which can be achieved through the use of catalytic cracking. This involves gaining a better understanding of the complex mechanism involving the catalyst by using model compounds. Once there is a clear understanding of the catalyst's role, an industrially viable catalyst and process can be developed to produce high-quality fuels that are cost competitive with oil-derived alternatives. Methanolysis is considered a promising new technology, which can be a sustainable alternative to traditional fuel sources. In addition to developing an optimal catalyst, work must be done to improve methane conversion and incorporate catalyst regeneration into the reaction.

2.2. Liquefaction

2.2.1. Mechanism

Liquefaction is the thermochemical conversion of biomass to primarily yield liquid fuel at moderate temperature and high pressures. This is a chemical reforming process of organic matters, usually with a high hydrogen partial pressure and catalysts to enhance the rate of reaction and/or to improve the selectivity of the product. This approach is summarized in **Figure 3**. In the process of liquefaction, the carbonaceous materials of biomass are ultimately converted into liquefied products through a series of complex physical and chemical reactions [18, 19]:

- Micellar-like substructures are formed by solvolysis;
- Smaller and soluble molecules are generated via depolymerization;
- New molecular rearrangements occur through thermal decomposition, including dehydration, decarboxylation, C-O and C-C bonds rupture.
- Hydrogenolysis and hydrogenation of functional groups.

http://dx.doi.org/10.5772/65826

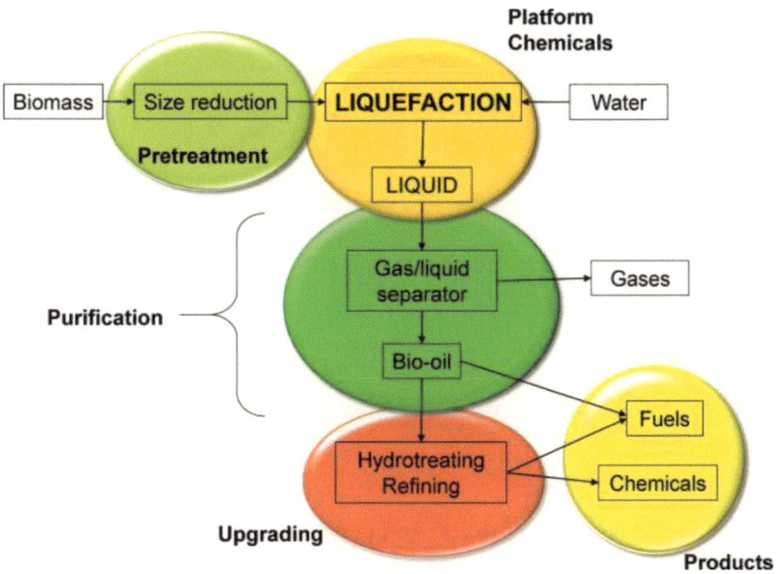

Figure 3. Process schematic of biomass liquefaction and upgrading [6].

Another remarkable feature of liquefaction is that it maximizes the liquid-yields with high quality. One important quality index of product oil is the H/C ratio, which can be increased by removing the oxygen content. The lower oxygen content makes the oil more chemically stable and requires less upgrading to the hydrocarbon product [20]. Generally, biomass contains typically 40–50 wt.% (DAF basis) of oxygen, which can be removed via dehydration and decarboxylation as the form of H_2O and CO_2, respectively. Thermodynamically, removal of H_2O and CO_2 from biomass can be considered as the best option to reduce the oxygen content. Removal of H_2O ultimately forms carbon as the remaining product such as charcoal while removal as CO_2 is prone to generate the product with high H/C ratio. Under hydrothermal upgrading reaction conditions, the decarboxylation selectivity is almost constant, 0.52 [21]. Dehydration and decarboxylation reactions are conceptually described as follows:

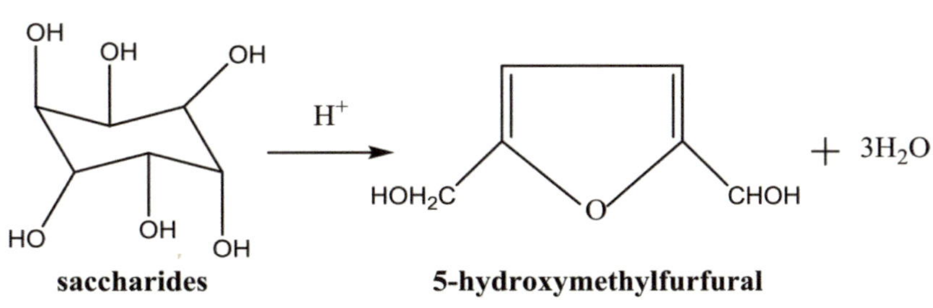

Dehydration of saccharides to 5-hydroxymethylfurfural [22]

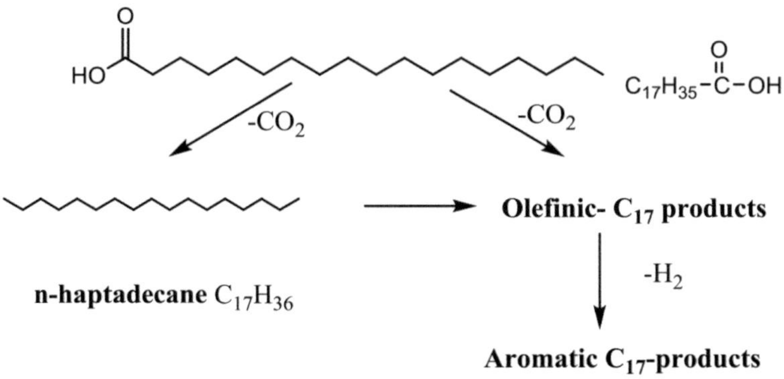

Simplified decarboxylation network of stearic acid to diesel products [23]

2.2.2. Conversion of biomass components

As we know, biomass is a broad definition and contains a wide range of substances with different compositions. The main components are categorized four groups: carbohydrates, lignin, protein and lipids. The protein and lipids are prevalent in the dried distiller grains with soluble (DDGS) and slaughterhouse waste [24, 25], which are not further discussed in this chapter. More concerns are focused on the conversion of carbohydrates and lignin.

2.2.2.1. Carbohydrates

The most abundant carbohydrates contained in the biomass are polysaccharides, cellulose, hemicellulose and starch. Carbohydrates first are hydrolyzed to glucose and other saccharides and further degraded into small molecule compounds at higher temperature. The hydrolysis rate varies with the components. Generally, the hydrolysis rate of hemicellulose and starch is much faster than cellulose, since the cellulose has relatively stable crystalline structure. The conversion of cellulose can achieve 100% at 280°C within 2 min. The addition of carbon dioxide increases cellulose hydrolysis rate, and this is most likely attributed to the formation of carbonic acid, acting as an acid catalyst. However, the beneficial effect is insignificant above 260°C [26]. Hemicelluloses are readily soluble and hydrolyzed above 180°C [27] and close to 100% conversion of hemicellulose is obtained for various wood and herbaceous biomass materials at 230°C within 2 min [28]. Several attempts have been made to reveal the general degradation mechanism of carbohydrates. Glucose is commonly chosen as the model compound because it is one of the main hydrolysates. Even though the employed models are different in details, the overall pattern is quite similar [29, 30]. An overall degradation route for glucose is presented in **Figure 4**.

2.2.2.2. Lignin

Lignin is an aromatic heteropolymer, which is relatively difficult to be degraded chemically or enzymatically [27]. The hydrothermal degradation of lignin can occur at high temperature of 350—400°C, and the main products are catechol, phenols and cresols [37, 38]. However, it can be catalytically hydrolyzed into various phenols and methoxy phenols by cleavage

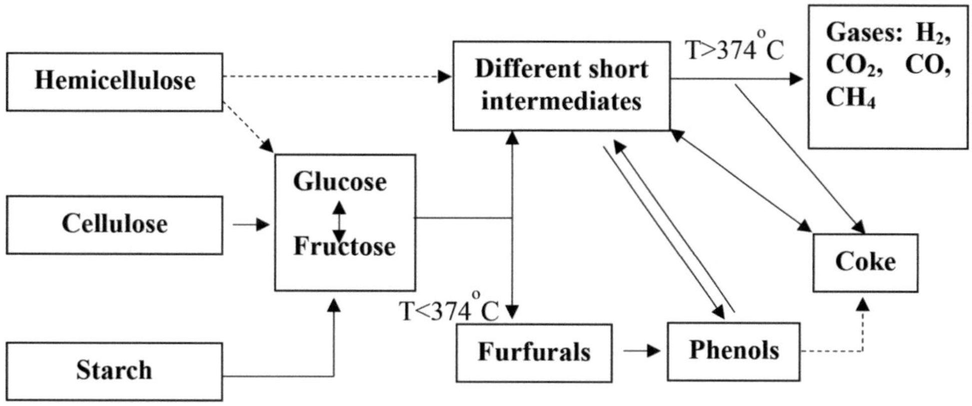

Figure 4. A simplified reaction mechanism for carbohydrate hydrolysis at sub- and supercritical conditions [69].

of ether-bonds under alkaline environment. The reaction temperature for hydrothermal processing of lignin is 200–300°C. The hydrothermal degradation pattern of lignin is shown in **Figure 5**.

2.2.3. *Impact of parameters*

Biomass liquefaction can be influenced by numerous parameters, including temperature, pressure, catalyst, residence time, heating rate, particle size, solvent, reducing agent and the type of biomass. In this chapter, the effect of temperature, pressure, catalyst and residence time on biomass liquefaction will be discussed separately.

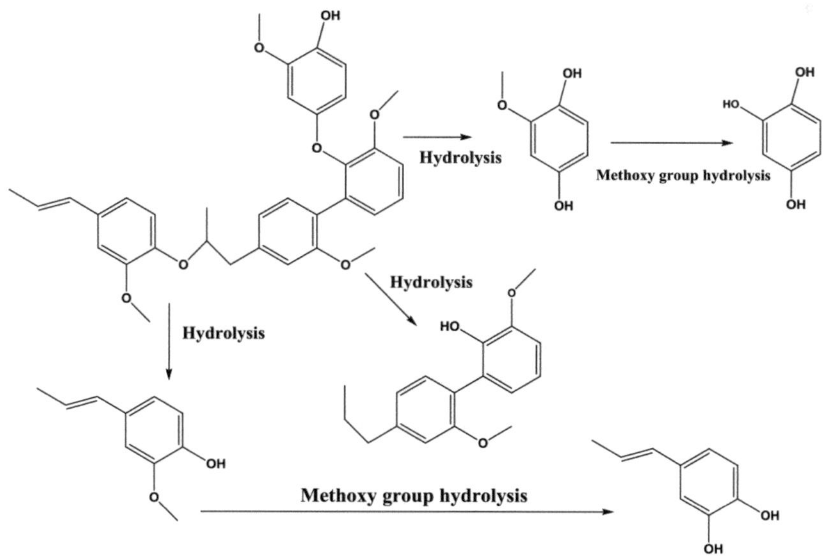

Figure 5. Simplified scheme of lignin degradation [39].

2.2.3.1. Temperature

The influence of temperature on the oil yield has been investigated by many researchers [31–36]. The results are illustrated in **Figure 6**. It is obvious that the maximum oil yield can be achieved in the intermediate temperature range. The reason is that temperature enhances the synergetic effect on the yield of liquids due to more biomass fragmentations generated with temperature. When temperature is sufficiently high to provide the activation energy for bond rupture, depolymerization occurs extensively, increasing more free radicals and fragmented species for further repolymerization at higher temperature. Therefore, the dominant reaction is depolymerization of biomass in initial stage, and repolymerization is prevailing in the later stages, and intermediate temperature is expected to yield higher amounts of bio-oil [40].

2.2.3.2. Pressure

Pressure is another factor affecting hydrothermal liquefaction (HTL). The hydrolysis rate and biomass dissolution are controlled by maintaining the pressure above the critical pressure of medium, which may enhance the favourable reaction pathways thermodynamically for production of liquid fuels or gas yield [40]. In addition, the density of medium solvent can be increased with pressure, which makes the medium more efficiently penetrates into molecules of biomass components, promoting the decomposition and extraction. However, the influence of pressure on the yield of oil or gas is negligible under supercritical conditions [41–44]. It is reported by Sangon et al. [41] that the oil yield was slightly increased with pressure increase from 7 to 12 MPa during the coal liquefaction in supercritical conditions. Kabyemela et al. [43, 44] observed that pressure rise had a certain degree of adverse effect on the degradation of glucose. The same phenomenon was witnessed in the process of hydrolysis and pyrolysis of cellobiose liquefaction at 400°C. This is because under the supercritical conditions, the increase in pressure results in the cage effect for C–C bonds, which inhibits the breakage of

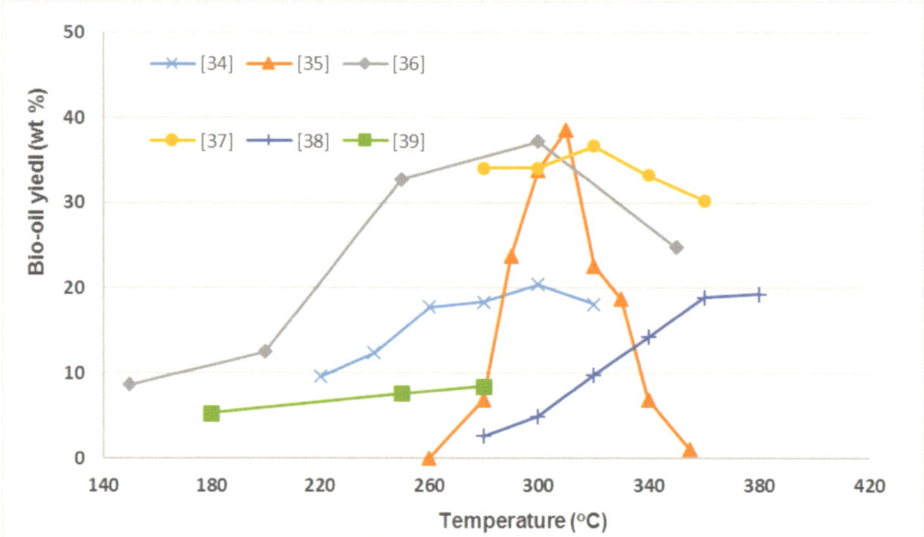

Figure 6. Oil yield dependence of temperature as reported by different research groups [31–36].

bonds and ends up with low degree of biomass fragmentation. Therefore, variation of pressure in supercritical region plays an insignificant role for overall oil yield.

2.2.3.3. *Catalyst*

Generally, homogeneous catalysts, such as alkali salts, organic and inorganic acids, are significantly more active than heterogeneous catalysts in the process of biomass liquefaction. Heterogeneous catalysts, such as Ni-containing catalysts, are frequently used in low-temperature water gasification of biomass. Homogeneous catalysts in the form of alkali salts have been commonly used in biomass liquefaction. For example, sodium carbonate has the catalytic effect of increasing oil yield and reducing char, and potassium carbonate can promote the water-gas shift reaction and lessen solid residue, while the catalysts sodium hydroxide and potassium hydroxide can enhance decarboxylization and increase isomerization of glucose. The heterogeneous catalysts Ni,Ru have the enhancement of H_2 and CH_4 yield in the hydrothermal conditions. According to the studies [45–48], the organic acid catalysts exhibit better catalytic activity than inorganic acid catalysts with respect to the solid residue content. Salts (phosphates, sulphates, chlorides, acetates, carbonates) don't have as good of catalytic performance as alkali (NaOH, KOH) at typical liquefaction temperatures.

2.2.3.4. *Residence time*

The effect of residence time on hydrothermal liquefaction is mainly demonstrated in two aspects: the overall conversion of biomass and the composition of products. According to the study of Boocock and Sherman [49], the bio-oil yield was reduced with increase in residence time. Other researchers [33, 36] witnessed the same results. It is observed by Yan et al. [50] that there is no significant increase in liquid yield for longer residence time. Therefore, the short residence time favours more liquid products in the liquefaction. There may be many reasons to explain the effect of residence time. One convincing explanation is that decrease in oil yield may be as a result of secondary and tertiary reactions for longer residence time when biomass conversion reached the saturation point.

In addition, the duration of liquefaction influences the composition of oil product. The significant difference between the decomposition products obtained for longer and shorter residence time was observed by Karagoz et al. [50]. For instance, some furan carboxaldehydes, benzoic acid and phthalate were gained for 60 min, which were not observed in the case of shorter residence time (15 min). Obviously, the preasphaltenes and asphaltene can be decomposed into lighter products for longer residence time, enhancing the yield of light oil and gases. Therefore, it is necessary to inhibit the further disintegration of lighter products in order to obtain high oil yield. One way to prevent oil decomposition is addition of reducing agents like tetraline, hydrogen and syngas. This strategy not only stabilizes the radicals and lighter products but also favours the high oil yield.

2.2.3.5. *Feedstock*

Theoretically, any biomass can be converted into bio-oil via liquefaction regardless of water content, including forestry, agriculture residues, sewage sludge and food waste. Due to the different behaviours of major biomass components (such as lignin, hemicellulose and cellulose) on the temperature variations, the liquid products and overall yield can be influenced significantly by the heterogeneity of biomass stock. More bio-oil can be obtained from feedstock with a higher content of cellulose and hemicellulose. In addition, the physical properties of bio-oil depend on the feedstock type. The bio-oil produced from liquefaction of the loosely structured biomass usually has poor quality and HHV and a low viscosity. This is because the oxygen and moisture contents are high in the liquid products. Therefore, modifications are necessary before the product can be used in engines, turbines and boilers. It is also unsuitable for long-term storage due to the complexity of the oil [51].

2.3. Gasification

Gasification is a high temperature partial oxidation process to optimize gas production, and it involves the extremely complex catalytic conversion of solid/liquid organic compounds into the gas or solid phase in the presence of a gasifying agent (air, steam, oxygen, CO_2). The gas phase, usually called "syngas", is a mixture of hydrogen, carbon monoxide, carbon dioxide, methane and small hydrocarbons. The solid phase, called "char", is actually composed of unconverted organic fraction, fixed carbon as well as ash. Gasification has been considered as a cost-effective way to increase the use of biomass for energy production, allowing widespread biomass utilization [52]. Especially in the beginning of this century, rising oil price and concerns over climate change resulted in the expansion of more advanced biomass gasification projects around the world. One typical biomass gasification process is shown in **Figure 7**.

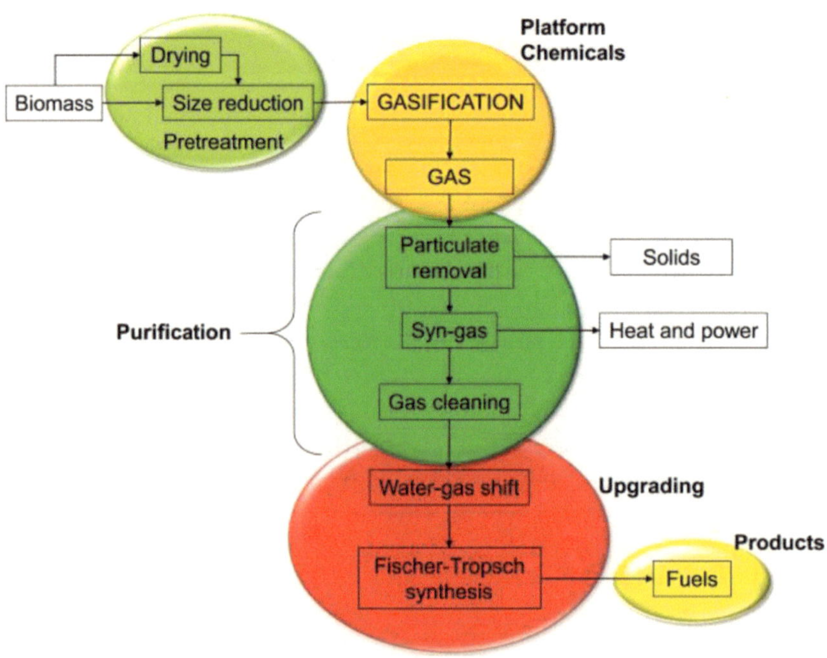

Figure 7. Process schematic of biomass gasification and upgrading [6].

http://dx.doi.org/10.5772/65826

2.3.1. Mechanism

It is believed that gasification involves several sequential and parallel reactions in an oxygen deficient environment. And the primary steps include drying, pyrolysis, reduction and oxidation, shown in **Figure 8**. Initially, the moisture of biomass is evaporated quickly during the drying, and this process is considered complete at 150°C. With the temperature increased, matrix carbonaceous materials, such cellulose and lignin, are decomposed into small species by cracking of chemical bonds within a temperature range of 250–700°C, which can be simply schematized with the following overall reaction:

$$\text{Biomass} \rightarrow H_2 + CO + CO_2 + \text{Hydrocarbons} + H_2O(g) + \text{Tar} + \text{Char}$$

Several complex phenomena, such as heat and mass transfer and reaction kinetics, are involved in this stage. The reaction kinetics is the determining step at low temperatures, while the heat or mass transfer becomes the limiting step at higher temperatures. Then the reduction stage comes into play. It involves all the products of the preceding stages. The final syngas is formed through the complicated reactions between gas mixture and char. The principal reactions include carbonation of char, water gas shift reaction, methanation, tar decomposition and steam reforming [1, 52].

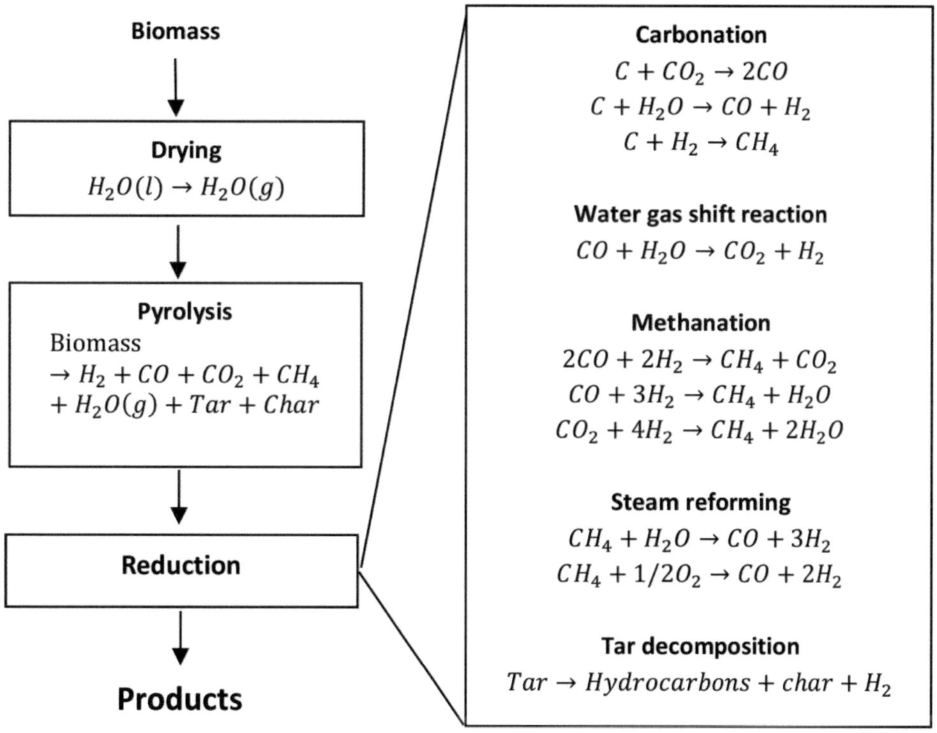

Figure 8. Main stages and the reactions involved in the gasification process [1, 52].

2.3.2. *Parametric impact*

There are many factors influencing the product distribution of biomass gasification, such as feedstock type, quality and moisture, particle size and density, operative conditions, the ratio of steam (or other gas) to biomass and catalysts.

2.3.2.1. *Particle size and density*

The smaller the particle size, the better would be the heat diffusion, which can make the temperature uniform throughout the particle, benefiting the reactions. The effect of particle size on reaction rate is more significant when there is a limitation on size above which heat transfer would be controlling [3]. In addition, the reduction in particle dimension can enhance the carbon conversion and hydrogen yield, improve the syngas efficiency and suppress the formation of tar and char [53, 54]. However, the particle size reduction requires intense energy [55].

Biomass is normally of low density due to the porous structure. The presence of the pores facilitates the heat transfer throughout the pellet, which in turn results in the homogeneous gasification and uniform product composition. As for the dense materials, there is a temperature difference between the exterior and interior of particle, leading to the simultaneous drying, pyrolysis and gasification. Accordingly, a non-homogeneous gas composition is produced.

2.3.2.2. *Operating conditions*

The operating conditions, including pressure, temperature, heating rate and other vital factors, have the potential to affect the gas yield and biomass conversion [54, 55]. The pressure is closely associated with bio-char reactivity and an increase in temperature would involve a faster heating rate, which generates a greater temperature difference over feedstock particles. While the fast and slower heating rate operations largely contribute to the design of gasifier and the desired products. The faster heating rate favours more gas production and less tar yield owing to the volatilization and degradation of the active tars by transforming them to the product gases. The slower heating rate contributes to lower gas yield and higher tar yield because of the recombination of lower-volatility hydrocarbons on the surface of the char particles.

2.3.2.3. *Steam-to-biomass ratio*

The ratio of steam to biomass (S/B) can determine the input energy requirements, outlet gas quality and yield of products. Increasing S/B ratio can positively enhance the reforming, cracking and water gas shift reactions, leading to higher hydrogen yield and consequently generating the syngas with high calorific content [56, 57]. However, there is a threshold limit beyond which increasing S/B will generate excess steam, resulting in the enthalpy loss and reduction in process efficiency [58]. This issue necessitates optimizing the ratio in steam biomass gasification.

http://dx.doi.org/10.5772/65826

2.3.2.4. Catalysts

Catalysts can provide an alternative lower-energy pathway for the reactions to follow. The catalysts used in the biomass gasification are categorized alkaline (Na, K) metal, alumina and zeolites, dolomites and limestones, Ni-based, Zn-based, as well as some rarer metals such as platinum- and ruthenium-based, specifically in relation with gasifier design or the type of biomass feed. Among them, alkaline metal oxides, dolomite and Ni-based catalysts have been testified to greatly promote the reformation reaction [59], and alumina silicates are found to enhance the char gasification effectively, whereas Ni-based catalysts are identified to facilitate the conversion of lighter hydrocarbons [60]. It is essential to develop more efficient and economical catalysts with the ability of enhancing the quality and yield of the desired product while minimizing the residual char and tar [1].

2.3.2.5. Feedstock.

In the case of gasification, due to the indispensable role these constituents play during the process of gasification, the proportion of cellulose and hemicellulose is closely related to the yield of the gaseous products, while the lignin has significant influence on the production of bio-oil. Generally, the higher the ratio of cellulose (including hemicellulose) to lignin in the feedstock, the higher the gaseous products yields. The moisture content of biomass affects the characteristics of the final gas products significantly. A feedstock with more than 30 wt% moisture would result in less gas production, higher tar content and more energy input. Gasification technology can be a good solution for energy production from renewable sources; however, the water and impurities in the gas products can be problematic in downstream FT processes, where a clean gas feed is required [6].

2.3.3. Application

The versatility of gasification manifests in that it can meet the demand for the electricity or thermal energy and industrial purposes. The products obtained from the biomass gasification are shown in **Figure 9**. The gas products, typically including H_2, CO, CO_2, CH_4, H_2O and other gaseous hydrocarbons, can be used to generate heat or electricity by direct firing in diesel engines, gas turbines and boilers. Alternatively, the gas mixture can be further reformed to produce usable fuels such as methanol or hydrogen, which can be transported with high energy densities, enabling the generation of electricity or heat to be centralized based on disperse gasification systems. Renewable biomass is considered as a potential material for gasification to produce syngas, hydrogen or other liquid fuels; however, the economics of current processes favour the use of light hydrocarbons (in natural gas) and coal owing to the characteristic of high oxygen content in the biomass.

On the other hand, bio-syngas originated from biomass gasification can act as the raw material for Fischer-Tropsch Synthesis (FTS) to produce green liquid fuel (diesel) or high value-added chemicals. It is needed that the ratio of H_2 to CO in syngas is adjusted to an appropriate range through water gas shift reaction prior to FTS, and then the syngas is catalytically converted to liquid fuels

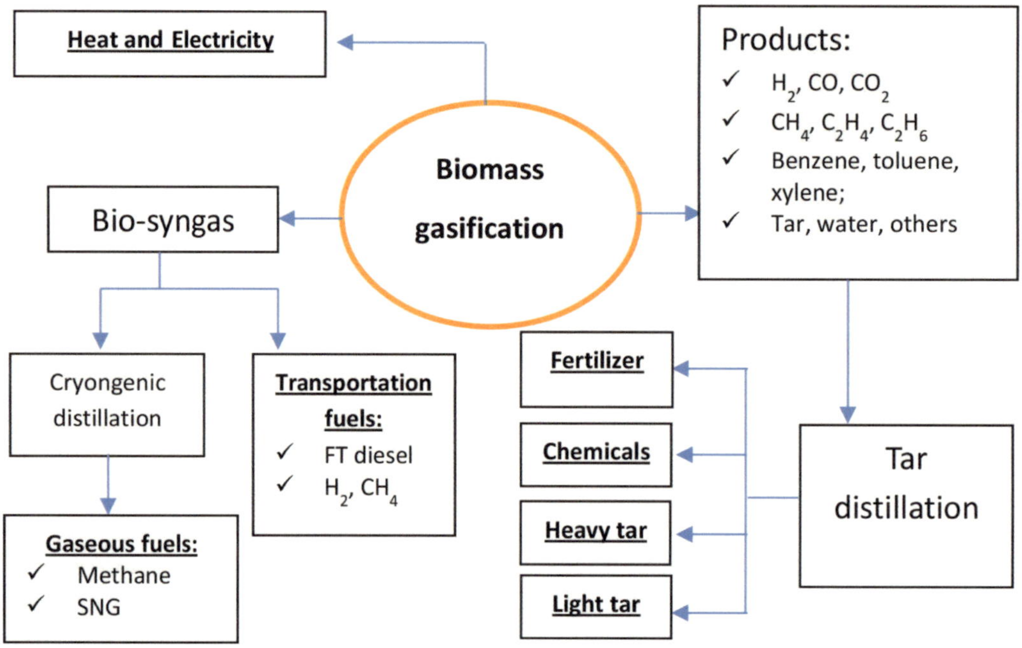

Figure 9. Products from gasification process [61].

and/or chemicals via FT process over Co-based or Fe-based catalysts. This type of synthetic FT diesel fuels has excellent quality. For instance, the emissions can be reduced significantly when fuels are employed in ignition combustion engines because of low content of nitrogenous and aromatic species. In addition, the FT fuels are sulphur-free, which is vital for lowering emissions of exhaust pollutants. Therefore, biomass gasification coupled with FT process provides a promising alternate for mitigating the burden on conventional transportation fuel; however, it requires significant development and scale-up efforts for commercial-scale installations [62].

The poly-generation concept has been developed with advancement in biomass gasification. For instance, co-generation of thermal power with electricity; poly-generation of heat, fertilizer and bio-char; poly-generation of heat, SNG/bio-fuels; and poly-generation of H_2 with heat and electricity. All these approaches not only optimize the thermal efficiency of the process but also provide flexibility and sustainability, thereby enhancing the economic advantage in the long run [3].

2.4. Biochemical transformation

Biochemical conversion of biomass involves the use of bacteria, microorganisms and enzymes to break down biomass into gaseous or liquid fuels, such as biogas or bioethanol. These processes are essentially microbic digestion and fermentation, which are most suitable for the high moisture herbaceous plants, marine crop and manure. The main products in the process are methanol, ethanol and gas mixture (CH_4, CO and CO_2) [63]. This biocatalytic process is considered to be a suitable alternative to chemical catalysis. Some important considerations when selecting the biocatalyst used for the upgrading of biomass are listed in **Table 3**.

(A) Source of biocatalyst	
Whole cell or free enzyme	Whole cells can also be used in permeabilized form
Wild type or mutant enzyme produced by rDNA technology	Some organisms are frequently used as the source for industrial enzymes
Refolded from IBs or use of IBs itself	Choice of the host is important. For enzymes where activity depends upon post-translational modification, higher organisms are used as host expression systems
(B) Free or immobilized form	
Soluble conjugates	For insoluble substrates (which is often the case with biomass), soluble conjugates or enzymes in membrane reactors are preferred
Carrier free or insoluble support	Enzyme aggregates like CLEA or CLEC have high volumetric activity
(C) Operational stability	This may be different from storage stability
(D) Reaction medium	The use of co-solvents is under exploited
(E) Normal or promiscuous activity	As more enzymes, engineered for better promiscuous activity, become available, this application may increase

Table 3. Summary of important parameters of biocatalysts [64].

2.5. Techno-economic analysis

Techno-economic analyses are essential for determining the economic viability of biomass conversion technologies in the research or development stage. On the basis of current performance and targets established for process improvements, evaluation of the production costs can allow us to assess the potential economic feasibility of a process configuration. These results are significant for investing in the technology with high potential for deployment success.

The economic feasibility of fast pyrolysis and upgrading facilities employing two bio-oil upgrading pathways have been evaluated based on the scale of 2000 metric tons/day and 65% biomass conversion in the literature [65]. The results showed that the hydrotreating followed by fluid catalytic cracking (FCC) pathway with hydrogen production via natural gas reforming is the best option with highest IRR (internal rate of return) of 13.3% when compared with single stage hydrotreating. According to the sensitivity analysis, product yield, fixed capital investment (FCI) and biomass cost have significant impacts on the facility IRR. Variation of 5% bio-oil yield can lead to product value (PV) of \$2.60–3.89/gal [66]. The hydrotreating/FCC pathway is proved to be of relatively low risk for project investment based on the Monte Carlo analysis.

Techno-economic analysis was implemented to assess the feasibility of developing a commercial large-scale woody biomass hydrothermal liquefaction (HTL) and upgrading plant with the scale of 2000 dry metric ton/day in the literature [67]. Two cases were involved: state-of-technology (SOT) and goal case considering future improvements. The annual production rate was estimated to be \$42.9 and \$69.9 million gallon gasoline-equivalent (GGE) for SOT and goal case, respectively. The minimum fuel-selling price was estimated to be \$4.44/GGE and \$2.52/GGE. Obviously, the liquid fuel cost of current HTL and upgrading technologies (SOT) is not competitive with the conventional petroleum-based gasoline. For advancing SOT to goal case, future improvements of HTL competitiveness should be focused on reducing organic loss to obtain higher final yield and lowering wastewater treatment cost. Based on the sensitivity analysis, the key factors that influence the production cost significantly are plant scale, yield, upgrading equipment cost and feedstock cost. Although the goal case technology is promising for future liquid fuel production via woody biomass HTL, there are still financial risks resulting from lack of knowledge and experience.

The techno-economic analysis of green fuel production via BG-FT (biomass gasification and Fischere Tropsch) with a scale of 2000 metric tons per day of corn stover was conducted in the literature [68]. Two scenarios are considered: high temperature (HT) gasification and low temperature (LT) gasification. The results showed that the total capital investment required for nth plant scenarios are \$610 million and \$500 million for HT and LT, respectively. Product value (PV) for two scenarios were estimated to be \$4.30/GGE and \$4.80/GGE, but the HT scenario was more desirable due to its higher carbon efficiency and subsequent higher fuel yield. Based on the sensitivity analysis, the capital investment and feedstock cost are chiefly responsible for affecting the PV, whereas the conversion of FT process, feedstock moisture and catalyst cost are least influential parameters.

3. Conclusion

Pyrolysis and liquefaction are inclined to obtain more liquid products (bio-oil), but the stability of bio-oil produced in the pyrolysis process is worse than that generated in the liquefaction process. However, the interest in liquefaction is considerably lower due to the reactors and feed systems, which are more complex and expensive than that for the pyrolysis process. Therefore, upgrading bio-oil by lowering the oxygen content, removing alkalis by means of hydrogenation and catalytic cracking are required for further application of pyrolysis conversion. Catalytic cracking pyrolysis gasification has shown promising application, since the syngas can be used for the production of methanol and hydrogen, which may have a future as fuel for transportation. The economic studies show biomass gasification plants can be as economical as conventional coal fired plants. Compared with chemical catalysts in the process of biomass conversion, biocatalysts have higher selectivity to products and cheaper capital investment. However, the practical industrial application of biocatalysts is limited due to slow conversion process, high dependence of environment and feedstock and difficult recycling of the biocatalysts.

It is noteworthy that one novel technology, achieved by a green catalysis research group at the University of Calgary, is catalytic coal or biomass upgrading using natural gas for liquid chemical or fuel productions. This process directly employs methane as reducing agent instead of hydrogen for removing oxygen from biomass under the facilitation of specially developed low-cost supported catalyst, leading to significantly reduced operation cost.

http://dx.doi.org/10.5772/65826

Acknowledgements

We gratefully acknowledge the financial supports from Natural Sciences and Engineering Research Council of Canada (NSERC, RGPIN/04385-2014).

Author details

Aiguo G. Wang, Danielle Austin and Hua Song*

*Address all correspondence to: sonh@ucalgary.ca

University of Calgary, Calgary, Canada

References

[1] Vineet S. S., Ming Z., Peter C., Joseph Y., Xia Z., Mohammad Z. M., Nilay S., Edward J. An overview of advances in biomass gasification. Energy & Environmental Science. 2016; 9; 2939-2977. DOI: 10.1039/c6ee00935b

[2] Kucuk M. M., Demirbas A. Biomass conversion process. Energy Conversion and Management. 1997;**38**(2):151–165.

[3] Kirubakaran V., Sivaramakrishnan V., Nalini R., Sekar T., Premalatha M., Subramaniam P. A review on gasification of biomass. Renewable and Sustainable Energy Reviews. 2009;**13**:179–186. DOI: 10.1016/j.rser.2007.07.001

[4] Peter M. Energy production from biomass (part 2): Conversion technologies. Bioresource Technology. 2002;**83**(1):47–54. DOI: 10.1016/S0960-8524(01)00118-3

[5] Ayhan D. Biomass resource facilities and biomass conversion processing for fuels and chemicals. Energy Conversion and Management. 2000;**42**(11):1357–1378. DOI: 10.1016/S0196-8904(00)00137-0

[6] Alonso D. M., Bond J. Q., Dumesic J. A. Catalytic conversion of biomass to biofuels. Green Chemistry. 2010;**12**(9):1493–1513. DOI: 10.1039/C004654J

[7] Bahng M., Mukarakate C., Robichaud D. J., Nimlos M. R. Current technologies for analysis of biomass thermochemical processing: A review. Analytica Chimica Acta. 2009;**651**:117–138. DOI: 10.1016/j.aca.2009.08.016

[8] Neves D., Thunman H., Matos A., Tarelho L., Gomez-Barea A. Characterization and prediction of biomass pyrolysis products. Progress in Energy and Combustion Science. 2011;**37**:611–630. DOI: 10.1016/j.pecs.2011.01.001

[9] Bridgwater A. V. Review of fast pyrolysis of biomass and product upgrading. Biomass and Bioenergy. 2012;**38**:68–94. DOI: 10.1016/j.biombioe.2011.01.048

[10] Bridgwater A. V., Meier D., Radlein D. An overview of fast pyrolysis of biomass. Organic Geochemistry. 1999;**30**:1479–1493. DOI: 10.1016/S0146-6380(99)00120-5

[11] Mettler M. S., Vlachos D. G., Dauenhauer P. J. Top ten fundamental challenges of biomass pyrolysis for biofuels. Energy and Environmental Science. 2012;**5**:7797–7809. DOI: 10.1039/c2ee21679e

[12] French R., Czernik S. Catalytic pyrolysis of biomass for biofuels production. Fuel Processing Technology. 2010;**91**:25–32. DOI: 10.1016/j.fuproc.2009.08.011

[13] Su-Ping Z., Yong-Jie Y., Zhengwei R., Tingchen L. Study of hydrodeoxygenation of bio-oil from the fast pyrolysis of biomass. Energy Sources. 2003;**25**:57–65. DOI: 10.1080/00908310290142118

[14] Corma A., Huber G. W., Sauvanaud L., O' Conner P. Processing biomass-derived oxygenates in the oil refinery: Catalytic cracking (FCC) reaction pathways and role of catalyst. Journal of Catalysis. 2007;**247**:307–327. DOI: 10.1016/j.jcat.2007.01.023

[15] Elliott D. C. Historical developments in hydroprocessing bio-oils. Energy & Fuels. 2007;**21**:1792–1815. DOI: 10.1021/ef070044u

[16] He P., Song H. Catalytic conversion of biomass by natural gas for oil quality upgrading. Industrial & Engineering Chemistry Research. 2014;**53**:15862–15870. DOI: 10.1021/ie502272j

[17] Xiao Y., He P., Cheng W., Liu J., Shan W., Song H. Converting solid wastes into liquid fuel using a novel methanolysis process. Waste Management. 2016;**49**:304–310. DOI: 10.1016/j.wasman.2015.12.017

[18] Esteban C., Ralph P. O. Fundamentals of thermochemical biomass conversion. In: R. P. Overend, T. A. Milne, L. K. Mudge, editors. Biomass Liquefaction: An Overview. 1st ed. New York: Springer Netherlands; 1985. pp. 967–1002. DOI: 10.1007/978-94-009-4932-4_54

[19] Zhang Y., Riskowski G., Funk T. Thermochemical conversion of swine manure [dissertation]. Urbana: University of Illinois at Urbana-Champaign; 1999. 46 p.

[20] Morf P. O. Secondary reactions of tar during thermochemical biomass conversion [dissertation]. Swiss: Federal Institute of Technology Zurich; 2001.

[21] Goudriaan F., van de Beld B. Thermal efficiency of the HTU process for biomass liquefaction. In: A. V. Bridgwater, editor. Progress in Thermochemical Biomass Conversion Conference; September 18–21; Tyrol, Austria; 2000. pp. 1312–1325.

[22] Feridoun S. A., Hiroyuki Y. Dehydration of fructose to 5-hydroxymethylfurfural in sub-critical water over heterogeneous zirconium phosphate catalysts. Carbohydrate Research. 2006;**341**(14):2379–2387. DOI: 10.1016/j.carres.2006.06.025.

[23] Snare M., Kubickova I., Maki-Arvela P. Production of diesel fuel from renewable feeds: Kinetics of ethyl stearate decarboxylation. Chemical Engineering Journal. 2007;**134**(1–3): 29–34. DOI: 10.1016/j.cej.2007.03.064

[24] Heddle J. F. Activated sludge treatment of slaughterhouse wastes with protein recovery. Water Research. 1979;**13**(7):581–584. DOI: 10.1016/0043-1354(79)90004-6

[25] Youngmi K., Nathan S., Mosiera R. H. Composition of corn dry-grind ethanol by-products: DDGS, wet cake, and thin stillage. Bioresource Technology. 2008;**99**(12):5165–5176. DOI: 10.1016/j.biortech.2007.09.028

[26] Tim R., Kaiyue L., Tobias A. Hydrolysis kinetics of biopolymers in subcritical water. The Journal of Supercritical Fluids. 2008;**46**(3):335–341. DOI: 10.1016/j.supflu.2007.09.037

[27] Ortwin B. Hydrothermal degradation of polymers derived from plants. Progress in Polymer Science. 1994;**19**(5):797–841. DOI: 10.1016/0079-6700(94)90033-7

[28] William S., Lai M., Antal M. J., Jr. Uncatalyzed solvolysis of whole biomass hemicellulose by hot compressed liquid water. Industrial & Engineering Chemistry Research. 1992;**31**(4):1157–1161. DOI: 10.1021/ie00004a026

[29] Andrea K., Palanikumar M., Franziska S. Influence of proteins on the hydrothermal gasification and liquefaction of biomass. 2. Model Compounds. Industrial & Engineering Chemistry Research. 2007;**46**(1):87–96. DOI: 10.1021/ie061047h

[30] Kruse A., Gawlik A. Biomass conversion in water at 330–410°C and 30–50 MPa. Identification of key compounds for indicating different chemical reaction pathways. Industrial & Engineering Chemistry Research. 2003;**42**(2):267–279. DOI: 10.1021/ie0202773

[31] Dong Z., Liang Z., Shicheng Z., Hongbo F., Jianmin C. Hydrothermal liquefaction of Macroalgae Enteromorpha prolifera to bio-oil. Energy Fuels. 2010;**24**:4054–4061. DOI: 10.1021/ef100151h

[32] Sudong Y., Ryan D., Matt H., Zhongchao T. Subcritical hydrothermal liquefaction of cattle manure to bio-oil: Effects of conversion parameters on bio-oil yield and characterization of bio-oil. Bioresource Technology. 2010;**101**(10):3657–3664. DOI: 10.1016/j.biortech.2009.12.058

[33] Motoyuki S., Hirokazu T., Katsumi H., Kiyoshi M. Hydrothermal liquefaction of plantation biomass with two kinds of wastewater from paper industry. Journal of Materials Science. 2008;**43**(7):2476–2486. DOI: 10.1007/s10853-007-2106-8

[34] Yixin Q., Xiaomin W., Chongli Z. Experimental study on the direct liquefaction of Cunninghamia lanceolata in water. Energy. 2003;**28**(7):597–606. DOI: 10.1016/S0360-5442(02)00178-0

[35] Shupinga Z., Yulong W., Mingde Y., Imdad K., Chun L., Junmao T. Production and characterization of bio-oil from hydrothermal liquefaction of microalgae Dunaliella tertiolecta cake. Energy. 2010;**35**(12):5406–5411. DOI: 10.1016/j.energy.2010.07.013

[36] Selhan K., Thallada B., Akinori M., Yusaku S., Md. Azhar U. Low-temperature hydrothermal treatment of biomass: Effect of reaction parameters on products and boiling point distributions. Energy Fuels. 2004;**18**(1):234–241. DOI: 10.1021/ef030133g

[37] Wahyudiono, Takayuki K., Mitsuru S. Decomposition of a lignin model compound under hydrothermal conditions. Chemistry Engineering and Technology. 2007;**30**(8):1113–1122. DOI: 10.1002/ceat.200700066.

[38] Bo Z., Hua-Jiang H., Ramaswamy S. Reaction kinetics of the hydrothermal treatment of lignin. Applied Biochemistry and Biotechnology. 2008;**147**(1):119–131. DOI: 10.1007/s12010-007-8070-6

[39] Aiguo L., YoonKook P., Zhiliang H. Product identification and distribution from hydrothermal conversion of Walnut Shells. Energy Fuels. 2006;**20**(2):446–454. DOI: 10.1021/ef050192p

[40] Javaid A., Nor A., Saidina A. A review on process conditions for optimum bio-oil yield in hydrothermal liquefaction of biomass. Renewable and Sustainable Energy Reviews. 2011;**15**(3):1615–1624. DOI: 10.1016/j.rser.2010.11.054

[41] Sangon S., Ratanavaraha S., Ngamprasertsith S., Prasassarakich P. Coal liquefaction using supercritical toluene–tetralin mixture in a semi-continuous reactor. Fuel Processing Technology. 2006;**87**(3):201–207. DOI: 10.1016/j.fuproc.2005.07.007

[42] Sascha R. A., Kersten Biljana P., Wolter P., Wim P. M., Van S. Gasification of model compounds and wood in hot compressed water. Industrial & Engineering Chemistry Research. 2006;**45**(12):4169–4177. DOI: 10.1021/ie0509490

[43] Kabyemela B. M., Takigawa M., Adschiri T., Malaluan R. M., Arai K. Mechanism and kinetics of cellobiose decomposition in sub- and supercritical water. Industrial & Engineering Chemistry Research. 1998;**37**(2):357–361. DOI: 10.1021/ie9704408

[44] Bernard M. K., Tadafumi A., Roberto M. M., Kunio A. Kinetics of glucose epimerization and decomposition in subcritical and supercritical water. Industrial & Engineering Chemistry Research. 1997;**36**(5):1552–1558. DOI: 10.1021/ie960250h

[45] ElGayar M., McAuliffe, Shellsol C. As a processing liquid in biomass liquefaction. Energy Sources. 1997;**19**(7):665-676. DOI: 10.1080/00908319708908881

[46] Mun S., Hassan E., Hassan M. Liquefaction of lignocellulosic biomass with dioxane/polar solvent mixtures in the presence of an acid catalyst. Journal of Industrial and Engineering Chemistry. 2004;**10**(3):473–477.

[47] Funazukuri T., Cho J., Wakao N. Effect of adding Na_2CO_3, HCl and or CO during liquefaction of lignin sulfonate with water. Fuel. 1990;**69**(10):1328–1329. DOI: 10.1016/0016-2361(90)90298-5

[48] Alma M., Maldas D., Shiraishi N. Liquefaction of several biomass wastes into phenol in the presence of various alkalis and metallic salts as catalysts. Journal of Polymer Engineering. 1998;**18**(3): 161-177.

[49] Boocock D. G. B. Further aspects of powdered poplar wood liquefaction by aqueous pyrolysis. The Canadian Journal of Chemical Engineering. 1985;**63**(4):627–631.

[50] Yongjie Y., Jie X., Tingchen L., Zhengwei R. Liquefaction of sawdust for liquid fuel. Fuel Processing Technology. 1999;**60**(2):135–143. DOI: 10.1016/S0378-3820(99)00026-0

[51] Akhtar J., Saidina Amin N. A review on process conditions for optimum bio-oil yield in hydrothermal liquefaction of biomass. Renewable and Sustainable Energy Reviews. 2011;**15**(3):1615–1624. DOI: 10.1016/j.rser.2010.11.054

[52] Antonio M., Simeone C., Dino M. Biomass gasification technology: The state of the art overview. Journal of Energy Chemistry. 2016;**25**(1):10–25. DOI: 10.1016/j.jechem.2015.11.005

[53] Lv P. M., Xiong Z., Chang J., Wu C. Z., Chen Y., Zhu J. X. An experimental study on biomass air-steam gasification in a fluidized bed. Bioresource Technology. 2004;**95**(1):95–101. DOI: 10.1016/j.biortech.2004.02.003

[54] Siyi L., Bo X., Xianjun G., Zhiquan H., Shiming L., Maoyun H. Hydrogen-rich gas from catalytic steam gasification of biomass in a fixed bed reactor: Influence of particle size on gasification performance. International Journal of Hydrogen Energy. 2009;**34**(3):1260–1264. DOI: 10.1016/j.ijhydene.2008.10.088

[55] De Lasa H., Salaices E., Mazumder J., Lucky R. Catalytic steam gasification of biomass: Catalysts, thermodynamics and kinetics. Chemical Review. 2011;**111**(9):5404–5433. DOI: 10.1021/cr200024w

[56] Prakash P., Sheeba K. N. Hydrogen production from steam gasification of biomass: Influence of process parameters on hydrogen yield: A review. Renewable Energy. 2014;**66**:570–579. DOI: 10.1016/j.renene.2013.12.025

[57] Ian N., Alberto O., Maria P. A., José C. Biomass gasification with air in an atmospheric bubbling fluidized bed: Effect of six operational variables on the quality of the produced raw gas. Industrial & Engineering Chemistry Research. 1996;**35**(7):2110–2120. DOI: 10.1021/ie9507540

[58] Shweta S., Pratik N. S. Air–steam biomass gasification: Experiments, modelling and simulation. Energy Conversion and Management. 2016;**110**:307–318. DOI: 10.1016/j.enconman.2015.12.030

[59] Meng N., Leung D. Y. C., Michael K. H., Leung K. S. An overview of hydrogen production from biomass. An overview of hydrogen production from biomass. Fuel Processing Technology. 2006;**86**(5):461–472. DOI: 10.1016/j.fuproc.2005.11.003

[60] Corte P., Lacoste C., Traverse J. P. Gasification and catalytic conversion of biomass by flash pyrolysis. Journal of Analytical and Applied Pyrolysis. 1985;**7**(4):323–335. DOI: 10.1016/0165-2370(85)80104-2

[61] Mustafa B., Mehmet B., Elif K., Havva B. Main routes for the thermo-conversion of biomass into fuels and chemicals. Energy Conversion and Management. 2009;**50**(12):3158–3168. DOI: 10.1016/j.enconman.2009.08.013

[62] Mustafa B., Mehmet B., Elif K., Havva B. Main routes for the thermo-conversion of biomass into fuels and chemicals. Part 1: Pyrolysis systems. Energy Conversion and Management. 2009;**50**(12):3147–3157. DOI: 10.1016/j.enconman.2009.08.014

[63] Mustafa B. Biomass energy and biochemical conversion processing for fuels and chemicals. Energy Sources, Part A: Recovery, Utilization, and Environmental Effects. 2006; 28(6): 517–525. DOI: 10.1080/009083190927994

[64] Mukherjee J., Gupta M. N. Biocatalysis for biomass valorization. Sustainable Chemical Process. 2015;3(1):3–7. DOI: 10.1186/s40508-015-0037-2

[65] Zhang Y., Brown T. R., Hu G., Brown R. C. Techno-economic analysis of two bio-oil upgrading pathways. Chemical Engineering Journal. 2013;**225**:895–904. DOI: 10.1016/j.cej.2013.01.030

[66] Wright M. M., Daugaard D. E., Satrio J. A., Brown R. C. Techno-economic analysis of biomass fast pyrolysis to transportation fuels. Fuel. 2010;**89**:S2–S10. DOI: 10.1016/j.fuel.2010.07.029

[67] Zhu Y., Biddy M. J., Jones S. B., Elliott D. C., Schmidt A. J. Techno-economic analysis of liquid fuel production from woody biomass via hydrothermal liquefaction (HTL) and upgrading. Applied Energy. 2014;**129**(15):384–394. DOI: 10.1016/j.apenergy.2014.03.053

[68] Swanson R. M., Platon A., Satrio J. A., Brown R. C. Techno-economic analysis of biomass-to-liquids production based on gasification. Fuel. 2010;**89**(1):S11–S19. DOI: 10.1016/j.fuel.2010.07.027

[69] Saqib Sohail T., Lasse R., Andreas R. Hydrothermal liquefaction of biomass: A review of subcritical water technologies. Energy. 2011;**36**(5):2328–2342. DOI: 10.1016/j.energy.2011.03.013

Chapter 11

Biomass Compositional Analysis for Conversion to Renewable Fuels and Chemicals

C. Luke Williams, Rachel M. Emerson and
Jaya Shankar Tumuluru

Additional information is available at the end of the chapter

http://dx.doi.org/10.5772/65777

Abstract

As the world continues to deplete its nonrenewable resources, there has begun a shift toward using renewable materials for the production of fuels and chemicals. Terrestrial biomass, as well as municipal solid wastes, provides renewable feedstocks for fuel and chemical production. However, one of the major challenges to using biomass as a feedstock for fuel and chemical production is the great amount of innate variability between different biomass types and within individual biomass species. This inconsistency arises from varied growth and harvesting conditions and presents challenges for conversion processes, which frequently require physically and chemically uniform materials. This chapter will examine intrinsic biomass compositional characteristics including cellulose, hemicellulose, lignin, extractives/volatiles, and ash for a wide array of biomass types. Additionally, extrinsic properties, such as moisture content and particle grind size, will be examined for their effect on biomass conversion to fuels using four major conversion processes: direct combustion, pyrolysis, hydrothermal liquefaction, and fermentation. A brief discussion on recent research for the production of building block chemicals from biomass will also be presented.

Keywords: biomass, composition, variability, renewable, fuels, chemicals

1. Introduction

CO_2 and other greenhouse gases are contributing to increased concerns about global warming. Biomass growth and utilization provides one solution to reduce greenhouse gas emissions and balance the ecosystem. For instance, carbon sequestered by the world's forests accounts for about 77% of terrestrial ecosystem [1]. In concert with fears about CO_2 production and global warming, concerns over dwindling and limited petroleum resources have given momentum to the production of renewable fuels, chemicals, and other materials from biomass [2]. In addition to providing carbon sequestration benefits for the world, biomass is also positioned to have a large impact on the domestic production of fuels and chemicals in the United States. According to the "Billion-Ton" Study [3] and its update [4] by the US Department of Energy (USDOE) and US Department of Agriculture (USDA), biomass has the potential to sustainably supply one third of the nation's petroleum consumption. This large available supply makes biomass one of the most abundant, inexpensive, and currently underutilized products from the agricultural and biorefinery industries [5]. While the first generation of biofuels is being produced using sugarcane in places such as Brazil, the second generation of biofuels will likely be derived from the over 1 billion tons of lignocellulosic biomass that is produced annually in the United States or from the estimated 10 to 50 billion tons of waste lignocellulose that is produced worldwide [3, 6]. The beginning of the second-generation enzymatic conversion of lignocellulosic material to fermentable sugars for fuel ethanol is becoming more economically viable in the United States [7] with companies such as DuPont [8] and POET [9] making gains in the production of second-generation cellulosic ethanol using agricultural wastes such as corn stover. In fact, to meet the US congressional mandate to manufacture 36 billion gallons of biofuels per year by 2022, several new types of biomass including energy crops, forest residues, and municipal wastes will have to be processed through modern conversion systems.

Besides just the production of fuel petroleum also produces a significant portion (basically all) of our plastics and other materials. The use of petroleum to produce chemical building blocks and materials has resulted in a global interest in using renewable bio-based polymers and composites derived from biomass to reduce our environmental impact [10, 11]. Additionally, from a broad energy standpoint, biomass has a significant advantage over renewable energy sources such as hydropower, wind, geothermal, and solar in that biomass is the only renewable energy source that can be turned directly into fuels and chemicals, as opposed to just generate electricity.

The output of these conversion systems, whether it is fuels or chemicals, is highly dependent on the quality of biomass input to the system. The quality of the biomass is dependent on inherent species variability, production conditions, and differing harvest, collection, and storage practices. Some of the most important parameters related to the biomass composition, in regard to the impact on biofuels production, are moisture content, ash content and speciation, carbohydrate distribution, and higher heating value. For example, moisture content impacts the storage, supply, and transportation of feedstock to biorefineries, and ash content often reduces oil yields in thermochemical conversion processes and, to a lesser extent, reduces the effectiveness of dilute alkali pretreatment in biochemical processes. Commercialization of biorefineries in the United States has resulted in an understanding of the importance of biomass quality (moisture, ash, and sugar content) and physical properties (particle size and shape), especially in regard to feed and handling issues. The overall objective of the present research is to understand the impact of the chemical and physical composition of various biomass feedstocks on the production of fuels and chemicals through a variety of conversion pathways. This work will focus on everything from the feeding and handling of biomass all the way through the effect of feedstock variability on the final oil product. Specific objectives

http://dx.doi.org/10.5772/65777

seek to understand chemical composition of woody and herbaceous crops, agricultural residues, and municipal solid wastes and suggest their suitability for different biofuel production conversion pathways. This research will also briefly touch on platform chemical production from biomass and consider methods for mitigating the problems associated with feedstock variability while converting biomass to fuels and chemicals.

2. Biomass compositional analysis

To effectively produce fuels and chemicals from biomass, it is critical to understand the composition of the feedstock material. The chemical composition of biomass, whether it is lignocellulosic or herbaceous, can be characterized by five primary components: cellulose, hemicellulose, lignin, extractives/volatiles, and ash. The most abundant biopolymer on earth, cellulose, is a polysaccharide of glucose monomers held together by β(1→4) linkages. These β(1→4) linkages are what make cellulose resistant to hydrolysis. The second major component of biomass, hemicellulose, is an amorphous heteropolymer comprised of several different carbohydrates including xylose, mannose, and glucose, among others. Due to its amorphous structure, hemicellulose is significantly more susceptible to hydrolysis than crystalline cellulose. Cellulose and hemicellulose, combined with the third major component of biomass, lignin, make up over 90% of lignocellulosic biomass and 80% of herbaceous biomass. Lignin is an intricate array of aromatic alcohols and is intertwined with the cellulose and hemicellulose fraction of the biomass structure. This interwoven nature of the lignin helps provide rigidity to lignocellulosic materials, such as trees.

The other minor components of biomass are extractives/volatiles and ash. While these components make up a smaller portion of the biomass composition, they can still have a major influence on what ends up being the optimal conversion process. The components comprising the extractives/volatiles include both water and ethanol solubles. Water-soluble compounds include nonstructural sugars and proteins, and ethanol-soluble components are typically represented by chlorophyll and waxes. Ash, which comprises the inorganic content in biomass, can be intrinsic to the biomass or added anthropogenically. Intrinsic ash includes material-like calcium and potassium ions, while anthropogenic ash is mostly silica (dirt) collected during harvest.

There is obviously significant compositional variation between different biomass types, but there is also a lot of variation within a single feedstock. This variation, while substantial across terrestrial feedstocks, varies even more widely when municipal solid wastes are included as renewable energy feedstocks. **Table 1** illustrates the large difference in composition across three broad categories of renewable feedstocks including lignocellulosic, herbaceous biomass, as well as municipal solid wastes. Algal biomass was not included in this study due to a lack of available data and the difficulty in obtaining consistent analysis methods across institutions [12].

As seen in **Table 1**, there exists a significant amount of variability in overall composition (i.e., cellulose, hemicellulose, and lignin) between different types of feedstocks. These

Feedstock composition	Woody	Herbaceous	Wastes
Proximate			
Volatiles (%)	84.0 (2.1)[193]	79.1 (5.8)[284]	76.7 (5.5)[21]
Ash (%)	1.3 (0.9)[193]	5.5 (3.2)[284]	6.6 (6.7)[21]
Fixed carbon (%)	14.7 (1.6)[193]	15.4 (4.0)[284]	14.8 (5.0)[21]
Ultimate			
Hydrogen (%)	6.0 (0.1)[192]	5.8 (0.3)[276]	5.9 (0.4)[21]
Carbon (%)	50.7 (4.71)[192]	47.4 (1.9)[276]	46.0 (4.0)[21]
Nitrogen (%)	0.32 (0.01)[192]	0.75 (0.49)[276]	1.3 (1.6)[21]
Oxygen (%)	41.9 (1.4)[134]	41.0 (2.4)[107]	38.3 (4.2)[7]
Sulfur (%)	0.03 (0.01)[135]	0.10 (0.32)[107]	0.15 (0.16)[7]
Structural			
Cellulose (%)	51.2 (8.7)[241]	32.1 (4.5)[2425]	28.4 (13.2)[27]
Hemicellulose (%)	21.0 (8.7)[241]	18.6 (3.4)[2425]	16.4 (5.5)[27]
Lignin (%)	26.1 (5.3)[241]	16.3 (3.3)[2425]	12.5 (2.7)[15]

Table 1. Feedstock compositions for woody, herbaceous, and waste materials; average (standard deviation)$^{\text{number of samples}}$.

differences are large enough that conversion reactors have to be operated under different conditions based on the type of material supplied to the conversion facility (such as lower pyrolysis temperatures for herbaceous feedstocks). Also, herbaceous feedstocks, in addition to having higher ash content, exhibit more variability in their composition of volatiles (and ash) than woody biomass.

While a high degree of variability is expected across broad categories such as lignocellulosic material and municipal solid waste (MSW), there also exists significant variability within individual feedstock categories. **Tables 2–4** highlight the differences within an individual feedstock category for lignocellulosic material, herbaceous material, and municipal solid waste, respectively.

While it is obvious that compositional differences can be stark between different biomass types, there is also a substantial compositional variability between different anatomical fractions of the same type of biomass. **Table 5** compiles information on the chemical composition of different plant fractions for woody biomass, corn, and wheat.

It can be seen that lignocellulosic biomass contains a large fraction of cellulose in the heartwood (shown by *whole tree*), while the *bark* contains a high percent lignin. In woody biomass, the extractives are fairly evenly distributed. Conversely, corn stover contains a majority of the extractives in the leaves and internodes (the links between different stalk segments). Taking advantage of processing a specific anatomical fraction could allow for greater control over product output by tailoring the composition of the reactor feed. Additionally, utilizing anatomical fractionation separation could increase the economic viability of a process by

http://dx.doi.org/10.5772/65777

Feedstock composition	Shrub willow	Hybrid poplar	Pine	Other softwoods	Other hardwoods
Proximate					
Volatiles (%)	84.7 (0.8)^{76}	84.0 (1.3)^{41}	83.5 (2.5)^{46}	81.3 (2.9)^{18}	85.1 (3.0)^{11}
Ash (%)	1.5 (0.4)^{76}	1.3 (0.5)^{41}	0.7 (0.6)^{46}	2.1 (2.0)^{18}	1.8 (1.2)^{11}
Fixed carbon (%)	13.8 (0.7)^{76}	14.6 (0.1)^{41}	15.7 (1.9)^{46}	16.5 (1.6)^{18}	13.1 (1.8)^{11}
Ultimate					
Hydrogen (%)	6.0 (0.2)^{76}	6.0 (0.1)^{41}	6.1 (0.1)^{45}	6.1 (0.1)^{18}	6.1 (0.1)^{11}
Carbon (%)	50.3 (0.9)^{76}	50.0 (1.1)^{41}	51.5 (1.0)^{45}	51.8 (0.9)^{18}	50.2 (0.5)^{11}
Nitrogen (%)	0.36 (0.10)^{76}	0.35 (0.17)^{41}	0.17 (0.12)^{45}	0.27 (0.21)^{18}	0.55 (0.49)^{11}
Oxygen (%)	42.6 (0.4)^{44}	42.8 (1.2)^{28}	41.4 (1.0)^{38}	39.7 (1.8)^{14}	41.1 (1.6)^{10}
Sulfur (%)	0.04(0.01)^{44}	0.03 (0.01)^{28}	0.02 (0.01)^{39}	0.03 (0.01)^{14}	0.05 (0.05)^{10}
Structural					
Cellulose (%)	–	43.8 (1.2)^{43}	47.4 (2.2)^{55}	42.1 (7.1)^{26}	50.8 (6.9)^{24}
Hemicellulose (%)	–	14.7 (0.1)^{43}	21.9 (4.9)^{55}	25.1 (5.2)^{26}	29.7 (4.3)^{24}
Lignin (%)	–	25.7 (0.3)^{43}	28.6 (0.7)^{55}	29.1 (1.7)^{26}	19.5 (4.1)^{24}

Table 2. Feedstock compositions for specific woody feedstocks; average (standard deviation)$^{\text{number of samples}}$.

Feedstock composition	Corn stover	Switchgrass	*Sorghum*	Energy cane (bagasse)	Mixed grasses	*Miscanthus*
Proximate						
Volatiles (%)	78.1 (5.0)^{50}	82.4 (4.1)^{43}	77.0 (3.7)^{44}	82.2 (1.9)^{48}	78.6 (2.8)^{47}	82.5 (3.5)^{35}
Ash (%)	6.3 (3.5)^{50}	4.0 (2.0)^{43}	7.2 (2.6)^{44}	3.4 (1.6)^{48}	6.6 (1.7)^{47}	2.6 (1.3)^{35}
Fixed carbon (%)	15.6 (4.4)^{50}	13.6 (3.0)^{43}	15.7 (2.3)^{44}	14.4 (1.0)^{48}	14.8 (2.4)^{47}	14.8 (2.9)^{35}
Ultimate						
Hydrogen (%)	5.7 (0.3)^{40}	5.9 (0.2)^{43}	5.7 (0.2)^{44}	6.1 (0.1)^{48}	5.8 (0.3)^{47}	5.8 (0.1)^{35}
Carbon (%)	47.1 (2.3)^{40}	47.1 (1.1)^{43}	46.4 (1.3)^{44}	48.8 (0.9)^{48}	47.6 (1.1)^{47}	48.9 (1.5)^{35}
Nitrogen (%)	0.63 (0.32)^{40}	0.60 (0.26)^{43}	1.04 (0.38)^{44}	0.43 (0.20)^{48}	1.38 (0.54)^{47}	0.35 (0.17)^{35}
Oxygen (%)	40.3 (2.2)^{39}	42.4 (2.3)^{42}	40.3 (0.6)^{3}	-	39.5 (0.7)^{2}	42.3 (1.1)^{4}
Sulfur (%)	0.14 (0.53)^{39}	0.06 (0.03)^{42}	0.11 (0.01)^{3}	-	0.12 (0.02)^{2}	0.04 (0.02)^{4}
Structural						20.1 (1.4)^{274}
Cellulose (%)	34.3 (2.5)^{251}	34.2 (2.7)^{348}	28.6 (2.6)^{488}	32.1 (3.2)^{479}	28.9 (2.9)^{465}	38.9 (3.2)^{274}
Hemicellulose (%)	20.7 (2.0)^{251}	21.9 (2.6)^{348}	15.4 (1.6)^{488}	19.5 (1.9)^{479}	16.7 (3.9)^{465}	20.1 (1.4)^{274}
Lignin (%)	15.2 (1.6)^{251}	19.2 (1.4)^{348}	12.2 (1.9)^{488}	16.3 (1.8)^{479}	15.7 (1.7)^{465}	21.1 (1.6)^{274}

Table 3. Feedstock compositions for specific herbaceous feedstocks; average (standard deviation)$^{\text{number of samples}}$.

Feedstock composition	MSW	C&D waste	Woody residues
Proximate			
Volatiles (%)	76.5 (1.1)[11]	76.5 (3.7)[9]	81.1 (2.4)[2]
Ash (%)	11.8 (5.2)[11]	0.8 (0.4)[9]	1.2 (0.3)[2]
Fixed carbon (%)	11.2 (5.2)[11]	18.9 (2.1)[9]	17.8 (2.0)[2]
Ultimate			
Hydrogen (%)	5.6 (0.4)[11]	6.2 (0.2)[9]	6.0 (0.0)[2]
Carbon (%)	43.3 (3.3)[11]	48.3 (1.2)[9]	52.5 (0.2)[2]
Nitrogen (%)	1.52 (1.72)[11]	1.09 (1.47)[9]	0.22 (0.06)[2]
Oxygen (%)	36.3 (4.8)[4]	42.4 (0.1)[2]	40.1 (0.6)[2]
Sulfur (%)	0.25 (0.14)[4]	0.02 (0.01)[2]	0.01 (0.01)[2]
Structural			
Cellulose (%)	28.4 (13.2)[15]	–	–
Hemicellulose (%)	16.4 (5.5)[15]	–	–
Lignin (%)	12.5 (2.7)[15]	–	–

Table 4. Feedstock compositions for specific waste feedstocks; average (standard deviation)$^{\text{number of samples}}$.

Structural component	Cellulose	Hemicellulose	Lignin	Extractives
Woody biomass (wt%–daf)[a]				
Whole tree	51.2	23.4	25.4	3.0
Bark	22.0	47.0	31.0	3.3
Twigs	15.4	62.3	22.3	1.6
Leaves	26.5	47.2	26.3	3.7
Corn (wt%–db)[b]				
Corn cobs	35.92	30.7	16.44	5.89
Corn leaves	34.33	22.77	13.99	10.54
Corn husk	37.73	31.18	10.52	5.80
Corn internodes	40.21	20.03	17.24	12.29
Wheat (wt%–db)[b]				
Internode 1	34.34	21.30	16.36	16.24
Internode 2	39.04	21.07	18.58	10.98
Internodes 3/4/5	38.92	21.56	19.50	9.67

Source: [a]Vassilev et al. [13]; [b]INL Library [14]; daf—dry ash-free, db—dry basis.

Table 5. Compositional variation with anatomical fraction for woody biomass, corn, and wheat.

http://dx.doi.org/10.5772/65777

extricating high value components. Profitable use of a coproduct is exemplified by the use of distiller's dried grains with solubles (DDGS) in ethanol production for high-protein animal feed [15, 16].

The majority of the data included in **Tables 1–5** can be found in Idaho National Laboratory's (INL) Bioenergy Feedstock Library. The woody materials in **Table 1** include a wide variety of softwoods, hardwoods, and other wood varieties making up around 23 different woody species. The herbaceous materials include those listed in **Table 3** along with sugarcane, sugarcane bagasse, and wheat. The waste materials from **Table 1** are all represented in **Table 4**. In **Table 4** municipal solid waste (MSW) includes fractions of paper, cardboard, and grass clippings. No food-based waste is currently accounted for by this data. The construction and demolition (C&D) waste included oriented strand board, particle board, and a variety of lumber conditions. Woody residues included forest thinning and logging residues.

All of the values reported for proximate and ultimate (reported on a dry basis) for **Tables 1–5** were collected at INL [17] and stored in the Bioenergy Feedstock Library [14]. The reported cellulose, hemicellulose, and lignin values were a combination of glucose (representing the cellulose fraction) and xylose, galactose, and arabinose (representing the hemicellulose) values measured using NREL's LAP determination of structural carbohydrates and lignin in biomass [18], glucose, and xylose values predicted using an NIR-based predictive models developed at NREL [19], and cellulose and hemicellulose values reported in literature are all reported on a dry basis. It should be noted that the value for volatiles in the previous tables is determined by heating samples to 950°C in an inert atmosphere. This value for volatile will therefore include all thermal decomposition products, in addition to molecules that could be removed without thermal decomposition.

Understanding the degree of biomass compositional variability is crucial to developing a robust conversion process. However, in addition to understanding compositional variability, it is useful to know where this variability originates. Kenney et al. have produced a thorough review discussing several sources of biomass variability [20]. Briefly, some of the major sources of biomass compositional variation derive from local agronomic conditions [21], drought [22], harvest season and year [23], and harvest method [24]. A further analysis of the sources of biomass variability and its impact on conversion processes has been compiled by Williams et al. [25].

3. Biomass conversion to renewable fuels and chemicals

As can be seen in **Tables 1–5**, biomass has a broad range of compositional variability, even within an individual feedstock. This variation has a substantial impact on biomass conversion to fuels and value-added chemicals that varies depending on the chosen conversion process. The following section investigates how feedstock quality impacts four common conversion processes: biochemical fermentation, direct combustion, pyrolysis, and hydrothermal liquefaction (HTL). General impacts of feedstock physical and chemical properties will be discussed before a more in-depth look at each conversion process.

The physical properties of biomass have a myriad of effects on its conversion to fuels and chemicals. Arguably, the two most important physical properties of biomass, regardless of conversion process, are particle size and moisture content. Practically all conversion methods require some degree of size reduction. Biochemical conversion processes can accept a greater range of particle sizes, and the final size needed tends to be dependent on the processing system utilized [26, 27]. On the thermochemical side, hydrothermal liquefaction is much more insensitive to particle size due to high heating rates in the liquid media [28], but a significant amount of size reduction is needed to pump biomass sludges in a continuous system [29]. Pyrolysis uses particles smaller than 0.5 mm because small particles decrease char yields and have higher heating rates [30]. Optimal combustion particle size is often larger and varies for different biomass types at approximately 6 mm for straw, 4 mm for *Miscanthus*, and 2–4 mm for wood [31]. While particle size is obviously important, others have argued that moisture content is likely the single most problematic property affecting feedstock supply and biorefining operations [20]. Moisture increases heating rates during steam pretreatment for biological conversion [32], reduces bio-oil quality and thermochemical conversion [33], and causes low thermal efficiency in combustion processes [34]. Aside from particle size and moisture content, other physical properties of interest include bulk density, elastic properties, and microstructure. Bulk density has a strong effect on transportation and handling costs (lower densities greatly increase transportation costs), and the elastic properties/microstructure can increase compressibility and interparticle interactions at constricted flow points such as hopper openings.

Biomass chemical properties also have a large influence on best conversion process and the quality of the final product. The three primary chemical components of interest in biomass conversion are ash content, volatiles, and lignin. High ash content generally has a negative effect on biomass conversion across the board by reducing the effectiveness of dilute acid pretreatment for biological processes [35] and increasing char yields and fouling in thermochemical processes such as HTL [36], pyrolysis [37], and combustion [38]. However, there exist several strategies for ash removal including leaching and air classification [39]. Volatiles are generally represented by light organic acids (such as acetic acid) and furans. The furan fraction of the volatiles can reduce fermentation efficiency in biological processes [40] and lower energy content and stability in bio-oils produced by thermochemical processes [41]. Lignin, on the other hand, can have a variety of effects on biomass conversion depending on the process chosen. Lignin generally has a negative effect on ethanol production by blocking enzyme access to cellulose [42] but can increase oil yields for pyrolysis [43] and heating values for combustion [34] during thermochemical conversion.

3.1. Biochemical conversion—fermentation

Ethanol production from biomass occurs via two primary steps: depolymerization of the cellulose and hemicellulose to fermentable sugars and fermentation of these sugars to ethanol. Biomass conversion to ethanol has been evaluated in many reviews [42–45] which vary focuses from pretreatment and enzymatic hydrolysis [42] to optimization of the cellulase enzyme for improving sugar conversion to ethanol [44] and evaluation of current and future economic aspects of fuel ethanol production [46]. This work will build upon these previous

http://dx.doi.org/10.5772/65777

reviews to explain how biomass compositional variability can influence fermentation processes for fuel production.

Mixed rangeland grasses are a prime example of a feedstock with high compositional variability. These grasses are an emerging alternative to traditional energy crops. Mixed rangeland grasses also preserve natural habitat and typically require less maintenance than traditional energy crops. However, the naturally high variability of these grasses can lead to reduced product yields in biochemical conversion processes. Adler et al. have shown that ethanol yield per unit area decreases as plant species diversity increases. Ethanol yields are maximized when there is increased targeted coverage of C_4 prairie grass energy crops, such as switchgrass, which sequester more carbon than typical C_3 conservation grassland varieties [47]. This preference for C_4 grasses illustrates how the production of ethanol using fermentation is typically much more dependent on biomass carbohydrate content. In fact, technoeconomic analysis has indicated that adjusting total carbohydrate content by 1% of total dry matter can change the minimum ethanol selling price (MESP) by $0.018/gal [46]. Given the compositional data above, fermentation is better matched to herbaceous crops than lignocellulosic material due to the higher carbohydrate content of grasses. Additionally, fermentation processes are typically more tolerant of the higher ash contents of herbaceous feedstocks [48]. However, it should be repeated that high alkali metal content from excess soil collected during harvest can increase acid neutralization during pretreatment and lower the xylan digestibility for corn stover, consequently lowering ethanol yields [35].

Despite compositional variability generally being a disadvantage in feedstock processing, there exists at least one aspect to variability that could be advantageous. Changes in structural carbohydrate content with anatomical fraction in corn stover significantly affect glucose yield. After hydrolysis, glucose concentration can be three times greater in the cobs, leaves, and husks than stalks [49]. Additionally, the corn cobs, leaves, and husks respond better than stocks to simultaneous saccharification and fermentation (SSF) despite having similar glucan levels [50]. Therefore, selective fermentation of specific anatomical fractions could increase process efficiency if a cost-effective separation process could be devised and there is a value-added coproduct that could be produced from the stalks. The advantage of separating biomass by anatomical fraction extends to other biomass types as well. For example, different fractions within wheat stover exhibit an almost 10% difference in glucan content, and some parts are much more susceptible to chemical saccharification [51].

3.2. Thermochemical conversion—general

While biomass as a feedstock exhibits a significant amount of compositional variability, illustrated in the tables above, the different options for thermochemical conversion are almost diverse. Thermochemical conversion operations utilize reactions using both solids (pyrolysis and combustion) and liquids (hydrothermal liquefaction). Products from thermochemical processes also span a wide range of states from solid (biochar), through liquid (bio-oil), all the way to gas (syngas). The wide variety of processing options and product outputs, along with short reaction times (on the order of seconds), allows thermochemical conversion operations to utilize a wide array of diverse process inputs.

3.3. Thermochemical conversion—combustion

The combustion of biomass, which is still common in developing countries, has been used for thousands of year to do everything from managing agricultural lands to producing heat and energy for industrial processes [52]. Currently, developed countries use nonrenewable fossil fuels such as oil, coal, and natural gas as a primary source of energy; however, these energy supplies could be depleted in the next 40–50 years [53]. In an effort to reduce the rate at which these nonrenewable resources are being depleted and reduce environmental impact, there is a shift toward the combustion of renewable biomass and other waste products (such as paper and plastics). Literature reviews focus on the combustion of biomass as an energy source both with [54–56] and without [30, 53, 57, 58] torrefaction as a pretreatment to improve combustion efficiencies and material grinding and storage properties. One of the major problems with combusting biomass in a traditional coal plant is slagging, a mineral buildup due to the higher ash content in biomass than in coal. This problem means that low-ash content biomass, such as woody feedstocks, is better to use than herbaceous materials (which have intrinsic ash contents about five times greater than woody materials) in combustion applications. While biomass combustion does present problems with slagging, it does have the benefit of reducing harmful greenhouse gas emissions as compared to coal [59], and the energy produced can be incorporated directly into the current energy grid without infrastructure changes.

3.4. Thermochemical conversion—pyrolysis

Pyrolysis is a thermochemical process that starts with a solid and can be tuned to produce either a solid (biochar) or a liquid (bio-oil). However, this chapter will focus on the production of bio-oil and the effects of biomass composition on the resulting oil yields and quality. Pyrolysis is performed at temperatures from 400 to 600°C [60] and often includes a catalyst with the aim of increasing the energy density of the product by removing oxygen (as water and volatiles) [61]. Pyrolysis of biomass to produce fuels has been thoroughly reviewed in the academic literature [33, 61–64].

The pyrolysis process is well suited for low-moisture-content material with low ash and high lignin content, meaning that pyrolysis processes favor lignocellulosic feedstocks. For example, lignin content increases the average molecular weight of resulting pyrolysis oil by 100 Da as lignin content rises from 5 to 15% [43]. The high ash content of herbaceous feedstock can decrease oil yields by 1–5% for every 1% increase in ash over an ash range of 1.5–7.5% [43]. In addition to decreasing oil yields, the alkali metals common in herbaceous crops can also have damaging effects on reactors and reduce catalyst lifetimes [65]. However, more recent studies have taken into account not only the production of pyrolysis oil but also the upgrading of that oil to the final fuel for a range of feedstocks including pines, poplars, switchgrass, and corn stover. In these integrated fast pyrolysis/hydrotreating studies, the effects of ash content on oil and upgraded fuel yields were relatively insignificant over a narrow ash range of 0.7–1.6% [66] (an ash range common for woody materials but low for grasses).

http://dx.doi.org/10.5772/65777

3.5. Thermochemical conversion—hydrothermal liquefaction (HTL)

Hydrothermal liquefaction (HTL) is a unique thermal conversion process that utilizes biomass and water slurries. This makes HTL particularly well suited to turning high water content material, such as algae, municipal solid wastes, or grasses into bio-based oils. Additionally, HTL bio-oils tend to be higher quality than pyrolysis oils because they have less oxygen. However, the oil yields for HTL are lower than pyrolysis and the oxygen content is still higher than crude oil [67]. Performing the dissolution of biomass in a water media also saves energy on drying the feedstock, and the high heat transfer rates in a liquid media reduce particle size reduction requirements [36]. HTL can operate over a wide range of temperatures (200–600°C) to create products that range from solid biochars to gases. Reaction temperatures from 200 to 275°C are suitable for solid production [68], while temperatures from 275 to 350°C produce liquid products, and temperatures above 400°C are suitable for gas production [36]. Due to the liquid nature of the reaction media and the high temperatures, these reactors often operate at high pressures (5–40 MPa) to keep the reaction media as a liquid or supercritical fluid. Since the operating conditions and products of hydrothermal reactors are so diverse, the reviews of this material span a wide range. Some reviews cover both sub- and supercritical temperature regimes, with an array of model compounds and biomass feedstocks, and product arrays including liquid bio-oils and gases [67]. Other reviews focus on narrower operating regimes and liquid products from a variety of feedstocks with both high and low ash content [28, 36] or simply the processing of lignin (which is usually considered a waste product) [69]. While the hydrothermal processing of biomass offers advantages in being more feedstock agnostic, it has drawbacks in high capital equipment cost due to the extreme operating conditions, high energy input to heat the water, and lower yields (even though the oil quality is generally high).

3.6. Chemical production

Aside from the production of biochar and bio-oil, hot liquid water can also be used to convert biomass into value-added chemicals. Luterbacher et al. have achieved a 65% yield of sugars from woody biomass and a 55% yield from switchgrass using a biphasic CO_2/H_2O system. This biphasic system improves process separations and can use larger particles (~1 cm) at a high solids loading (40 wt%) [70]. The targeted production of sugars from biomass, instead of a bio-oil destined for fuel blending, could facilitate the production of high-value chemicals and materials. For instance, biomass-derived sugars can be used to make renewable plastics by producing *p*-xylene [71–73]. The conversion of biomass-derived cellulose to *p*-xylene could take place using a scheme such as the one in **Figure 1**. In this scheme cellulose is converted to *p*-xylene in a four-step process: step one uses a biphasic CO_2/H_2O system to convert biomass into sugars [70], step two isomerizes glucose to fructose [74], step three converts fructose 2,5-dimethylfuran (DMF) [75], and step four converts DMF to *p*-xylene [71]. The final step of converting *p*-xylene to polyethylene terephthalate (PET) would take place in a typical refinery because this renewable *p*-xylene would act as a standard drop in feedstock.

The production of chemicals from biomass has the potential to produce a wide array of drop in building blocks. The top twelve most promising drop in chemical building blocks can be

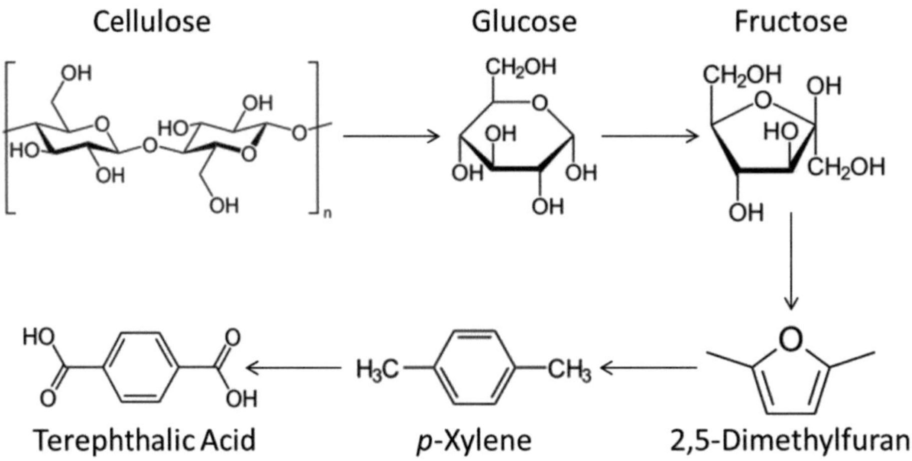

Figure 1. A representative pathway for conversion of cellulose to terephthalic acid through the transformation of cellulose-derived sugars to furans.

found in the Department of Energy's report on Top Value-Added Chemicals from Biomass [76]. This report lists several chemicals that could be made from biomass with an emphasis on the conversion of sugars to building block chemicals and the conversion of these building block chemicals to intermediates. After examining both biochemical and thermochemical pathways, it was noted that biochemical pathways focused on the conversion of sugars to building block chemicals, and thermochemical pathways dominated the conversion of building block chemicals to final products.

4. Pretreatments to improve biomass feedstock chemical composition and suggestions for optimal biomass conversion pathways

Raw herbaceous biomass has a chemical composition which is low in carbon content and high in oxygen, volatiles, and ash; is high in moisture; and has low energy content. This combination of properties does not make herbaceous crops suitable for thermochemical applications such as gasification, pyrolysis, and co-firing [77]. The shortcoming of many types of raw biomass, in terms of chemical and physical properties, can be overcome by pretreatment to produce a conversion-ready feedstock. Currently, there exist a variety of pretreatment methods including pelletization, air classification, dry torrefaction, hydrothermal carbonization, steam explosion, ionic liquid dissolution, acid and alkali leaching, and ammonia fiber expansion (AFEX). These pretreatment techniques are being looked at to improve biomass quality to produce a conversion-ready feedstock for both thermochemical and biochemical applications [25, 78]. Pretreatment can reduce biomass chemical and physical heterogeneity and lessen problems in (a) conversion applications (removing using air classification to remove ash prior to co-firing biomass could reduce slagging), (b) supply chain logistics (pelletizing biomass reduces transportation costs), (c) operational constraints (certain forms of pretreatment allow for utilization of coal infrastructure for feeding, milling, etc. of biomass, without

http://dx.doi.org/10.5772/65777

costly modifications or installation of separate processing lines), and (d) technical constraints (e.g., reduction of corrosion due to biomass washing).

Pretreatment for the optimization of chemical production from biomass is very much in its infancy. However, it is a safe bet that pretreatment will be required to get a consistent product, given that specialty chemicals require a much higher purity than the fuels currently being produced. Current research is ongoing for the production of many different value-added chemicals such as *p*-xylene [71], dimethylfuran [75], and levulinic acid [79] to name just a few, but at this point, all of these studies start with pure feedstocks, such as cellulose, and not biomass. To move the industry, past fuels to value-added chemicals will require a greater understanding of how biomass composition effects its conversion to fuels and chemicals.

As the previous pages have illustrated, the transformation of biomass to fuels and chemicals can take place over a wide variety of pathways with numerous influences from the biomass composition. These conversion pathways can be generally grouped as either biochemical or thermochemical. A greatly simplified process diagram for the production of renewable liquid fuels and chemicals from biomass can be seen in **Figure 2**. This figure groups feedstocks with their most likely conversion pathway based on the previous discussion regarding biomass composition.

Given the current variability in biomass resources, it is apparent that conversion technology will have to be tailored to regional renewable supply, be it lignocellulosic, herbaceous, a municipal solid waste stream, or algae. Given the high ash content of herbaceous biomass and the high water content of some municipal solid wastes, it is likely that these streams will be destined for use in biochemical pathways to produce sugars through enzymatic hydrolysis or methane using anaerobic digestion. However, there is also a chance that these materials could be passed through the thermochemical process of hydrothermal liquefaction to produce oils or undergo a more mild hydrothermal treatment to produce a platform chemical stream based on biomass-derived sugars. The abundant lignocellulosic biomass will likely be converted to bio-oil or energy using a thermochemical process such as pyrolysis or combustion, respectively. Thermochemical processes make use of lignocellulosic feedstocks in part due to their low ash content and because a high lignin content is unsuitable for enzymatic digestion in biochemical fermentation.

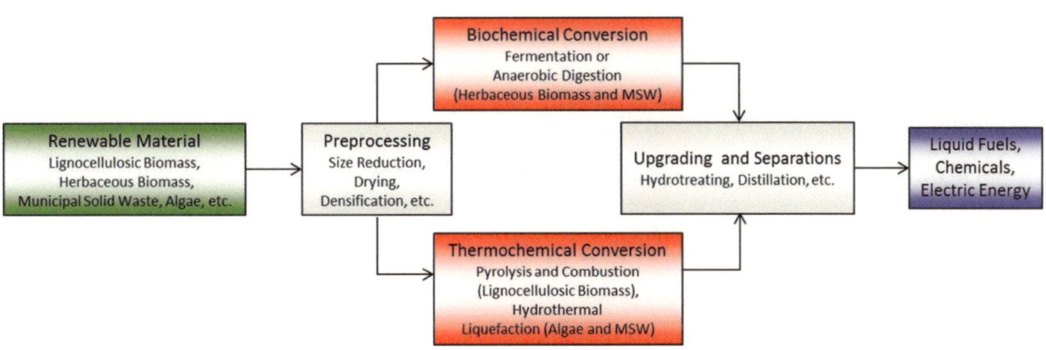

Figure 2. Broad scheme for conversion of renewable material to fuels, chemicals, and energy.

5. Conclusions

The large degree of variability between biomass resources, both currently available and emerging, is a significant barrier to the utilization of biomass as a feedstock for fuel and chemical production. The impacts of physical characteristics such as moisture content and particle size, as well as chemical characteristics such as ash content, extractives/volatiles, and lignin, all play varying, and intricate, roles during biomass conversion. Adding to the complexity of this system is the fact that, in addition to a myriad of compositionally diverse feedstocks, there also exist numerous conversion pathways to the final fuel or chemical products. To alleviate this problem, it will be necessary to develop techniques to reduce biomass variability and develop a consistent, conversion-ready feedstock for biorenewable fuel and chemical production.

Financial and competing interests disclosure

The US government retains, and the publisher, by accepting the article for publication, acknowledges that the US government has a nonexclusive, paid-up, irrevocable, worldwide license to publish or reproduce the published form of this manuscript, or allow others to do so, for US government purposes. The authors have no other relevant affiliations or financial involvement with any organization or entity with a financial interest in or financial conflict with the subject matter or materials discussed in the manuscript apart from those disclosed. No writing assistance was utilized in the production of this manuscript.

Acknowledgements

This research was supported by the US Department of Energy under the Department of Energy Idaho Operations Office Contract No. DE-AC07-05ID14517.

Author details

C. Luke Williams*, Rachel M. Emerson and Jaya Shankar Tumuluru

* Address all correspondence to: luke.williams@inl.gov

Idaho National Laboratory, Idaho Falls, ID, USA

References

[1] Noble IR, Bolin B, Ravindranath NH, Verardo DJ, Dokken DJ. Land Use, Land Use Change, and Forestry. Cambridge University Press: Cambridge England; 2000. p. 375

[2] Chu S, Majumdar A (2012) Opportunities and challenges for a sustainable energy future. Nature 488 (7411):294–303

[3] Perlack RD, Wright LL, Turhollow AF, Graham RL, Stokes BJ, Erbach DC. Biomass as Feedstock for a Bioenergy and Bioproducts Industry: The Technical Feasibility of a Billion- Ton Annual Supply. DTIC Document. April 2005. 78p. DOE/GO-102005-2135. ORNL/TM-2005/66

[4] Perlack RD, Eaton LM, Turhollow Jr AF, Langholtz MH, Brandt CC, Downing ME, Graham RL, Wright LL, Kavkewitz JM, Shamey AM (2011) U.S. Billion-Ton Update: Biomass Supply for a Bioenergy and Bioproducts Industry. August 2011. 195 p. DOE/EE-0363

[5] Brennan L, Owende P (2010) Biofuels from microalgae—a review of technologies for production, processing, and extractions of biofuels and co-products. Renewable and Sustainable Energy Reviews 14 (2):557–577

[6] Lynd LR, Cushman JH, Nichols RJ, Wyman CE (1991) Fuel ethanol from cellulosic biomass. Science (Washington) 251 (4999):1318–1323

[7] Humbird D, Davis R, Tao L, Kinchin C, Hsu D, Aden A, Schoen P, Lukas J, Olthof B, Worley M (2011) Process design and economics for biochemical conversion of lignocellulosic biomass to ethanol. National Renewable Energy Laboratory; March 2011. 147 p. TP-5100-47764

[8] DuPont (2011) DuPont industrial biosciences announces new leader and name for cellulosic ethanol business. http://www2.dupont.com/BioFuel/en_US/whats_new/DuPont_Cellulosic_Ethanol_111511.html. Accessed 03-11-2015

[9] Poet-DSM (2013) POET and DSM to make advanced biofuels a reality by 2013. http://poet-dsm.com/news/poet-and-dsm.aspx. Accessed 03/30/2015

[10] Nagarajan V, Mohanty AK, Misra M (2013) Sustainable green composites: value addition to agricultural residues and perennial grasses. ACS Sustainable Chemistry & Engineering 1 (3):325–333

[11] Ragauskas AJ, Williams CK, Davison BH, Britovsek G, Cairney J, Eckert CA, Frederick WJ, Jr., Hallett JP, Leak DJ, Liotta CL, Mielenz JR, Murphy R, Templer R, Tschaplinski T (2006) The path forward for biofuels and biomaterials. Science 311 (5760):484–489. doi:10.1126/science.1114736

[12] Laurens LML, Dempster TA, Jones HDT, Wolfrum EJ, Van Wychen S, McAllister JSP, Rencenberger M, Parchert KJ, Gloe LM (2012) Algal biomass constituent analysis: method uncertainties and investigation of the underlying measuring chemistries. Analytical Chemistry 84 (4):1879–1887. doi:10.1021/ac202668c

[13] Vassilev SV, Baxter D, Andersen LK, Vassileva CG, Morgan TJ (2012) An overview of the organic and inorganic phase composition of biomass. Fuel 94:1–33. doi:10.1016/j.fuel.2011.09.030

[14] INL (2015) DOE biomass feedstock library. bioenergy.inl.gov. Accessed 3-10-2015

[15] Wyman CE, Goodman BJ (1993) Biotechnology for production of fuels, chemicals, and materials from biomass. Applied Biochemistry and Biotechnology 39 (1):41–59

[16] Dale BE, Allen MS, Laser M, Lynd LR (2009) Protein feeds coproduction in biomass conversion to fuels and chemicals. Biofuels, Bioproducts and Biorefining 3 (2):219–230

[17] ASTM D3172-13, standard practice for proximate analysis of coal and coke, ASTM International, West Conshohocken, PA, 2013, www.astm.org

[18] Sluiter A, Hames B, Ruiz R, Scarlata C, Sluiter J, Templeton D, Crocker D (2008) Determination of structural carbohydrates and lignin in biomass. Laboratory analytical procedure. April 2008. 15 p. NREL/TP-510-42618

[19] Wolfrum EJ, Sluiter AD (2009) Improved multivariate calibration models for corn stover feedstock and dilute-acid pretreated corn stover. Cellulose 16 (4):567–576

[20] Kenney KL, Smith WA, Gresham GL, Westover TL (2013) Understanding biomass feedstock variability. Biofuels 4 (1):111–127

[21] Templeton DW, Sluiter AD, Hayward TK, Hames BR, Thomas SR (2009) Assessing corn stover composition and sources of variability via NIRS. Cellulose 16 (4):621–639. doi:10.1007/s10570-009-9325-x

[22] Emerson R, Hoover A, Ray A, Lacey J, Cortez M, Payne C, Karlen D, Birrell S, Laird D, Kallenbach R, Egenolf J, Sousek M, Voigt T (2014) Drought effects on composition and yield for corn stover, mixed grasses, and Miscanthusas bioenergy feedstocks. Biofuels 5 (3):275–291. doi:10.1080/17597269.2014.913904

[23] Lemus R, Brummer EC, Moore KJ, Molstad NE, Burras CL, Barker MF (2002) Biomass yield and quality of 20 switchgrass populations in southern Iowa, USA. Biomass and Bioenergy 23 (6):433–442

[24] Eisenbies MH, Volk TA, Posselius J, Shi S, Patel A (2014) Quality and variability of commercial- scale short rotation willow biomass harvested using a single pass cut-and-chip forage harvester. BioEnergy Research.(8):546. doi:10.1007/s12155-014-9540-7

[25] Williams CL, Westover TL, Emerson RM, Tumuluru JS, Li C (2015) Sources of biomass feedstock variability and the potential impact on biofuels production. BioEnergy Research (9):1–14

[26] Dibble CJ, Shatova TA, Jorgenson JL, Stickel JJ (2011) Particle morphology characterization and manipulation in biomass slurries and the effect on rheological properties and enzymatic conversion. Biotechnology Progress 27 (6):1751–1759. doi:10.1002/btpr.669

[27] Van Walsum GP, Allen S, Spencer M, Laser M, Antal M, Jr., Lynd L (1996) Conversion of lignocellulosics pretreated with liquid hot water to ethanol. In: Wyman C, Davison B (eds) Seventeenth symposium on biotechnology for fuels and chemicals, vol 57/58. ABAB Symposium. Humana Press: New York USA pp 157–170. doi:10.1007/978-1-4612-0223-3_14

[28] Akhtar J, Amin NAS (2011) A review on process conditions for optimum bio-oil yield in hydrothermal liquefaction of biomass. Renewable and Sustainable Energy Reviews 15 (3):1615–1624. doi:10.1016/j.rser.2010.11.054

[29] Jazrawi C, Biller P, Ross AB, Montoya A, Maschmeyer T, Haynes BS (2013) Pilot plant testing of continuous hydrothermal liquefaction of microalgae. Algal Research 2 (3):268–277. doi:10.1016/j.algal.2013.04.006

[30] Demirbas A (2004) Effects of temperature and particle size on bio-char yield from pyrolysis of agricultural residues. Journal of Analytical and Applied Pyrolysis 72 (2):243–248. doi:10.1016/j.jaap.2004.07.003

[31] Spliethoff H, Hein K (1998) Effect of co-combustion of biomass on emissions in pulverized fuel furnaces. Fuel Processing Technology 54 (1):189–205

[32] Brownell HH, Yu EKC, Saddler JN (1986) Steam-explosion pretreatment of wood: effect of chip size, acid, moisture content and pressure drop. Biotechnology and Bioengineering 28 (6):792–801. doi:10.1002/bit.260280604

[33] Bridgwater A, Meier D, Radlein D (1999) An overview of fast pyrolysis of biomass. Organic Geochemistry 30 (12):1479–1493

[34] Jenkins B, Baxter L, Miles Jr T, Miles T (1998) Combustion properties of biomass. Fuel Processing Technology 54 (1):17–46

[35] Weiss ND, Farmer JD, Schell DJ (2010) Impact of corn stover composition on hemicellulose conversion during dilute acid pretreatment and enzymatic cellulose digestibility of the pretreated solids. Bioresource Technology 101 (2):674–678. doi:http://dx.doi.org/10.1016/j.biortech.2009.08.082

[36] Toor SS, Rosendahl L, Rudolf A (2011) Hydrothermal liquefaction of biomass: a review of subcritical water technologies. Energy 36 (5):2328–2342. doi:10.1016/j.energy.2011.03.013

[37] Tumuluru JS, Hess JR, Boardman RD, Wright CT, Westover TL (2012) Formulation, pretreatment, and densification options to improve biomass specifications for co-firing high percentages with coal. Industrial Biotechnology. 8 (3):113–132

[38] Jenkins BM, Bakker RR, Wei JB (1996) On the properties of washed straw. Biomass and Bioenergy 10 (4):177–200. doi:http://dx.doi.org/10.1016/0961-9534(95)00058-5

[39] Lacey JA, Aston JE, Westover TL, Cherry RS, Thompson DN (2015) Removal of introduced inorganic content from chipped forest residues via air classification. Fuel 160:265–273

[40] Palmqvist E, Hahn-Hägerdal B (2000) Fermentation of lignocellulosic hydrolysates. I: inhibition and detoxification. Bioresource Technology 74 (1):17–24. doi:http://dx.doi.org/10.1016/S0960-8524(99)00160-1

[41] Carpenter D, Westover TL, Czernik S, Jablonski W (2014) Biomass feedstocks for renewable fuel production: a review of the impacts of feedstock and pretreatment on the yield and product distribution of fast pyrolysis bio-oils and vapors. Green Chemistry 16 (2):384. doi:10.1039/c3gc41631c

[42] Sun Y, Cheng J (2002) Hydrolysis of lignocellulosic materials for ethanol production: a review. Bioresource Technology 83 (1):1–11

[43] Fahmi R, Bridgwater AV, Donnison I, Yates N, Jones JM (2008) The effect of lignin and inorganic species in biomass on pyrolysis oil yields, quality and stability. Fuel 87 (7):1230–1240. doi:10.1016/j.fuel.2007.07.026

[44] Lin Y, Tanaka S (2006) Ethanol fermentation from biomass resources: current state and prospects. Applied Microbiology and Biotechnology 69 (6):627–642. doi:10.1007/s00253-005-0229-x

[45] Bothast RJ, Schlicher MA (2005) Biotechnological processes for conversion of corn into ethanol. Applied Microbiology and Biotechnology 67 (1):19–25. doi:10.1007/s00253-004-1819-8

[46] Ruth MF, Thomas SR The effect of corn stover composition on ethanol process economics. In: 25th Symposium on Biotechnology for Fuels and Chemicals, 2003. Humana Press Breckenridge, CO.

[47] Adler PR, Sanderson MA, Weimer PJ, Vogel KP (2009) Plant species composition and biofuel yields of conservation grasslands. Ecological Applications 19 (8):2202–2209. doi:10.1890/07-2094.1

[48] McKendry P (2002) Energy production from biomass (part 1): overview of biomass. Bioresource Technology 83 (1):37–46. doi:http://dx.doi.org/10.1016/S0960-8524(01)00118-3

[49] Montross MD, Crofcheck CL (2004) Effect of stover fraction and storage method on glucose production during enzymatic hydrolysis. Bioresource Technology 92 (3):269–274. doi:http://dx.doi.org/10.1016/j.biortech.2003.09.007

[50] Duguid KB, Montross MD, Radtke CW, Crofcheck CL, Wendt LM, Shearer SA (2009) Effect of anatomical fractionation on the enzymatic hydrolysis of acid and alkaline pretreated corn stover. Bioresource Technology 100 (21):5189–5195. doi:http://dx.doi.org/10.1016/j.biortech.2009.03.082

[51] Duguid KB, Montross MD, Radtke CW, Crofcheck CL, Shearer SA, Hoskinson RL (2007) Screening for sugar and ethanol processing characteristics from anatomical fractions of wheat stover. Biomass and Bioenergy 31 (8):585–592. doi:10.1016/j.biombioe.2007.03.002

[52] Andreae MO (1991) Biomass burning: its history, use, and distribution and its impact on environmental quality and global climate, in Global Biomass Burning: Atmospheric, Climatic and Biospheric Implications,Cambridge MA: MIT Press; 1991 p[3–21]

[53] Saidur R, Abdelaziz EA, Demirbas A, Hossain MS, Mekhilef S (2011) A review on biomass as a fuel for boilers. Renewable and Sustainable Energy Reviews 15 (5):2262–2289. doi:10.1016/j.rser.2011.02.015

[54] Shankar Tumuluru J, Sokhansanj S, Hess JR, Wright CT, Boardman RD (2011) REVIEW: a review on biomass torrefaction process and product properties for energy applications. Industrial Biotechnology 7 (5):384–401

[55] Tumuluru JS, Wright CT, Hess JR, Kenney KL (2011) A review of biomass densification systems to develop uniform feedstock commodities for bioenergy application. Biofuels, Bioproducts and Biorefining 5 (6):683–707. doi:10.1002/bbb.324

[56] Tumuluru J, Tabil L, Song Y, Iroba K, Meda V (2014) Impact of process conditions on the density and durability of wheat, oat, canola, and barley straw briquettes. BioEnergy Research (8):1–14

[57] Demirbas A (2004) Combustion characteristics of different biomass fuels. Progress in Energy and Combustion Science 30 (2):219–230. doi:10.1016/j.pecs.2003.10.004

[58] Demirbas A (2005) Potential applications of renewable energy sources, biomass combustion problems in boiler power systems and combustion related environmental issues. Progress in Energy and Combustion Science 31 (2):171–192. doi:10.1016/j.pecs.2005.02.002

[59] Veijonen K, Vainikka P, Järvinen T, Alakangas E (2003) Biomass co-firing–an efficient way to reduce greenhouse gas emissions. A report from European Bioenergy Networks (EUBIONET),VTT Processes, Altener Industries, Espoo, Finland (March 2003)

[60] Carlson TR, Vispute TP, Huber GW (2008) Green gasoline by catalytic fast pyrolysis of solid biomass derived compounds. ChemSusChem 1 (5):397–400

[61] Czernik S, Bridgwater A (2004) Overview of applications of biomass fast pyrolysis oil. Energy & Fuels 18 (2):590–598

[62] Bridgwater A (1999) Principles and practice of biomass fast pyrolysis processes for liquids. Journal of Analytical and Applied Pyrolysis 51 (1):3–22

[63] Bridgwater A, Peacocke G (2000) Fast pyrolysis processes for biomass. Renewable and Sustainable Energy Reviews 4 (1):1–73

[64] Bridgwater AV (2012) Review of fast pyrolysis of biomass and product upgrading. Biomass and Bioenergy 38:68–94. doi:10.1016/j.biombioe.2011.01.048

[65] Davidsson KO, Pettersson JBC, Nilsson R (2002) Fertiliser influence on alkali release during straw pyrolysis. Fuel 81 (3):259–262

[66] Howe DT, Westover T, Carpenter DL, Santosa D, Emerson R, Deutch S, Starace A, Kutnyakov I, Lukins C (2015) Field-to-fuel performance testing of lignocellulosic feedstocks: an integrated study of the fast pyrolysis/hydrotreating pathway. Energy & Fuels 29 (5):3188–3197

[67] Peterson AA, Vogel F, Lachance RP, Fröling M, Antal Jr MJ, Tester JW (2008) Thermochemical biofuel production in hydrothermal media: a review of sub-and supercritical water technologies. Energy & Environmental Science 1 (1):32–65

[68] Reza MT, Lynam JG, Uddin MH, Coronella CJ (2013) Hydrothermal carbonization: fate of inorganics. Biomass and Bioenergy 49:86–94. doi:http://dx.doi.org/10.1016/j.biombioe.2012.12.004

[69] Kang S, Li X, Fan J, Chang J (2013) Hydrothermal conversion of lignin: a review. Renewable and Sustainable Energy Reviews 27:546–558. doi:10.1016/j.rser.2013.07.013

[70] Luterbacher JS, Tester JW, Walker LP (2012) Two-temperature stage biphasic CO_2–H_2O pretreatment of lignocellulosic biomass at high solid loadings. Biotechnology and Bioengineering 109 (6):1499–1507

[71] Williams CL, Chang C-C, Do P, Nikbin N, Caratzoulas S, Vlachos DG, Lobo RF, Fan W, Dauenhauer PJ (2012) Cycloaddition of biomass-derived furans for catalytic production of renewable *p*-xylene. ACS Catalysis 2 (6):935–939. doi:10.1021/cs300011a

[72] Chang C-C, Green SK, Williams CL, Dauenhauer PJ, Fan W (2014) Ultra-selective cycloaddition of dimethylfuran for renewable *p*-xylene with H-BEA. Green Chemistry 16 (2):585–588. doi:10.1039/c3gc40740c

[73] Williams CL, Vinter KP, Patet RE, Chang C-C, Nikbin N, Feng S, Wiatrowski M, Caratzoulas S, Fan W, Vlachos DG, Dauenhauer PJ (2016) Inhibition of xylene isomerization in the production of renewable aromatic chemicals from biomass-derived furans. ACS Catalysis. doi:10.1021/acscatal.5b02329

[74] Moliner M, Román-Leshkov Y, Davis ME (2010) Tin-containing zeolites are highly active catalysts for the isomerization of glucose in water. Proceedings of the National Academy of Sciences 107 (14):6164–6168

[75] Roman-Leshkov Y, Barrett CJ, Liu ZY, Dumesic JA (2007) Production of dimethylfuran for liquid fuels from biomass-derived carbohydrates. Nature 447 (7147):982–985. doi:http://www.nature.com/nature/journal/v447/n7147/suppinfo/nature05923_S1.html

[76] Werpy T, Petersen G, Aden A, Bozell J, Holladay J, White J, Manheim A, Eliot D, Lasure L, Jones S (2004) Top value added chemicals from biomass. Volume 1—Results of screening for potential candidates from sugars and synthesis gas. Aug. 2004. 76 p. ADA436528

[77] Tumuluru JS (2015) Comparison of chemical composition and energy property of torrefied switchgrass and corn stover. Frontiers in Energy Research 3:46

[78] Bonner IJ, Thompson DN, Plummer M, Dee M, Tumuluru JS, Pace D, et al. Impact of ammonia fiber expansion (AFEX) pretreatment on energy consumption during drying, grinding, and pelletization of corn stover. Drying Technol. 2016 Jan;34(11):1319-1329

[79] Weingarten R, Conner WC, Huber GW (2012) Production of levulinic acid from cellulose by hydrothermal decomposition combined with aqueous phase dehydration with a solid acid catalyst. Energy & Environmental Science 5 (6):7559–7574

Chapter 12

Modeling the Calorific Value of Biomass from Fruit Trees Using Elemental Analysis Data

Borja Velázquez-Martí, Isabel López-Cortés,
Domingo Salazar-Hernández and
Ángel Jesús Callejón-Ferre

Additional information is available at the end of the chapter

http://dx.doi.org/10.5772/65276

Abstract

Pruning of fruit trees produces a great quantity of biomass each year that can be used for energy production. For this purpose, it is necessary to carry out an energy characterization of these pruned wastes, where the determination of heating value is significant. This value is usually measured by an adiabatic or isoperibolic calorimeter, which causes high economic costs and wastes time. The present study is focused on the development of indirect models for heating value prediction of biomass from orange trees *Citrus × sinensis* Osbeck, almond trees *Prunus dulcis* (Mill) D.A. Webb, and olive trees *Olea europaea* L. from an elemental analysis in order to reduce the time of determination as well as the economic costs. Residual biomass was classified and characterized according to CEN regulations such as received, without drying. Also, moisture content wet basis, bark ratio, density, heating value, and elemental composition (carbon, hydrogen, nitrogen, and sulfur) were measured. The influence of these variables on the heating value was analyzed. Finally, mathematical models were developed to predict this value for this studied species. These models showed coefficients of determination between 0.83 and 0.97, being suitable for industrial use.

Keywords: bioenergy, economical studies, wood residues, higher heating value

1. Introduction

A number of researchers have published mathematical models to predict the higher heating value of different biomass materials from the concentration of the main elements present, such as percentage of carbon, percentage of hydrogen, and percentage of nitrogen together with others [1–6]. On the other hand, other models have used proximate analysis [7–10] or structural analysis [9–11]. Indirect calculation of the higher heating value by means of these types of models is justified by the expensive cost of the use of calorimeters [3, 12]. The aim of this chapter is to compare the resources used in direct heat value determination with that used in indirect calculation from elemental analysis by means of prediction models in three common lignocellulosic materials coming from pruning Mediterranean fruit trees. In this work, predictive models of the heating value of biomass from pruning of *Citrus* × *sinensis* Osbeck (orange tree), *P. dulcis* (Mill.) D.A. Webb (almond tree) and *O. europaea* L. (olive tree) have been developed; the calorific value determined directly using the isoperibolic calorimeter was compared with that determined from the elemental composition of materials. The influence of the percentage of leaves, wood moisture content, bark percentage, and percentage content of C, H, N, and S was analyzed.

The heat value is an invariable parameter for a material with constant composition, defined by the empirical form $CH_wO_xN_yS_z$, where w is the number of moles of hydrogen per mole of carbon, x is the number of moles of oxygen per mole of carbon, y is the number of moles of nitrogen per mole of carbon, and z is the number of moles of sulfur per mole of carbon. The moles of each element are obtained by multiplication of the sample mass with its ratio and dividing by the atomic weight of each element. The values w, x, y, and z are obtained from the division of the moles of each element in the sample by the moles of carbon.

The capacity of retaining water in the biomass, caused by its porosity, must be considered in the measurement of the calorific value. The moisture content in the material changes its molecular formula, and therefore the gravimetric percentages of C, H, O, and N. For this reason, standards to determine the calorific value for a particular material, such as UNE 164001:2005 EX [13], require the material to be obtained in the anhydrous state, or with a known moisture content.

The problem in industrial applications is that the biomass materials received for the combustion in boilers show variation in their moisture content, they are sometimes mixed with leaves and even with other materials. Under these conditions, the heat properties are not constant. In order to measure the calorific value instantly, regression models are analyzed in this paper from the percentage contents of C, H, and N of the materials as received. Currently, there are devices capable of measuring these elements in 5 min [5, 6].

When there is variability in the composition, and uncertainty of the type of materials or proportion of the mixture thereof, the mathematical models for the indirect determination of the calorific value are only applicable in the scope where they were developed. This uncertainty is quite common in industry. Thus researchers, such as Francis and Lloyd [14], Ebeling and Jenkins [15], and Kathiravale et al. [16], have provided models for different types of mixtures. However, none of them showed an economical study to justify the use of the model. In this chapter, the development of models specific for the studied material is shown; in addition, an analysis was carried out to compare the cost of direct determination with an indirect mathematical model from elemental analysis.

http://dx.doi.org/10.5772/65276

In industrial facilities, it is very difficult to find biomass received for combustion without some moisture. Because of processes increase production costs, they are rarely used in the production of energy wood. On the other hand, open air drying rarely decreases the moisture content below 20% [17, 18]. Moreover, it is normal that industrial boilers do not work with a well-defined type of material but with mixtures of different types of biomass. These reasons make the composition of the biomass used in industry variable that directly affects the expected calorific value. So, calorific determination before the introduction of the materials in the boiler is useful to understand the energy performance of the combustion. If this direct determination by the bomb calorimeter is more expensive than the indirect determination from their composition, developing predictive models is fully justified. This is studied in this work.

2. Materials and methods

2.1. Vegetal material

The species studied in this work were *Citrus × sinensis* Osbeck (orange tree), *P. dulcis* (Mill) D.A. Webb (almond tree), and *O. europaea* L. (olive tree). These three species are widely grown in the Mediterranean region; they represent 10% of the total cultivated area in Spain [19].

The studied orange variety was "Valencia Late". This is one of the latest varieties most cultivated in Spain. The harvesting begins from March, and it is usually pruned in summer after harvesting. Its main use is fresh consumption due to its high juice content. It also has high chances of industrialization [20].

The almond variety studied was "Blaquerna". It is a self-fertile cultivar that is currently replacing other varieties of lower production. Almond pruning can be annual or biennial according to the development of the tree [21]. During the first 4 years, winter pruning is carried out focused on defining the architecture of the vegetation. From the fourth production year, pruning is performed to remove unproductive branches and to improve the fruit quality [22].

The studied olive tree variety was "Villalonga", which is the variety most widely used in Valencia, with a total cultivation area of 23,550 ha. Its main use is in the manufacture of oil and it is also used for fresh consumption [23].

To define these raw materials as biomass for biofuel, the standard UNE EN 14961-1:2011[24] was used. According to this standard, the analyzed samples in this work were classified by their origin and sources of solid biofuels: 1. Wood biomass → 1.1. Wood biomass from forest or plantation → 1.1.4. Residues from cuttings → 1.1.4.1. Fresh/green, hardwood (including leaves). Therefore, it would be "cutting hardwood waste" (1.1.4.1). Following the mentioned standard, the specifications of the properties of the studied biofuel are defined in **Table 1**.

Origin: Cutting hardwood waste (1.1.4.1)		Wood biomass (1.1) - *Citrus × sinensis* Osbeck - *Prunus dulcis* (Mill.) D.A. Webb - *Olea europaea* L.
Marketed form		Wood logs, firewood
Sizes (cm)		
Length (*L*) cm		
Citrus × sinensis Osbeck *L* 100 (max 100 cm ± 5 cm)	*Prunus dulcis* (Mill.) D.A. Webb *L* 100+ (max 153 cm)	*Olea europaea* L. *L* 100+ (max 182 cm)
Diameter (*D*) cm		
Citrus × sinensis Osbeck *D*10 (2 cm ≤ *D* ≤ 10 cm)	*Prunus dulcis* (Mill.) D.A. Webb *D*10 (2 cm ≤ *D* ≤ 10 cm)	*Olea europaea* L. *D*10 (2 cm ≤ *D* ≤ 10 cm)
Moisture content, M (% as it is received) CEN/TS 15149-1, CEN/TS 15149-2		
Citrus × sinensis Osbeck M 45 (≤45%)	*Prunus dulcis* (Mill.) D.A. Webb M 35 (≤35%)	*Olea europaea* L. M 40 (≤40%)
Volume or weight, m^3, kg, loose or piled as received		15–20 kg of each specied.
Volume ratio of split logs		Trunk without cuttings, whole branches
Cut surface[a]		Surface smooth and regular cut
Rust and rot		None of the samples has mold

[a] The use of chainsaw is considered to produce a smooth and regular surface.

Table 1. Specification of the properties of the energy wood pieces (EN 14961-1:2011) [22].

2.2. Sample preparation

The branches of each species, obtained from pruning, were divided into six size classes (0–1, 1–2, 2–3, 3–4, 4–5, and >5 cm). Five samples of each class were taken for analysis; therefore, it resulted in 30 samples per species. For sample preparation, the methods defined by the UNE-EN 14780:2012 [25] were followed. The main purpose of sample preparation was to reduce the size of the branches in test portions with the same initial composition, being representative of the original sample. Initial wet basis moisture content average in wood was about 42.24%. In **Figure 1**, the process of the preparation of the sample is shown. First step was to separate the leaves and wood of the 30 branches of each species arrived at the laboratory (**Figure 2a–c**). These leaves were crushed with hammer mill and stored in airtight jars with identification labels. On the other hand, the wood was milled until their sizes were lower than 3 mm. Special care was taken to avoid loss of fine particles and moisture during milling and other operations. Average wet basis moisture content in tested particles was 29.85%. The devices used for the sample preparation were as follows:

- Manual saw. In order to prevent the moisture loss at the border, cuttings of the central part was used for the analytical determinations. The pieces obtained using this device are shown in **Figure 2(d)**.

- Hammer mill crusher of stainless steel, equipped with a 3 mm screen.

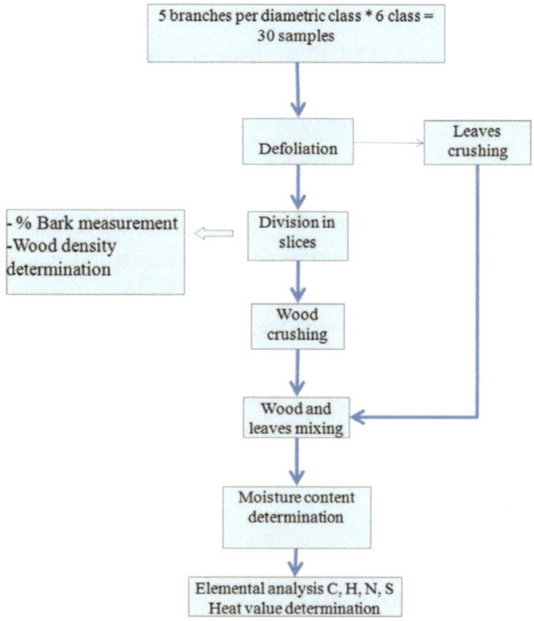

Figure 1. Sample preparation process.

Figure 2. Images of sample preparation process.

Once the samples were crushed, leaves and wood were mixed in defined proportions in each group of size class. The different proportions analyzed were from 10% weight leaves/weight wood in the first sample, 20% weight leaves/weight wood in the first sample in the second, etc. up to 50% of mixture leaves/wood to the fifth sample. So, the effect of leaves on the calorific value was measured. This is very important in this study because pruned material used in boilers has always got a high percentage of leaves, and this influences the moisture content and composition. In many publications, the calorific value calculated referred to that of completely dry and bare material, but this condition is far from the actual applications. In this work, the analysis was focused on samples with variable elemental composition which is usually obtained due to variation in the percentage of leaf and bark content (caused by different size class) and the moisture content (obtained after outdoor drying). For this reason, neither the wood nor the leaf fractions were dried rigorously, but they only experience the natural loss of moisture content during the transport and storage prior to analysis.

2.3. Measurement process

Higher heating value, wet basis moisture content, and elemental composition (C, H, and N) of each sample were measured. For this, standards shown in **Table 2** were used.

Reference of standard	Standard
CEN/TS 14779	Solid biofuels—sampling—methods for preparing sampling plans and sampling certificates
CEN/TS 14780	Solid biofuels—methods for sample preparation
EN 14774-3	Solid biofuels—determination of moisture content—Stufe drying method part 3 moisture content analysis for overall sample analysis
EN 14918	Solid biofuels—determination of calorific value
CEN/TS 15104	Solid biofuels—determination of total carbon, hydrogen, and nitrogen—instrumental methods
CEN/TS 15290	Solid biofuels—determination of major elements
EN 14961-1	Solid biofuels—fuel specifications and classes—part 1: general requirements

Table 2. Methods used to sample and measure parameters.

The calorific value was measured by means of a LECO AC-500 isoperibolic calorimeter. Before analyzing the samples, the calorimeter was calibrated by the combustion of a reactive standard of a known calorific power (benzoic acid, 1 g). Subsequently, each sample was prepared with a mass between 0.1 and 1 g. This was introduced into a combustion vessel where a fuse wire caused ignition. Note that 10 ml of distilled water was added. Then, combustion vessel was sealed, and oxygen with a pressure of 3000 kPa was introduced inside the calorimeter . This container was placed in a bucket of water which was surrounded by an insulating layer to maintain a constant temperature. During analysis, the water temperature was measured by an electronic thermometer with an accuracy of 1/10,000 degree. In order to control energy exchange, the temperatures of the cuvette and the insulating layer were continuously monitored. With this, the device applies a correction to the result.The water temperature was monitored by a microprocessor in every 6 seconds. The difference between the water temperature before ignition and after ignition was processed through the calorimeter software, for obtaining the calorific value.

The weight percentage of carbon, hydrogen, and nitrogen was measured by means of a LECO TruSpec CHN analyzer. According to EN 14918, samples between 0.1 and 1 g were weighed. Then, they were wrapped in titanium sheets that is completely inorganic. These were placed in a feeding carrousel. The analysis cycle consists of three phases: purging, combustion, and analysis. In the purge phase, the sample is casted into the load compartment, which is sealed and atmospheric gases are removed. In the second phase, the sample is casted into a compartment at 950°C, and oxygen is injected for rapid and complete combustion. The gases pass through an afterburner at 850°C to oxidize and remove particles. The combustion gases are collected in the ballast (gas collection vessel). During the analysis phase, combustion gases are homogenized in the ballast. Subsequently, an aliquot of 3 cm^3 is purged with helium through infrared detectors of CO_2 and H_2O. Another gas loop aliquot is transported through hot copper to remove O_2 and transform the NO_X to N_2. Then, in order to remove CO_2 and H_2O, they are allowed to flow through the tubes containing Anhydrone Lecosorb. The N content is determined on a thermal conductivity cell. The results are shown as percent or kg/mg.

In order to calculate the cost of analysis all inputs were counted. Market prices of nine enterprises were checked. Consumables, labor maintenance, and amortization were considered. Cost of technical labor was estimated in 20 €/h. Residual value of the device was considered to be 10% of investment. Time of analysis was measured.

2.4. Percentage of bark

The percentage of bark in the branches was calculated after their classification according to six diameter classes: 0–1, 1–2, 2–3, 3–4, 5–6, and >6 cm:. The diameter influences the bark ratio [26]. For each diameter, class five samples were taken, so 30 samples in total were analyzed in each species. In each branch, diameter with bark and diameter without bark were measured using of a digital caliper with 0.01 mm accuracy, as shown in **Figure 3**. From these diameters, the percentage of bark was calculated by Eq. (1) [26], where *Bark* (%) is the percentage of bark; ϕ_{cc} is the diameter over bark; ϕ_{sc} is the diameter without bark:

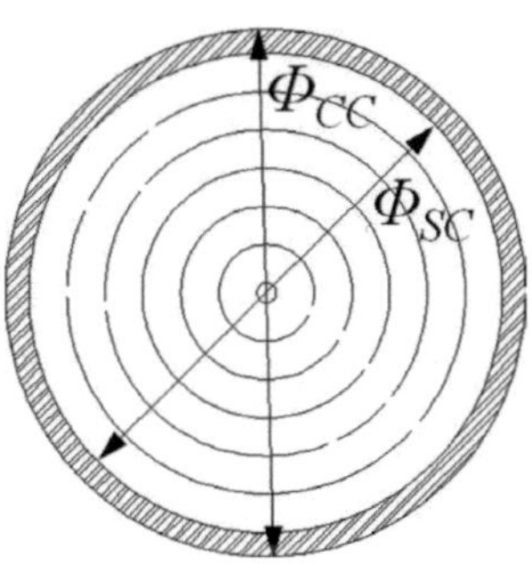

Figure 3. Measurements to calculate the percentage of bark.

$$\text{Bark (\%)} = \frac{\varphi_{\text{cc}}^2 - \varphi_{\text{sc}}^2}{\varphi_{\text{cc}}^2} \cdot 100 \tag{1}$$

2.5. Determination of wood density

Wood density is expressed as the mass of dry wood per unit volume. To calculate the density, first the samples were immersed in a beaker with water, calibrated to 250 mL. The volume was measured by the difference between the water level before and after immersing. Then the dry weight of the samples was determined; for this purpose, samples were placed in a drying oven at a constant temperature of 105 ± 2°C for 24 h. Mean and standard deviation for the densities are obtained by Eq. (2) [26]:

$$\rho_{\text{m}} = \frac{P_{\text{s}}}{V_{\text{v}}} \tag{2}$$

where ρ_{m} is the wood density (g·cm^{-3}), P_{s} is the dry weight of the sample (g), and V_{v} is the volume of the sample (cm^3).

2.6. Prediction model of heat value

To obtain predictive models by regression up to three variables (C, H, and N) have been used. For evaluating the models, the coefficient of determination (R^2), root mean square of the errors (RMS), and mean absolute error (MAE) were obtained. The coefficient of determination, denoted by R^2, is a number that indicates the proportion of the variance in the dependent variable that is predictable from the independent variable. RMS represents the sample standard deviation of the differences between predicted values and observed values. The mean absolute error is an average of the absolute errors $|e_i| = |f_i - y_i|$, where f_i is the prediction and y_i is the true value. The model with the best fit had highest R^2, minimum MAE, and RMS.

For all equations, 30 data were used to develop the models, and another 15 independent data were used for validation. The statistical program used was Statgraphics Centurion XV©14, even for the calculation of the significance of the variables of the mathematical prediction models by the beta coefficients and Student's *t*-test. The data observed in the new experiments and predicted by the model were compared with paired-sample test based on the *t*-Distribution.

3. Results and discussion

Several tests were initially applied to check the normal distribution of the data, such as Shapiro-Wilk test [27, 28], Anderson-Darling test, and the Lilliefors test [29, 30]. In **Table 3**, statistical description of the studied variables is shown for each species. It is observed that the coefficient of skewness and kurtosis are between −2 and +2. This indicates that the observed samples are normally distributed.

Variable	Species	Average	Standard deviation	Coef. skewness	Coef. kurtosis	Max.	Min.
HHV (kJ/kg)	Almond	15840.63	234.752	−0.775	−0.435	17458.54	13744.85
	Orange	12653.35	230.03	−0.23	−1.044	14108.26	10781.85
	Olive	15234.97	243.7	−0.281	−0.717	16861.92	13354.64
C %	Almond	40.026	2.261	−0.382	−0.683	44.050	35.300
	Orange	31.192	2.414	−0.274	−1.158	34.800	26.650
	Olive	37.974	2.359	0.192	−0.706	41.800	33.600
H %	Almond	5.661	0.740	1.740	−0.239	7.410	4.875
	Orange	7.179	0.291	−0.284	−0.485	7.590	6.540
	Olive	6.911	0.275	−0.429	−1.061	7.285	6.390
N %	Almond	1.132	0.535	0.147	−1.415	1.935	0.349
	Orange	0.430	0.143	1.383	0.474	0.800	0.222
	Olive	0.549	0.178	−0.792	−0.871	0.812	0.219
S %	Almond	0.128	0.024	0.705	−0.055	0.182	0.088
	Orange	0.061	0.054	1.802	−0.898	0.166	0.006
	Olive	0.031	0.013	1.100	−0.031	0.063	0.015
Moisture content %	Almond	19.009	4.870	0.967	−0.392	30.314	12.291
	Orange	34.679	4.496	−0.457	−1.084	41.094	26.770
	Olive	31.085	7.659	−0.020	−0.637	46.070	18.490
Bark %	Almond	24.741	6.485	0.548	−1.070	36.282	13.830
	Orange	15.038	3.688	1.936	0.176	23.960	10.135
	Olive	16.469	4.297	1.586	1.235	28.260	10.500
Leaves %	Almond	30.231	14.434	0.235	−1.346	50	10
	Orange	30.417	14.590	−0.156	−1.325	50	10
	Olive	31.905	14.703	−0.483	−1.202	50	10

HHV, high heat value; C, carbon; H, hydrogen; N, nitrogen; S, sulfur.

Table 3. Statistical analysis of the studied variables in each species.

The calorific values are clearly influenced by the moisture content and the leaf content in the sample. The obtained calorific value for the three species was between 12 and 16 MJ/kg, which were relatively lower than values cited in the literature. This is because, in this work, we have studied samples of diverse diameters with different percentage of bark, without any drying process, and mixtures of wood and leaves in different proportions, so smaller high heat values (HHVs) were obtained. For example, González et al. [31] gave values of HHV for biomass from orange tree pruning as 16–18 MJ/kg. Yin [32] also analyzed mixtures of biomass and got HHV values 18 MJ/kg, but all these were measured on a dry condition basis.

To compare the calorific value of the three species studied, analysis of variance was performed. In **Figure 4**, LSD intervals are shown at 95% confidence level. It can be seen that the

HHV of almond and olive trees were similar, but that of orange was significantly lower, which may be due to the characteristics of wood, leaves, and bark with moisture.

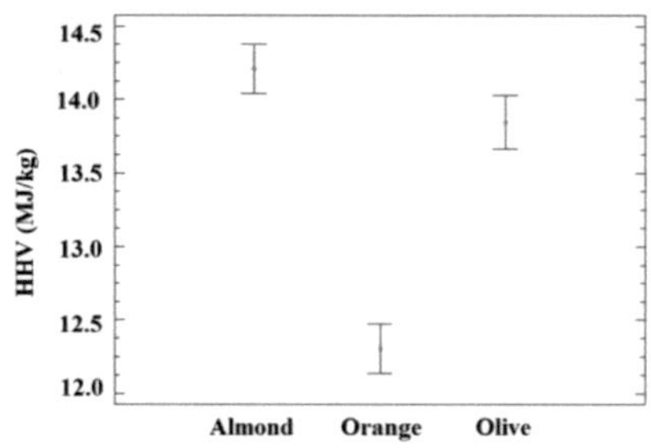

Figure 4. Intervals LSD for the calorific value of the species studied at 95% level of confidence.

Prunus dulcis	C	H	N	S	% leaves	% bark	% moisture content	HHV (kJ/kg)
C		−0.576*	0.1515	−0.3196	0.2535	−0.0076	−0.9844*	0.9739*
H	−0.576*		0.1712	0.397*	−0.0914	−0.1514	0.6393*	−0.5449*
N	0.1515	0.1712		0.4904*	0.8992*	−0.2699	−0.2055	0.2928
S	−0.3196	0.397*	0.4904*		0.3773	−0.3122	0.2873	−0.2427
% leaves	0.2535	−0.0914	0.8992*	0.3773		−0.09	−0.301	−0.3658
% bark	−0.0076	−0.1514	−0.2699	−0.3122	−0.09		0.0612	−0.0737
% moisture	−0.9844*	0.6393*	−0.2055	0.2873	−0.301	0.0612		−0.9786*
HVV (kJ/kg)	0.9739*	−0.5449*	0.2928	−0.2427	0.3658	−0.0737	−0.9786*	
Citrus × sinensis	C	H	N	S	% leaves	% bark	% moisture content	HHV (kJ/kg)
C		−0.8699*	−0.1595	0.5819*	−0.4233*	−0.5698*	−0.9551*	0.916*
H	−0.8699*		−0.1615	−0.5637*	0.165	0.4785*	0.8811*	−0.8249*
N	−0.1595	−0.1615		0.2159	0.7504*	0.2578	−0.0793	−0.1708
S	0.5819*	−0.5637*	0.2159		−0.173	−0.3339	−0.6484*	0.6292*
% leaves	−0.4233*	0.165	0.7504*	−0.173		0.1614	0.2708	−0.3111
% bark	−0.5698*	0.4785*	0.2578	−0.3339	0.1614		0.4213*	−0.7104*
% moisture	−0.9551*	0.8811*	−0.0793	−0.6484*	0.2708	0.4213*		−0.8464*
HVV (kJ/kg)	0.916*	−0.8249*	−0.1708	0.6292*	−0.3111	−0.7104*	−0.8464*	
Olea europaea	C	H	N	S	% leaves	% bark	% moisture content	HHV (kJ/kg)
C		−0.2761	0.8297*	0.0164	0.7945*	0.3936	−0.9715*	0.9752*
H	−0.2761		0.1347	0.2282	0.0321	−0.3109	0.3303	−0.279
N	0.8297*	0.1347		0.0287	0.9104*	0.1444	−0.7907*	0.8261*
S	0.0164	0.2282	0.0287		0.0894	0.0508	0.0596	−0.0737
% leaves	0.7945*	0.0321	0.9104*	0.0894		0.0024	−0.727*	−0.7955*
% bark	0.3936	−0.3109	0.1444	0.0508	0.0024		−0.4414*	−0.421
% moisture	−0.9715*	0.3303	−0.7907*	0.0596	−0.727*	−0.4414*		−0.9411*
HVV (kJ/kg)	0.9752*	−0.279	0.8261*	−0.0737	0.7955*	0.421	−0.9411*	

C, carbon; H, hydrogen; N, nitrogen; S, sulfur; HHV, calorific value.
* Pairs of variables with p-values below 0.05 (95% of significance level).

Table 4. Correlation coefficients of Pearson between the studied variables per species.

Table 4 shows the correlation between analyzed variables. Significant negative influence of the percentage of moisture in the calorific value (HHV) is observed, with −0.97 Pearson coefficient. On the other hand, it is noted that a higher percentage of C increases the HHV (+0.97). It is observed that the percentage of H is associated with the highest moisture content (+0.63) and obviously decreases the percentage of carbon in the sample and the calorific value (−0.54). It is observed that the percentages of sulfur, leaves, and bark do not have a clear influence on the calorific value of almonds but have a negative influence on the olive and orange trees.

Specie	Model	R^2 adj. (%)	RMS	MAE
Mixing three species	HHV = 717.79 + 380.85·C + 7.61% leaves−14.00% bark	97.352	278.77	195.74
	HHV = 863.61 + 383.93·C − 15.257% bark	96.976	297.90	218.00
	HHV =769.58 + 371.80·C + 8.02% leaves	97.147	289.35	201.07
	HHV = 928.66 + 374.30·C	96.733	309.62	223.64
Prunus dulcis	HHV = 7624.11 + 237.76·C + 228.75·N − 83.37%w	97.131	166.21	125.38
	HHV = −1024.04 + 412.99·C + 272.54·N	96.743	177.08	133.28
	HHV = −1106.84 + 422.77·C	94.630	227.38	173.87
Citrus × sinensis	HHV = 4323.54 + 301.46·C − 72.74% bark	88.127	331.30	250.40
	HHV = 1254.39 + 365.79·C	83.17	394.46	303.26
Olea europaea	HHV = −790.92 + 421.37·C	94.858	231.17	118.44

HHV, high heat value (kJ/kg); MAE, mean absolute error; C, carbon; N, nitrogen; %w, moisture content in wet basis.

Table 5. Prediction models proposed for the indirect calculation of calorific value.

Calorific models from elemental analyses are proposed in **Table 5**. All models show a high R^2 and relatively low standard error (RMS) and mean absolute error (MAE). All models are considered valid for calculating the calorific value, as an alternative to the calorimeter, reducing the time and cost of analysis. The p-values for all explanatory variables were less than 0.05. It is observed that the difference between the simplest models, whose explanatory variable is only carbon, and the more complicated is very small. Therefore, we recommend using the simplest, common to all species Eq. (3). In the variance analysis of the regression models, the p-value was less than 0.01 for all variables. This means that there is a significant relationship between variables and the volume for a confidence level of 99%:

$$HHV = 28.66 + 374.30 \cdot C \tag{3}$$

Vargas-Moreno et al. [4] conducted a review of models to predict the calorific value for different biomass materials based on elemental analyses. Most of these models reviewed gave determination coefficients between 0.8 and 0.99. For this reason, the correlation coefficients obtained in this work, higher than 0.8, were considered very acceptable (**Table 4**). However, on the other hand, Acda [33] proposed models with higher R^2 for different materials, but their models were obtained for dry materials and without leaves, whereas the models obtained in this study were obtained from mixtures of wood and leaves, and different moisture content, as received. Therefore, they are more applicable in actual situations in industries, where the material cannot be usually dried completely and they have different leaf percentage.

Velázquez-Martí et al. [5] already applied this method to obtain prediction models to predict a high heat value on lignocellulosic waste materials from urban tree pruning. In these works, the determination coefficients obtained are similar to those of the present work, between 0.7 and 0.9. This means that the presence of moisture causes decrease in accuracy of the calorific value calculated by prediction models.

It is observed that a higher carbon content (C) provides bigger calorific values. This coincides with many other studies [2, 34]. It is also shown in **Table 4** with a Pearson correlation coefficient of 0.97 between HHV and C. Because of this, all proposed equations for predicting the gross calorific value (**Table 5**), whether univariate or multivariate, present the variable C.

The sulfur (S) is not high in all species studied (**Table 3**), so it would not be a problem in biomass combustion boilers caused by this element. As it can be seen, in olive and citrus wood the S and N contents are lower than the limits established by the standard EN 14691-part 4 [35], which fix the conditions for chips used as biofuels in %N < 1% and %S < 0.1%. Almond has not got values excessive high. This allows preparing mixtures with materials with low N and S contents.

This discrepancy may be due to the extractive substances of plant biomass (sugars, tannins, sterols, fatty acids, resin acids, oligomeric terpenics, hydrocarbons, etc.) that influence the gross calorific value. Their contents depend on the species, the part of the plant, season, and the growing stage, among other factors [36]. Another explanation could be the botanical family of the studied species [37].

Once calculated using the prediction models, we proceeded to the validation of these data by applying them to new samples, whose observations were compared with the predicted values.

3.1. Comparison cost between direct and indirect measurement of HHV

Measurements with the isoperibolic LECO AC-500 calorimeter showed that the time per analyzed sample was 15 min, which includes sample preparation. As consumables, 6 L of pure oxygen measured at standard conditions (25°C and 1 atm of pressure) and a fuse for ignition of the sample are used. **Table 6** shows the cost of reagents. Device depreciation and calibration cost are respectively shown in **Tables 7** and **8**. The total costs for different number of analyzed samples are shown in **Table 9**.

Consumable	Price per consumable pack		Consumption per sample		Analysis cost (€/analysis)
Standard for calibration ref. 502-208-T Ac. Benzoic	160.0	€/100 tablets	5 tablets/calibration		1.60
Spike for combustion ref. 502-815 Mineral oil	175.7	€/118 mL	0.25	mL	
Fuse for ignition ref. 502-462	38.2	€/375 fuse wire	1	Ud	0.10
Rent gas tank O_2	94.4	€/year			0.019
Gas O_2					2.25
Maintenance	1200	€/year			0.24
Technical labor	20	€/h	20	min/sample	6.67
			100	min/calibration	

Table 6. Cost of the reagents for isoperibolic LECO AC500 calorimeter (12/03/2013).

Acquisition cost	50,000	€
Useful life	10	
Residual value 10%	5000	€
Depreciation cost	0.9	€/sample

Table 7. Depreciation cost per sample of LECO AC-500 calorimeter.

	n	Time (min/analysis)	Total time (min)	Cost (€)
Number of analysis	5	20	100	33.33
Standard for calibration ref. 502-208-T Ac. Benzoic	5			8.00
Fuse wire, ref. 502–462	5			0.51
Gas O_2				11.25
Rent gas tanks + maintenance				1.294
			Total	54.39

n, number of analysis.

Table 8. Calibration cost of LECO AC-500 calorimeter.

Sample number	Number of calibration	Technical labor time (h)	Technical labor cost (€)	Gasses (€)	Wire fuse (€)	Total cost (€)	Cost/sample (€/sample)
1	1	2.00	6.67	1.71	0.10	**64.03**	64.03
2	1	2.33	13.33	3.43	0.20	**73.67**	36.83
3	1	2.67	20.00	5.14	0.30	**83.31**	27.77
4	1	3.00	26.67	6.86	0.40	**92.95**	23.24
5	1	3.33	33.33	8.57	0.50	**102.59**	20.52
6	1	3.67	40.00	10.29	0.60	**112.23**	18.70
7	1	4.00	46.67	12.00	0.70	**121.87**	17.41
8	1	4.33	53.33	13.71	0.80	**131.51**	16.44
9	1	4.67	60.00	15.43	0.90	**141.15**	15.68
10	1	5.00	66.67	17.14	1.00	**150.78**	15.08
11	1	5.33	73.33	18.86	1.10	**160.42**	14.58
12	1	5.67	80.00	20.57	1.20	**170.06**	14.17
13	1	6.00	86.67	22.29	1.30	**179.70**	13.82
14	1	6.33	93.33	24.00	1.40	**189.34**	13.52
15	1	6.67	100.00	25.71	1.50	**198.98**	13.27
16	1	7.00	106.67	27.43	1.60	**208.62**	13.04
17	1	7.33	113.33	29.14	1.70	**218.26**	12.84
18	1	7.67	120.00	30.86	1.80	**227.90**	12.66
19	1	8.00	126.67	32.57	1.90	**237.54**	12.50
20	1	8.33	133.33	34.29	2.00	**247.18**	12.36
21	1	8.67	140.00	36.00	2.10	**256.82**	12.23
22	1	9.00	146.67	37.71	2.20	**266.46**	12.11
23	1	9.33	153.33	39.43	2.30	**276.10**	12.00
24	1	9.67	160.00	41.14	2.40	**285.74**	11.91
25	1	10.00	166.67	42.86	2.50	**295.38**	11.82

Table 9. Total costs for different number of samples analyzed by LECO AC-500 calorimeter.

http://dx.doi.org/10.5772/65276

The time spent for elemental analysis using a TruSpec CHN analyzer is 5 min. The costs of the reagents, calibration, and amortization are respectively shown in **Tables 10–12**.

Consumable	Price per consumable pack		Consumption per sample		Analysis cost (€/analysis)
Standard for calibration EDTA ref. 502-092	44.8	€/50 g	0.2 × 8	g/calibration	0.18
Tin Foil for solid samples (Large Tin Foil ref. 502-397-400)	65.5	€/400 tin foil	1	Unit/sample	0.16
Rent tank O_2	94.4	€/year			0.019
Rent tank N_2	47.2	€/year			0.009
Rent tank He	47.2	€/year			0.009
Rent tank compressed air	47.2	€/year			0.009
Gas O_2	450.0	€/tank	1	Tank/200 analysis	2.25
Gas N_2	342.9	€/tank	1	Tank/200 analysis	1.71
Gas He	342.9	€/tank	1	Tank/200 analysis	1.71
Compressed air	126.2	€/year			0.03
Maintenance	1200	€/year			0.24
Technical labor	20	€/h	5	min/simple	3.33

Table 10. Cost of the reagents for elemental LECO TRUSPEC CHNS analyzer (12/03/2013) (supposing 5000 analysis per year).

		n	Time (min/analysis)	Total time (min)	Cost
Time of calibration calculation	Blanc test	15	5	75	
	Standard analysis	8	5	40	
				115	38.33 €
Standard for calibration	EDTA ref. 502-092	8			1.43 €
Tin foil for solid samples	Large Tin Foil ref. 502-397-400	8			1.31 €
Gases		8			18.00 €
Rent gas tanks + maintenance		8			2.30 €
				Total	61.37 €

n, number of analysis.

Table 11. Calibration cost of LECO Truspec CHNS analyzer.

Acquisition cost	**125,000**	**€**
Useful life	10	
Residual value 10%	12,500	€
Depreciation cost	2.25	€/sample

Table 12. Depreciation cost per simple of LECO Truspec CHNS analyzer.

In **Table 13**, the total costs of the elemental analysis with the analyzer LECO CHNS TruSpec are shown. When **Tables 9** and **13** are compared, it can be seen that the cost for determining the calorific value indirectly from elemental analysis is 23% cheaper than the direct measurement with AC500 LECO isoperibolic calorimeter for 25 samples. Moreover, time of determination is lower. The possibility to calculate the calorific value of a substance from its elemental composition reduces investment to a single computer, instead of two. This is very important in laboratories with limited resources of small and medium enterprises.

No. analysis	**Calibration**	**Calibration cost**	**Analysis time (h)**	**Technician labor**	**Gases**	**Tin foil**	**Total**	**Cost/sample (€/sample)**
1	1	61.37	2.00	1.67	2.25	0.16	**67.99**	67.99
2	1	61.37	2.08	3.33	4.50	0.33	**74.61**	37.30
3	1	61.37	2.17	5.00	6.75	0.49	**81.23**	27.08
4	1	61.37	2.25	6.67	9.00	0.66	**87.85**	21.96
5	1	61.37	2.33	8.33	11.25	0.82	**94.46**	18.89
6	1	61.37	2.42	10.00	13.50	0.98	**101.08**	16.85
7	1	61.37	2.50	11.67	15.75	1.15	**107.70**	15.39
8	1	61.37	2.58	13.33	18.00	1.31	**114.32**	14.29
9	1	61.37	2.67	15.00	20.25	1.47	**120.93**	13.44
10	1	61.37	2.75	16.67	22.50	1.64	**127.55**	12.76
11	1	61.37	2.83	18.33	24.75	1.80	**134.17**	12.20
12	1	61.37	2.92	20.00	27.00	1.97	**140.79**	11.73
13	1	61.37	3.00	21.67	29.25	2.13	**147.40**	11.34
14	1	61.37	3.08	23.33	31.50	2.29	**154.02**	11.00
15	1	61.37	3.17	25.00	33.75	2.46	**160.64**	10.71
16	1	61.37	3.25	26.67	36.00	2.62	**167.26**	10.45
17	1	61.37	3.33	28.33	38.25	2.78	**173.87**	10.23
18	1	61.37	3.42	30.00	40.50	2.95	**180.49**	10.03

No. analysis	Calibration	Calibration cost	Analysis time (h)	Technician labor	Gases	Tin foil	Total	Cost/sample (€/sample)
19	1	61.37	3.50	31.67	42.75	3.11	**187.11**	9.85
20	1	61.37	3.58	33.33	45.00	3.28	**193.73**	9.69
21	1	61.37	3.67	35.00	47.25	3.44	**200.34**	9.54
22	1	61.37	3.75	36.67	49.50	3.60	**206.96**	9.41
23	1	61.37	3.83	38.33	51.75	3.77	**213.58**	9.29
24	1	61.37	3.92	40.00	54.00	3.93	**220.20**	9.17
25	1	61.37	4.00	41.67	56.25	4.09	**226.81**	9.07
26	1	61.37	4.08	43.33	58.50	4.26	**233.43**	8.98
27	1	61.37	4.17	45.00	60.75	4.42	**240.05**	8.89
28	1	61.37	4.25	46.67	63.00	4.59	**246.67**	8.81
29	1	61.37	4.33	48.33	65.25	4.75	**253.29**	8.73
30	1	61.37	4.42	50.00	67.50	4.91	**259.90**	8.66
31	1	61.37	4.50	51.67	69.75	5.08	**266.52**	8.60
32	1	61.37	4.58	53.33	72.00	5.24	**273.14**	8.54
33	1	61.37	4.67	55.00	74.25	5.40	**279.76**	8.48
34	1	61.37	4.75	56.67	76.50	5.57	**286.37**	8.42
35	1	61.37	4.83	58.33	78.75	5.73	**292.99**	8.37
36	1	61.37	4.92	60.00	81.00	5.90	**299.61**	8.32
37	1	61.37	5.00	61.67	83.25	6.06	**306.23**	8.28
38	1	61.37	5.08	63.33	85.50	6.22	**312.84**	8.23
39	1	61.37	5.17	65.00	87.75	6.39	**319.46**	8.19
40	1	61.37	5.25	66.67	90.00	6.55	**326.08**	8.15
41	1	61.37	5.33	68.33	92.25	6.71	**332.70**	8.11
42	1	61.37	5.42	70.00	94.50	6.88	**339.31**	8.08
43	1	61.37	5.50	71.67	96.75	7.04	**345.93**	8.04
44	1	61.37	5.58	73.33	99.00	7.21	**352.55**	8.01
45	1	61.37	5.67	75.00	101.25	7.37	**359.17**	7.98
46	1	61.37	5.75	76.67	103.50	7.53	**365.78**	7.95
47	1	61.37	5.83	78.33	105.75	7.70	**372.40**	7.92
48	1	61.37	5.92	80.00	108.00	7.86	**379.02**	7.90
49	1	61.37	6.00	81.67	110.25	8.02	**385.64**	7.87
50	1	61.37	6.08	83.33	112.50	8.19	**392.26**	7.85

Table 13. Total cost of analysis by LECO Truspec CHNS analyzer.

4. Conclusions

The advantage of using indirect methods for determining the heat value based on regression models from the analysis of C, H, and N elements has been proven this paper. Time to determine the high heat value indirectly is 40% lower than the time taken using the calorimeter directly. The cost of indirect method is 23% cheaper. Along with the cost savings, reduced analysis time is associated with a lower environmental impact linked to the reagents used.

It is proved that a higher carbon content (C) provides bigger calorific values. However, hydrogen (H) has a negative influence, with a negative Pearson's coefficient. This is due to the fact that hydrogen is associated with the water content. As it is known, the moisture content decreases the high heat value of the biomass, therefore, hydrogen presents inverse proportionality with the heat obtained from combustion. The colinearity between the moisture content and the hydrogen ratio justifies that both were rarely considered in the same model to predict the heat value.

In this paper, models for determining the calorific value in samples of *Citrus × sinensis* Osbeck (orange tree), *P. dulcis* (Mill) D.A. Webb (almond tree), and *O. europaea* L. (olive tree) are proposed using an elemental analysis. The accuracy is high, obtaining coefficients of determination higher than 0.95, an average error of 223.64 kJ/kg, and a RMS of 309.62 kJ/kg.

According to the thermochemical characterization of plum wood, the residual biomass from pruning can be used as chips for bioenergy. These species did not have significant differences in C and H composition, between 30 and 40% C, between 5 and 7% H. However, small differences exist with respect to N and S. Olive and citrus wood have S and N contents lower than the limits established by the standard EN 14691-part 4 [32], which fix the conditions for chips used as biofuels in $\%N < 1\%$ and $\%S < 0.1\%$. Almond has not got values excessive high but still it exceeds. This leads to the preparation of mixtures with materials with low N and S contents to decrease their content.

Author details

Borja Velázquez-Martí[1*], Isabel López-Cortés[2], Domingo Salazar-Hernández[2] and Ángel Jesús Callejón-Ferre[3]

*Address all correspondence to: borvemar@dmta.upv.es

1 Department of Rural and Food Engineering, Universitat Politècnica de València, Valencia, Spain

2 Department of Vegetal Production, Universitat Politècnica de València, Valencia, Spain

3 Department of Engineering, University of Almeria, Almería, Spain

References

[1] Friedl A, Padouvas E, Rotter H, Varmuza K. Prediction of heating values of biomass fuel from elemental composition. Anal Chim Acta. 2005;544:191–198. DOI: 10.1016/j.aca.2005.01.0

[2] Sheng C, Azevedo JLT. Estimating the higher heating value of biomass fuels from basic analysis data. Biomass Bioenergy. 2005;28(5):499–507. DOI: 10.1016/j.biombiec.2004.11.008

[3] Callejón-Ferre AJ, Velázquez-Martí B, López-Martinez JA, Manzano-Agugliaro F. Greenhouse crop residues: energy potential and models for prediction of their higher heating value. Renew Sust Energ Rev. 2011;15(2):948–955. DOI: 10.1016/j.rser.2010.11.01

[4] Vargas-Moreno JM, Callejón-Ferre AJ, Pérez-Alonso J, Velázquez-Martí B. A review of the mathematical models for predicting the heating value of biomass materials. Renew Sust Energ Rev. 2012;16:3065–3083. DOI: 10.1016/j.rser.2012.02.054

[5] Velázquez-Martí B, Sajdak M, López-Cortés I, Callejón-Ferre AJ. Wood characterization for energy application proceeding from pruning *Morus alba* L., *Platanus hispanica* Münchh. and *Sophora japonica* L. in urban áreas. Renew Energy. 2014;62:478–483. DOI: 10.1016/j.renene.2013.08.010

[6] Pérez-Arévalo JJ, Callejón-Ferre AJ, Velázquez-Martí B, Suárez-Medina MD. Prediction models based on higher heating value from the elemental analysis of neem, mango, avocado, banana, and carob trees in Guayas (Ecuador). J Renew Sustain Ener. 2015;7:053122.

[7] Saidur R, Abdelaziz EA, Demirbaş A, Hossain MS, Mekhilef S. A review on biomass as a fuel for boilers. Renew Sust Energ Rev. 2011;15(5):2262–2289. DOI: 10.1016/j.rser.2011.02.015

[8] Khan AA, Jonga WD, Jansens PJ, Spliethoff H. Biomass combustion in fluidized bed boilers: potential problems and remedies. Fuel Process Technol. 2009;90:21–50. DOI: 10.1016/j.fuproc.2008.07.012

[9] Telmo C, Lousada J, Moreira N. Proximate analysis, backwards stepwise regression between gross calorific value, ultimate and chemical analysis of wood. Bioresource Technol. 2010;101(11):3808–3815. DOI: 10.1016/j.biortech.2010.01.021

[10] Demirbas A, Demirbas AH. Estimating the calorific values of lignocellulosic fuels. Energ Explor Exploit. 2004;22:135–143. DOI: 10.1260/0144598041475198.

[11] Demirbas A. Relationship between heating value and lignin, moisture, ash and extractive contents of biomass fuels. Energy Explor Exploit. 2002;20:105–111. DOI: 10.1260/014459802760170420

[12] Callejón-Ferre AJ, Carreño-Sánchez J, Suárez-Medina FJ, Pérez-Alonso J, Velázquez-Martí B . Prediction models for higher heating value based on the structural analysis

of the biomass of plant remains from the greenhouses of Almería. Fuel. 2014;116:377–387. DOI: 10.1016/j.fuel.2013.08.023

[13] UNE-EN 164001:2005 EX. Solids Biofuels. Method for the detemination of calorific value. AENOR: Madrid. Spain; 2005.

[14] Francis HE, Lloyd WG. Predicting heating value from elemental composition. J Coal Qual. 1983;2(2):21–25.

[15] Ebeling JM, Jenkins BM. Physical and chemical properties of biomass. Trans ASAE. 1998;28:898–902.

[16] Kathiravale S, Yunus MNM, Sopian K, Samsuddin AH, Rahman RA. Modeling the heating value of municipal solid waste. Fuel. 2003;82(9):1119–1125. DOI: 10.1016/S0016-2361(03)00009-7

[17] Velázquez-Martí B, Fernández-González E, López-Cortes I, Salazar-Hernández DM. Quantification of the residual biomass obtained from pruning of vineyards in Mediterranean area. Biomass Bioenergy. 2011;35(3):3453–3464. DOI: 10.1016/j.biombioe.2011.04.009

[18] Velázquez-Martí B, Fernández-González E, López-Cortes I, Salazar-Hernández DM. Quantification of the residual biomass obtained from pruning of trees in Mediterranean olive groves. Biomass Bioenergy. 2011;35(2):3208–3217. DOI: 10.1016/j.biombioe.2011.04.042

[19] INE. Instituto Nacional de Estadística. Censo Agrario 2009 [Internet]. 2009 [Updated: 2014]. Available from: http://www.ine.es/censoagrario/censoag.htm

[20] Avanza MM, Bramardi SJ, Mazza SM. Statistical models to describe the fruit growth pattern in sweet orange "Valencia late". Spanish J Agric Res. 2008;6(4):577–585.

[21] Velázquez-Martí B, Fernández-González E, López-Cortes I, Salazar-Hernández DM. Quantification of the residual biomass obtained from pruning of trees in Mediterranean almond groves. Renew Energy. 2011;36:621–626. DOI: 10.1016/j.renene.2010.08.008

[22] López-Cortés I, Salazar-García DC, Salazar DM, Velázquez-Martí B. Fatty Acid, Sterol and Polyphenol content in most Monovarietal Oils of the Spanish East Areas. In: De Leonardis A. (eds.). Virgin olive oil. Production, composition, uses and benefit for the man. New York: Nova Publishers; 2014.

[23] Barranco D, Rallo L. Olive cultivars in Spain. Horttechnology. 2000;10(1):107–110.

[24] UNE-EN 14961-1:2011. Biocombustibles sólidos. Especificaciones y clases de combustibles. Parte 1: Requisitos generales. AENOR: Madrid, Spain; 2011.

[25] UNE-EN 14780:2012. Biocombustibles sólidos. Preparación de muestras. AENOR: Madrid, Spain; 2012.

[26] Husch B, Beers TW, Kershaw JA Jr, editors. Forest Mensuration. John Wiley & Sons. Inc.: Hoboken, NJ; 2003.

[27] Shapiro SS, Francia RS. An approximate analysis of variance test for normality. J Amer Statist Assoc. 1972; 67: 215–216.

[28] Shapiro SS, Wilk MB. An analysis of variance test for normality (complete samples). Biometrika. 1965;52:591–611.

[29] Sheskin DJ. Handbook of Parametric and Nonparametric Statistical Procedures: Fifth Edition. London: Chapman & Hall/CRC Taylor & Francis Group 2011.

[30] Razali NM, Wah YB. Power Comparisons of Shapiro-Wilk, Kolmogorov-Smirnov, Lilliefors and Anderson-Darling Tests. J Stat Model Anal. 2011;2(1):21–33.

[31] González Z, Rosal A, Requejo A, Rodríguez A. Production of pulp and energy using orange tree prunings. Bioresource Technol. 2011;102(19):9330–9334. DOI: 10.1016/j.biortech.2011.07.088

[32] Yin CY. Prediction of higher heating values of biomass from proximate and ultimate analyses. Fuel. 2011;90(3):1128–1132. DOI: 10.1016/j.fuel.2010.11.031

[33] Acda MN. Physico-chemical properties of wood pellets from coppice of short rotation tropical hardwoods. Fuel. 2015;160:531–533. doi:10.1016/j.fuel.2015.08.018

[34] Erol M, Haykiri-Acma H, Kücükbayrak S. Calorific value estimation of biomass from their proximate analysis data. Renew Energy. 2010;35:170–173. DOI:10.1016/j.renene.2009.05.008

[35] EN 14691-part 4. Wood chips—product standard for non-industrial use. CEN-UNE, Madrid, Spain, 2011.

[36] Nhuchhen DR. Prediction of carbon, hydrogen, and oxygen compositions of raw and torrefied biomass using proximate analysis. Fuel. 2016;180:348–356. doi: 10.1016/j.fuel.2016.04.058

[37] Veiga JPS, Valle TL, Feltran JC, Bizzo WA. Characterization and productivity of cassava waste and its use as an energy source. Renew Energy. 2016;93:691–699. doi:10.1016/j.renene.2016.02.078

Chapter 13

Microalgal Biomass: A Biorefinery Approach

Luis C. Fernández Linares,

Kevin Á. González Falfán and Citlally Ramírez-López

Additional information is available at the end of the chapter

http://dx.doi.org/10.5772/65827

Abstract

The biorefinery concept has been identified as the most promising way to create a biomass-based industry, which can be defined as the sustainable biomass processing to obtain energy biofuels and high-value products through processes and equipment for biomass. Microalgae can be used as an efficient and economically viable biorefinery feedstock; microalgae could be used in different areas such as human and animal nutrition, nutraceutical and therapeutic products, fertilizers, plastics, isoprenes and biofuels and also in the treatment of wastewaters and CO_2 capture. Microalgae biomass can be used for biofuel production, such as bioelectricity, methane, biohydrogen, bioethanol and biodiesel. In this chapter, an overview of the factors that affect the production of the microalgal biomass yield and value-added by-products production is presented. Likewise, we present the results of the microalgal culture of *Scenedesmus* sp. SCX2 performed in semicontinuous culture (SCC), in 2000 raceway ponds employing Bold's Basal Medium (BBM), under greenhouse conditions. Over the SCC, we monitored biomass concentration, lipid, protein, pigments and sugar production, light irradiance, culture and greenhouse temperature and nitrate concentration in the medium.

Keywords: biomass, microalgae, *Scenedesmus* sp. SCX2, biorefinery, biofuels

1. Introduction

Microalgae have the ability to convert solar energy into chemical forms though photosynthesis, in which CO_2 is captured from the atmosphere and oxygen is produced during the process, furthermore, microalgae offers the potential to produce high-value compounds such as nutraceuticals and compounds for human consumption, and also low-value products as feed, biofertilizers and bioenergy, as well as wastewater treatment, nitrogen fixing and CO_2

mitigation [1, 2]. Microalgae have been taken into consideration as a feedstock for renewable biofuels production, such as bioelectricity, methane produced by anaerobic digestion of the algal biomass, biohydrogen produced under anaerobic conditions, bioethanol (sugar fermentation) and biodiesel derived from microalgal oil [1, 3–5]. Microalgae offer, compared to other renewable sources, higher growth rates and require less water than terrestrial crops [6]; microalgae have a very short harvesting cycle (1–10 days) compared to other land-based feedstock (which are harvested once or twice a year) [7]; algae cultures do not compete with food or arable lands [8]; their phototrophic growth has a favorable environmental impact, since the CO_2 that is released to the atmosphere during hydrocarbons combustion is recycled by microalgae in photosynthetic processes (100 biomass tons require about 183 CO_2 tons) [9]. Microalgae have higher photosynthetic efficiency than crops. Therefore, microalgae biomass cultivation will help in reducing the CO_2 in the atmosphere at a faster rate than land-based crops. Phototrophic metabolism of microalgae is mainly influenced by light irradiance, culture temperature and nutrients supply. However, in large-scale open systems as raceways, it is practically impossible to control diurnal fluctuations of irradiance and temperature [10, 11]. However, raceways are the most commonly employed technology for microalgae biomass production due to their low cost of construction and operation. Microalgae biomass production for obtaining just biofuels is not economically feasible, for which it is necessary to use high-productivity and economic reactors, as well as the use of economic and available nutrients; likewise, the maximum valorization of biomass, including the carbon fixation and wastewater treatment by integrating the concept of industrial ecology. All previous allows the conceptualization of microalgal biorefinery.

This chapter addresses the microalgal biomass production from a biorefinery point of view and describes the improvements of the microalgal biomass production in open systems raceways (RW) of 2000 L capacity operated in semicontinuous culture under greenhouse conditions.

2. Biorefineries

In a conventional refinery, the raw oil undergoes several processes after which the products are used for transportation, electricity and the further production of more complex and valuable chemicals. On the other hand, the biorefineries allow the utilization of raw materials that may be widely available at low cost and sustainable for the production of high-value products that may be considered in terms of their energetic potential. Within a biorefinery there is a wide variety of configurations and generation of several end products. An appropriate definition of biorefinery is as follows: the sustainable processing of the biomass in a wide commercial spectrum of products and energy [12]. Other definition is sustainable biomass processing to obtain energy biofuels and high-value products through processes and equipment for biomass transformation [13]. The biorefinery concept has been identified as the most promising way to create a biomass-based industry.

A biorefinery can utilize all kinds of biomass such as wood (straw and bark), agricultural crops, organic waste (derived from vegetables and animals), industrial and domestic wastes and aquatic biomass (algae and microalgae from sea and freshwater) [14]. The microalgae present a high potential for the biodiesel production compared to the terrestrial biomass [15]. The impact on the utilization of the soil for the microalgal production is significantly lower than the generated by the agricultural biomass based on corn, canola and switchgrass [16]. The versatility of the microalgal metabolism allows the application in different areas such as human and animal nutrition, nutraceutical and therapeutic products, fertilizers, plastics, isoprenes, biofuels and also in the treatment of wastewaters and CO_2 capture [17, 18]. Algae can be used as an efficient and economically viable biorefinery feedstock. An efficient biorefinery using algae can only be constructed through its integration with other industries [19].

2.1. Microalgae as raw material for a biorefinery implementation

Algae are a polyphyletic group of organisms (including organisms that do not have the same origin, but are multiple, independent evolutionary lines) that comprise both, unicellular and multicellular forms and both, prokaryotes and eukaryotes, which are able to capture light energy through pigments (such as chlorophylls, carotenoids, anthocyanins and phycoerythrinsphycobilins) that are necessary to carry out photosynthesis [20, 21]. The classification of algae has five main branches: chromista (brown algae, golden brown algae and diatoms), red algae, dinoflagellates, euglenids and green algae [21]. Microalgae can be eukaryotic (Chlorophyta, Rhodophyta and Bacillariophyta) or prokaryotic (Cyanophyta) [3]. Microalgae are a large group of microscopic algae that are considered primary producers on a global scale, and are involved in all marine and freshwater ecosystems [15, 22]. Microalgae are the largest autotrophic microorganisms of plant life taxa in the world. The biodiversity of microalgae is enormous and they represent an almost untapped resource. It has been estimated that about 20,000–800,000 species exist of which about 40–50,000 species are described [23, 24]. The biomass produces three major biochemical components by *de novo* synthesis consisting of lipids, proteins, carbohydrates, pigments and others.

The proposal to use microalgae for biofuels production is not new and was first suggested in the 1940s–1950s. In the 1970s, the large-scale cultivation of microalgae for producing sustainable liquid fuels was previously investigated and in the 1990s extensive research was carried out, especially at the Solar Energy Research Institute in Golden, Colorado, USA [3, 4, 9, 25].

Specifically, microalgae production offers the potential for the production of high-value compounds, products for human and animal nutrition, bioactive compounds, fluorescent pigments, stable-isotope biochemicals, drug screening, microalgal recombinant proteins, cosmetics, biofertilizers and wastewater treatment, nitrogen fixing, CO_2 mitigation and bioenergy production [2, 9, 21, 25–28]. In this manner, these microscopic organisms have been taken into consideration as a feedstock for the production of renewable biofuels such as bioelectricity, methane produced by anaerobic digestion of the algal biomass, biohydrogen produced under anaerobic conditions, bioethanol (sugar fermentation) and biodiesel derived from microalgal oil [3, 4, 25, 27, 29, 30]. The environmental factors such as illumination, temperature, the amount of nutrients and the salinity affect the photosynthesis and the productivity of the cellular biomass by influencing the pathway and the activity of the cellular metabolism [31, 32].

2.2. Open pond culture systems

Microalgae can be grown in open ponds or in closed systems named photobioreactors. The use of open systems to produce biofuel feedstock is 2.5-times less expensive than using photobioreactors [33]. For that reason, open systems are currently the preferred option for large-scale bioprocessing. There are many types of open ponds used for microalgae cultivation such as raceway, shallow ponds or circular ponds. Raceway pond is the most common open system due to its potential to produce large quantity of microalgae. In a raceway pond system, the cultures are circulated around a race track by using paddle wheels. This will keep the microalgae suspended in the medium and allow the utilization of CO_2 from the atmosphere; CO_2 can also be injected into the medium. The pond is shallow to allow the light penetration into the pond, which in turn maximizes the light exposure of the microalgae for photosynthesis. Open systems make it possible to reduce production costs, this is mainly due to the greater investment required for closed photobioreactors, which increases the depreciation costs (making up as much as 78% of the total microalgae production cost) and also due to the greater power consumption of closed photobioreactors, which likewise increases the operational costs [34]. By using open reactors both the depreciation and the operating costs are reduced. At the same scale and under the same conditions as those considered for tubular photobioreactors, the microalgae production and the reactor costs are reduced almost 4 and 20 fold, respectively, for raceway reactors [35].

The temperature and light irradiance variations are practically impossible to control in outdoor cultures [10]. Such fluctuations can affect microalgal growth at the point to collapse the whole culture [36]. Furthermore, contamination by predators, protozoa, ciliates and other fast growing heterotrophic bacteria have restricted the commercial production of algae in open systems to only those organisms that can grow under specific conditions such as high pH [37], hyper-saline cultures [26] or where suitable irradiance is available. For all these reasons, a crucial factor for the success in culturing open systems is to choose a microalgal strain with the ability to grow in a selective medium, in outdoor environmental conditions with relatively constant productivities [38, 39].

The operational mode of the open systems is important to be considered. The main cultivation modes are in batch, fed-batch, continuous and semicontinuous culture (SCC). The SCC consists of periodic withdrawals of volume from the culture and the removed volume is reinstated with fresh medium. This culture mode is propitious for long-term outdoor cultures, because the growth is maintained in log phase, it avoids the shading effect caused by high cell density [40] and allows the inoculum and biomass availability during each harvesting cycle [41].

SCC cultures have been studied for wastewater removal [42], CO_2 fixation [43, 44] and to enhance biomass productivity in several microalgae species [40, 45, 46], but little work has been published on long-term pilot-scale open systems concerning its effect on the fatty acids methyl esters (FAMES) content, which are the feedstock for microbial biodiesel production.

http://dx.doi.org/10.5772/65827

2.3. Environmental factors and their effect on the microalgal growth

The microalgal growth depends on the availability and efficient utilization of light. A lack of an adequate illumination exerts a negative effect on the productivity, growth and other parameters such as lipids content. The microalgal may be in the respiration phase in the absence of light, in the limiting phase when the illumination is not enough, in a saturation phase if there is a maximum photosynthetic yield or they may be in an inhibition phase due to the suppression of the photosynthetic activity caused by an excess of light (photoinhibition) [47]. High temperatures and concentrations of dissolved O_2, low CO_2 concentrations, as well as a high intensity in illumination and a high pH in the medium, all promote an increase in the photorespiration [48].

The illumination can be supplied to the microalgal culture either by the sun light or by using artificial light such as fluorescent lamps, LEDs, optic fibers, etc. The sources of artificial light may be modified for a specific wavelength emission that leads to metabolic and physiological changes in the microalgae. For a large scale microalgal culture, the utilization of the sun light is recommended in order to operate at a low cost that allows a positive energetic balance. The utilization of artificial sources of light may be necessary for the production of high-value products, but in general, artificial light at large scale should be avoided [49].

2.3.1. Temperature

An adequate temperature is necessary for a microalgal photosynthetic activity in which there are no modifications either in the biochemistry or the physiology of the microalgae. The optimum temperature range at which the microalgal growth is between 20 and 25°C for mesophilic microalgal. The thermophilic strains can be cultured at 40°C and the psychrophilic strains can grow at 17°C. The computational modeling designed to estimate the effect of the temperature on the microalgal growth demonstrates that the temperature control at large scale is even more expensive due to the energetic requirement to control the temperature than the energy that eventually could be produced from the microalgal production (e.g. this is the case in the biodiesel production) [38]. The temperature is one of the most difficult factors to be optimized in the open culture systems due to climatic variations and the seasonal changes. This fact may cause a possible decrease in productivity [50].

2.3.2. Nutrients

A culture medium for microalgal culturing must contain all of the following: a carbon source, which makes up to 50% of the microalgal biomass. Nitrogen (N), which corresponds to approximately 5–10% dry weight basis of the biomass. Phosphorous (P), which is a component of the DNA, RNA, ATP and the cell membrane. Sulfur (S), which is a component in certain amino acids, vitamins and sulfolipids and is involved in the synthesis of proteins. Potassium (K), which is a cofactor for several enzymes and plays an important role in the osmotic regulation and the synthesis of proteins. Magnesium (Mg), which is found in the core of the chlorophyll molecule andiron (Fe), which is important during the assimilation of nitrogen and is part of the cytochromes. The trace metals, salts and organic components in addition to the other elements mentioned above can be manipulated in order to modify the production of certain desired metabolites [51]. The utilization of wastewater that is rich in N and P as culture medium for microalgae is an economical alternative. Wastewaters can come of sectors such as domestic, commercial-service, industrial and agricultural [52]. One of the advantages that wastewater offers as a culture medium is the possibility of recovery and utilization of inorganic compounds simultaneous to the production of recycling or groundwater recharge water streams [53].

2.3.3. Salinity

An excess of sodium ions (Na^+) leads to a saline stress, whereas an osmotic stress may imply the presence or absence of sodium ions. An accumulation of Na^+ and chloride (Cl^-) ions causes an ionic imbalance in the cell. This decreases the capability to absorb minerals such as K, calcium (Ca^{2+}) and manganese (Mn^{2+}) [54]. Under a saline stress, the microalgae present change their metabolism in order to protect themselves from the osmotic damage. The salt concentration that can be tolerated by microalgae, bacteria and cyanobacteria is up to 1.7 M in marine medium [55].

2.4. Selection of the microalgal strain

The selection of the microalgal strain is of upmost importance because each strain has different and specific features. Microalgal strains may be acquired through a specialized collection or they can be isolated from extreme environments. The species isolated from local places present a better adaptation capability and often times can be adapted and grown under adverse and changing conditions and therefore, they tend to survive and grow under stressful conditions even better than the microalgae coming from specialized collections [56]. The biomass of the microalgae can be used in several forms. Biogas can be obtained from the anaerobic digestion of the biomass and the lipids are treated for a transesterification in order to produce biodiesel. The carbohydrates can be used to produce ethanol or butanol, whereas the biomass can be treated thermochemically or by gasification to obtain bio-oil [57]. The decision on the type of microalgae to use depends on its composition and the type of biofuel and other desired valuable products that need to be produced or that are of interest (**Table 1**).

Microalgal species	Mode of cultivation	BR	Biomass productivity (mg L^{-1} d^{-1})	Carbohydrates (%)	Proteins (%)	Lipids (%)	Ref.
Stichococcus sp.	Batch	Erlenmeyer flasks	167.00	40.63	26.25	11.56	[58]
Scenedesmus obliquus	Batch	1-L glass vessel	840.57	46.65	–	22.4	[59]
Scenedesmus obliquus	Semibatch	1-L glass vessel	883.80	52.90	–	–	[60]
Scenedesmus bijugatus	SCC	Vertical tubular	260.00	26.00	–	24	[61]
Chlorella vulgaris	Semibatch	Vertical tubular	268.10	–	52.2	–	[62]

Microalgal species	Mode of cultivation	BR	Biomass productivity (mg L^{-1} d^{-1})	Carbohydrates (%)	Proteins (%)	Lipids (%)	Ref.
Chlorella sp.	Batch	Vertical tubular	154.48	–	–	21.27	[63]
Chlorella ellipsoidea	Batch	Bubble column	31.55	–	41.75	–	[64]
Chlorella variabilis	Batch	Raceway	113.33	–	–	10	[65]
Chaetoceros muelleri	–	Fiberglass cylinders	–	12.7–19.3*	46.9–64.4†	22.1–33.2‡	[66]
Graesiella sp.	SCC	Raceway	36.00	–	–	33.4	[39]

*Winter, 18–19.3; spring, 15.2–15; summer, 12.7–13.6%.
†Winter, 49.3–46.9; spring, 54.2–56.3; summer, 59.2–64.4%.
‡Winter, 30.3–33.2; spring, 29.5–28.2; summer, 24, 22.1%.

Table 1. Microalgal strains for the production of biofuels and high-value products.

In this work, the marine microalgae *Scenedesmus* sp. SCX2 was grown in 2000 L raceway ponds under greenhouse conditions and was used to study the effect of semicontinuous culture mode during summer and autumn seasons on biomass, protein, pigments, sugar and lipid production.

3. Semicontinuous culture of *Scenedesmus* sp. SCX2 in RW under greenhouse conditions

3.1. Materials and methods

The cultivation system of the microalgae *Scenedesmus* sp. SCX2 was performed in semicontinuous culture in 2000 L raceway ponds employing Bold's basal medium (BBM), for 123 days (eight harvesting cycles) during summer and autumn seasons both under greenhouse conditions. Over the SCC, we monitored biomass concentration, lipid, protein, pigments and sugar production, light irradiance, culture temperature, greenhouse temperature and nitrate and ammonium concentrations in the medium as described in the following sections.

3.1.1. Microalgal strain and growth conditions

Scenedesmus sp. SCX2 strain was acquired from the Centro de Investigación Científica y de Educación Superior de Ensenada (CICESE). Prior to use it, the strain was grown autotrophically and axenically in BBM medium in a 1 L photobioreactor (PBR) with a working volume of 0.9 L, and bubbled with air at a constant flow rate of 0.5 vvm and then in 20 L PBR. The PBR was illuminated with a light intensity of 180 μmol photons m^{-2} s^{-1} on the surface of the reactor by using cool white fluorescent tubes, with a photoperiod of 12 h of light:12 h of darkness and the temperature was maintained at 24 ± 1°C.

3.1.2. Raceway pond cultures

The open channel raceway reactors dimensions are: two 3.1 m channels, each of 0.72 m wide and connected by 180° bends at both ends to give a total surface area of 6.1 m^2. The raceways were constructed with fiberglass and the culture was circulated and mixed by a two paddle wheel system at a superficial flow velocity of 0.3 ms^{-1}. The operation volume was 1400 L, with a culture medium depth of 0.24 m. The cultures were conducted under greenhouse conditions. The light irradiance and greenhouse temperature were recorded with a data logger T&D RTR-500; the culture temperature was recorded with a data logger T&D RT-200. Additionally, temperature, pH, turbidity, NO_3^-, NH_4^+, conductivity and dissolved oxygen were continuously determined with HYDROLAB DS 5 (Hach) probe. The harvesting cycles were carried out on day 14 or 16, and to do so, half of the culture volume was removed and replaced with the same volume of fresh medium. Water losses due to evaporation were quantified and replaced periodically. The experiments were conducted under greenhouse conditions in Mexico City, Mexico (19.513°N, 99.126°W).

3.2. Analytical methods

3.2.1. Biomass determination

Biomass was determined by absorbance at 600 nm and dry biomass gravimetrically with a thermobalance. The sample was filtered through a glass microfiber membrane (Ahlstrom, 4.7 cm diameter, 0.7 μm pore size) to constant weight.

3.2.2. Determination of nitrate concentration

The nitrate concentration in the culture medium was determined as in Ref. [67] by using 0.5 mL of culture supernatant obtained after centrifugation of a microalgal culture sample at 3823×*g* for 10 min, then this supernatant was placed in a test tube and dried at 100°C for 12 h in an oven Riossa H62. A 0.5 mL portion of diphenyl sulfonic acid was added to the dry sample followed by the addition in four pulses of 2.2 mL of 12N KOH during constant mixing. A mix of 100 μL of supernatant with 900 μL of distilled water was read at 410 nm in a spectrophotometer GENESYS 10S UV-vis (Thermo Scientific, USA). A standard curve with the culture medium was prepared within a range of 0–500 mg L^{-1} of NO_3^- as $NaNO_3$, to obtain the nitrate concentration.

3.2.3. Quantitative lipid determination and FAMEs determination

Cell disruption, lipid extraction, quantitative lipid and lipid composition, determined as fatty acid methyl esters (FAMEs) by gas chromatography after direct lipid transesterification, were realized as in Ref. [68].

3.2.4. Pigment content

Carotenoids and chlorophylls were measured according to the protocols and the equations obtained by Ref. [69]. Culture samples of 1.5 mL were centrifuged for 5 min at 17,000×*g*, after which the supernatant was removed. The cells were suspended in 1.5 mL of methanol and incubated at 45°C in the dark during 30 min. The tubes were centrifuged (17,000×*g*/5 min) and the methanol extract was transferred into plastic cuvettes for measurement at 470, 653, 666 and 750 nm. Carotenoids and chlorophyll A (Chl*a*) and B (Chl*b*) concentrations were estimated according to the following equations (Eqs. 1–4):

$$\text{chl}a\left[\mu\text{g mL}^{-1}\right] = 15.65\left(\text{ABS}_{666}\right) - 7.34\left(\text{ABS}_{653}\right), \tag{1}$$

$$\text{chl}b\left[\mu\text{g mL}^{-1}\right] = 27.05\left(\text{ABS}_{653}\right) - 11.21\left(\text{ABS}_{666}\right), \tag{2}$$

$$\text{Total carotenoids}\left[\mu\text{g mL}^{-1}\right] = \frac{1000\left(\text{ABS}_{470}\right) - 2.86\left(\text{chl}a\right) - 129.2\left(\text{chl}b\right)}{221}, \tag{3}$$

$$\text{Total pigments}\left[\mu\text{g mL}^{-1}\right] = \text{chl}a + \text{chl}b + \text{total carotenoids}. \tag{4}$$

3.2.5. Total sugars

A sample of 5 mg of dried algae was added to 1 mL of 2 N HCl. Hydrolysis was achieved by heating this mixture at 100°C for 120 min. Then total sugars were quantified by the Dubois method [70].

3.2.6. Proteins

A sample of 5 mg of dried algae was added to 1 mL of 1 N NaOH. Hydrolysis was achieved by heating this mixture at 100°C for 120 min. Then proteins were quantified by the Bradford method [71].

All the samples were analyzed in triplicate. The average productivity referenced in [72] was calculated for biomass, lipid, total pigment and sugar and protein content, at the end of every cycle of culture. Productivity is defined as the milligram of product per liter of media per day.

3.3. Results and discussion

The aim of implementing an autotrophic microalgal culture in outdoor conditions was to take advantage of solar energy to convert CO_2 into organic compounds through photosynthesis and then to produce biomass, lipids, proteins, pigments and sugars. The *Scenedesmus* sp. SCX2 culture was performed using BBM medium, under semicontinuous mode, into a raceway reactor of 2000 L. The first culture cycle was considered as an adaptation period. The major growth was observed in the second and third cycles; then the growth was decreasing gradually in subsequent cycles (**Figure 1**). This was certainly due to the weather conditions; in our research group we have observed an increase in biomass with the sequenced cycles [73]. In the second and third cycles, the volumetric biomass productivities of 53 and 5 mg L^{-1} d^{-1} were obtained, respectively, without significant differences between them. These were 15 and 11% higher than volumetric biomass productivity achieved in cycle 1 (**Figure 2**) and up to 50% higher than volumetric biomass productivity obtained in the last two cycles. Ref. [36] reported a volumetric biomass productivities of 30–260 mg L^{-1} d^{-1} with *Scenedesmus* sp. and all productivities obtained in this work are within this range. Areal biomass productivities obtained at the end of the second and third cycles were 12 and 11.5 g m^{-2} d^{-1}, respectively, which were 40% higher than that reported for *Chlorophyta* sp. and similar to the areal biomass productivity obtained with *Chlorella* sp. [74], both microalgae grown in semicontinuous mode in raceways (0.986 m^2), under greenhouse conditions. In addition, higher productivities of this work were half of that obtained by Ref. [35] with *Scenedesmus* sp. cultivated in semicontinuous mode in a raceway of 32 m^2 and CO_2-enriched aeration.

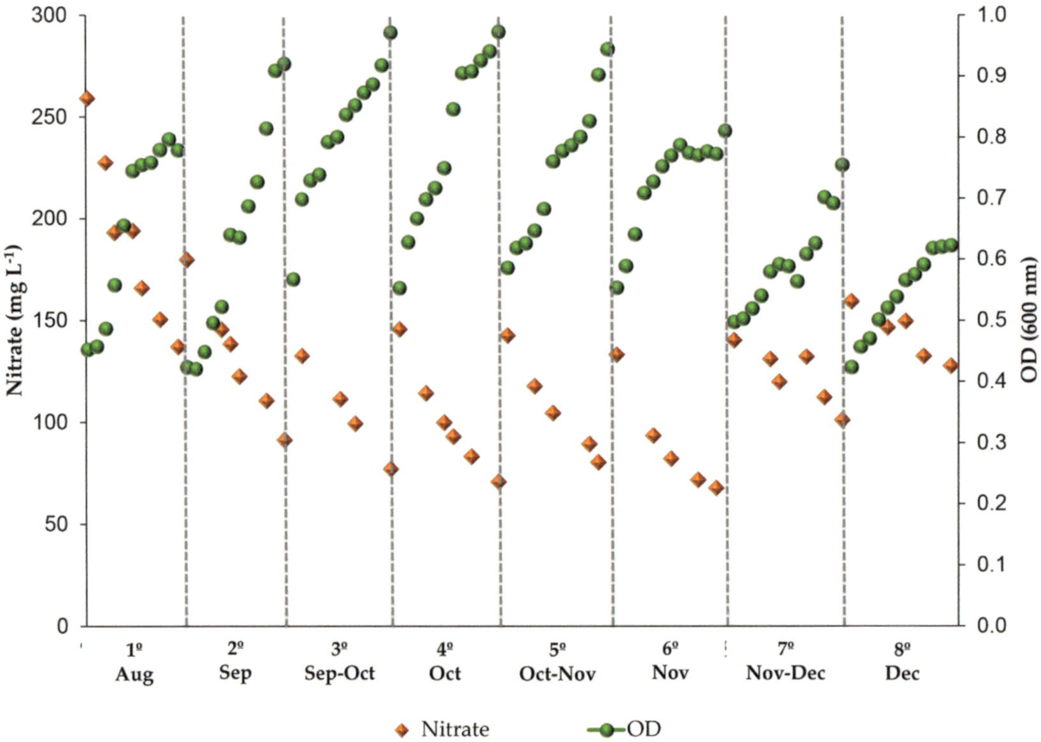

Figure 1. *Scenedesmus* sp. SCX2 growth and nitrate consumption during the eight cycles of culturing under semicontinuous mode in a raceway of 2 m^3 with 1.4 m^3 of medium, under greenhouse conditions.

The first two cycles were developed in the late summer while the other cycles were developed in the autumn, which explains the gradual decrease in both irradiance culture temperatures (**Table 2**); which decreased the biomass productivities (**Figure 2**). Temperatures of the culture of this work were within the range reported by Ref. [75] for the culturing of *Scenedesmus* sp. between 10 and 30°C.

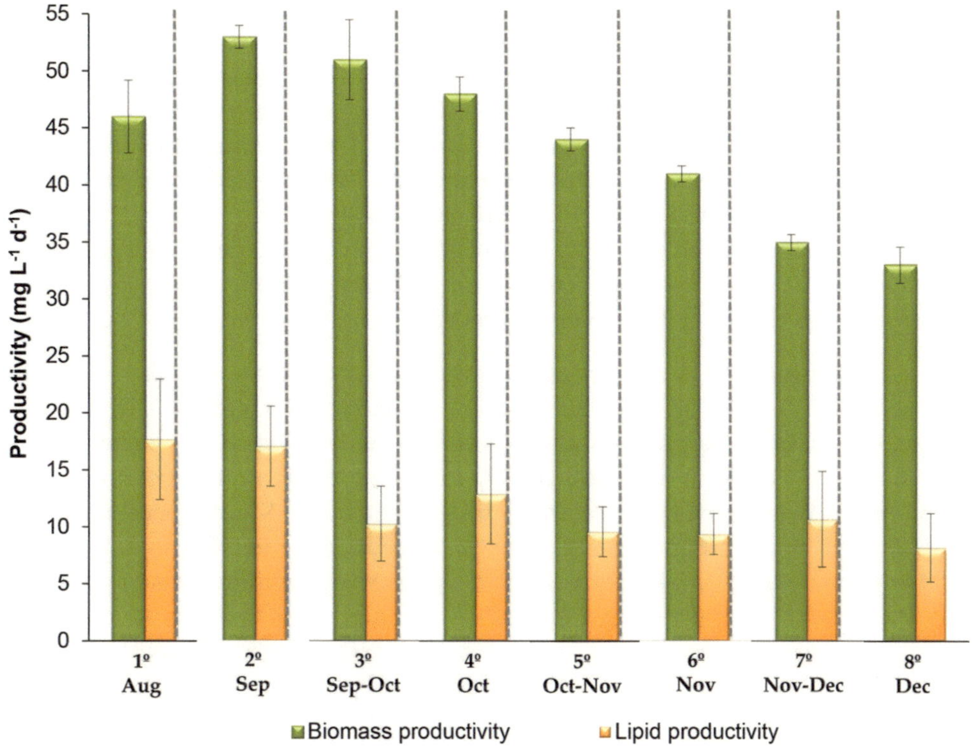

Figure 2. Biomass and lipid productivities from *Scenedesmus* sp. SCX2 cultivated during the eight cycles of culturing under semicontinuous mode in a raceway of 2 m^3 with 1.4 m^3 of medium, under greenhouse conditions.

Cycle	Month (2015)	Maximal irradiance (μE m^{-2} s^{-1})	Average temperature (°C)	Protein productivity (mg L^{-1} d^{-1})	Sugar productivity (mg L^{-1} d^{-1})	Pigment productivity (mg L^{-1} d^{-1})
1	Aug	1251.1	22.0	2.5 ± 0.02	19.2 ± 2.3	1.1 ± 0.07
2	Sep	1279.3	24.7	3.0 ± 0.01	14.8 ± 2.5	1.4 ± 0.11
3	Sep–Oct	1076.3	18.9	2.9 ± 0.01	11.9 ± 1.0	1.3 ± 0.17
4	Oct	1053.5	18.9	2.4 ± 0.02	10.8 ± 0.7	0.8 ± 0.00
5	Oct–Nov	1090.4	18.5	1.1 ± 0.02	12.2 ± 3.5	0.8 ± 0.03
6	Nov	1066.8	18.5	0.8 ± 0.02	9.0 ± 0.9	0.5 ± 0.07
7	Nov–Dec	894.3	17.0	0.8 ± 0.01	11.3 ± 2.0	0.5 ± 0.03
8	Dec	893.2	15.6	0.8 ± 0.01	5.6 ± 0.3	0.4 ± 0.01

Table 2. Irradiances, temperatures and productivities corresponding to the eight cycles of *Scenedesmus* sp. SCX2 culture in the semicontinuous mode.

Lipid production is very important, because lipids are the raw material for biodiesel synthesis by the transesterification reaction. Higher lipid productivities were obtained in the first and second cycles with 17.7 and 17.1 mg L^{-1} d^{-1} (**Figure 2**). The maximum lipid content in the biomass was obtained in the first cycle with 38.4%, which was threefold higher than lipid content obtained by *Chlorella variabilis* grown in a raceway with working volume 450 L [65] and 30% higher than that reported by [76], who cultivated *Scenedesmus* sp. in Erlenmeyer flasks with an operation volume of 800 mL and irradiance of 400 μE m^{-2} s^{-1} at 23°C, using culture media supplemented with nitrogen and phosphate sources, and even vitamins, in different concentrations. In the other culture cycles, the lipid productivities were lower without significant differences among them, with an average lipid productivity of 10.9 mg L^{-1} d^{-1}.

After carbon, nitrogen is the most important nutrient for the microalgae, which is assimilated in the form of NO_3^- or NH_4^+ [77–79]. During the semicontinuous mode, the NO_3^- was always present in the culture medium, and it was not fully assimilated in any cycle (**Figure 1**). Approximately 50% of the initially added NO_3^- for each cycle was consumed in the first six cycles; therefore, none of the cycles reached nitrogen limitation. Nitrate concentration never was less than 68 mg L^{-1} (**Figure 1**). In general, no effect between NO_3^- uptake and lipid productivity was observed, insomuch as the NO_3^- uptake was practically the same in each cycle and the lipid productivity decreased as the semicontinuous culture was progressing. The most significant change was in the last two cycles, when the NO_3^- uptake by *Scenedesmus* sp. was only 44% (**Figure 1**) and the lower productivities of biomass, lipids (**Figure 2**), proteins, total pigments and sugars were obtained (**Table 2**).

The pH of the culture is influenced by several factors such as algal productivity, photosynthesis and ionic composition of the culture medium [79, 80]. The pH is one of the most important factors in the microalgal culture because it determines the solubility and availability of CO_2 and essential nutrients, which has a significant impact on the microalgal metabolism [81]. Generally, a proportional relationship between biomass and pH was observed. In the first three cycles, a pH range of 9.1–11.3 was obtained, which is an advantage insomuch as with high pH values. Only few microorganisms can grow in these pH ranges and a non-contaminated microalgal culture could be guaranteed; however, with pH values above 8.5 the availability of inorganic carbon (CO_2 and HCO_3^-) is limited and the growth of microalgae is affected [82]. In this work, higher pH values (around 11) were achieved in the first two cycles; nevertheless, the pH had no negative effect on microalgal growth. From the third cycle forward, the pH oscillated in a range of 8.7–10.5, which allowed a greater availability of CO_2; however, it was not reflected in the biomass increase.

In regard to the protein content, it was proportional to the biomass content. The higher protein productivities were obtained in the first four cycles; the maximum productivity was achieved in the second cycle with 3.0 mg L^{-1} d^{-1}. From the fifth cycle forward, the protein productivity decreased 61% and in the last two cycles it decreased until 75% from the productivity reached in the second cycle (**Table 2**). It was report that *Scenedesmus dimorphus* has a protein content from 8 to 18% (dry weight basis) [83], and in this work the highest protein content was 5.6%, this was achieved at the end of the second cycle. Ref. [84] obtained six and four times more protein with *Scenedesmus* sp., respectively, that protein content achieved in this work, at the end of the second cycle. *Scenedesmus* sp. was cultured in asbestos tanks, with an operation volume of 150 L and natural lighting, using fertilizer (Nitrofoska Foliar) and residual water fish.

The maximum total sugar productivity (19.2 mg L^{-1} d^{-1}) was obtained in the first cycle (**Table 2**). Total sugar productivities of the other cycles were declining gradually over time. It has been reported that *Scenedesmus* sp. can present a total sugar content of 50% (dry weight basis) [85], similar to what it has been achieved in this work (41.7%). Total pigments presented a similar behavior to the microalgal biomass, insomuch as the first three cycles presented higher productivities. The highest total pigments productivity was achieved in the second cycle (1.4 mg L^{-1} d^{-1}). For the fourth cycle, the total pigment content decreased by 40% and continued to decrease during the subsequent cycles to reach a productivity of 0.4 mg L^{-1} d^{-1} in the eighth cycle (**Table 2**).

The most abundant FAMES were palmitic acid (C16:0), stearic acid (C18:0) and palmitoleic acid (C16:1 Δ9). In several studies C18:0 is usually found in low concentration in microalgae and C16:0, C18:1 and C18:3 are the major fatty acids [86–88].

3.4. Conclusions

Higher productivities of biomass, lipids, proteins, total sugars and pigments were obtained in the first two and three cycles of cultivation, which covered a period of culturing between late summer and early autumn, when the irradiation and temperatures stimulated the *Scenedesmus* sp. SCX2 growth. A *Scenedesmus* sp. SCX2 culture was maintained under semicontinuous mode during 123 days (eight cycles) without NO_3^- limitation, with an average volumetric biomass productivity of 44 mg L^{-1} d^{-1}. The maximum content of protein, sugars, lipids and pigments was 5.6, 41.7, 38.4 and 2.4%, respectively. The lipid productivity decreased after the third cycle by almost 42%. The lowest lipid productivity was obtained in the eighth cycle and decreased by 53% compared with the highest lipid productivity. The total sugar productivity varied throughout of the experiment and decreased by 70% in the eighth cycle, compared with the highest productivity obtained in the first cycle. The maximum productivities of lipids and total sugars were obtained in the first cycle at 22°C and an irradiance of 1251.1 μE m^{-2} s^{-1}. The frequency of renewal of the semicontinuous culture can be reduced to increase the productivities and diminishing the contamination of the culture due to other microorganisms.

Acknowledgements

The authors are grateful for the financial support provided by National Science and Technology Council (CONACYT-Mexico) (Grant 247402), the Secretary for Research and Posgraduated studies—Instituto Politécnico Nacional (SIP-IPN, Grants 20151343, 20152207, and 20161677). The authors would also like to thank Claudia Guerrero Barajas for her participation in reviewing the paper.

Author details

Luis C. Fernández Linares*, Kevin Á. González Falfán and Citlally Ramírez-López

*Address all correspondence to: lfernand36@gmail.com

Professional Interdisciplinary Biotechnology Unit of the National Polytechnic Institute, Mexico City, Mexico

References

[1] Hadj-Romdhane F, Jaouen P, Pruvost J, Grizeau D, Van Vooren G, Bourseau P. Development and validation of a minimal growth medium for recycling Chlorella vulgaris culture. Bioresour Technol. 2012 [cited 2016 Jul 14];123. Available from: http://www.ncbi.nlm.nih.gov/pubmed/22940343

[2] Mendoza JL, Granados MR, de Godos I, Acién FG, Molina E, Heaven S, et al. Oxygen transfer and evolution in microalgal culture in open raceways. Bioresour Technol. 2013;137:188–195.

[3] Aguirre A-M, Bassi A, Saxena P. Engineering challenges in biodiesel production from microalgae. Crit Rev Biotechnol [Internet]. 2012 [cited 2016 Apr 11];8551:1–16. Available from: http://www.ncbi.nlm.nih.gov/pubmed/22804334

[4] Fon Sing S, Isdepsky A, Borowitzka MA, Reza Moheimani N. Production of biofuels from microalgae. Mitig Adapt Strategies Glob Chang. 2013 [cited 2016 Apr 11];18(1): 47–72. Available from: http://link.springer.com/10.1007/s11027-011-9294-x

[5] Ramanna L, Guldhe A, Rawat I, Bux F. The optimization of biomass and lipid yields of Chlorella sorokiniana when using wastewater supplemented with different nitrogen sources. Bioresour Technol. 2014;168:127–135.

[6] Balat M. Potential alternatives to edible oils for biodiesel production—A review of current work. Energy Convers Manag. 2011; 52, 1479–1492.

[7] Harun R, Singh M, Forde GM, Danquah MK. Bioprocess engineering of microalgae to produce a variety of consumer products. Renew Sustain Energy Rev. 2010;14(3):1037–1047.

[8] Demirbas A, Fatih Demirbas M. Importance of algae oil as a source of biodiesel. Energy Convers Manag [Internet]. 2011 [cited 2016 Jul 18];52(1):163–170. Available from: http://dx.doi.org/10.1016/j.enconman.2010.06.055

[9] Yusuf C, Chisti Y. Biodiesel from microalgae. Biotechnol Adv [Internet]. 2007 [cited 2016 Apr 11];25(3):294–306. Available from: http://www.ncbi.nlm.nih.gov/pubmed/1735 0212

[10] Borowitzka MA, Moheimani NR. Open pond culture systems. In: Algae for Biofuels and Energy, (Ed). Borowitzka MA, Moheimani NR. Dordrecht: Springer Netherlands; 2013. 133–152. Available from: http://link.springer.com/10.1007/978-94-007-5479-9_8

[11] Molina Grima E, Acién Fernández FG, Camacho FG, Chisti Y. Photobioreactors: light regime, mass transfer, and scaleup. J Biotechnol 1999;70:231–247.

[12] International Energy Agency. Biorefinery Concept [Internet]. 2016 [cited 2016 Apr 4]. Available from: http://www.iea.org/topics/renewables/

[13] González-Delgado Á-D, Kafarov V. Microalgae based biorefinery: issues to consider. CT&F. 2011;4(4):5–22.

[14] IEA, Bioenergy Secretariat. Biorefineries: adding value to the sustainable utilisation of biomass. 2009. 1–16. Available from: http://www.ieabioenergy.com/wp-content/uploads/2013/10/Task-42-Booklet.pdf

[15] Posten C, Schaub G. Microalgae and terrestrial biomass as source for fuels: a process view. J Biotechnol. 2009;142(1):64–69.

[16] Clarens AF, Resurreccion EP, White MA, Colosi LM. Environmental life cycle comparison of algae to other bioenergy feedstocks. Environ Sci Technol [Internet]. 2010 [cited 2016 Apr 6];44(5):1813–1819. Available from: http://pubs.acs.org/doi/abs/10.1021/es902838n

[17] Gouveia L. From tiny microalgae to huge biorefineries. Oceanography [Internet]. 2014;2(1):2–9. Available from: http://www.esciencecentral.org/journals/from-tiny-microalgae-to-huge-biorefineries-2332-2632-2-120.php?aid=24524

[18] Milano J, Ong HC, Masjuki HH, Chong WT, Lam MK, Loh PK, et al. Microalgae biofuels as an alternative to fossil fuel for power generation. Renew Sustain Energy Rev 2016;58:180–197.

[19] Trivedi J, Aila M, Bangwal DP, Kaul S, Garg MO. Algae based biorefinery—How to make sense? Renew Sustain Energy Rev. 2015;47:295–307.

[20] Barsanti L, Gualteri P. Algae: Anatomy, Biochemistry and Biotechnology. CRC Press; Taylor & Francis Group, Boca Raton, FL, USA; 2014.

[21] Salis A, Nicolò M, Guglielmino S, Solinas V. Biodiesel from microalgae. In: Handbook of Hydrocarbon and Lipid Microbiology Protocols. McGenity, Terry J., Timmis, Kenneth N., Nogales, Balbina (Eds). Springer-Verlag; Berlin Germany, 2010; 2827–2839.

[22] Hernandez J-P, de-Bashan LE, Rodriguez DJ, Rodriguez Y, Bashan Y, et al. Growth promotion of the freshwater microalga Chlorella vulgaris by the nitrogen-fixing, plant growth-promoting bacterium Bacillus pumilus from arid zone soils. Eur J Soil Biol [Internet]. 2009 [cited 2016 Apr 11];45(1):88–93. Available from: http://linkinghub.elsevier.com/retrieve/pii/S1164556308001052

[23] Oncel S, Microalgae for a macroenergy world. Renew Sustain Energy Rev. 2013; 26:241–264.

[24] Hu Q, Sommerfeld M, Jarvis E, Ghirardi M, Posewitz M, Seibert M, et al. Microalgal triacylglycerols as feedstocks for biofuel production: perspectives and advances. Plant J [Internet]. 2008 [cited 2016 Apr 17];54(4):621–639. Available from: http://www.ncbi.nlm.nih.gov/pubmed/18476868

[25] Schenk PM, Thomas-Hall SR, Stephens E, Marx UC, Mussgnug JH, Posten C, et al. Second generation biofuels: high-efficiency microalgae for biodiesel production. BioEnergy Res [Internet]. 2008 [cited 2016 Apr 11];1(1):20–43. Available from: http://link.springer.com/10.1007/s12155-008-9008-8

[26] Das P, Aziz SS, Obbard JP. Two phase microalgae growth in the open system for enhanced lipid productivity. Renew Energy. 2011;36(9):2524–2528.

[27] Hadj-Romdhane F, Zheng X, Jaouen P, Pruvost J, Grizeau D, Croué JP, et al. The culture of Chlorella vulgaris in a recycled supernatant: effects on biomass production and medium quality. Bioresour Technol 2013;132:285–292.

[28] Mendoza JLL, Granados MRR, de Godos I, Acién FG, Molina E, Banks C, et al. Fluid-dynamic characterization of real-scale raceway reactors for microalgae production. Biomass and Bioenergy. 2013;54:267–275.

[29] Feng Y, Li C, Zhang D. Lipid production of *Chlorella vulgaris* cultured in artificial wastewater medium. Bioresour Technol. 2011;102(1):101–105.

[30] Mutanda T, Ramesh D, Karthikeyan S, Kumari S, Anandraj A, Bux F. Bioprospecting for hyper-lipid producing microalgal strains for sustainable biofuel production. Bioresour Technol [Internet]. 2011 [cited 2016 Apr 11];102(1):57–70. Available from: http://www.ncbi.nlm.nih.gov/pubmed/20624676

[31] He Q. Microalgae as Platforms for Recombinant Proteins. In: Handbook of Microalgal Culture. Biotechnology and Applied Phycology, A. Richmond (Ed.), Blackwell Science, Oxford, 2004; pp. 472.

[32] Suganya T, Varman M, Masjuki HH, Renganathan S. Macroalgae and microalgae as a potential source for commercial applications along with biofuels production: a biorefinery approach. Renew Sustain Energy Rev 2016;55:909–941.

[33] Richardson JW, Johnson MD, Outlaw JL. Economic comparison of open pond raceways to photo bio-reactors for profitable production of algae for transportation fuels in the Southwest. Algal Res [Internet]. 2012 [cited 2016 Jul 18];1(1):93–100. Available from: http://dx.doi.org/10.1016/j.algal.2012.04.001

[34] Acién FG, Fernández JM, Magán JJ, Molina E. Production cost of a real microalgae production plant and strategies to reduce it. Biotechnol Adv [Internet]. 2012;30(6):1344–1353. Available from: http://dx.doi.org/10.1016/j.biotechadv.2012.02.005

[35] Morales-Amaral M del M, Gómez-Serrano C, Acién FG, Fernández-Sevilla JM, Molina-Grima E. Outdoor production of *Scenedesmus* sp. in thin-layer and raceway reactors using centrate from anaerobic digestion as the sole nutrient source. Algal Res. 2015;12:99–108.

[36] Mata TM, Martins AA, Caetano NS. Microalgae for biodiesel production and other applications: a review. Renew Sustain Energy Rev. 2010;14(1):217–232.

[37] Pawlowski A, Mendoza JL, Guzmán JL, Berenguel M, Acién FG, Dormido S. Selective pH and dissolved oxygen control strategy for a raceway reactor within an event-based approach. Control Eng Pract [Internet]. 2015;44:209–218. Available from: http://www.scopus.com/inward/record.url?eid=2-s2.0-84939644511&partnerID=tZOtx3y1

[38] Ras M, Steyer J-PP, Bernard O. Temperature effect on microalgae: a crucial factor for outdoor production. Rev Environ Sci Biotechnol. 2013;12(2):153–164.

[39] Wen X, Du K, Wang Z, Peng X, Luo L, Tao H, et al. Effective cultivation of microalgae for biofuel production: a pilot-scale evaluation of a novel oleaginous microalga *Graesiella* sp. WBG-1. Biotechnol Biofuels [Internet]. 2016 [cited 2016 Aug 26];9(1):123. Available from: http://www.ncbi.nlm.nih.gov/pubmed/27303444

[40] Ho S-H, Chen C-NN, Lai Y-Y, Lu W-B, Chang J-S. Exploring the high lipid production potential of a thermotolerant microalga using statistical optimization and semi-continuous cultivation. Bioresour Technol [Internet]. 2014 [cited 2016 Jul 18];163:128–135. Available from: http://linkinghub.elsevier.com/retrieve/pii/S0960852414005306

[41] Fábregas J, Patiño M, Arredondo-Vega BO, Tobar JL, Otero A. Renewal rate and nutrient concentration as tools to modify productivity and biochemical composition of cyclostat cultures of the marine microalga Dunaliella tertiolecta. Appl Microbiol Biotechnol [Internet]. 1995 [cited 2016 Jul 18];44(3–4):287–292. Available from: http://link.springer.com/10.1007/BF00169918

[42] Zhou W, Min M, Li Y, Hu B, Ma X, Cheng Y, et al. A hetero-photoautotrophic two-stage cultivation process to improve wastewater nutrient removal and enhance algal lipid accumulation. Bioresour Technol 2012;110:448–455.

[43] Bao P, Huang H, Hu Z-Y, Häggblom MM, Zhu Y-G. Impact of temperature, CO_2 fixation and nitrate reduction on selenium reduction, by a paddy soil Clostridium strain. J Appl Microbiol [Internet]. 2013 [cited 2016 Jul 18];114(3):703–712. Available from: http://doi.wiley.com/10.1111/jam.12084

[44] Ho S-H, Chen C-Y, Chang J-S. Effect of light intensity and nitrogen starvation on CO_2 fixation and lipid/carbohydrate production of an indigenous microalga Scenedesmus obliquus CNW-N. Bioresour Technol 2012;113:244–252.

[45] Chiu SY, Kao CY, Tsai MT, Ong SC, Chen CH, Lin CS. Lipid accumulation and CO_2 utilization of Nannochloropsis oculata in response to CO_2 aeration. Bioresour Technol

[Internet]. 2009;100(2):833–838. Available from: http://dx.doi.org/10.1016/j.biortech.2008.06.061

[46] Radmann EM, Reinehr CO, Costa JAV. Optimization of the repeated batch cultivation of microalga Spirulina platensis in open raceway ponds. Aquaculture. 2007;265(1–4): 118–126.

[47] Tebbani S, Titica M, Ifrim G, Caraman S. Control of the light-to-microalgae ratio in a photobioreactor. 2014 18th International Conference on System Theory, Control and Computing (ICSTCC) [Internet]. IEEE; 2014 [cited 2016 Jul 18]. 393–398. Available from: http://ieeexplore.ieee.org/lpdocs/epic03/wrapper.htm?arnumber=6982448

[48] Parsons TR, Maita Y, Lalli CM. A manual of chemical & biological methods for seawater analysis. Oxford: Pergamon Press, 1984; pp.135–139.

[49] Blanken W, Cuaresma M, Wijffels RH, Janssen M. Cultivation of microalgae on artificial light comes at a cost. Algal [Internet]. 2013;2(4):333–340. Available from: http://dx.doi.org/10.1016/j.algal.2013.09.004

[50] Griffiths M. Microalgal Cultivation Reactor Systems. In: Biotechnological Applications of Microalgae: Biodiesel and Value-Added Products. Bux F, (Ed). CRC Press; Taylor & Francis Group, Boca Raton, FL, USA; 2013; pp. 51–76.

[51] Sydney E, Novak A, Cesar de Carvalho C, Soccol C. Chapter 4 – Respirometric Balance and Carbon Fixation of Industrially Important Algae. In: Biofuels from Algae. Pandey A, Lee D, Chisti Y, Soccol SR, (Eds). Elsevier, Burlington, MA, USA; 2014; pp. 67–84.

[52] UN-Water. Wastewater management—A UN-water analytical brief [Internet]. 2012 [cited 2016 Aug 28]. Available from: http://www.unwater.org/fileadmin/user_upload/unwater_new/docs/UN-Water_Analytical_Brief_Wastewater_Management.pdf

[53] Dibenedetto A. Production of aquatic biomass and extraction of Bio-oil. In: Biorefinery: From Biomass to Chemicals and Fuels, Aresta M, Dibenedetto A, Dumeignil F, (Eds). Walter de Gruyter GmbH & Co. Germany; 2015; pp. 81–100.

[54] Sudhir P, Murthy SDS. Effects of salt stress on basic processes of photosynthesis. Photosynthetica [Internet]. 2004 [cited 2016 Apr 11];42(4):481–486. Available from: http://link.springer.com/10.1007/S11099-005-0001-6

[55] Yen H-W, Hu I-C, Chen C-H, Chang J-S. Design of Photobioreactors for Algal Cultivation. In: Biofuels from Algae. Pandey A, Lee D, Chisti Y, Soccol SR, (Eds). Elsevier, Burlington, MA, USA; 2014; pp. 23–46.

[56] Duong VT, Li Y, Nowak E, Schenk PM. Microalgae isolation and selection for prospective biodiesel production. Energies [Internet]. 2012 [cited 2016 Apr 11];5:1835–1849. Available from: www.mdpi.com/journal/energies

[57] Hall P, Gifford J. Bioenergy Options for New Zealand [Internet]. Richardson M, editor. Scion, Energy Group, New Zealand, 2008 [cited 2016 Apr 11]. Available from: https://

www.niwa.co.nz/sites/niwa.co.nz/files/import/attachments/Situation_Analysis_-_Bioenergy_Options.pdf

[58] Karapatsia A, Penloglou G, Chatzidoukas C, Kiparissides C. An experimental investigation of *Stichococcus* sp.cultivation conditions for optimal co-production of carbohydrates, proteins and lipids following a biore fi nery concept. Biomass and Bioenergy [Internet]. 2016 [cited 2016 Apr 9]; Available from: http://dx.doi.org/10.1016/j.biombioe.2016.01.009

[59] Ho S-HH, Chen C-YY, Chang J-SS. Effect of light intensity and nitrogen starvation on CO_2 fixation and lipid/carbohydrate production of an indigenous microalga Scenedesmus obliquus CNW—N. Bioresour Technol [Internet]. 2012 [cited 2016 Jul 18];113:244–252. Available from: http://dx.doi.org/10.1016/j.biortech.2011.11.133

[60] Ho SH, Kondo A, Hasunuma T, Chang JS. Engineering strategies for improving the CO_2 fixation and carbohydrate productivity of Scenedesmus obliquus CNW-N used for bioethanol fermentation. Bioresour Technol [Internet]. 2013;143:163–171. Available from: http://dx.doi.org/10.1016/j.biortech.2013.05.043

[61] Ashokkumar V, Salam Z, Tiwari ON, Chinnasamy S, Mohammed S, Nasir F, et al. An integrated approach for biodiesel and bioethanol production from Scenedesmus bijugatus cultivated in a vertical tubular photobioreactor. Energy Convers Manag [Internet]. 2015;101:778–786. Available from: http://dx.doi.org/10.1016/j.enconman.2015.06.006

[62] Chen C, Chang Y, Chang H. Outdoor cultivation of Chlorella vulgaris FSP-E in vertical tubular-type photobioreactors for microalgal protein production. Algal [Internet]. 2016;13:264–270. Available from: http://dx.doi.org/10.1016/j.algal.2015.12.006

[63] Zhou X, Xia L, Ge H, Zhang D, Hu C. Feasibility of biodiesel production by microalgae *Chlorella* sp. (FACHB-1748) under outdoor conditions. Bioresour Technol [Internet]. 2013;138:131–135. Available from: http://dx.doi.org/10.1016/j.biortech.2013.03.169

[64] Wang SK, Hu YR, Wang F, Stiles AR, Liu CZ. Scale-up cultivation of Chlorella ellipsoidea from indoor to outdoor in bubble column bioreactors. Bioresour Technol 2014;156:117–122.

[65] Bhowmick G De, Subramanian G, Mishra S, Sen R. Raceway pond cultivation of a marine microalga of Indian origin for biomass and lipid production: a case study. Algal [Internet]. 2014;6(PB):201–209. Available from: http://dx.doi.org/10.1016/j.algal.2014.07.005

[66] Voltolina D, Sánchez-Saavedra M del P, Torres-Rodríguez LM. Outdoor mass microalgae production in Bahia Kino, Sonora, NW Mexico. Aquac Eng. 2008;38(2):93–96.

[67] Keeney DR, Nelson DW. Nitrogen-inorganic forms. Black CA, editor. Methods of Soil Analysis Part 2: Agronomy. Madison, Wisconsin, U.S.A.: American Society of Agronomy; 1982. 643–698.

[68] Ramírez-López C, Chairez I, Fernández-Linares L. A novel culture medium designed for the simultaneous enhancement of biomass and lipid production by Chlorella vulgaris UTEX 26. Bioresour Technol. 2016;212:207–216.

[69] Wellburn AR. The espectral determination of chlorophylls, a and b, as well as total carotenoids, using various solvents with spectrophotometers of different resolution. J Plant Physiol. 1994;144:307–313.

[70] DuBois M, Gilles KA, Hamilton JK, Rebers PA, Smith F. Colorimetric method for determination of sugars and related substances. Anal Chem [Internet]. 1956 [cited 2016 Jul 21];28(3):350–356. Available from: http://pubs.acs.org/doi/abs/10.1021/ac60111a017

[71] Bradford MM. A rapid and sensitive method for the quantitation of microgram quantities of protein utilizing the principle of protein-dye binding. Anal Biochem [Internet]. 1976 [cited 2016 Jul 21];72(1–2):248–254. Available from: http://linkinghub.elsevier.com/retrieve/pii/0003269776905273

[72] Griffiths MJ, van Hille RP, Harrison STL. The effect of nitrogen limitation on lipid productivity and cell composition in Chlorella vulgaris. Appl Microbiol Biotechnol [Internet]. 2014 [cited 2016 Aug 27];98(5):2345–2356. Available from: http://link.springer.com/10.1007/s00253-013-5442-4

[73] Millán-Oropeza A, Fernández-Linares L. Biomass and lipid production from Nannochloropsis oculata growth in raceway ponds operated in sequential batch mode under greenhouse conditions. Environ Sci Pollut Res [Internet]. 2016 [cited 2016 Jul 21];1–9. Available from: http://link.springer.com/10.1007/s11356-016-7013-6

[74] Hase R, Oikawa H, Sasao C, Morita M, Watanabe Y. Photosynthetic production of microalgal biomass in a raceway system under greenhouse conditions in Sendai City. J Biosci Bioeng [Internet]. 2000 [cited 2016 Jul 21];89(2):157–163. Available from: http://linkinghub.elsevier.com/retrieve/pii/S1389172300887307

[75] Singh SP, Singh P. Effect of temperature and light on the growth of algae species: a review. Renew Sustain Energy Rev. 2015;50:431–444.

[76] Hakalin NLS, Paz AP, Aranda DAG, Moraes LMP. Enhancement of cell growth and lipid content of a freshwater microalga *Scenedesmus* sp. by optimizing nitrogen, phosphorus and vitamin concentrations for biodiesel production. Nat Sci [Internet]. 2014 [cited 2016 Jul 21];6(12):1044–1054. Available from: http://www.scirp.org/journal/PaperDownload.aspx?DOI=10.4236/ns.2014.612095

[77] Grobbelaar JU. Algal nutrition-mineral nutrition. In: Handbook of Microalgal Culture. Biotechnology and Applied Phycology, A. Richmond (Ed.), Blackwell Science, Oxford, 2004; [cited 2016 Jul 21]; 95–115. Available from: http://doi.wiley.com/10.1002/9780470995280.ch6

[78] Abdel-Raouf N, Al-Homaidan AA, Ibraheem IBM. Microalgae and wastewater treatment. Saudi J Biol Sci. 2012;19(3):257–275.

[79] García LM. Elimination of CO2 with autochthonous microalgae. (Ph D Thesis). University of León, Spain; 2008.

[80] Park JBKBK, Craggs RJJ, Shilton ANN. Wastewater treatment high rate algal ponds for biofuel production. Bioresour Technol [Internet]. 2011 [cited 2016 Jul 21];102(1):35–42. Available from: http://dx.doi.org/10.1016/j.biortech.2010.06.158

[81] Chenl CY, Durbin EG. Effects of pH on the growth and carbon uptake of marine phytoplankton. Mar. Ecol. Prog. Ser. 1994;109: 83–94.

[82] Ugwu CUU, Aoyagi H, Uchiyama H. Photobioreactors for mass cultivation of algae. Bioresour Technol. 2008;99(10):4021–4028.

[83] Balat M, Balat H. Progress in biodiesel processing. Appl Energy [Internet]. 2010 [cited 2016 Jul 21];87(6):1815–1835. Available from: http://dx.doi.org/10.1016/j.apenergy.2010.01.012

[84] Andrade RCE, Vera BAL, Cárdenas LCH, Morales AED. Biomass production of the microalgae *Scenedesmus sp.* Using sewage from fishmongers. Rev Técnica la Fac Ing Univ del Zulia. 2009; 32(2):126–134.

[85] Ueda R, Hirayama S, Sugata K, Nakayama H. Process for the production of ethanol from microalgae. 1996; Patent US 5578472 A. https://www.google.com/patents/US5578472

[86] He Q, Yang H, Wu L, Hu C. Effect of light intensity on physiological changes, carbon allocation and neutral lipid accumulation in oleaginous microalgae. Bioresour Technol 2015;191:219–228.

[87] Toledo-Cervantes A, Morales M, Novelo E, Revah S. Carbon dioxide fixation and lipid storage by Scenedesmus obtusiusculus. Bioresour Technol. 2013;130:652–658.

[88] Liu J, Huang J, Sun Z, Zhong Y, Jiang Y, Chen F. Differential lipid and fatty acid profiles of photoautotrophic and heterotrophic Chlorella zofingiensis: assessment of algal oils for biodiesel production. Bioresour Technol [Internet]. 2011 [cited 2016 Jul 22];102(1): 106–110. Available from: http://dx.doi.org/10.1016/j.biortech.2010.06.017

Chapter 14

Biomass Production on Reclaimed Areas Tailing Ponds

Martin Bosák

Additional information is available at the end of the chapter

http://dx.doi.org/10.5772/65829

Abstract

This chapter presents the results of a multiannual systematic research and development of essentially new environmental safety technology of overlapping tailing pond modelled in terms of Vojany thermal power plant (EVO), Slovakia. Re-cultivated tailing area can be used to produce biomass (Swedish willow) and this biomass is used again for subsequent incineration with coal. Laboratory and small plot experiments conducted directly in the tailing pond area resulted in the development of another new dimension of environmental technology of decontamination of tailing ponds. This technology connects technical, safety, economical and environmental effects for biomass production.

Keywords: biomass, willow, tailing pond, re-cultivation, energy

1. Introduction

Nowadays the world is facing a global problem, which has an effect of plants on the environment. Slag (dross ashes mixture) and fly ash are formed by the combustion of coal. In the wider vicinity of power plants, it is a waste that burdens the environment. The problem of landfill sites must be effectively addressed because they are in terms with the landscape stability and severe environmental and safety problems.

There is a real danger of breaking up the dam especially, in the case of dumping large quantities of such wastes in an area of huge tailings ponds. Therefore, it can have serious consequences on the population, components of the environment and property, in general. The tailings ponds are stores of very fine waste containing significant amounts of water. Their mobility from the pond is high, if they are released. Subsequently, they can have trans-boundary impacts, also impact on the landscape areas and protected areas of European importance because they migrate long distances, particularly over the surface water flows.

Especially by using environmentally sound technologies, it is necessary to ensure adequate management of these wastes, for security and long-term stability of tailing ponds.

2. Tailing ponds

2.1. Tailing ponds in the Middle Europe

Industrial production generates some kind of waste (by-product) toxic substance, which contaminates the sites and often degrades the surroundings of human living including air, ground water and surface.

From the viewpoint of environmental safety of deposited materials, there are 56 tailing ponds of various levels and types in Slovakia (**Figure 1**). They are in different stages of their existence or life cycles (**Tables 1–3**). They contain mainly wastes from coal gangue, ore processing products (floating sludge), heating plants (slag, ash) and power plants. Safe closing or re-cultivation of tailing ponds is a current problem in the Slovak Republic and also for the environmental safety in Europe because generally the tailing ponds, as watery building constructions, are considered immensely dangerous objects, in terms of environment [1].

Types of tailing ponds in Slovakia:

- 15 with ash material,
- 27 with ore and
- 14 others (industrial).

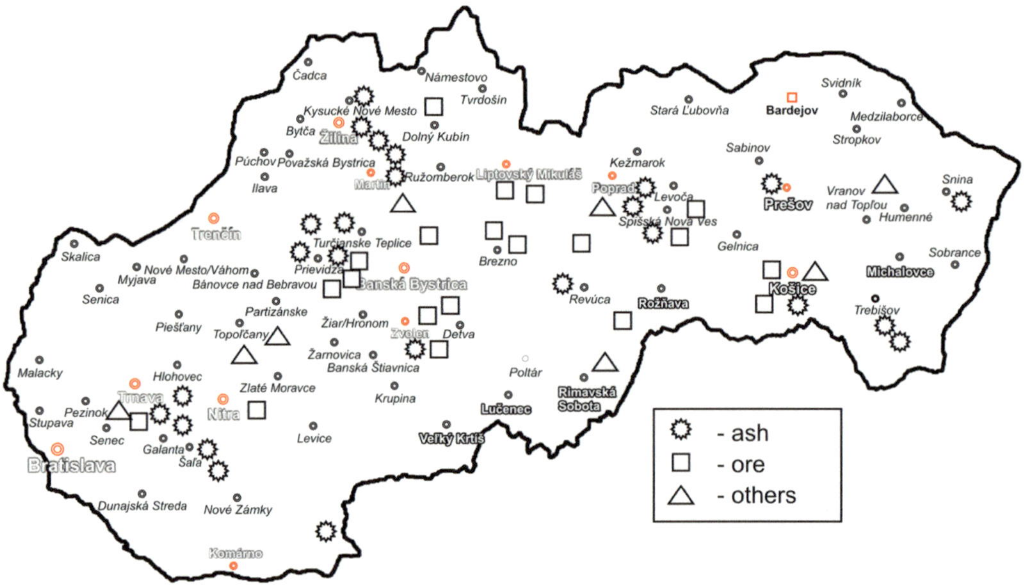

Figure 1. Registered tailing ponds in Slovakia.

No.	Name	Place	District	Category
1.	**Tailings pond ENO— new**	Zemianske Kostol'any	Prievidza	I
2.	**Tailings pond ENO —old**	Zemianske Kostol'any	Prievidza	II
3.	**Tailings pond ENO**	Bystričany —Chalmová	Prievidza	II
4.	**EVO Vojany**	Vojany - Drahnov	Mihalovce	II
5.	**Tailings pond KAPPAs.**	Štúrovo- časť Obid	Nové Zámky	II
6.	**Martin—old tailings pond**	Martin	Martin	II
7.	**Martin—new tailings pond**	Bystrička	Martin	II
8.	**Tailings pond Poša**	Poša-Nižný Hrabovec	Vranov n. T.	II
9.	**Tailings pond Snina**	Snina	Snina	II
10.	**Tailings pond DUSLO**	Tmovec n. Váhom,	Šal'a	II
11.	**Tailings pond Žilina**	Bytčica	Žilina	II
12.	**Tailings pond Košice**	Krásna nad Hornádom	Košice	III
13.	**Sered'**	Dolná Streda	Galanta	III
14.	**Zvolen**	Zvolen	Zvolen	III
15.	**Ash tailings pond**	Homé Opatovce	Žiar n. H.	III

Table 1. The list of ash tailing ponds in Slovakia.

No.	Name	Place	District	Category
1.	**Hačava**	Hačava	Rimavská Sobota	II
2.	**Hodruša Hámre**	Hodruša Hámre	Ziar nad Hronom	II
3.	**Jelšava**	Jelšava	Rožňava	II
4.	**Nižná Slaná**	Nižná Slaná	Rožňava	II
5.	**Rudňany**	Závadka	Spišská Nová Ves	II
6.	**Sedem žien**	Banská Belá	Žiar n. Hronom	II
7.	**Tailings pond Slovinky**	Slovinky	Spišská Nová Ves	II
8.	**Baňa Cígel' II.**	Sebedražie	Prievidza	III
9.	**Dúbrava 01**	Dúbrava	Liptovský Mikuláš	III
10.	**Dúbrava 02**	Dúbrava	Liptovský Mikuláš	III
11.	**Dúbrava 03**	Liptovský Mikuláš	Liptovský Mikuláš	III
12.	**Žiar nad Hrouom**	Žiar n. H.	Žiar n. H.	III
13.	**Košice – Bankov**	Košice	Košice	III
14.	**Lintych**	B. Štiavnica	B. Štiavnica	III
15.	**Pezinok**	Pezinok	Pezinok	III
16.	**Podrečanv**	Podrečany	Lučenec	III

No.	Name	Place	District	Category
17.	**Smolník**	Smolník	Spišská Nová Ves	III
18.	**Široká**	Široká	Dolný Kubín	IV
19.	**Baňa Cígel' I.**	Sebedražie	Prievidza	IV
20.	**Košice Bankov**	Košice	Košice	IV
21.	**Horná Ves (Kremnica)**	Horaá Ves	Žiar nad Hronom	IV
22.	**Hronský Beňadik,**	Hronský Beňadik	Nová Baňa	IV
23.	**Lubeník**	Jelšava	Rožňava	IV
24.	**Pezinok**	Pezinok	Pezinok	IV
25.	**Rožňava**	Rožňava	Rožňava	IV
26.	**Sered'**	Sered'	Galanta	IV
27.	Špania dolina	Špania dolina	B. Bystrica	IV

Table 2. The list of ore tailing ponds in Slovakia.

No.	Name	Place	District	Category
1.	**Čifáre**	Čifáre	Nitra	II
2.	**Bukocel**	Hencovce	Vranov n. T.	III
3.	**Dubová**	Dubová	B. Bystrica	III
4.	**Novácke odkalisko 7**	Nováky	Prievidza	III
5.	**Stabilizačný násyp Handlová**	Handlová	Prievidza	III
6.	**Odkalisko ČOV Sokol'anv**	Sokol'any-Bočiar	Košice	IV
7.	**Fámeš**	Pastuchov	Hlohovec	IV
8.	**Plešivec- Gemerská Hôrka**	Plešivec	Rožňava	IV
9.	**Nádrže oceliarenských kalov**	Vel'ká Ida	Košice	IV
10.	**Mokrá halda**	Vel'ká Ida	Košice	IV
11.	**Novácke odkalisko 6**	Nováky	Prievidza	IV
12.	**Sal'a RSTO**	Sal'a	Galanta	IV
13.	**Šulekovo**	Šulekovo	Trnava	IV
14.	**Veronika**	Dežerice	Topol'čany	IV

Table 3. The list of industrial tailing ponds in Slovakia.

According to the summary records of water cannons, 28 tailings of I–IV categories were located in the Czech Republic as on 1.1.2014, all listed in the following **Table 4**.

Not only red mud is produced in tailings, but there are also ash and uranium tailings in Hungary. Some of them are already being re-cultivated and prepared for liquidation. In Hungary, there are 20 tailing ponds as characterised below, in **Table 5**.

http://dx.doi.org/10.5772/65829

No.	Name	Place	Category
1.	**Hodějovice**	České Budějovice	III
2.	**Mydlovary**	České Budějovice	III
3.	**Zbrod North 1/4**	Hodonín	III
4.	**Nové Chalupy**	Karlovy Vary	III
5.	**Tailing ponds Π.**	Ostrov	III
6.	**Dolní Radechová**	Náchod	III
7.	**Debrné**	Trutnov	III
8.	**Odkaliště TDK IV/3**	Trutnov	III
9.	**Stráž p. Ralskem**	Česká Lipa	II
10.	**Dřiteč**	Pardubice	III
11.	**Lhotka**	Pardubice	II
12..	**Semtín no. 7**	Pardubice	II
13.	**Chvaletice I.**	Přelouč	III
14.	**Božkov**	Plzeň	III
15.	**Panský les**	Mělník	II
16.	**Odkaliště Spolana**	Neratovice	III
17.	**Bvtíz**	Přibram	III
18.	**Rýzmburk**	Vlašim	III
19.	**Ušák**	Kadaň	II
20.	**SEPAP no. 4**	Litoměňce	III
21.	**Třískohipy**	Louny	III
22.	**Barbora III.**	Ústi nad Labem	III
23.	**Užin - old tailing ponds**	Ústi nad Labem	III
24.	**Dolni Rožinka**	Bystřie nad Pemštejnem	II
25.	**Zlatkov**	Bystřice nad Pemštejnem	II
26.	**Tailing ponds Synthesia a.s.**	Pardubice	IV
27.	**Tailing ponds**	Mladá Boleslav	IV
28.	**Ústí – new tailing ponds**	Ústi nad lab em	IV

Table 4. The list of registered tailing ponds in the Czech Republic.

On the presented maps (**Figures 1–3.**) are shown tailing ponds of middle European countries—Slovakia, the Czech Republic and Hungary. Even though Slovakia is the smallest country, it has the most tailing ponds (56) compared to the Czech Republic (28) and Hungary (20). Of course, this is also closely related with 'energy policy' of individual countries.

No.	Name	Type
1.	**Pellérd**	Uranium
2.	**Ajka**	Red mud
3.	**Almásfüzítő**	Red mud
4.	**Kurityán**	Red mud
5.	**Mosonmagyaróvár**	Red mud
6.	**Neszmély**	Red mud
7.	**Bokod**	Ash
8.	**Borsodnádasd**	Ash
9.	**Borsodszirák**	Ash
10.	**Estergom/Dorog**	Ash
11.	**Dunaújváros**	Ash
12.	**Gyöngyösoroszi**	Ash
13.	**Inota**	Ash
14.	**Kazincbarcika**	Ash
15.	**Múcsony**	Ash
16.	**Pécs**	Ash
17.	**Tatabány/Bánhida**	Ash
18.	**Tiszapalkonya**	Ash
19.	**Tiszaújváros**	Ash
20.	**Visonta**	Ash

Table 5. The list of tailing ponds in Hungary.

It is shown on the maps that locations of tailing ponds in Slovakia and the Czech Republic are distributed almost equally on the whole area of the countries. In Hungary, tailing ponds are placed mostly in the northern part of the country, near the border with the Slovakia.

The present tailing ponds are still considered environmentally dangerous and are also expensive to maintain; therefore, nowadays, a higher attention is given to them in Slovakia. This is demonstrated by the example. There was an accident in the Hungarian village Ajka, where on 4 October 2010, the dam pond broke after heavy rains. Subsequently, more than 0.7 million cubic meters of toxic mud struck in seven towns and villages and red sludge flooded the neighbourhood. The environmental disaster has claimed up to 10 human victims and over 150 people were injured and it destroyed dozens of homes. Other accidents at tailings ponds that are fatal and environmental devastation are shown in **Table 6**. Therefore, the discussion about closing these ponds is very important.

http://dx.doi.org/10.5772/65829

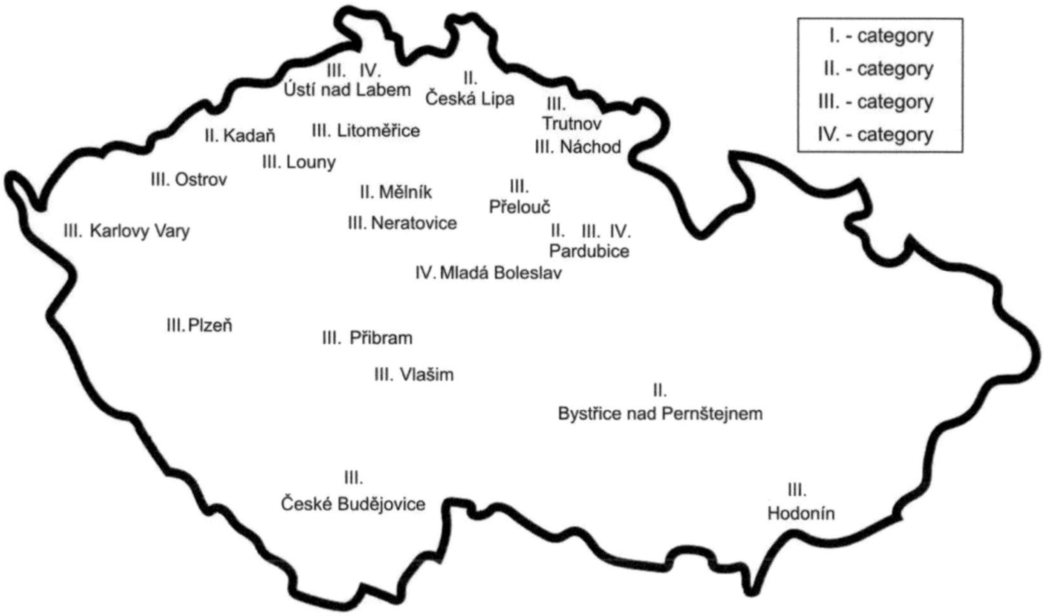

Figure 2. Registered tailing ponds in Czech Republic.

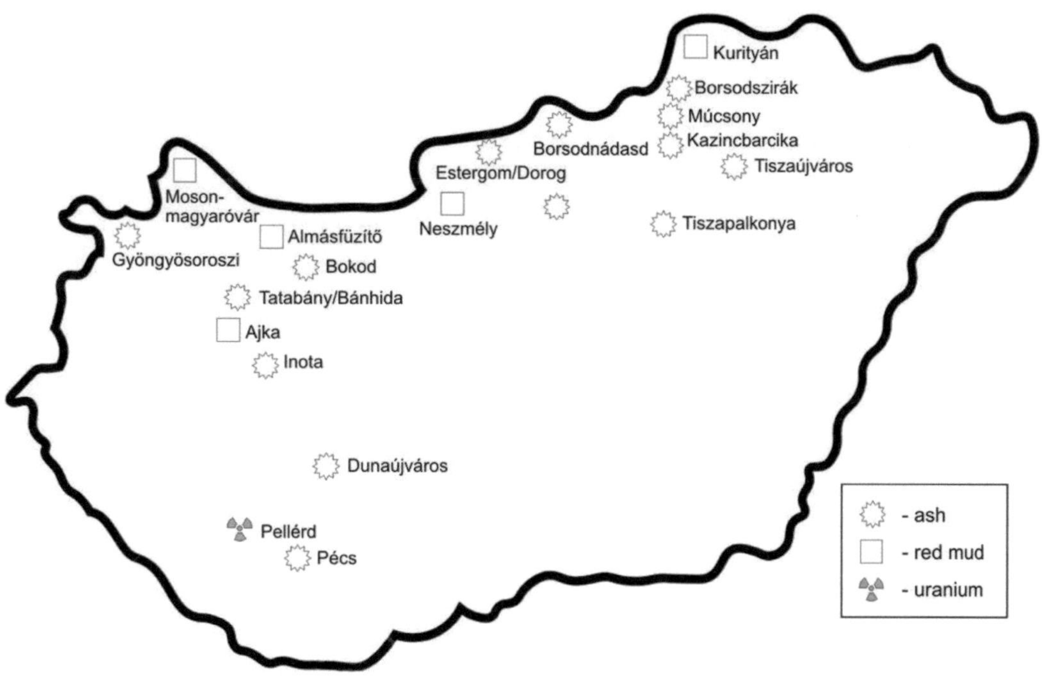

Figure 3. Registered tailing ponds in Hungary.

Town (Country)	Date	Number of death	Type of pond
Zemianske Kostol'any (Slovakia)	26.5.1965	4	Ashes from heat power plant
Stava (Italy)	19.7.1985	268	Fluorite sludge
Harmony (South Africa Republic)	6.2.1994	10	Cyanide pond
Placer (Philippines)	2.9.1995	12	Sludge
Ajka (Hungary)	4.10.2010	10	Red sludge

Table 6. Examples of tailing ponds from the world of accidents resulting in death.

Figure 4. Filling tailing pond.

For member countries, the European Union currently allocates huge funds in development projects. This is to prevent and rectify environmental damage; hence, the restoration and rehabilitation of tailing ponds dross ashes biological mixtures are important [1].

2.2. Tailing pond of EVO Vojany

The biggest fossil fuel plant in Slovakia is Vojany power plant, where mainly semi-anthracite coal from Ukraine and Russia is used as the fuel. Currently, for disposal of waste products from coal combustion the plant operates two facilities:

http://dx.doi.org/10.5772/65829

- dump with stabilisation material tailing,
- tailing ponds with dross ashes mixture.

Coal combustion products that are hydraulically transported are stored with stabilisation material in the tailing pond with two cassettes of dross ash mixtures (cassette no. 1 is already closed) (**Figure 4**). Stabilisation material is a by-product of the desulphurisation of power plant technology and combustion processes [2].

Safety and operation oversight within the relevant legislation of tailing pond of EVO plant Vojany are needed because it is a water work. On the verge of PLA Latorica, on the left bank of the

Area of base	Cassette No. 1—47.2 ha
	Cassette No. 2—48.1 ha
Overall capacity of stock ash	Cassette No. 1—7,580,000 m^3
	Cassette No. 2—5,760,000 m^3
Dispozition capacity	Cassette No. 1—full
	Cassette No. 2—850.000 m^3

Table 7. Base parameters of tailing pond.

Figure 5. Place cassette no. 1 on pond in scale 1:10,000.

river Laborec, it was built in 1965 to store dross ash mixture and is located in the administrative area of village Drahňov and Vojany. It is bounded on all sides by high grass-covered embankments. Two separate approximately similar cassettes create the tailing pond (**Table 7**):

- cassette No. 1–29 ha (with dam 47.2 ha), (**Figure 5**) and
- cassette No. 2–27 ha.

Cassettes are separated by dividing dam, which originally had the function of the peripheral dam cassette No. 1. This means that the area to be addressed after the final shutdown of the pond is about 56 ha [3].

3. Experiments and methods of research

This chapter presents development of original remediation technologies for unconventional tailing pond dross ashes mixture disposal and results of the experiments in the research. This technology uses structured layers of land, soil and stabilisation material. The reason is to replace previous legislative solution by the overlapping of hydrofilm material and drainage system.

3.1. Experiments of containers

The experiment simulating any large-scale use of this new non-traditional, uncertified practice or technology was based on the verification of replacement waterproofing properties of the stabiliser. The purpose of verification of experiments was to assess the possibility of using a stable material due to its potential to prevent solidification of penetrating rain water into the lower layers of the pond, with the risk of another accident.

A mixture of grass varieties that are resistant to typical and local conditions is used for covering of the energy crop with the future consideration of using pond grown plants as biomass for co-incineration with coal in power plants. The cultivation of fast growing Swedish willow with respect to its root system was experimentally verified. Therefore it was used in this experiment and the subsoil thickness was of 500 mm.

The laboratory experiment was set up with the following procedure:

- the stabilisation material of thickness of 0, 50, 100, 150, 200, 250, 300, 350, 400, 450 and 500 mm, was kept in the bottom of a 2 × 11 container with a size of 1000 mm × 1000 mm × 1000 mm,
- the second layer reflects the subsoil profile of the rehabilitated land and was then deposited in the soil with subsoil thicknesses of 300 mm for grass and 500 mm for Swedish willow (**Table 8**),
- like the subsoil, even topsoil profile describes the reclaimed area, the last build up layer of topsoil with a thickness of 200 mm is uniformly used for all variants,
- a part of containers was located in areas with variable weather conditions (the first half) and the other part is placed under the local climate conditions (the second half) (**Figures 6** and **7**) [4].

http://dx.doi.org/10.5772/65829

	Subsoil	Topsoil	Soil together
Grass	300 mm	200 mm	500 mm
Willow	500 mm	200 mm	700 mm

Table 8. Requirements for growing.

Figure 6. Containers in terms of controllable.

3.1.1. Stabilisation material: its analysis and usage

We provide an analysis report of hazardous substances contained in the stabilised in the following tables and **Figure 8** show the structured layers in individual containers, which appeared in the experiment. The used stabilisation material has pH 8.45 and a conductivity of 38.7 (**Tables 9** and **10**).

Every container was filled with layers of stabilisation material, subsoil and topsoil according to their draft. The first container was filled without a layer of stabiliser, as a control container—nothing else has been done. Afterwards, sufficiently compacted individual layers were filled to bring them closer to actual conditions [4].

3.1.2. Experimental method

The half of containers were located in areas under local climatic conditions and the second half of containers were placed in areas where weather conditions were controlled.

Figure 7. Containers under conditions consistent with local climate.

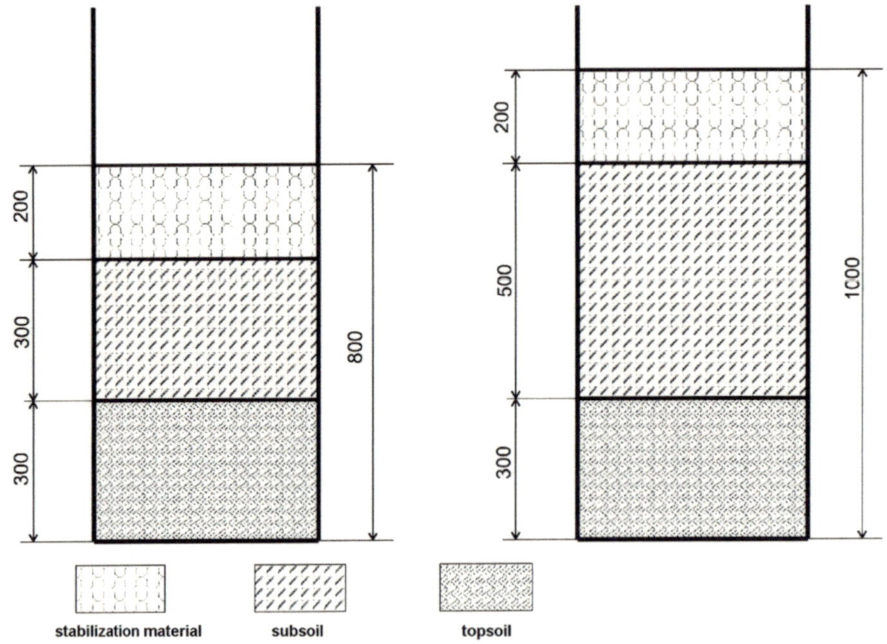

Figure 8. Diagram of the experimental variations of the experiment at 300 mm stabiliser.

http://dx.doi.org/10.5772/65829

	Indicator	Abbreviation	Value
1	Total organic carbon	TOC	**105,100**
2	Benzene, toluene, ethylbenzene and xylenes	BTEX	**<0.001**
3	Polychlorinated biphenyls, seven members	PCB	**<0.01**
4	Mineral oil (C10–C40)	NEL	**<1**
5	Polvcyclic aromatic hydrocarbons	PAH	**<0.05**

Table 9. Analysis of stabilisation material based on based on regulation of Ministry of Environment of SR No. 599/2005 Z.z.

	Indicator	Abbreviation	Value [mg]
1	Arsenic	As	**0.02**
2	Barium	Ba	**0.58**
3	Cadmium	Cd	**<0.003**
4	Total chromium	Cr	**0.03**
5	Copper	Cu	**0.03**
6	Mercury	Hg	**0.002**
7	Molybdenum	Mo	**0.05**
8	Nickel	Ni	**<0.02**
9	Lead	Pb	**0.05**
10	Antimony	Sb	**0.01**
11	Selenium	Se	**<0.01**
12	Zinc	Zn	**0.10**
13	Chlorides	cl^-	**31.7**
14	Fluorides	F^-	**2.2**
15	Sulphates	SO_4^{2-}	**1490**
16	Phenol index	FNI	**<0.3**
17	Dissolved organic carbon	DOC	**243**
18	Total solubles	CL	**3240**

Table 10. Analysis of water extract of stabilisation material based on regulation of Ministry of Environment of SR No. 599/2005 Z.z.

Reflecting the maximum monthly average for last 50 years indoor watering and simulation of precipitation in real outdoor was applied with water. Data on rainfall was used from the Slovak Hydrometeorological Institute, Regional Centre of Kosice from stations in Michalovce, Somotor and Milhostov ,which are the closest to the tailing pond.

Except for the rainy year 2010, when the monthly average rose to 85 mm/month, the long-term measurements of rainfall in the area show that the average monthly values range from 40 to 50 mm/month [4].

3.1.3. *Results of experiment*

According to different climatic conditions, the results of the experiment can be divided into two groups, with respect only to the quantity of water received:

3.1.3.1. *Containers under conditions consistent with local climate–natural amount of rainwater*

Rainfall was observed in containers on average 43 mm/month, which represents 1.43 mm of rainfall per day during the period January 2011–December 2011 (**Table 11**). The rainwater do not get through the layers, but in one container that contained stabilisation material, soil absorbed them.

Station	Month												Total	Average per month
	I	II	III	IV	V	VI	VII	VIII	IX	X	XI	XII		
Michalovce	40	6	30	17	30	92	180	27	51	16	1	77	567	47
Milhostov	28	4	31	14	46	112	166	11	41	14	1	58	525	44
Somotor	30	9	36	21	28	92	115	9	49	18	3	76	484	40
Average														**44**

Table 11. The rainfall [mm] of nearby stations in 2011.

3.1.3.2. *Containers with controllable received amount of water (pouring)*

The containers with variable conditions were simulated with extreme daily rainfall amounts of water, minimum 50 mm/day, that is to say, more than 7 times the maximum daily average precipitation. In May 2010, a long-term, extremely high quantity of rainwater were measured—the calculated average of monthly rainfall was 208 mm, it means that the average rainfall of the day is 6.9 mm (**Table 12**).

A part of the precipitation is absorbed by each layer in the container and seepage water accumulated the discharge outlet in the prepared containers and continuously measured.

Station	Year	Month	Rainfall [mm]
Michalovce	2011	VII	180
Milhostov	2010	V	219
Somotor	2010	V	226
Max. average			**208**

Table 12. Maximum of rainfall [mm] of nearby stations.

In order to determine the waterproofing ability of the individual layers for carrying out experiments, the maximum long-term nature of rainfall in the area of the tailing pond was taken as the basis of the daily rainfall amounts of water in the controlled.

In the following **Figures 9** and **10** are shown information depending on the amount of water tightness where the thickness of stabiliser and the subsoil are 300 and 500 mm, respectively.

Alternative use of 170 and 230 mm of stabilisation material, depending on the thickness of subsoil differences for willows and seed grass, resulted in implementation of laboratory experiments, leading to establish the maximum daily amount of rainwater at 100 mm in containers (70-times more than the average daily value). The root system of plants planted on the surface will gradually pump saturated water to the topsoil and subsoil.

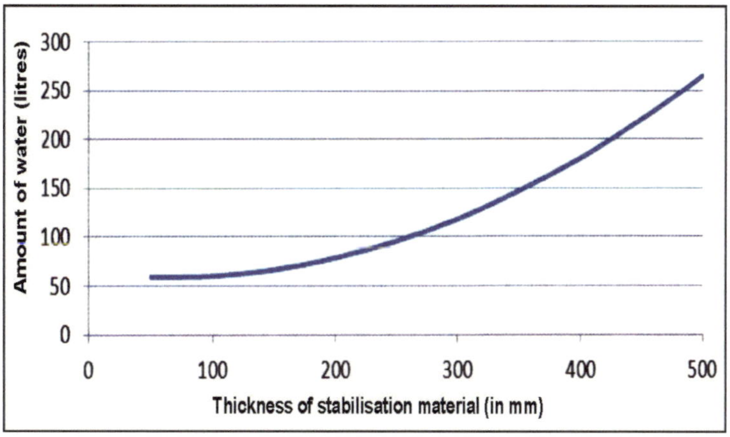

Figure 9. Dependence of the thickness of stabiliser and the amount of water tightness in thickness of the subsoil 300 mm.

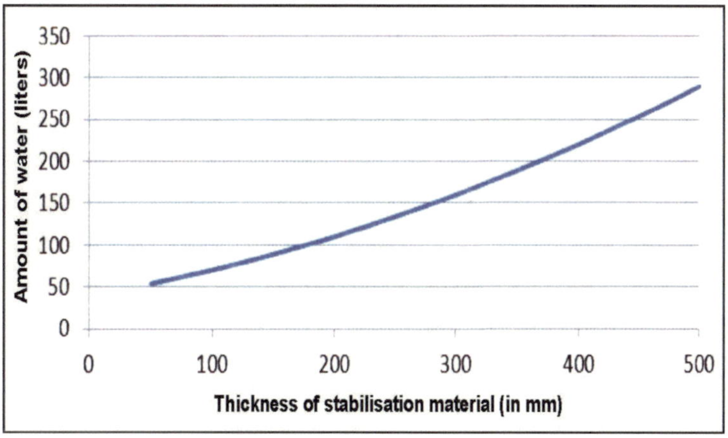

Figure 10. Dependence of the thickness of stabiliser and the amount of water tightness in thickness of the subsoil 500 mm.

3.2. Willow cultivation on tailings pond

3.2.1. Small plot trials on tailings pond

The second phase of experiment was to implement small plot trials on the surface of tailing where four large-sized parcels of 7 m × 20 m in a given structure have been built (**Tables 13** and **14**).

Parcel	Stabilisation material	Subsoil	Topsoil	Together
1st parcel	300 mm	500 mm	200 mm	1000 mm
2nd parcel	500 mm	500 mm	200 mm	1200 mm
3rd parcel	-	500 mm	200 mm	700 mm
4th parcel	-	500 mm	200 mm	700 mm

Table 13. Structure of parcels for willow.

Parcel	Stabilisation material	Subsoil	Topsoil	Together
1st parcel	300 mm	300 mm	200 mm	800 mm
2nd parcel	500 mm	300 mm	200 mm	1000 mm
3rd parcel	-	300 mm	200 mm	500 mm
4th parcel	-	300 mm	200 mm	500 mm

Table 14. Structure of parcels for grass.

The possibility cultivation of willow and grass cultivation in the closed tailing ponds in reality during the vegetation seasons are tested the proposed alternatives of the covering layer for the biological re-cultivation of the tailing ponds from the viewpoint of water permeability under natural conditions of atmospheric impacts (**Figures 11** and **12**). The results of experiment are

Figure 11. Experimental parcels on tailings pond.

Figure 12. Grass mixtures rapid and prosoil.

supplemented with 7 years of practical knowledge about the cultivation of Swedish willow on the territory of 15 ha in the nearby town of Kežmarok.

The average root striking of plant slips after the willow planting in two research stations of the Slovak University of Agriculture in Nitra was in the range between 66.91–89.51% and 45.67–91.35%. The planting was implemented in 2011.

Under suitable conditions, it is possible to achieve a root striking of more than 90% according to Dawson statement (2007). High root striking percentage of slips is inevitable for the optimal structure of vegetation and for optimal crops [5]. The number of rooted pieces can be influenced by the planting way of slips during the founding of commercial plantations. Better root striking in the case of planting of slips horizontally to the soil surface in comparison with the classical planting vertically to the soil surface was identified by Lowthe-Thomas et al. This planting method at the same time can significantly reduce the cost of planting [6].

We have achieved an average root striking of 91.76% (**Table 15**) in our experiment. The conditions for willow cultivation (structured layers of the stabilized stabilised waste, subsoil layer and arable layer) were suitably prepared. We can conclude from this result that, it is feasible to produce biomass directly in the power plant or in the tailing ponds which is 2 km away from the plant.

Plot	% of rooted plants
1.	88.72
2.	92.63
3.	91.14
4.	94.57
Average	**91.76**

Table 15. Number of rooted pieces of the Swedish willow in the experiment.

The optimal total quantity of precipitation in summer months should achieve 300 mm and during the total vegetation season it should achieve 550 mm [7]. Besides the extremely rainy year 2010, the measured values for the whole vegetation season were lower than 570 mm.

The precipitation data have been acquired from the three nearest monitoring stations to the tailing ponds (**Table 16**). The monitoring stations belong to relevant regional centre of the Slovak Hydro-Meteorological Institute.

Station	2009	2010	2011	2012	2013	2014	2015
Milhostov	585	892	526	496	529	567	556
Michalovce	636	929	567	635	578	625	593
Somotor	590	1030	486	481	522	520	537
Average	604	950	527	537	543	570	562

Table 16. The rainfall of nearby stations.

Swedish willow was planted in the spring of 2011. As most of the shoots were damaged, it was necessary to carry out planting of new shoots of Swedish willow in spring 2012. Rodents damaged willow root systems and most of them were completely destroyed by the spring of 2013. Subsequently the tree was replanted and fenced and rodent repellers were installed as well. Since the Swedish willow has been gradually replanted, various heights of tree shoots can be found on experimental fields. It is possible to achieve more than 90% of embeddedness of the Sweden tree shoots under appropriate conditions. High embeddedness of shoots is essential for achieving the optimal harvest. The method of planting Swedish willow cuttings might influence the number of in-rooted units. Worse embeddedness of shoots is monitored during planting cuttings vertically to the soil surface than planting the cuttings horizontally to the soil surface. We can conclude that conditions for growing Swedish willow were suitably prepared because in this experiment the average embeddedness of shoots was up to 92% (**Table 17, Figure 13**).

Approximately 580–600 mm of rainfall is the optimal precipitation value for the entire growing season. The willow can produce large amounts of biomass at this level. The rainiest year was 2010, as shown in **Table 13**, according to the atmospheric precipitation of individual

Season	Hight of willow
2012 – June	120 – 640 mm
2013 – April	1100 – 1750 mm
2014 – May	1900 – 2450 mm
2015 – April	2800 – 3200 mm

Table 17. Height of Swedish willow.

Figure 13. Swedish willow on 4th parcel.

periods. The total amount of rainfall was below the required level and did not exceed 550 mm in the next 3 years. As a result, Swedish willows have been drying out.

At Vojany thermal power plant through small plot trials continues the general process of experimental testing of the possibility of re-cultivating cinder/ash mixture tailings. The biomass yield of mowed grass is corresponding with the expected values which are increasing every year. It would be appropriate if the annual rainfall total was around 600 mm in terms of growing willow. We might conclude that total rainfall for the last period was below the long-term average. Because of lower-than-average precipitation during the year, the newly planted willow took a much lesser extent than initially expected.

3.2.2. Results of experiment

Upon analysing the individual components of experimental plots it was shown that at the edges of the experimental plots the formation of continual solidified layer of stabilisation material was not observed. The coherent layer of solidified stabilisation material was formatted in cuts, which was made closer to the centre of the plot. This situation could significantly affect the rainfall, whose intensity was at a time of experimentation very low. On the edges of the plots rainfalls withered quickly, more than the centre of the plots, in which the water was maintained for longer. In the edge stabilisation material was loose, but gradually towards the centre were produced visible larger chunks of hardened stabilisation material **Figure 14**.

From the plots were taken samples of manually cut root systems of Swedish willow (**Figure 15**). Selected root systems had different length, branching and direction depending on structured underlay of various plots. We can conclude, that in terms of length, the root system corresponds to the length of the part above the topsoil (mutually correlated). Another factor that cannot be ignored is, that the root system of Swedish willow did not crush the layer of stabilisation material in the vertical direction.

Figure 14. Solidified layer of stabilisation material.

Figure 15. Horizontal root system of Swedish willow above stabilisation material.

The experiments resulted in the application of new environmental recycling technology by new remediation technologies unconventional pond dross ashes mixture by using structured layers of stabilisation material, soil and land. Stabilisation material is the by-product of the desulphurisation process of power plant in combustion processes. The uniqueness new environmental recycling technology lies in the presented technology solutions, where a waste product of energy combustion processes will be used in another defused form.

During the growing season, the proposed real alternatives to the coating of bio-remediation of the pond in terms of water permeability under natural conditions and atmospheric effects were tested. It also verifies the best type of plants grown in the creation of sanitation, security conditions, with respect to the possibility of their further use as a co-incineration of biomass with coal in the same technology plant.

The final results showed that stabilisation layer prevented penetration of water into the lower layers of the tailing pond. The solution thus achieves a synergistic environmental-security effect [8].

3.3. Combustion of biomass

According to EU legislative from year 2008 EU is to provide 20% energy from renewable sources which is the energy sector goal for 2020 [8]. Co-incineration of biomass can be one of the ways to achieve this goal. Under the shared combustion of biomass and coal, there is a reduction, specifically, partial elimination of the environmental impact due to low content of nitrogen and sulphur in the biomass, resulting in a reduction of CO, NO*x* and SO_2, as well as reducing emissions of heavy metals [9].

One of the most promising methods for the provision of energy production in general is co-incineration of biomass energy willows and other biomass plant with coal in the near future and in the present [10]. It failed to apply in Slovakia and at global level, despite the availability, environmental and technological benefits announced by this system on a larger scale. Increased costs associated with the production and logistics of biomass assurance seems to be the biggest problem. Co-incineration of biomass with coal significantly increases the clean energy ratio. It is defined as the ratio of produced electricity to the total consumption of fossil energy. It is primarily reducing the greenhouse gas emissions from mining, transportation and combustion of coal when substituting a certain percentage of coal with biomass [11]. In practice, co-incineration of willow chips with wood wastes have been used in biomass power stations in Sweden. In the local energy supply, it plays an important role. Study of environmental impact of coal combustion from a power plant have been carried out in Poland and also in a number of countries [12].

The power plant Vojany started to implement them as well in collaboration with this author, in 2009 in line with the above trends. Project combustion of black coal with biomass in fluidised boilers was realised in the form of a scientific research complex. The research includes the provision of (growing) plant biomass in the pond area surrounding the facility gaining self-made slag-ash mixture. The first positive results in the reduction of emissions was produced by co-incineration of biomass, mainly wood chips mixed with black coal in a 4%

ratio by 40 kg per MWh produced, and operational savings associated with the consumption of limestone, creation and disposal of ash, water consumption and steam. The project's next phase was experimentally realised with the co-incineration of biomass with a value 7% and then 15%. The surroundings of the facility has good power potential of the fast-growing energy crops—willow was showed in research focused on the potential provision of biomass. In connection with this research specific purposeful cultivation of biomass was initiated in the large pond of about 56 ha that contained slag-ash mixture. One of most burdened areas in eastern Slovakia is the area of Vojany [13].

3.3.1. *Co-incineration of wood chips with coal*

In 2007, co-incineration of biomass in EVO boilers started. Forest biomass has been chosen for the first tests of wood chips. To achieve the same thermal input it was needed to deliver six cubic meters of biomass to the boiler, instead of one cubic of black coal, due to different heating value, calorific value and density of black coal and wood chips. It could not be replaced by any amount of coal with biomass because wood chips have a density of 0.3 t/m^3 and calorific value 10 MJ/kg and coal has a density of about 1 t/m^3 and heating value of 25 MJ/kg.

The added mixture of wood chips containing about 4% of the heat energy mix wood chips—coal, does not negatively affect dynamic characteristics of the power plant units—it was proved from calculations based on the time. It was also proved by tests in 2007 that it is possible to smoothly combust biomass (wood chip) in fluidised boilers in the power plant of Vojany. It turned out that wood chips have a positive impact on the boiler combustion mode, because of a higher proportion of volatile matter and a lower ignite temperature than coal. The resuls were a decrease in the concentration of carbon monoxide in the exhaust gas and more efficient combustion of irradiated fuel [14].

The launch of scientific-research activities aimed at complex solution for the issue of environmental energy-biomass combustion optimisation processes was initiated because of these partial positive results.

Replacement of a share of combusted black coal in thermal power EVO with biomass-based fuels was carried with priority to maximise the reduction of emissions especially carbon-sulphur oxides, by providing the required energy performance and therefore ultimately in increasing competitiveness improving economic indicators, manufacturing and energy-production companies.

The co-incineration of black semi-anthracite coal and wood biomass in fluidised FK5 boilers in the examined facility was performed as the experiment.

The wood chips ranged from 8.0 to 8.65 MJ/kg and the black coal average heating values in the experiment ranged from 25.4 to 28.1 MJ/kg. Graphically illustrated in **Figure 16** is the dependency of the heating value of wood chips on its moisture.

From different forms of concentrations of pollutants (the individual VOP) it is significant that:

- with increased performance there was a decrease in the concentration of carbon monoxide in co-incineration ratio of wood chips,

- during testing, legislative allowed emission limits for SO_2 were preserved and there has been an increase in performance and in values and
- other values of VOP were as well below the individual emission limits.

At the same time the efficiency of boiler and co-incineration of coal and the biomass, are presented in **Table 18**.

The co-incineration of a mixture of wood chips and coal led indeed into a slight reduction in boiler efficiency, which is negligible compared to the achieved environmental-safety effects—it can be stated when taking into account the partially different characteristics specifically, quality of supplied and combusted coal and wood chips.

The implementation of the I phase in plant biomass co-incineration in the power plant Vojany at the block no. 6 began in July 2009 based on these tests, and partial results of experimentation.

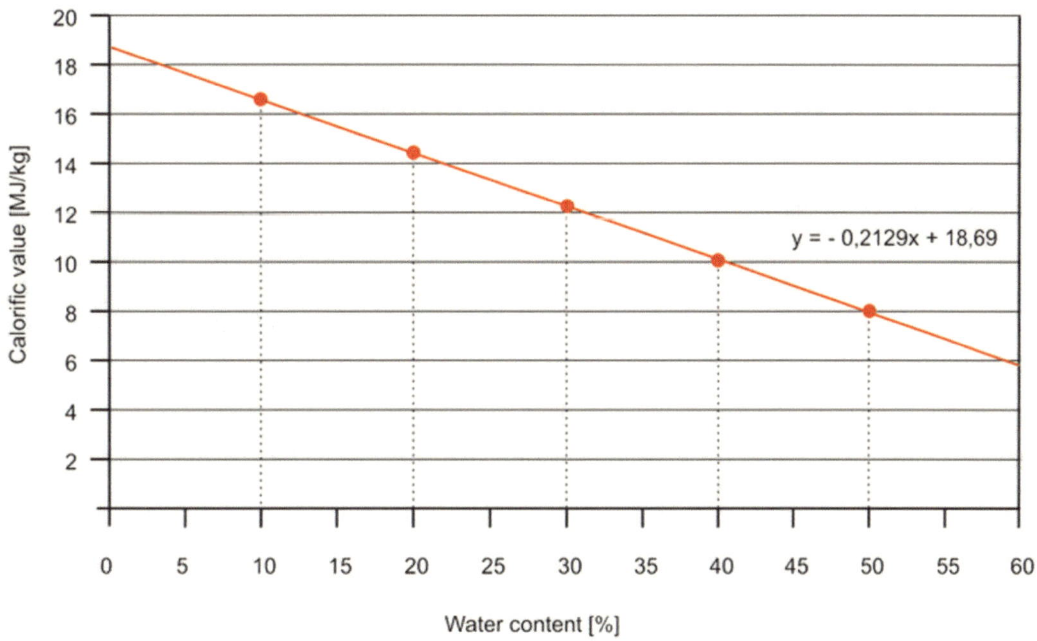

Figure 16. Dependency of the calorific value of wood chips on the water content.

Electrical performance	Black coal	Share of wood chip 1.91%	Share of wood chip 3.91%
[MW]	[%]	[%]	[%]
66	93.81	93.47	92.30
88	93.55	93.32	92.91
110	93.52	92.57	93.20

Table 18. The effectiveness of fluidised boiler K5 in Vojany.

A landfill with open capacity of 400 tons adjusted with transport (conveyor belts) and technology (crusher-sorter) biomass was built and customised to support experiments.

A rated power of conveyor belts was set to properly balance the mutual ratio and a mechanised system was used for wood chips, to transport them over conveyor belt and coal through the other one. The use of raw wood chip (hardwood and softwood) and its share was gradually increased to 5.3%.

Co-incineration of a better quality of the combusted wood chips with a higher calorific value than previously considered biomass was achieved with a share of 5.3%, not by increasing the weight. This means that projected calorific value changed from 9.5 MJ/kg to over 11 MJ/kg. This ratio has proved to be the best possible, to maintain the maximum dynamic properties of the boiler and in the execution of transporting the fuel mixture into the boiler. The implementation of II stage of biomass project co-incineration, which consisted of the construction of independent mechanised access to the boiler especially for biomass was determined in relation to the achieved results.

Table 19 shows that the combustion of 1 ton of biomass eliminated approximately one ton of carbon dioxide emissions and capacities of coal lines remained clear and the new path was used to transport higher volume of wood chips into the boiler (**Figure 17**).

3.3.2. Results of experiment

The production of biomass and co-incineration of coal in fluidised boilers, of thermal power stations SE, a.s. as well as other power plants is need in terms of environmental benefits. By burning coal and biomass is reflected in a significant reduction in emissions and production of solid waste.

The facility is obtaining its biomass currently from six local suppliers. EVO facility is a promising purchaser and will be their regular customer in the future as well because the surrounding wetlands around the facility provide good conditions for rapidly growing trees (poplar, willow). There is potential for increasing employment in the region for people with lower qualifications with the cultivation of biomass in the surrounding area of the plant or in the wetlands.

Year	Wood chip share [t]	CO_2 – eliminated [t]
2009	8,310	10,487
2010	21,443	27,061
2011	24,099	30,413
2012	26,917	33,969
2013	60,794	92,954
2014	48,752	84,899

Table 19. The amount of CO_2 saved by co-incineration of biomass.

http://dx.doi.org/10.5772/65829

Figure 17. Biomass—landfill and transport.

4. Summary

The research carried out in the tailing pond EVO Vojany showed that on the tailing pond can be planted Swedish willow, as a source of biomass, while the by-product (waste) of the desulphurisation of power plant technology, combustion processes, can be used as the stabilisation material, making it possible to be reused on reclaimed areas of tailing ponds.

Using biomass has a positive impact on the environment and it is environmentally adequate way of power generation. We developed our knowledge with this new technology with 7 years of experience in cooperation with the EVO. About 80 tons of wood chips are co-incinerated based on daily experiment based model. The European Union's commitment to continually increase the use of renewable energy of 20% by 2020—there are contributing implementations of research that results into practice.

Acknowledgements

In this chapter are the results of the project implementation VEGA No. 1/0936/15 Economics and Environmental Studies and experimental verification of the possibility of reclaiming tailings ash mixture in SE–EVO Vojany.

Author details

Martin Bosák

Address all correspondence to: martin.bosak@euke.sk

University of Economics in Bratislava, Košice, Slovak Republic

References

[1] Bosák M, Hajduová Z, Majerník M vodohospodárske dielo, Andrejovský P. Experimental-Energy Combustion of Biomass Combined with Coal in Thermal Power Plants. Polish Journal of Environmental Studies. Olsztyn, 2015, vol. 24, no. 4, pp. 1517–1523.

[2] Majerník M, Bosák M, Andrejkovič M, Hajduová Z, Turisová R. Prevention of Environmental Risks of Tailings Ponds, 12th International Multidisciplinary Scientific GeoConference SGEM 2012, Albena Resort, 2012, vol. 5, pp. 619–626.

[3] Hronec O, Vilček J, Majerník M, Tkáč M. Ecological stabilization tailings pond in EVO Vojany - study – štúdia, EU PHF Košice, 2009, 106 p.

[4] Tkáč M, Majerník M, Hronec O, Vilček J, Bosák M. Experimental verification options reclamation tailings pond in SE - EVO Vojany, EU PHF Košice, 2011, 53 p.

[5] Dawson WM. Short Rotation Coppice Willow – Best Practice Guidelines. Renew Project, 2007, 48 p.

[6] Lowthe-Thomas SC, Slater FM, Randerson PF. Reducing the Establishments Costs of Short Rotation Willow Coppice (SRC). Biomass and Bioenergy, 2010, vol. 34, no. 5, pp. 677–686.

[7] Hall RL. Short Rotation Coppice for Energy Production: Hydrological Guidelines. Report. Crown Copyright, 2003, 21 p.

[8] Bosák M, Tarča A, Majerník M, Tomčíková M. Application of New Environmental Recycling Technology. In: SGEM 2016, 16th International Multidisciplinary Scientific Geoconference, Albena, 2016,

[9] Heller M.C, Keoleian GA, Mann MK, Volk TA. Life Cycle Energy and Environmental Benefits of Generating Electricity from Willow Biomass. Renewable Energy, 2004, vol. 29, no. 7, pp. 1023–1042.

[10] McCormick K, Kåberger T. Key barriers for bioenergy in Europe: Economic conditions, Know-How and Institutional Capacity and Supply Chain Coordination. Biomass and Bioenergy, 2007, vol. 31, no. 7, pp. 443–452.

[11] Keoleian GA, Volk TA. Renewable Energy from Willow Biomass Crops. Life Cycle Energy, Environmental and Economics Performance. Critical Reviews in Plant Sciences, 2005, vol. 24, no. 5–6, pp. 386–406 .

[12] Kalembkiewicz J, Chmielarz U. Effects of Biomass Co-Combustion with Coal on Functional Speciation and Mobility of Heavy Metals in Industrial Ash. Polish Journal of Environmental Studies, 2013, vol. 22, no. 3, p. 741.

[13] Vilček J, Hronec O, Tomáš J. Risk Elements in Soils of Burdened Areas of Eastern Slovakia, Polish Journal of Environmental Studies, 2012, vol. 21, no. 5, 1429 p.

[14] Vaszily M. Three Years Experience. Slovenská energetika, Slovenské elektrárne, a. s., 2012, vol. 37, no. 6, pp. 3–5.

Chapter 15

Biomass Blending and Densification: Impacts on Feedstock Supply and Biochemical Conversion Performance

Allison E. Ray, Chenlin Li, Vicki S. Thompson,
Dayna L. Daubaras, Nicholas J. Nagle and
Damon S. Hartley

Additional information is available at the end of the chapter

http://dx.doi.org/10.5772/67207

Abstract

The success of lignocellulosic biofuels and biochemical industries depends on an economic and reliable supply of high-quality biomass. However, research and development efforts have been historically focused on the utilization of agriculturally derived cellulosic feedstocks, without considerations of their low energy density, high variations in compositions and potential supply risks in terms of availability and affordability. This chapter demonstrated a strategy of feedstock blending and densification to address the supply chain challenges. Blending takes advantage of low-cost feedstock to avoid the prohibitive costs incurred through reliance on a single feedstock resource, while densification produces feedstocks with increased bulk density and desirable feed handling properties, as well as reduced transportation cost. We also review recent research on the blending and densification dealing with various types of feedstocks with a focus on the impacts of these preprocessing steps on biochemical conversion, that is, various thermochemical pretreatment chemistries and enzymatic hydrolysis, into fermentable sugars for biofuel production.

Keywords: blending, densification, conversion performance, advanced feedstock supply system, preprocessing

1. Introduction

Global demands for energy, finite petroleum reserves, and growing concerns over climate change have prompted considerable interest in lignocellulosic biomass as a sustainable alternative to fossil-derived sources for the production of transportation fuels. The Renewable

Fuel Standard (RFS2) [1] mandates the use of 36 billion gallons of renewable fuels by 2022 under the U.S. Energy Independence and Security Act (EISA) of 2007 [2]. Biomass availability and quantity pose significant barriers to the realization of large-scale production of lignocellulose-derived biofuels. The U.S. Department of Energy's (DOE) 2016 Billion Ton Report has projected the potential for more than one billion tons of biomass in the form of agricultural, forestry, waste, and algal materials capable of displacing approximately 30% of U.S. petroleum consumption without adverse environmental effects or negative impacts to production of food and agricultural products [3].

The conversion of biomass into affordable bio-based fuels and chemicals aims to displace all of the products currently made from a barrel of oil. Research and development efforts focused on the production of bio-derived hydrocarbon fuels and products seek to mobilize the bioeconomy in order to diversify energy resources that enable energy production. However, development of biomass as a sustainable energy resource for fuels and chemicals will require advances aimed at solving logistical challenges to ensure a cost-effective and consistent feedstock supply to the biorefinery [4–7]. Efficient utilization of the available resources for biofuels production requires considerations of supply chain cost, feedstock quality and conversion performance that dictates overall process economics. Logistical operations like harvest, collection, preprocessing, storage, and transportation have a significant impact on biomass availability and feedstock cost and quality [6, 8]. The large-scale deployment of lignocellulosic biomass for energy production has been severely limited by the high cost associated with the feedstock supply chain and technology barriers in conversion to fuel [8–10].

Initial development of the biofuels industry has centered around high-productive, single-resource areas that rely on sufficient quantity to enable selection and sourcing of suitable materials for conversion processes. However, as the bioeconomy grows and production moves away from highly productive, resource-rich areas, the impact of the spatial and temporal variability inherent to biomass feedstocks [6] cannot be managed solely by passive means in order to meet requirements for quality and quantity. The expansion of the industry will necessitate the adoption of "advanced" concepts within the supply system in order to meet cost, quality, and quantity requirements.

In addition, the "conventional system" currently employed by the cellulosic biofuel industry relies on a vertically integrated feedstock supply system where a single biomass feedstock is procured through contracts with local farmers, harvested and stored locally, and delivered in a low-density, baled format to the conversion facility [7]. This system has been demonstrated to work in high-yield regions, such as the U.S. Corn Belt; however, recent analyses have shown that conventional systems fail to meet feedstock cost targets outside of highly productive regions [11]. Realization of large-scale production of lignocellulosic biofuels will require modification to the current system in order to enable a consistent, cost-effective, and continuous supply of biomass to the biorefinery [10]. In comparison, the advanced feedstock supply system (AFSS) employs a wide range of preprocessing techniques, such as feedstock blending and densification in distributed biomass depots, and shows great promise for enabling improvements in handling and quality, consistency and uniformity, facilitating access to resources, and stability in storage [4, 7, 12, 13].

http://dx.doi.org/10.5772/67207

Biorefineries that rely upon a single feedstock to meet tonnage requirements are vulnerable to significant risks, in terms of both availability and affordability. Diversification of biomass supply has the potential to reduce risk [7], in some cases by as much as 80% [14], while enabling the lowest delivery cost [15]. Achieving a continuous, year-round supply of a single biomass resource is unlikely given the seasonal availability of most agricultural crops. Furthermore, climate change poses an inevitable risk to biomass supply systems for a developing bioenergy industry. Langholtz et al. [16] highlight the risk of extreme weather events to the bioenergy supply chain that are certain to cause reductions in feedstock production and increased price for agricultural commodities and biofuels. Other work has shown that drought has a significant, negative impact on biomass quality, in addition to biomass production yields [17]. Recent studies suggest a blended feedstock strategy to enable supply chain resilience may provide a solution to reliance on a single biomass resource [10, 18].

Low-density biomass feedstocks also pose a significant challenge to supply chain operations that translate to difficulties in storage, transportation, handling, and feeding [19], which hinder the large-scale use of biomass for biofuel production. Large volumes of low-density feedstocks require more resources for transportation and shipping. The size of the transportation resources needed to reach the 2050 target of 50% reduction of greenhouse gas via biofuels, biopower, and bioproducts exceeds the resources used to move the 2010 world grain and oil seed resources by 6- to 10-fold [20]. Densification processes, such as pelletization would increase the bulk and energy density of raw biomass, improve stability during storage and handling, create flowable feedstocks that are compatible with existing handling systems, and improve transport efficiency and cost [19].

The use of blended and densified feedstocks in conversion pathways instead of conventionally ground biomass from a single source addresses several challenges in the current biomass supply chain, including availability, transportation, storage, cost, quality, and supply variability [7, 19, 21–23]. This chapter provides a glimpse into the potential for preprocessing options, for example, blending and densification, to provide benefits to both biomass cost and conversion.

2. Addressing feedstock supply challenges

2.1. Feedstock blending strategy

A promising strategy to reduce supply risk is to blend different biomass feedstocks. Blending has been used by many industries (e.g., coal and animal feed) to affect the quality of the feedstock [24]. In the coal industry, different grades of coal are blended in order to meet emission targets and minimize ash production during power generation [25–27]. In the animal feed industry, a variety of feedstocks are blended to meet the desired nutritional requirements for a specific target animal [28]. Similarly, the concept of blending can be extended to the biofuel and bioproducts industry.

Formulating a designed feedstock through blending and other preprocessing methods allows low-cost and typically low-quality biomass to be blended with biomass of higher cost and typically higher quality to achieve the specifications required at the in-feed of a conversion facility (note that different conversion processes may require different specifications, and the cost required to meet those specifications will vary). The use of low-cost biomass allows the supply chain to implement additional preprocessing technologies that actively control feedstock quality, while also bringing more biomass into the system. This analysis and design approach is referred to as the "least-cost formulation" strategy [29]. In addition, recent work has shown that blended biomass feedstocks demonstrate improved flowability behavior [30], suggesting the potential for blending to extend benefits from the supply chain to feeding systems at the refinery.

The farmgate price is used to describe the economic availability of biomass resources and includes the cost of production and harvest [3]. **Figures 1** and **2** represent the cost of corn stover as a function of availability by state or region; these figures illustrate the increase in farmgate price with increasing demand. By blending feedstocks, the biorefinery can take advantage of the lower end of each supply curve to reduce cost. For example, **Figure 3** shows supply curves for switchgrass and corn stover from a 12-county region in northwestern Kansas, approximately 90 miles by 120 miles in size. In this region, only 700,000 tons per year of switchgrass (red curve) are available at $50/ton which could not support a biorefinery (capacity of ≥800,000 tons per year). There is sufficient corn stover to supply 1.6 million tons but at a farmgate price of $58/dry ton (blue curve). Thus, the strategy of combining the two feedstocks (green curve) shows that 1.6 million tons could be supplied to a biorefinery for a lower farmgate price of $48/dry ton.

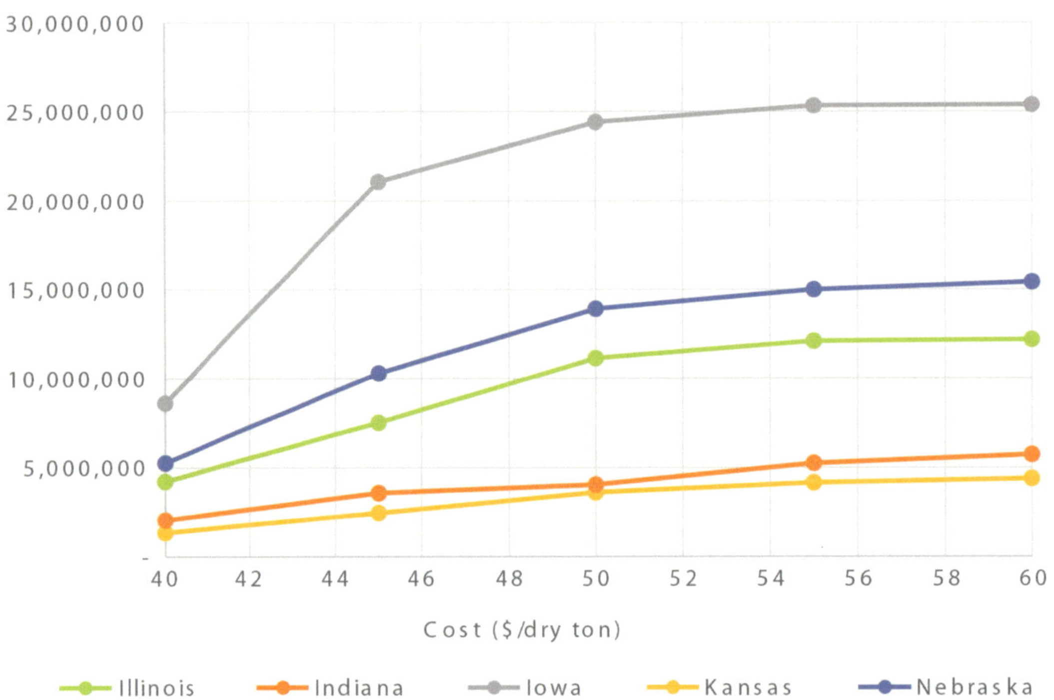

Figure 1. Corn stover availability by state as a function of farmgate price [33]. Availability and cost data assume base case corn stover yields for 2015. The data for each state is the sum of available corn stover for each county at a given farm gate price.

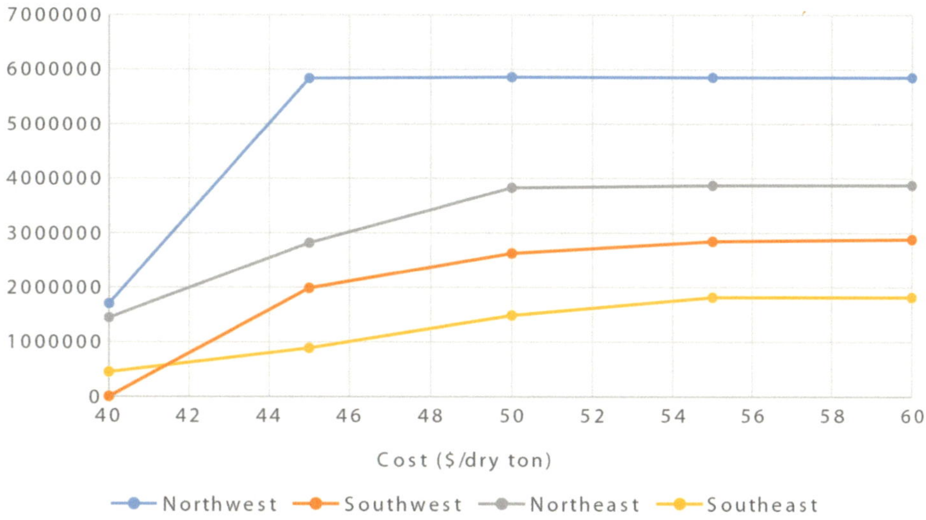

Figure 2. Feedstock supply curves for various locations in Iowa [33]. Availability and cost assume base case corn stover yields for 2015 for 61 counties in Iowa.

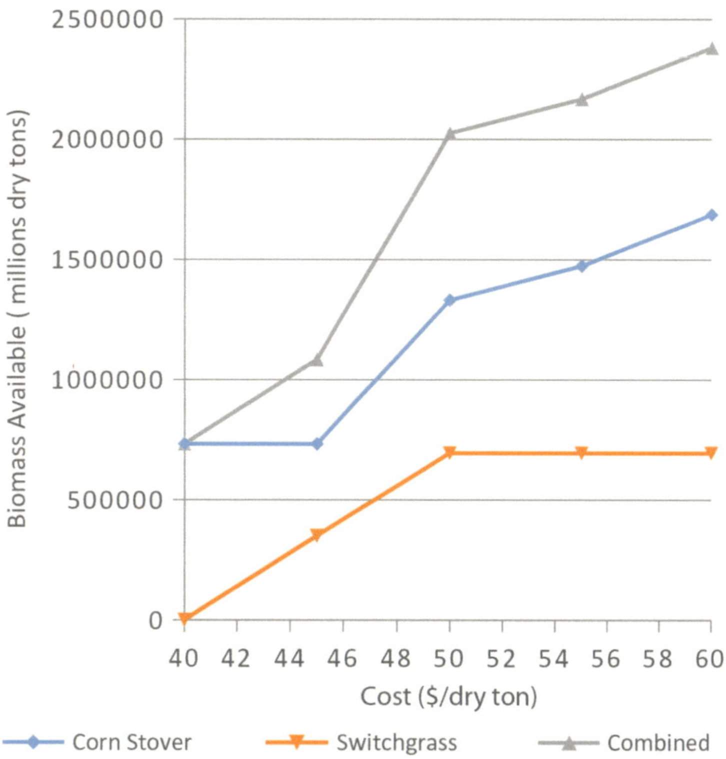

Figure 3. Biomass availability in northwestern Kansas as a function of farmgate price; prices and quantities shown are for 2015 and assume base case yields [34]. Supply curves illustrate biomass availability as a function of cost for a 12-county region in northwestern Kansas, with an approximate area of 90 miles × 120 miles. Curves are shown for corn stover (blue), switchgrass (red), and combined for a blended supply of corn stover and switchgrass (green).

Feedstock blending allows a biorefinery to utilize less of a single and expensive biomass type by collecting a variety of biomass (e.g., corn stover, switchgrass, sorghum, yard waste) and effectively moving down the cost versus supply curve and paying a lower average price for each feedstock. This does not change the supply versus cost curves for each resource; instead, it describes a system where purchasers are using a combination of least-cost resources and blending them to meet feedstock specifications for a subsequent biomass conversion process [29]. Costs may be further reduced by contraction of the draw radius for material collection, which reduces transportation cost. Feedstock formulation enabled through blending and other preprocessing strategies allows low-cost, low-quality biomass to be blended with higher cost and higher quality to achieve the in-feed specifications at the conversion facility. Blending feedstocks of differing quality results in a feedstock that has properties representative of the proportions of the materials that were blended together. Final price and quality are basically a weighted average of the price and quality of the components. It is important to realize a balance must be maintained and cost benefits may be not be linearly related to quality impacts. For many feedstock blends, there is likely a threshold quality level that cannot be surpassed to realize equal economic benefit. Biomass quality is a key consideration when analyzing biomass cost and availability. In combination with densification, wider sourcing areas can be tapped (including resources that are considered stranded using conventional supply systems).

Combining different biomass resources into the supply system also creates cost benefits by reducing overall grower payments [12]. The blended feedstock strategy relies on the availability of multiple feedstock resources that can be blended in an economical supply radius [31], which, in turn, decreases grower payment by reducing the required amount of any single biomass resource. In this manner, blending has the potential to expand the regionally available, biomass resource pool to include feedstocks of marginal quality at lower cost. In addition, a blended strategy offers the potential for feedstock quality upgrades and reduced variability [6, 21]. Blending high-quality feedstocks with low-cost, low-quality feedstocks is a strategy that can be used to meet quality specifications [21] at the biorefinery, in addition to achieving volume and cost targets in the supply chain [32]. An analysis by Maung and colleagues [18] has shown that a multi-crop cellulosic feedstock strategy lowers transportation costs compared to reliance on a single resource. Additionally, sourcing multiple feedstocks for cellulosic biofuel production mitigates supply risks associated with policies that govern crop residue removal. Further, Maung et al. suggest that a multi-feedstock strategy enhances understanding of the links between environmental policy, economies of density, economics of geography, transportation, risk and diversification in the biomass feedstock supply chain.

2.2. Densification

Reducing transportation costs while producing feedstocks with desirable (and consistent) physical properties such as increased bulk density and enhanced handling and processing characteristics requires densification of low-bulk density biomass. Commodity production for renewable fuels and chemicals requires large-scale biomass resources managed through AFSS and distributed biomass depots. These depots can provide feedstock stability, size reduction, and managed moisture [20]. Distributed biomass depots can reduce transportation and shipping costs

http://dx.doi.org/10.5772/67207

and improve feedstock stability and consistency by employing strategies such as size reduction, moisture management, blending, and densification. This allows greater access to stranded feedstocks and can reduce grower payment through feedstock blending [4, 11, 35]. Reducing transportation costs while producing feedstocks with desirable (and consistent) physical properties such as increased bulk density and enhanced handling and processing characteristics requires densification of low-bulk density biomass.

Pelleted biomass is produced from raw, ground material that is conditioned with heat and/or moisture, compacted, and extruded through a die [2, 3]. The economics and physical properties of densified biomass formats produced from agricultural residues have been explored in several studies [36–39]. Pelleting of biomass can increase unit density of raw biomass resources by as much as 10-fold [19], resulting in a flowable and durable product that is compatible with existing biomass supply system infrastructure. It has been shown that activation of the natural binders in biomass, such as lignin, through combined moisture and temperature effects during the process of densification is key to the development of particle-particle bonding that is required for durability [9]. The extent of lignification contributes significantly to biomass recalcitrance [4], and lignin alteration during the process of densification may impact biomass reactivity to pretreatment and enzymatic hydrolysis [12].

Currently, there are many types of densification systems available: pellet mills, piston and roller presses, tabletizers, and extruders [19]. Pelleting of biomass can increase unit density of raw biomass resources by as much as 10-fold [1], producing a uniform, durable product with free-flowing characteristics that may be more compatible with biorefinery operations. The economics and physical properties of densified biomass formats produced from agricultural residues have been explored in several studies [2, 9–11]. Industrial pelleting has developed into a well-established process using wood and wood chips. Global pellet demand has reached 23-million metric tons [40].

3. Impact of blending and densification on pretreatment processes in biochemical conversion pathways

Thermochemical pretreatment processes are used in biochemical conversion pathways to facilitate enzymatic access to cellulose and enable conversion of complex carbohydrate polymers into fermentable sugars. These promising processes include ammonia fiber expansion (AFEX), dilute acid, alkaline and ionic liquid (IL) pretreatment technologies. Specifically, AFEX is a physicochemical pretreatment process performed under high pressure (200–400 psi) and moderate temperature (80–150°C) with concentrated ammonia for a brief residence time (5–30 min) before pressure release [41]. AFEX pretreatment facilitates enzymatic access to cellulose by breaking down the cellulose crystalline structure and depolymerizing the lignin. Dilute-acid (DA) pretreatment relies on the combined effect of dilute sulfuric acid (0.25–2 wt.%), temperature (140–200°C), and time (seconds to minutes) to solubilize hemicellulose and improve enzymatic accessibility to cellulose [42, 43]. Alkaline pretreatment technologies focus on lignin solubilization and deacetylation under relative mild conditions (60–180°C) with NaOH or ammonium hydroxide (i.e., soaking in aqueous ammonia, namely SAA) as catalyst [44–46]. Recently, ILs are receiving significant attention as a class of novel environmental benign "green solvents" to dissolve and disrupt the biomass cell wall, reduce cellulose crystallinity and lignin content, and increase the porosity and surface area for enhanced enzymatic digestibility [26, 47–52]. In addition, this pretreatment technique shows great capability of fractionating wide range of feedstocks [50, 52, 53].

Although significant efforts have been focused on pretreatment of single lignocellulosic biomass in loose and ground format, recently researchers started to look into the application of biomass blending and densification for biochemical conversion into fermentable sugars. The details of biomass blending and quality improvement, characteristics of various densification formats of diverse feedstock types, and their impacts on conversion performance are discussed below.

3.1. Impact of biomass blending

Feedstock blending is one approach offering promising solution to overcome current challenges on biomass supply such as significant compositional variations [21, 22]. Therefore, it is imperative to develop conversion technologies that can process blended biomass feedstocks with minimal negative impact in terms of overall performance of the relevant biochemical pathway unit operations: pretreatment, fermentable sugar production, fermentation, and fuel titers.

Ionic liquid (IL) pretreatment has shown uniqueness in efficiently handling wide range of feedstocks; thus, this technology was investigated on the feasibility to process mixed feedstocks. It was firstly demonstrated in a US patent that the two or more feedstocks, including softwood, hardwood, grass, agricultural residues, and byproducts, can be combined for IL pretreatment with equivalent sugar conversion in comparison with single feedstocks [54]. Shi et al. evaluated the efficiency of feedstock blending along with the densification coupled with IL pretreatment to address the issues of feedstock diversity and compositional variations [55]. The IL 1-ethyl-3-methylimmidizolium acetate can process mixtures of pine, eucalyptus, switchgrass, and corn stover (in 1:1:1:1 ratios) and result in fast saccharification by reaching 90% digestibility within 24 h, which is comparable to any single feedstock type among the four starting biomass materials [52, 55]. A continuation study was further performed to investigate the IL pretreatment of the same mixture of four biomass in both flour and pellet formats, in comparison with dilute acid (DA) and soaking in aqueous ammonia (SAA) pretreatment methods, for simultaneous saccharification and fermentation into advanced biofuel isopentenol [26]. Their results show significant variations on the chemical composition, crystallinity, and enzymatic digestibility of the pretreated feedstock across the three different pretreatment technologies studied. IL pretreatment liberated the highest sugar titers from mixed biomass either in flour or pellets and is capable of handling mixed feedstocks with equal efficiency, and thus outperformed DA and SAA pretreatment methods which are more effective in pretreating herbaceous biomass feedstock and less effective in woody biomass for the mixed feedstock utilization. The high sugar production from IL process in turn led to the highest isopentenol titers in fermentation as compared to DA and SAA pretreatments.

While these three studies focused on the blends of various feedstock types that are agriculturally derived, researchers also looked into the utilization of the municipal solid waste as blending agent with lignocellulose to provide lower cost of biorefinery feedstock inputs [51]. The MSW/corn stover blends (ratio varying from 1:1 to 1:9 on the dry weight basis) went through two types of IL pretreatment for sugar conversion, one is pretreatment by IL 1-ethyl-3-methylimidazolium acetate followed by enzymatic hydrolysis, and another is enzyme-free acidolysis in IL 1-ethyl-3-methylimidazolium chloride with addition of mineral acid. Both processes show promising sugar conversion with glucose yield over 80% and xylose yield over 75%, suggesting the great potential to use MSW for biofuel production while maintaining performance and lowering cost.

Since the data from these four studies of biomass blends were obtained at low solid loading and milliliter level of operations, which cannot be directly transferred to industrially relevant scales, Li et al. performed the process scale-up and integration of IL1-ethyl-3-methylimidazolium acetate pretreatment on herbaceous (switchgrass) and woody (eucalyptus) blends (1:1 ratio) by 30-fold at 10% solid loading [50]. In comparison with single feedstocks, this biomass blend recovered similar yields of glucan, xylan, and lignin as switchgrass and eucalyptus at 6-L scale operation. The pretreated mixed feedstock was further enzymetically hydrolyzed at 2-L scale with 96% sugar yield [50, 56]. Additionally, the same group also investigated the scale-up of IL acidolysis using 1-ethyl-3-methylimidazolium chloride and mineral acid on MSW/corn stover blends and obtained sugar conversion yields that are comparable to small-scale studies [51, 53, 57]. These results indicate that mixed feedstocks, either agriculturally derived or MSW blended, are viable and valuable resources to consider when assessing biomass availability and affordability demands of the biorefineries. These initial scale-up evaluations demonstrate that IL-based pretreatment is feedstock agnostic, and there is no fundamental issues in terms of performance associated with the larger operations. This early-stage, 6-L scale-up process development integrates the unit operations of pretreatment, homogenization, continuous washing/separation, and product recovery for simplified feedstock handling, reduced water consumption and mitigation of IL inhibition, all of which can be further connected with downstream microbial fermentation for advanced biofuel production.

A few studies have examined the impact of blended or mixed biomass feedstocks on sugar yields from biochemical conversion using other pretreatment technologies. Karki et al. [58] reported on the enzymatic hydrolysis of mixtures of switchgrass and tall wheatgrass following dilute-acid and aqueous ammonia pretreatments. Switchgrass and tall wheatgrass were similar in composition before and after dilute-acid pretreatment, although tall wheatgrass had significantly higher glucose yields from enzymatic hydrolysis. Mixtures of the two species produced glucose yields that were higher than switchgrass and lower than tall wheatgrass following dilute-acid pretreatment and enzymatic hydrolysis. This study also demonstrated hydrolysis yields for mixtures could be predicted based on results of the individual components.

Brodeur-Campbell et al. [59] reported on the effects of biomass mixtures on sugar recovery from combined dilute-acid pretreatment and enzymatic hydrolysis. Aspen, a hardwood species that is suitable for efficient biochemical processing, was chosen as a model species in this study. Balsam, representing a high-lignin, softwood species, and switchgrass, a herbaceous energy crop with high ash, were chosen for comparative studies using 1:1 mixtures of aspen:balsam and aspen:switchgrass. No synergistic or antagonistic effects were identified in this study for three different pretreatment severities and three enzyme loadings examined. Again, total sugar recoveries for mixtures could be predicted by linear interpolation (±4%) from sugar yields of the pure biomass species. Similarly, Wolfrum et al. examined the effect of blending combined with densification on sugar yields from blends of corn stover, switchgrass, and *Miscanthus* biomass feedstocks [60, 61]; results showed the pelleting had a slightly positive, although not significant effect on total sugar yield. As in the previous studies, sugar yields could be predicted with reasonable accuracy from knowledge of the pure biomass feedstocks.

These studies demonstrate the efficient conversion of blended feedstocks to fermentable sugars and highlight the great potential for blending to expand the available resources for biofuel production. Biomass blending strategy certainly provides equivalent conversion performance as compared with single feedstock, in addition to its economic benefits toward the future development.

3.2. Impact of biomass densification

Lignocellulosic biomass with low bulk and energy density requires relatively high energy to transport, store, and distribute the feedstock from the field to the biorefinery gate for conversion, and the loose ground materials often pose problems of material feeding and handling in the reactors. Biomass densification typically involves exposing the biomass to elevated pressures and temperatures to remove excess water and compress the biomass. This process acts as a mild thermochemical pretreatment and can also impact the composition and structure of the biomass [55]. Several densification forms have been demonstrated recently, and this section reviews and compares the impact of densification on various thermochemical pretreatment.

3.2.1. Pellets

Recently, a growing body of literature has assessed the impact of pelletizing herbaceous and woody materials on the bioconversion process when combined various pretreatment technologies. Pelleted biomass is produced from raw, ground material that is conditioned with heat and/or moisture, compacted, and extruded through a die [2, 3]. It has been shown that activation of the natural binders in biomass, such as lignin, through combined moisture and temperature effects during the process of densification is key to the development of particle-particle bonding that is required for durability [9]. The extent of lignification contributes significantly to biomass recalcitrance [4], and lignin alteration during the process of densification may impact biomass reactivity to pretreatment and enzymatic hydrolysis [12].

Published reports evaluating the impact of pelletization on the bioconversion of corn stover, sorghum, wheat straw, big bluestem grasses, softwood, and switchgrass have shown positive trends using lower severity alkaline pretreatment. Similar or slightly higher sugar release and ethanol yield were observed in the pelleted format when compared to the nonpelleted format after pretreatment and enzymatic hydrolysis. Guragain et al. [62] evaluated the effect of alkaline pretreatment on sugar release and ethanol production in pelleted and nonpelleted wheat straw, corn stover, big bluestem, and sorghum stalk; mass recovery after alkali pretreatment increased by 14, 11, 2, and 5%, respectively, compared to nonpelleted biomass. Volumetric sugar production increased for all feedstocks except sorghum, although final sugar yields were not significantly different between the pelleted and non-pelleted biomass. Nahar and Pryor [63] reported that combining pelleting and pretreatment with SAA treatment reduced cellulase loading to achieve 90% glucose yield at 10 FPU per g glucan in switchgrass. Pelleting the switchgrass did not affect the feedstock composition compared to the non-pelleted switchgrass. Hoover et al. [64] evaluated the effect of physical properties resulting from pelleting AFEX-pretreated corn stover. Comparing grind size, die speed, and preheating on pellet properties on the sugar release after enzymatic hydrolysis showed the following: Die speed had no effect on sugar yield, while a larger grind size (4 mm vs. 6 mm) had a similar or lower effect on sugar yields after enzymatic hydrolysis. Overall, pelleting AFEX-treated biomass increased or had no effect on sugar yields at low or high ammonia loadings. Bals et al. [65] tested the susceptibility of AFEX-treated, corn stover pellets to enzymatic hydrolysis at high solids loading (18–36%). Pelletization slightly increased the initial rate of hydrolysis relative to raw biomass, enabled mixing, and resulted in higher glucose yields at 18% solids loading relative to unpelletized biomass (68% vs. 61%). Similarly, Rijal et al. [66] demonstrated that DA-treated switchgrass did not impact glucose yield in the finer ground, and pelleted materials compared to the native material. However, glucose yields from aqueous ammonia pretreatment, followed by enzymatic hydrolysis, were higher for both powder and pelleted materials compared to the native material. Glucose yield for the DA- and SAA-treated and pelleted switchgrass was 98 and 79%, respectively.

Theerarattananoon et al. [67] evaluated the impact of pelleting conditions on sugar release and chemical composition of corn stover, wheat straw, sorghum stalk, and big bluestem grass. Dilute-acid pretreatment and subsequent enzymatic hydrolysis increased glucan content in the pretreated solids compared to the nonpelleted companion feedstock for corn stover, wheat straw, and big bluestem prairie grass. Glucan content in the pretreated pelleted sorghum stalks was slightly less than nonpelleted sorghum stalks. Enzymatic hydrolysis results suggested that pelleting increased cellulose yield for all feedstocks. While wheat straw had the highest cellulose yield (94.1%), Ray et al. [68] evaluated the impact of densification on the bioconversion of corn stover, ground and pelleted format. The low solids dilute-acid pretreatment resulted in higher theoretical ethanol yields from the pelleted versus the non-pelleted format of 84 and 69%, respectively. Pelleted and ground corn stover was pretreated at higher solids loading at multiple pretreatment severities and showed slightly increased reactivity across three of the five severities tested.

Similar to other pretreatment technologies, conversion of biomass feedstocks with low energy and bulk density using ILs is not an economic process. To address this issue, Shi et al. investigated and compared the IL pretreatment of switchgrass, lodgepole pine, corn stover, and eucalyptus in both flour and densified pellet formats with 1-ethyl-3-methylimmidizolium acetate at 160°C and 10% solid loading for 3 h [55]. There was no significant difference between the physio-chemical properties, that is, composition and cellulose crystallinity, of the pretreated flour and pellets. The subsequent enzymatic digestibility results show that sugar yields from both formats reach 90% conversion within 24 h, suggesting densifying a wide range of feedstocks may be a competitive solution with no significant adverse impacts, provided that they are coupled with the appropriate conversion technology. Although significant improvements in terms of IL cost and recycling need to be resolved before this technology is commercially viable, biomass densification certainly provides the economic benefits toward the future development.

3.2.2. ComPAKo briquettes

Additional studies have been performed quantifying ethanol yields from densified AFEX-pretreated corn stover, switchgrass, and prairie cordgrass. Rijal et al. [69] studied the effect of initial particle size (2, 4, 8 mm) and densification on ethanol production. They employed a novel densification method, ComPAKo that uses a gear, mesh system to produce compacted biomass briquettes (1 inch × 0.5 inch × 4 inch). The ComPAKo equipment operates at lower temperatures (30–60°C) and pressures, lowering energy costs. Also, the capital investment for ComPAKo is less than half that of a pellet mill. The bulk density of the briquettes ranged between 380 and 460 kg/m^3 with moisture content of 11–15%. The AFEX-pretreated biomass was used for both simultaneous saccharification fermentation (SSF) at 4% glucan loading and separate hydrolysis with fermentation (SHF) at 1% glucan loading with an enzyme loading of 15 FPU and 64 CBU/g of glucan for hydrolysis. Results demonstrated that 2-mm densified corn stover briquettes yielded approximately 5% higher ethanol than 8-mm densified material. They also showed that grinding the densified 8-mm briquettes to 2 mm prior to SSF studies did not result in significant ethanol yield differences, but the 2-mm densified corn stover showed 4% higher yield than post-grinding the 8-mm briquettes prior to hydrolysis. The ethanol yields from the SSF did not significantly differ for the AFEX-treated corn stover or switchgrass when compared to the densified AFEX-treated material, but they noted a negative impact for the prairie cordgrass densified material by 16%. This was attributed to the observation that densified AFEX-treated prairie cordgrass was stronger and did not break apart during mixing and hydrolysis. Upon grinding of the AFEX-treated densified prairie cordgrass, the ethanol yields were 35% less than with the nondensified material, indicating that prairie cordgrass densification is not beneficial. The negative impacts of densification on AFEX-treated prairie cordgrass may be attributed to amount or structure of lignin in this feedstock. Sugar yields during SHF were not impacted for the corn stover or switchgrass densified material, but they were significantly diminished for prairie cordgrass. However, when comparing SHF ethanol yields, switchgrass densified material gave significantly lower yields, while yields from densified corn stover were only slightly higher, but the densified prairie cordgrass produced higher yields than either corn stover or switchgrass. The results support AFEX as an effective pretreatment technology for ComPAKo densification processes, thus reducing the need for additional particle size reduction for effective hydrolysis. These technologies however will produce different sugar and ethanol yields dependent on feedstock choice and subsequent hydrolysis and fermentation. AFEX-treated densified corn stover yielded the better quality briquettes of the three biomass types tested in the AFEX-treated, ComPAKo processes.

Biersbach et al. [70] also studied ethanol yields from briquettes of AFEX-treated corn stover, switchgrass, and prairie cordgrass and assessed the impact enzyme loading has during SSF and SHF, and they tested storage of these densified materials. They used the ComPAKo method to produce briquettes of 1–2 cm with a bulk density range between 380 and 460 kg/m^3 and moisture content of 11–15%. Ethanol yield was improved for all AFEX-pretreated biomass tested regardless of enzyme dose or fermentation regimen (SSF or SHF). They found that ComPAKo densified AFEX-treated biomass did not consistently have an impact on ethanol yields in most

http://dx.doi.org/10.5772/67207

of the conditions they tested, but in three of the corn stover tested conditions, densification increased ethanol yields up to 13%. For experiments using switchgrass and prairie cordgrass densified material, the densification caused 7 and 22% reduction in ethanol yields, respectively. They concluded that the ethanol yield differences of the various feedstocks could be attributed to the glucan content and pretreatment efficiency. They also found that the higher enzyme dose (15 FPU Spezyme CP, 64 CBU Novozyme 188) during enzymatic hydrolysis generally increased ethanol yields in the range of 18–317% for SHF and 28–62.5% for SSF, dependent on the feedstock. When the AFEX-treated densified briquettes were stored for 6 months, there was an increase in ethanol yield of 12–17%, with the exception of the prairie cordgrass which gave a 55% reduction in ethanol yield when SHF was performed, but not with SSF.

3.2.3. *Extrusion pelleting*

Extrusion pelleting is another densification technology that Sundaram and Muthukumarappan [71] used to evaluate AFEX-pretreated corn stover, switchgrass, and prairie cordgrass. They tested the effects of various parameters during laboratory-scale single-screw extrusion pelleting and the impact of those parameters on pellet bulk density, hardness, and sugar recovery from enzymatic hydrolysis. The parameters tested included moisture content (5, 10, and 15%), hammer mill particle size (2, 4, and 8 mm), and extrusion barrel temperature (75, 100, and 125°C). In general, the bulk density of the AFEX-treated biomass particles decreased as the particle size increased, and the bulk density increased with increasing the moisture content. Similar to other studies, the AFEX-pretreated material increased the pellet bulk density for each feedstock (650.6 kg/m^3 for corn stover, 680.1 kg/m^3 for prairie cord grass, and 627.7 kg/m^3 for switchgrass) compared to untreated material (453.0, 463.2, and 433.9 kg/m^3, respectively). The moisture content significantly impacted pellet bulk density with higher moisture content causing an increase. However, particle size of AFEX-pretreated material had no impact on pellet bulk density, but it inversely affected the untreated pellets; likewise extrusion temperature did not significantly impact AFEX-pretreated pellet bulk density but did negatively impact the untreated material. Pellet hardness was also determined for AFEX-pretreated pellets of corn stover, switchgrass, and prairie cordgrass with maximum hardness values of 2342.8, 2424.3, and 1298.6 N for each feedstock, respectively. The hardness of the AFEX-treated pellets was not significantly different at different barrel temperatures, indicating that good quality pellets can be achieved at 75°C, thus reducing costs. Moisture content correlated with pellet hardness for treated and untreated materials which is typical for extrusion pelleting and in combination with moisture content, particle size impacted pellet hardness, with 2 and 4 mm particles yielding maximum hardness. The percent glucose released form AFEX-pretreated pellets ranged from 88.9 to 94.9% for corn stover, 90.1 to 94.9% for prairie cord grass, and 87.0 to 92.9% for switchgrass. These glucose yields were 1.6, 2.1 and 2.3 fold higher than those from untreated pellets, respectively and xylose yields increased 1.6, 1.4, and 2.0 fold for AFEX-treated pellets compared to untreated pellets, respectively. Neither glucose yields nor xylose yields were significantly impacted by the extrusion temperatures or the particle sizes tested during extrusion pelleting, again indicating a low temperature of 75°C can be used to achieve quality pellets for conversion. Finally, the results show the extrusion pelleting process can be performed at low temperatures and larger particle size without significantly impacting sugar yields, thus reducing pellet processing costs.

These key findings suggest that densification of biomass does not negatively affect its composition and downstream conversion and may actually increase bioconversion or perhaps reduce the requirements for a given conversion level. However, many of these evaluations involving herbaceous feedstocks were conducted under low-solids, non-mixed, and batch conditions, which make extrapolations to more process-relevant conditions difficult.

4. Summary

This book chapter evaluates the potential of preprocessing options, that is, blending and densification, for uniform, consistent, quality-controlled, and cost-effective feedstock development, and reviews their impacts on feedstock supply chain logistic and downstream conversion performance. The use of blended and densified feedstocks in conversion pathways instead of conventionally ground biomass from a single source addresses several challenges in the current biomass supply chain, including availability, transportation, storage, cost, quality, and supply variability. Review and summary of recent research further demonstrate that a biomass blending strategy provides an efficient way to meet quality and conversion performance specifications in comparison with the conversion of single feedstock. Densified formats can perform equivalent to non-densified formats in terms of sugar and ethanol biochemical conversion performance. Both blending and densification provide great promise to enable more cost-effective downstream processing.

Acknowledgements

This work is supported by the U.S. Department of Energy, Office of Energy Efficiency and Renewable Energy, Bioenergy Technologies Office, under DOE Idaho Operations Office Contract DE-AC07-05ID14517. The authors would like to extend thanks to AAE for thoughtful review of this chapter.

Author details

Allison E. Ray[1*], Chenlin Li[1], Vicki S. Thompson[1], Dayna L. Daubaras[1], Nicholas J. Nagle[2] and Damon S. Hartley[1]

*Address all correspondence to: allison.ray@inl.gov

1 Idaho National Laboratory, Idaho Falls, ID, USA

2 National Renewable Energy Laboratory, Golden, CO, USA

References

[1] USEPA, March 17, 2016 [cited 2016 July 14]; Available from: https://www.epa.gov/renewable-fuel-standard-program.

[2] EISA, *Energy Independence and Security Act (EISA) of 2007*. 2007 January 4, 2007 [cited 2012 September 27, 2012]; Available from: http://www.gpo.gov/fdsys/pkg/BILLS-110hr6enr/pdf/BILLS-110hr6enr.pdf.

[3] USDOE, *2016 Billion-ton report: advancing domestic resources for a thriving bioeconomy*, in M.H. Langholtz, B.J. Stokes, and L.M. Eaton, Editors, Economic Availability of Feedstocks, vol 1. 2016, Oak Ridge National Laboratory: Oak Ridge, TN, USA.

[4] Hess, J.R., et al., *Commodity-scale production of an infrastructure-compatible bulk solid from herbaceous lignocellulosic biomass*, in Uniform-Format Bioenergy Feedstock Supply System Design Report. 2009, Idaho National Laboratory: Idaho Falls, ID.

[5] Hess, J.R., C.T. Wright, and K.L. Kenney, *Cellulosic biomass feedstocks and logistics for ethanol production*. Biofuels, Bioproducts and Biorefining, 2007. **1**(3): pp. 181–190.

[6] Kenney, K.L., et al., *Understanding biomass feedstock variability*. Biofuels, 2013. **4**(1): pp. 111–127.

[7] Lamers, P., et al., *Strategic supply system design—a holistic evaluation of operational and production cost for a biorefinery supply chain*. Biofuels, Bioproducts and Biorefining, 2015. **9**(6): pp. 648–660.

[8] Rentizelas, A.A., I.P. Tatsiopoulos, and A. Tolis, *An optimization model for multi-biomass tri-generation energy supply*. Biomass and Bioenergy, 2009. **33**(2): pp. 223–233.

[9] Banerjee, S., et al., *Commercializing lignocellulosic bioethanol: technology bottlenecks and possible remedies*. Biofuels, Bioproducts and Biorefining, 2010. **4**(1): pp. 77–93.

[10] Oke, M.A., M.S.M. Annuar, and K. Simarani, *Mixed feedstock approach to lignocellulosic ethanol production—prospects and limitations*. Bioenergy Research, 2016: pp. 1–15.

[11] Argo, A.M., et al., *Investigation of biochemical biorefinery sizing and environmental sustainability impacts for conventional bale system and advanced uniform biomass logistics designs*. Biofuels, Bioproducts and Biorefining, 2013. **7**(3): pp. 282–302.

[12] Jacobson, J., et al., *Techno-economic analysis of a biomass depot*. 2014, Idaho National Laboratory: Idaho Falls, ID.

[13] Searcy, E. and J.R. Hess, Uniform-Format Feedstock Supply System: A Commodity-Scale Design to Produce an Infrastructure-Compatible Biocrude from Lignocellulosic Biomass. 2010, Idaho National Laboratory: Idaho Falls, ID.

[14] Hansen, J., J. Jacobson, and M. Roni. Quantifying Supply Risk at a Cellulosic Biorefinery. in 33rd International Conference of the System Dynamics Society. 2015. Cambridge, MA, USA.

[15] Sultana, A. and A. Kumar, *Optimal configuration and combination of multiple lignocellulosic biomass feedstocks delivery to a biorefinery*. Bioresource Technology, 2011. **102**(21): pp. 9947–9956.

[16] Langholtz, M., et al., *Climate risk management for the U.S. cellulosic biofuels supply chain*. Climate Risk Management, 2014. **3**: pp. 96–115.

[17] Emerson, R., et al., *Drought effects on composition and yield for corn stover, mixed grasses, and Miscanthus as bioenergy feedstocks*. Biofuels, 2014. **5**(3): pp. 275–291.

[18] Maung, T.A., et al., *The logistics of supplying single vs. multi-crop cellulosic feedstocks to a biorefinery in southeast North Dakota*. Applied Energy, 2013. **109**: pp. 229–238.

[19] Tumuluru, J.S., et al., *A review of biomass densification systems to develop uniform feedstock commodities for bioenergy application*. Biofuels, Bioproducts and Biorefining, 2011. **5**(6): pp. 683–707.

[20] Richard, T.L., *Challenges in scaling up biofuels infrastructure*. Science, 2010. **329**(5993): pp. 793–796.

[21] Williams, C.L., et al., *Sources of biomass feedstock variability and the potential impact on biofuels production*. Bioenergy Research, 2016. **9**(1): pp. 1–14.

[22] Li, C., et al., *Impact of feedstock quality and variation on biochemical and thermochemical conversion*. Renewable and Sustainable Energy Reviews, 2016. **65**: pp. 525-536..

[23] Tumuluru, J.S., et al., *Formulation, pretreatment, and densification options to improve biomass specifications for co-firing high percentages with coal*. Industrial Biotechnology, 2012. **8**(3): pp. 113–132.

[24] Hill, L.D., Grain Grades and Standards: Historical Issues Shaping the Future. 1990, Chicago, IL, USA: University of Illinois Press.

[25] Sami, M., K. Annamalai, and M. Wooldridge, *Co-firing of coal and biomass fuel blends*. Progress in Energy and Combustion Science, 2001. **27**(2): pp. 171–214.

[26] Shi, J., et al., *Impact of pretreatment technologies on saccharification and isopentenol fermentation of mixed lignocellulosic feedstocks*. BioEnergy Research, 2015. **8**(3): pp. 1004–1013.

[27] Boavida, D., et al., *A study on coal blending for reducing NOx and N_2O levels during fluidized bed combustion*. Clean Air, 2012. **5**: pp. 175–191.

[28] Reddy, D.V. and N. Krishna, *Precision animal nutrition: a tool for economic and eco-friendly animal production in ruminants*. Livestock Research for Rural Development, 2009. **21**(3): p. 36.

[29] USDOE, Bioenergy Technologies Office Multi-Year Program Plan. 2016: Washington, D.C.

[30] Crawford, N.C., et al., *Evaluating the pelletization of "pure" and blended lignocellulosic biomass feedstocks*. Fuel Processing Technology, 2015. **140**: pp. 46–56.

[31] Kenney, K.L., et al., Feedstock Supply System Design and Economics for Conversion of Lignocellulosic Biomass to Hydrocarbon Fuels. Conversion Pathway: Biological Conversion of Sugars to Hydrocarbons: The 2017 Design Case. 2013, Idaho National Laboratory: Idaho Falls, ID.

[32] Thompson, V.S., et al., *Assessment of municipal solid waste as a blend feedstock to lower biomass feedstock costs*, in 36th Symposium on Biotechnology for Fuels and Chemicals. 2014: Clearwater Beach, FL.

[33] USDOE, *Bioenergy knowledge discovery framework*. 2016 [cited 2015 November 4, 2015]; Available from: www.bioenergykdf.net.

[34] USDOE, *Bioenergy knowledge discovery framework*. 2015 [cited 2015 November 4, 2015]; Available from: www.bioenergykdf.net.

[35] Lamers, P., et al., *Techno-economic analysis of decentralized biomass processing depots*. Bioresource Technology, 2015. **194**: pp. 205–213.

[36] Kaliyan, N. and R.V. *Morey, Natural binders and solid bridge type binding mechanisms in briquettes and pellets made from corn stover and switchgrass*. Bioresource Technology, 2010. **101**(3): pp. 1082–1090.

[37] Larsson, S.H., et al., *High quality biofuel pellet production from pre-compacted low density raw materials*. Bioresource Technology, 2008. **99**(15): pp. 7176–7182.

[38] Mani, S., L.G. Tabil, and S. Sokhansanj, *Effects of compressive force, particle size and moisture content on mechanical properties of biomass pellets from grasses*. Biomass and Bioenergy, 2006. **30**(7): pp. 648–654.

[39] Mani, S., L.G. Tabil, and S. Sokhansanj, *Specific energy requirement for compacting corn stover.* Bioresource Technology, 2006. **97**(12): pp. 1420–1426.

[40] Greene, J., *Q1 2015 Recap: Global Wood Pellet Demand Creates US Opportunities*. 2015 April 14, 2015 [cited 2016 September 11, 2016]; Available from: http://blog.forest2market.com/wood-pellet-demand-creates-opportunity.

[41] Bals, B., et al., *Ammonia fiber expansion (AFEX) treatment of eleven different forages: Improvements to fiber digestibility in vitro*. Animal Feed Science and Technology, 2010. **155**(2–4): pp. 147–155.

[42] Himmel, M.E., et al., *Biomass recalcitrance: engineering plants and enzymes for biofuels production*. Science, 2007. **315**(5813): pp. 804–807.

[43] Weiss, N.D., J.D. Farmer, and D.J. Schell, *Impact of corn stover composition on hemicellulose conversion during dilute acid pretreatment and enzymatic cellulose digestibility of the pretreated solids*. Bioresource Technology, 2010. **101**(2): pp. 674–678.

[44] Chen, Y., et al., *Understanding of alkaline pretreatment parameters for corn stover enzymatic saccharification*. Biotechnology for Biofuels, 2013. **6**(1): pp. 1–10.

[45] Chen, X., et al., *A highly efficient dilute alkali deacetylation and mechanical (disc) refining process for the conversion of renewable biomass to lower cost sugars*. Biotechnology for Biofuels, 2014. **7**(1): pp. 1–12.

[46] Tao, L., et al., *Process and technoeconomic analysis of leading pretreatment technologies for lignocellulosic ethanol production using switchgrass*. Bioresource Technology, 2011. **102**(24): pp. 11105–11114.

[47] Li, C.L., et al., *Comparison of dilute acid and ionic liquid pretreatment of switchgrass: Biomass recalcitrance, delignification and enzymatic saccharification*. Bioresource Technology, 2010. **101**(13): pp. 4900–4906.

[48] Passos, H., M.G. Freire, and J.A.P. Coutinho, *Ionic liquid solutions as extractive solvents for value-added compounds from biomass.* Green Chemistry, 2014. **16**(12): pp. 4786–4815.

[49] Brandt, A., et al., *Deconstruction of lignocellulosic biomass with ionic liquids.* Green Chemistry, 2013. **15**(3): pp. 550–583.

[50] Li, C., et al., *Scale-up of ionic liquid-based fractionation of single and mixed feedstocks.* BioEnergy Research, 2015: pp. 1–10.

[51] Sun, N., et al., *Blending municipal solid waste with corn stover for sugar production using ionic liquid process.* Bioresource Technology, 2015. **186**: pp. 200–206.

[52] Li, C.L., et al., *Comparing the recalcitrance of eucalyptus, pine, and switchgrass using ionic liquid and dilute acid pretreatments.* Bioenergy Research, 2013. **6**(1): pp. 14–23.

[53] Li, C., et al., *Scale-up and process integration of sugar production by acidolysis of municipal solid waste/corn stover blends in ionic liquids.* Biotechnol Biofuels 2017. In Press.

[54] Singh, S., et al., *Mixed feedstocks processing using an ionic liquid.* 2013, US20130183739 A1.

[55] Shi, J., et al., *Impact of mixed feedstocks and feedstock densification on ionic liquid pretreatment efficiency.* Biofuels, 2013. **4**(1): pp. 63–72.

[56] Tanjore, D., et al., *Resolving process scale-up issues of ionic liquid pretreatment and saccharification of biomass to monomeric sugars,* in 35th Symposium on Biotechnology for Fuels and Chemicals. 2013: Portland, Oregon.

[57] Li, C., et al., *Scale-up and process integration of sugar production by acidolysis of single and mixed feedstocks in ionic liquids,* in 37th Symposium on Biotechnology for Fuels and Chemicals. 2015: San Diego, CA.

[58] Karki, B., N. Nahar, and S.W. Pryor, *Enzymatic hydrolysis of switchgrass and tall wheatgrass mixtures using dilute sulfuric acid and aqueous ammonia pretreatments.* Biological Engineering Transactions, 2011. **3**(3): pp. 163–171.

[59] Brodeur-Campbell, M., J. Klinger, and D. Shonnard, *Feedstock mixture effects on sugar monomer recovery following dilute acid pretreatment and enzymatic hydrolysis.* Bioresource Technology, 2012. **116**: pp. 320–326.

[60] Wolfrum, E., et al., *The effect of feedstock densification on structural sugar release and yield in and biofuel feedstock and feedstock blends,* in Symposium on Biotechnology for Fuels and Chemicals. 2015: San Diego, CA.

[61] Wolfrum, E.J., et al., *The effect of biomass densification on structural sugar release and yield in biofuel feedstock and feedstock blends.* BioEnergy Research, 2017. Accepted.

[62] Guragain, Y.N., et al., *Evaluation of pelleting as a pre-processing step for effective biomass deconstruction and fermentation.* Biochemical Engineering Journal, 2013. **77**: pp. 198–207.

[63] Nahar, N. and S.W. Pryor, *Reduced pretreatment severity and enzyme loading enabled through switchgrass pelleting.* Biomass and Bioenergy, 2014. **67**: pp. 46–52.

http://dx.doi.org/10.5772/67207

[64] Hoover, A.N., et al., *Effect of pelleting process variables on physical properties and sugar yields of ammonia fiber expansion pretreated corn stover.* Bioresource Technology, 2014. **164**: pp. 128–135.

[65] Bals, B.D., et al., *Enzymatic hydrolysis of pelletized AFEX™-treated corn stover at high solid loadings.* Biotechnology and Bioengineering, 2014. **111**(2): pp. 264–271.

[66] Rijal, B., et al., *Combined effect of pelleting and pretreatment on enzymatic hydrolysis of switchgrass.* Bioresource Technology, 2012. **116**: pp. 34–41.

[67] Theerarattananoon, K., et al., *Effects of the pelleting conditions on chemical composition and sugar yield of corn stover, big bluestem, wheat straw, and sorghum stalk pellets.* Bioprocess and Biosystems Engineering, 2012. **35**(4): pp. 615–623.

[68] Ray, A.E., et al., *Effect of pelleting on the recalcitrance and bioconversion of dilute-acid pretreated corn stover under low- and high-solids conditions.* Biofuels, 2013. **4**(3): pp. 271–284.

[69] Rijal, B., et al., *Effect of initial particle size and densification on afex-pretreated biomass for ethanol production.* Applied Biochemistry and Biotechnology, 2014. **174**(2): pp. 845–854.

[70] Biersbach, G., et al., *Effects of enzyme loading, densification, and storage on AFEX-pretreated biomass for ethanol production.* Applied Biochemistry and Biotechnology, 2015. **177**(7): pp. 1530–1540.

[71] Sundaram, V. and K. Muthukumarappan, *Impact of AFEX™ pretreatment and extrusion pelleting on pellet physical properties and sugar recovery from corn stover, prairie cord grass, and switchgrass.* Applied Biochemistry and Biotechnology, 2016. **179**(2): p. 202–219.

Chapter 16

Metal Removal by Seaweed Biomass

Claudia Ortiz-Calderon, Héctor Cid Silva and
Daniel Barros Vásquez

Additional information is available at the end of the chapter

http://dx.doi.org/10.5772/65682

Abstract

Environmental metal pollution is a serious public problem, and it has become an issue leading to research in the effluent remediation area. Techniques involving biosorption processes have been found to be promising due to the low cost of nonliving biomaterials, which have the potential to adsorb metal ions from wastewaters. One of the most promising types of biomasses to be used as biosorbents is the seaweed biomass, particularly from brown algae. The biosorption capability of the seaweed biomass relies on their cell wall chemical composition, mainly composed of alginates and fucoidans, molecules with a high presence of functional groups that interact with metal ions. This book chapter focuses on the use of seaweed biomass for metal biosorption and the chemical basis underlying the process. The current state of the commercial status of biosorption technology based on seaweed biomass is presented. Examples of complementary uses of the algae biomass other than industrial wastewater cleaning processes are presented, and the potential reuse of the biomass after the biosorption focused on biofuel production is discussed.

Keywords: seaweed biomass, metal removal, biosorption, biosorbent, wastewater treatment

1. Introduction

Environmental metal pollution is a serious public problem, and it has become an issue, leading to research in the effluent remediation area. Many techniques have been reported for removing metals from solutions, such as chemical precipitation, adsorption, ion exchange, filtration, chemical oxidation or reduction, electrochemical treatment, membrane processes, and evaporation. It has been found that these methods are limited, because of high operational costs, especially when the initial metal ion concentrations are at the range of 10-100 mg/L [1]. Hence, techniques involving biosorption processes have been found to be promising, due to the low cost of nonliving biomaterials, which have the potential to adsorb metal ions from wastewater.

The biosorption processes occur when metal ions interact with the functional groups present in biopolymers that are part of the biomass. Chemical groups such as amide, hydroxyl, carboxylate, sulfonate, phosphate, and amino are responsible for the quantitative adsorption of metals [2]. Several interaction mechanisms such as complexation, coordination, chelation, ion adsorption, cation exchange, and microprecipitation have been proposed as the participants in the metal biosorption processes [3].

A wide variety of biomasses has been found to be capable of sequestering metal ions from dilute solutions. An interesting approach is the use of the nonliving forms of the biomaterials because they do not need nutrition for the maintenance and avoid metal toxicity problems [4]. One of the most promising types of biomasses suitable for their use as biosorbents is marine algal biomass (seaweeds), which exhibit a high abundance in the oceans [5].

The biosorption capability of algae biomass is mostly related to their cell wall chemical composition, which exhibits a fiber-like structure and an amorphous embedding matrix of polysaccharides such as alginates and fucoidan [6]. In brown algae, alginates have a high affinity for divalent cations and sulfated polysaccharides give account of the uptake of trivalent cations [7]. The physical and chemical nature of the interaction between the metals and the functional groups present in the biomass has been intensively studied, in order to develop technologies for the sequestration of metals to clean, or to recover, valuable metals from industrial effluents [5, 8, 9].

This book chapter focuses on metal biosorption by seaweed biomass and the chemical interactions between the functional groups of this biomass and the cations. To the end, the potential uses of algae biomass in industrial wastewater cleaning processes and its potential reuse are highlighted.

2. Seaweed biomass

Algae are autotrophic organisms that contain chlorophyll and carry out oxygenic photosynthesis; they are widely distributed and have great diversity. Algae do not represent a formal taxonomic group of organisms, but a highly heterogeneous collection of organisms of different evolutionary lineages and high genetic diversity, which is reflected in the huge diversity that algae in morphological terms, ultrastructure, ecological, biochemical, and physiological [10].

Macroalgae, or seaweed, are a group of fast-growing aquatic organisms including about 9000 species. They are commonly classified into three groups according to the color of the thallus, which correspond to the Chlorophyta (green algae), Rhodophyta (red algae), and Heterokontophyta phylum, class Phaeophyceae (brown algae) [11] (**Figure 1**).

The taxonomic classification of these organisms involves much more than this simple designation and is performed considering a combination of features, including the nature of photosynthetic pigments; polymers present in the cell wall and cellular organization. Today,

http://dx.doi.org/10.5772/65682

Figure 1. Brown algae *Durvillaea antarctica*. Left, a specimen freshly collected from the coastal line. Right, a sample of milled *Durvillaea antarctica* biomass with a size of 500–1000 μm.

thanks to molecular systematics, a good progress has been made in the classification of these organisms, solving the problem of underestimation of diversity when considering only morphological characters [12]. There is great interest in the commercial use of the chemical constituents present in the seaweeds, in the field of energy production, agriculture, food, environmental, and pharmaceutical industry. The global harvest seaweed for food and algal products (e.g., Agar, alginates, and carrageenan) exceeds 3 million tons per year, with a potential harvest estimated at 2.6 million tons for red algae and 16 million tons brown algae [13]. Of particular interest is the use of seaweed dead biomass as biosorbent of heavy metals in solution. Multiple studies have shown a high sorption capacity and selectivity for different metal cations attributed to the polysaccharides present in their cell walls [4, 5, 8, 9, 14–18]. The basic organization of their cell walls comprises a fibril skeleton mainly composed of cellulose, and an amorphous matrix of sulfated galactans constituted by carrageenans and agar in red algae and alginates or alginic acid and fucoidan in brown algae (**Table 1**). Studies to assess

Division	Common name	Pigments	Storage product	Cell wall
Chlorophyta	Green algae	Chlorophyll *a, b*; α-, β- and γ-carotenes, and several xanthophylls	Starch (amylose and amylopectin) (oil in some)	Cellulose (β-1,4-glucopyroside), hydroxyproline, glucosides, xylans, and mannans
Phaeophyta	Brown algae	Chlorophyll *a, c*; β-carotene and fucoxanthin and several other xanthophylls	Laminaran (β-1,3-lucopyranoside, predominantly); mannitol	Cellulose, alginic, acid, and sulfated, mucopolysaccharides (fucoidan)
Rhodophyta	Red algae	Chlorophyll *a* (*d* in some *Florideophyceae*); R- and C-phycocyanin, allophycocyanin; R- and B-phycoerythrin. α- and β-carotene and several xanthophylls	Floridean starch (amylopectin-like)	Cellulose, xylans, several sulfated polysaccharides (galactans) calcification in some; alginate in *Corallinaceae*

Table 1. Physical and chemical characteristics of three algal divisions. Adapted from [2].

the biosorption (mass of metal adsorbed by mass of biosorbent) of different metals (Pb, Cu, Zn, and Cd) by seaweed biomasses have shown that the higher sorption capacity is exhibited by brown algae [5, 8, 9, 18].

Phaeophyceae is a large and diverse class of species ranging from small filamentous algae to highly complex organisms up to 60 m long. This class presents 2046 species organized into 20 orders [19]. It is believed that *Phaeophyceae* emerged between 150 and 200 million years ago in a secondary endosymbiosis event, in which a red algae was captured by an ancestral protist [12, 20, 21]. Its characteristic color is given by large quantities of fucoxanthin present in their chloroplasts. The most important algae orders from the point of view of the biosorption are Fucales and Laminariales, presenting species with a greater structural complexity. The components in their cell walls and reserve polymers have been reported as the main cause of its great capacity for immobilization of metal cations. The cell wall structure varies according to the species, age of the population, climate, and geographical location. Fibril wall structure, which provides structural support, is composed of cellulose-like plants, although displayed in a lower portion, occupying from 1 to 8% of its dry weight [21]. Two groups of anionic polysaccharides that provide strength and flexibility to the cell wall are alginates (between 10 and 40% dry weight) and fucoidan (between 5 and 20% dry weight). Mannitol and laminarans represent energy backup products. Other components such as proteins and phlorotannins also contribute but in a lesser way to the biosorption, representing about 5% of the biomass dry weight (**Figure 2**).

The three major components of the cell wall extracellular matrix of brown algae, cellulose, alginic acids, and polymers like mannuronic and guluronic acids, are complexed with light

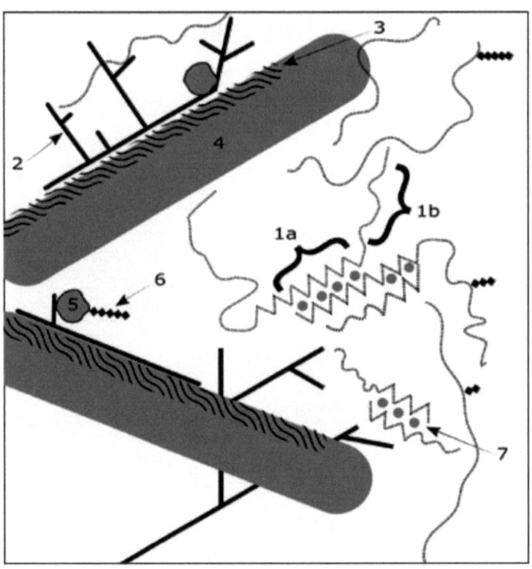

Figure 2. Schematic representation of the main components present in the cell wall of brown algae. Alginates rich in α-1,4-L-guluronate (1a); alginates rich in β-1,4-D-mannuronate (1b); fucoidans (2); hemicellulose (3); cellulose (4); proteins (5); phlorotannins (6); Ca^{2+} (7). Adapted from [2].

http://dx.doi.org/10.5772/65682

metals such as sodium, potassium, magnesium, and calcium, and other polysaccharides [19]. Alginates and sulfated polysaccharides have been reported as the predominant molecular components with reactive groups in brown algae [18]. Biosorption of heavy metals involves several mechanisms that differ qualitatively and quantitatively depending of the chemical species used, the origin of the biomass, and its processing procedure such as reinforcement by crosslinking [20]. Algae biomass possesses several chemical groups that can attract and sequester metals: acetamide, amine, amide, sulfhydryl, sulfate, and carboxyl [2]. This chemical diversity originates a combination of mechanisms for the capture of the metals, including electrostatic attraction, complexation, ion exchange, covalent binding, van der Waals attraction, adsorption, and microprecipitation [3].

Alginates are a family of linear polysaccharides, consisting of two uronic acids units: β-1,4-D-mannuronate (M) and α-1,4-L-guluronate (G). These units are arranged in homopolymer blocks of M, homopolymer blocks of G, and/or heteropolymer blocks of M and G (**Figure 1**). The relative abundance of M and G blocks in the macromolecular structure determines structural properties and affinity of alginates for divalent cations. The affinity of some divalent metal cations varies with M:G ratio [2, 6, 22]. Studies have shown that the affinity of the alginates for cations such as Pb, Cu, Cd, Zn, and Ca increases with a higher content of guluronic acid [23, 24]. The high specificity for divalent cations is explained by the structure of "zig-zag" formed by homopolymers of guluronic acid, which stabilizes the Ca^{2+} and other divalent cations easily (**Figure 1**) [2, 22]. Alginates fibers are able to adopt an ordered conformation in solution through dimerization of homopolymeric regions of guluronic acid, in the presence of calcium or other divalent cations, as they are filled with carboxylic groups and other electronegative oxygen atoms. This description is known as the model of "egg box" [25]. The carboxyl groups are the most abundant functional group in brown algae, determined by the percentage of quantifiable sites by titration, reaching about 70%. Furthermore, most of the metal cations of interest show high sequestering at pH near to the dissociation constant (pK_a) of carboxylic acids [2].

Fucoidans are branched sulfated polysaccharides mainly constituted by α-L-fucose, uronic acids, and a small portion of galactose, xylose, arabinose and/or mannose, glucose, and sometimes proteins, presenting an extremely variable molecular weight. They are presented in the form of homopolymers or homofucans called fucans, or heteropolymers called fucoidans. The sulfonate groups in the fucoidans are the second functional group in abundance in brown seaweed, and its role could become prominent, if the binding of the metal occurs at a low pH [2].

Mannitol is a compound derived from monomeric D-mannose present in all brown algae, which can represent up to 30% of the biomass dry weight [2]. The second largest reserve products are laminarans and polysaccharides, which are composed of (1, 3)-β-D-glucans. They consist of residues (1, 3)-β-D-glucopyranose with some 6-O-branches of the main chain. Two types of laminaran chains exist: M, with a mannitol monomer attached to the reducing end, and G, with a glucose monomer attached to the reducing end. All the polysaccharides present hydroxyl groups, but these are less abundant and only are negatively charged at pH above 10, playing a secondary role at low pH [2].

3. Seaweed biomass and metal biosorption

Biosorption of heavy metal ions in wastewater using algae can be ecologically safer, cheap, and efficient. Algae can be used for sorption of toxic and radioactive metal ions [26] and also to recover metal ions like gold and silver [27].

The biosorption of heavy metal ions by seaweed biomass may occur by different mechanisms such as ion exchange, complex formation, and electrostatic interaction [28], being ion exchange the most important [29]. Polysaccharides and proteins present in the algae cell walls provide the metal-binding sites [30]. The sorption capacity of a seaweed cell surface to a specific ion depends on several factors such as the amount of functional groups in the algae matrix, the coordination number of the metal ion to be sorbed, the accessibility of binding groups for metal ions, the complex formation, affinity constants of the metal with the functional group, and the chemical state of these sites [31]. Considering the heterogeneity of the cell wall composition in different seaweed species, the capacity of metal biosorption by the algal strains will vary. For instance, brown algae with alginate in their cell wall composition have a high biosorption affinity for lead ions [5]. Alginate polymers are the primary responsible for heavy metal ions sorption in brown algae, and their capacity to bind the metal directly depends on the number of binding sites on this polymer [32]. In a second place, fucoidans play a key role for heavy metal sequestration.

The functional groups present in the brown and green algae cell wall matrixes, such as carboxyl, hydroxyl, sulfate, phosphate, and amine groups, play a dominant role in the metal binding [30]. The presence of various functional groups and their complexation with heavy metals during biosorption process can be studied by using spectroscopic techniques, such as FT-IR and XPS [33]. The X-ray absorption fine structure spectroscopy and quantum chemistry calculation are also an experimental approach to explain the biosorption mechanisms [34]. An interesting methodology to determine the contribution of different functional groups in the metal adsorption is the derivatization of functional groups, like the pretreatment of the seaweed biomass with methanol in acid media or with propylene oxide, which blocks the action of the carboxyl groups in the biomass [35]. In *Sargassum fluitans*, after 4 h of treatment of the biomass with propylene oxide, about 50% of reduction in the biosorption capacity of Cd and Pb ions was observed [35]. Cid et al. [36] found a 43.3% of decrease in copper biosorption by esterifying biomass of the species *Durvillaea antarctica* using methanol in acid media. The presence of carboxyl, hydroxyl, sulfate, phosphate, and amine functional groups has been found to play a dominant role for the algae-metal interaction by an ion-exchange mechanism that occurs between heavy metals and intrinsic light metals ions such as Ca^{2+} and Mg^{2+}, and Na^{+} and K^{+} [37]. A summary of representative data for efficiency of copper removal for different types of algal biomass is presented (**Table 2**).

Heavy metal ion uptake by algal biomass can be enhanced by physical or chemical treatments that modify the seaweed cell surface structure and provide additional binding sites [32, 41–43]. Physical treatments such as heating/boiling, freezing, crushing, and drying usually lead to an enhanced level of metal ion biosorption. These treatments provide more surface area to increase the biosorption capacity [41] and release cell contents that might bind to

Brown algae specie	pH	q_m (mmol Cu g^{-1} biomass)	References
Ascophyllum nodosum	4.0	0.91	Romera et al. [5]
Durvillaea antarctica	5.0	1.73	Cid et al. [36]
Fucus serratus	5.5	1.60	Ahmady-Asbchin et al. [37]
Fucus spiralis	4.0	1.10	Romera et al. [5]
Fucus vesiculosus	5.0	1.66	Mata et al. [28]
Padina sp.	5.0	1.14	Sheng et al. [18]
Sargassum sp.	5.0	0.99	Sheng et al. [18]
Sargassum sp.	5.5	1.13	Karthikeyan et al. [38]
Sargassum filipendula	4.5	0.89	Davis et al. [39]
Sargassum filipendula	4.5	1.32	Kleinübing et al. [40]
Sargassum fluitans	4.5	0.80	Davis et al. [39]
Sargassum vulgarie	4.5	0.93	Davis et al. [39]

Table 2. Comparison of Cu(II) biosorption maximum capacities of different types of brown algae. Adapted from [36]

metal ions. The most common algal pretreatments are $CaCl_2$, formaldehyde, glutaraldehyde, NaOH, and HCl. Pretreatment with $CaCl_2$ causes calcium binding to alginate that plays an important role in ion exchange [42]. Formaldehyde and glutaraldehyde strengthen the cross-linking between hydroxyl groups and amino groups [43]. NaOH increases the electrostatic interactions of metal ion cations and provides optimum conditions for ion exchange, while HCl replaces light metal ions with a proton and also dissolves polysaccharides of cell wall [32], or denatures proteins, increasing the binding sites for the biosorption process.

4. Industrial uses of seaweed biomass

Over the past four decades, much effort had been devoted to identify readily available non-living seaweed biomass, capable of effectively removing heavy metals, with good hydro-dynamic capacities, physicochemical stability, and with the possibility to enhance their capacities to obtain biosorbents. After years of experimentation on hundreds of raw seaweed biomass for biosorption of heavy metals under different conditions, the optimum conditions for the biosorption process at bench scale have been stated for many seaweed biomasses. This research has conducted the efforts to the development of biosorption technologies for industrial applications, considering the volumes and the complex composition of different wastewaters.

One of the most important issues to consider is the biomass organic leaching phenomenon that is produced by the contact of the biomass with the water to be treated, liberating fractions of the biomass, biopolymers, and another chemical compounds. Organic leaching is an important factor to minimize, because it adds organic pollution to the treated water and generates an important biomass loss, resulting in a decrease on the availability of biomass for the next cycle of biosorption [24]. Also, organic leaching provokes hydraulic problems in column systems, because the biomass tends to accumulate at the exit of the packed columns, generating a clot that impairs the normal flow of the treated water passing by the biosorbent bed and generates elevated levels of TDS [44, 45]. The problems of excessive leaching and swelling can be minimized through proper engineering procedures, but the costs and practicability of these procedures are of concern. To control the swelling in seaweed biomass, Chu and Hashim [46] employed polyvinyl alcohol to immobilize biomass of *Sargassum* to construct packed columns. Omar et al. [47] immobilized *Enteromorpha torta* biomass using calcium alginate and solutions of calcium chloride for the biosorption of Cesium-134 on packed column beads. Other authors used calcium or magnesium salts to improve rheological properties of the biomass by the crosslinking generated by calcium ions and the alginate fibers [2], and used this material to make columns beds [48–52]. Other procedures applied to seaweeds biomass to improve the material properties include protonation treatments with different acids: crosslinking with formaldehyde, glutaraldehyde, bis(ethenyl)sulfone, 1-chloro-2,3-epoxypropane and polyethylene imine [14, 53], thermal treatment [54], and chemical modifications of functional groups, among others [55]. It is important to highlight that the costs associated with the raw biomass process and the ultimate disposal of the derived biomass waste should be considered for a viability of installation of the process. One simple way to prevent swelling of seaweeds in packed columns is to mix seaweed biomass with an inert and stable materials such as sand. This will improve the porosity inside the column and thereby enhance the solute flow pattern [56]. The modification of the surfaces of these biosorbents for further removal of other contaminants, for an integral application on wastewater treatment, is rather challenging.

Another important issue to be considered when using biosorbents to treat metal polluted wastewaters is the complexity of the solution, because it can affect the biosorption process by competition for the exchange sites by cationic chemical species other than metals. Vijayaraghavan et al. [57] studied the nickel biosorption capacities on *Sargassum wightii* in aqueous solutions and residual electroplating solutions, finding that the complex nature of electroplating solutions negatively affected the metal biosorption performance, because of the competition with other ions. Patrón-Prado et al. [58] studied the biosorption of cadmium and copper by *Sargassum sinicola* in aqueous solutions and in saline wastewaters, finding that the salinity of the solution caused a reduction in Cd biosorption from 89 to 5.8%. At the same time, the authors found a clear antagonist effect between both metals in solution. The same antagonist effect was reported on other seaweed species [59, 60].

Also, it is well known that most of the seaweed sorbents have poor affinity for anions such as nitrate, sulfate, and phosphate, due to the predominant anionic sites in their surface. These anions are common in effluents and if not removed may lead to eutrophication and other undesirable effects on the environment [61]. Alginates, one of the major constituents of the seaweed biomass, can be chemically modified to remove anionic contaminants from water solutions and can be used to encapsulate materials such as magnetite, leading to the formation of a multifunctional sorbent that has magnetic properties and can remove both cationic heavy metal ions such as copper ions and anionic contaminants, like arsenic [62].

Alginates can be cross-linked by addition of alkaline metals as Ca or Mg [63], resulting in a encapsulation of raw biomass, improving their biosorption capacities, and generating a solid biosorbent with a better hydrodynamic performance. Mata et al. [15] determined the effect of the immobilization of *Fucus vesiculosus* with alginate xerogels in the biosorption of Cd, Cu, and Pb. The immobilization increased the kinetic uptake and intraparticle diffusion rates for the three metals. Song et al. [64] evaluated the recovery of Li, Sr, and La by Ca-alginate beads at different physicochemical conditions, finding the best performance for Sr and La.

The feasibility of the biosorption process to reduce toxic metals presents some limitations, and a fully understanding of the process in the context of a reactor system is necessary. Engineering considerations are crucial when a seaweed-based biosorption system is designed and developed. In general, biosorption systems using dry seaweed biomass correspond to a solid-liquid contact process. This process ideally implies several cycles of biosorption and desorption stages. The effluents to be treated make contact with the biomass in a batch, semi-continuous or continuous flow system. Banks [65] describes different types of reactors with potential use in biosorption system designs: the conventional stirred tank reactors, packed bed reactors, expanded bed reactors, fluidized bed reactors, and airlift reactors, depending on the final result and the type of effluent to be treated.

Despite the considerable progress in the understanding of seaweed biomass interactions with heavy metals made over four decades of continuous research, most seaweed biosorption processes are still at bench scale. Some proposed processes based on biosorption have been patented for commercial applications, some of them at pilot scale and some at commercial scale, mainly represented by units that were constructed in Canada and USA during the 1990s [66]. Thus, only a few industrial processes or products based on biosorption technology have been implemented, especially if we refer only to seaweed applications. A search in WIPO Patentscope (http://patentscope.wipo.int/search/en/search.jsf) only shows 29 patents related to biosorption and heavy metals removal, and only four consider seaweed biomass in the development. Pohl [67] patented a preparation of a biosorbent based on brown algae, consisting in a method to prepare the raw material for biosorption of heavy metals and hydrocarbons to finally obtain a milled dry seaweed biosorbent material. Other inventions include directly applications of seaweed biomass. For example, Volesky and Kuyucak developed a method for the biosorption of gold using seaweed biomass of *Sargassum genus* [68]. In the market, there is a very limited offer on biosorption technologies. Actually, only five products offer a commercial solution to remove pollutants from water. These commercial technologies include AMT-BIOCLAIM™ from Advanced Mineral Technologies, Inc. (AMT), a sorbent obtained from industrial fermentation process by *Bacillus subtilis* [69, 70]. Bio-Recovery System, Inc. (Las Cruces, USA) developed a microalgae biosorbent named as AlgaSORB™, which consists of *Chlorella vulgaris* biomass immobilized on silica gel polymer matrix, to treat heavy metal ions from diluted solutions (1–100 mg g^{-1}). Metagener and RAHCO Bio-Beads have developed two commercial biosorbents, which could be used as effective materials for removal of heavy metal ions from wastewaters (mainly electroplating and mining industry) [1, 71]. Particularly interesting is the invention BV-SORBEX™, by B.V. SORBEX, Inc. (Montreal, Canada), that offers a family of commercial sorbents based on metal-binding biomass, including different-sized powders and granules made of different types of biomass, including seaweeds [72]. These biosorbents remove a wide range of metal ions from diluted or concentrated solutions with high efficiency (over 99%) in a wide range of pH values and aqueous system conditions [73]. However, these products have not been commercially successful [74, 75]. One reason is related to the lack of a full understanding of the mechanisms, kinetics, and thermodynamics of the biosorption process [76]. Another reason is related to the existence of competing technologies based on physical and chemical treatments such as ion exchange, activated carbon, chemical precipitation, oxidation/reduction methods, electrocoagulation, electrodialysis, ultrafiltration, reverse osmosis, and solvent extraction [77]. However, two major arguments that support the biosorption as a convenient cleanup technology are the low cost of the biosorbents and the constant increase in the demands on environmental regulatory standards [61]. Paradoxically, processes to improve the biosorbent performance not only increase the final price of the product but also raise questions about the toxicity and environmental hazards of the final product. All these issues create a need for other commercially attractive applications of biosorption technology. The main suggestion is the use of biosorption for the recovery of precious metals [61, 75]. Future directions may also include a most integrated solution addressed to clean up industrial effluents containing multiple pollutants and the applications of hybrid technologies between traditional and biosorbent technology. Unfortunately, the increasingly published output on the biosorption field do not reflect an improved knowledge of the process, nor aid to commercial exploitation [78]. Biosorptive processes may contribute to primary or secondary biological water treatments including domestic, municipal and industrial wastes, and in some circumstances, solid wastes. Biosorption technology is still on a developing stage, and commercial success will depend on a better understanding of the process, governed by a pragmatic rationale of its commercial development and potential applications.

The final cost of biosorption treatment certainly involves the harvest, transportation, and processing of the biomass, together with the control of optimal conditions of the process, the regeneration of the biosorbent, and the final disposing of the biomass. Other costs usually not discussed are the capital expenses and plant operation costs, because they depend on the design of the treatment plant and the nature of wastewaters to be treated. Because many species of seaweed are valuable for the production of molecules with nutritional value, cosmetic applications among other uses, the use of residual dead biomass is convenient. Preparation of biosorbents is usually a major cost associated with the biosorption process, and biomass preprocessing is necessary to guarantee a good performance of the biosorbent. Thus, much attention should be taken with the estimation of costs, including the final disposal of the residues. Once a biosorbent life cycle ends, the ultimate disposal should be addressed. Landfilling the biomass, chemical or thermal destruction techniques seems an easy way to manage waste biosorbents, procedures that are not cheap or environmentally friendly. Used biosorbents can also be reused for other applications. Therefore, once heavy metal ions are completely removed from the used seaweeds by a demineralization process, they can be used for other applications.

http://dx.doi.org/10.5772/65682

Because the sorption technology based on biomasses has not been fully developed at industrial level, there are scarce data about the technical-economical evaluation of industrial biosorption applications. Some calculations predict that the prices of a biosorbent system represent about a tenth of the price of resins [79]. A comparison between classic methods for metal sorption and biosorption techniques can be summarized considering advantages and disadvantages of different technological approaches (**Table 3**). From the analysis, it can be concluded that biosorption is a clean technology, reducing the amount of solid wastes generated, gives the possibility to recover various metals, and has low operational cost in terms of energy consumption.

Metal adsorption technology	Advantages	Disadvantages	References
Adsorption	High spectrum of pollutants High capacity and fast kinetics Potentially selective	Performance depends on the type of adsorbent Needs chemical or thermal treatment for better performance	Aderhold et al. [80]
Biosorption	High spectrum of pollutants High capacity and fast kinetics Obtained directly from the nature or from industrial waste (low cost)	Poorly selective without chemical modifications	Davis et al. [2]
Chemical precipitation	High spectrum of metals Low cost of implementation Easy operation	High sludge formation implies cost on maintenance and sludge disposal	Rubio et al. [81]
Ion exchange	Metal selective High regeneration capacity	High initial capital and maintenance cost Limited pH tolerance	Rubio et al. [81]
Coagulation/flocculation	Capacity to reduce biological viability. Good dewatering performance	High cost on chemicals High costs on sludge management	Rubio et al. [81]
Electrochemical precipitation	Applicable for the treatment of heavily contaminated wastes Easy to operate.	High initial capital, high operation cost, and maintenance Need of power energy	Qin et al. [82]
Membrane filtration	Low solid waste generation Low chemical consumption Small space requirement Metal selective method	Temperature instability Low stability at low pH High initial capital, high cost on operation, and maintenance Easy fouling generation and flow stacks	Madaeni and Mansourpanah [83]

Table 3. Comparison of conventional technologies and biosorption for the treatment of metal contaminated waters.

The reuse of seaweed biomass after the remediation process to obtain other products with commercial value is an approach to make more attractive the biosorption process. The usability of the seaweed biomass after the biosorption has several productive destinies. Seaweed biomass residues traditionally are burned, disposed in landfills, or confined, but also can be used to obtain a final valuable product. A new niche of application of the passive biosorption process of metals is the enrichment of biomass with microelements to be used as biological feed supplements and/or fertilizers. Dietary supplements obtained by this way have already demonstrated good results on animals [84, 85]. Micronutrients, such as Cu(II), Mn(II), Zn(II), and Co(III), are usually targets for biosorption on wastewater remediation processes, and a biomass that incorporates these elements allows obtaining a microelements enriched biological material. Seaweed biomass is rich in many nutritional elements such as carbohydrates, proteins, microelements, and polyphenols, representing a natural fertilizer, but is not a full complement to the all requirements to amend poor soils. Different seaweed biomasses were experimented for microelemental enrichment via biosorption, obtaining material to be used as feed additives, to supplement livestock diet [86–88]. It was observed that for all the studied seaweed, the smaller the content of the microelement in the natural biomass, the higher the enrichment coefficient of the biomass reached. Certainly, the difficulty to integrate a wastewater bioremediation process to recover metals and the use of the biomass as a fertilizer or as a feed supplement is that the microelements present in the wastewater are at minimal quantities. Also, the possibility of contamination of the biomass with undesirable elements has to be avoided. Another interestingly approach is the potential use of biomass for biogas production. Seaweed biomass can be anaerobically digested for the production of methane. Nkemka and Murto [89] experimented with the demineralization of seaweed biomass prior fermentative processes, obtaining an efficient production of biogas and at the same time, a residue to be used as a fertilizer. A good demineralization process can produce a useful biomass for composting soils [90]. The same concept would be applied to obtain bio-oils from residual biomass. Diaz-Vazquez et al. [91] evaluated the demineralization of *Sargassum sp.* biomass by several methods before a hydrothermal liquefaction process (HTL) of the biomass to increase the yield of bio-oil production. Algae biomass has a higher energy content than most of the lignocellulosic biomass used for biofuel production [92]. In contrast to the high lignocellulosic contents of terrestrial plants, seaweeds are primarily composed of elastic polysaccharides such as alginic acid, laminarin, carrageenan, and agarose that make them a more suitable feedstock for thermochemical conversion processes. The HTL process converts complex polymers present in the biomass into simpler molecules that can be converted to bio-oils. One major limitation on the HTL process is the high ash content present on seaweed biomass, which reduces the yields an quality of the generated bio-oils, restricting their use in direct combustion or gasification processes [93, 94], so the biomass has to be pretreated to reduce the ash content. Usually, the pretreatment process is carried out with mineral acids like sulfuric acid, nitric acid, or organic acids as acetic acid and citric acid usually in a rate of 10% (w/w) between biomass and acid [56]. The selection of a specific acid for demineralization relies on the mechanisms of removal, the mechanical stability of the biomass, and the final use of the recycled biomass [56]. A proper demineralization process allows the incorporation of seaweed biomass previously used in biosorption processes into a new productive cycle (**Figure 3**). Once the seaweed biomass has been used for the treatment of metal polluted effluents, the treated effluent is recovered, and the seaweed biomass loaded with the metal can be demineralized in situ. By this procedure, a metal-enriched solution can be recovered, and the biomass can be harvested from the container and be submitted to a secondary treatment for further productive processes (**Figure 3**).

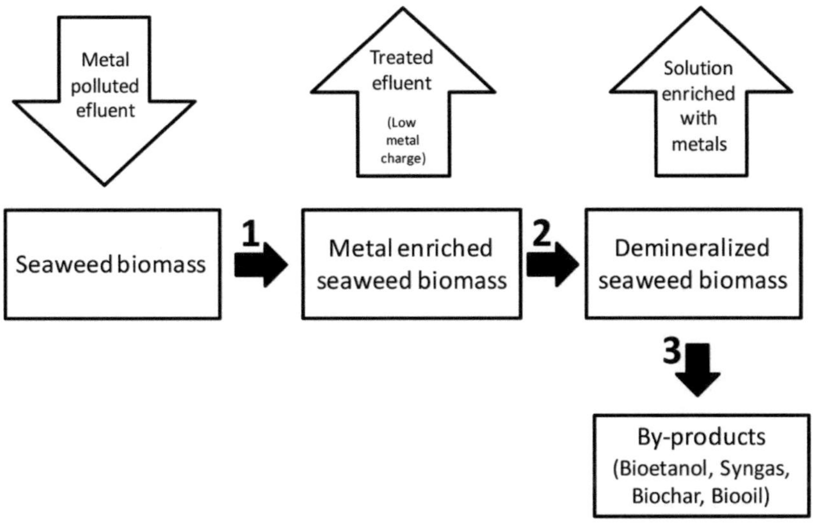

Figure 3. Simplified scheme for industrial use of seaweed biomass. (1) Biosorption process; (2) demineralization process; (3) secondary treatment such as pyrolysis, fermentation, HTL, gasification, among others.

5. Conclusions

A wide variety of biomasses has been evaluated for the sequestration of metal ions from solutions. An interesting approach is the use of the nonliving forms of the biomasses because they do not need nutrition for the maintenance and do not present the problem generated with the toxicity of the metals on living organisms. The biosorption capability of algae biomass is mostly related to the cell wall chemical composition, that is, a fiber-like structure, and an amorphous matrix embedding with polysaccharides such as alginates and fucoidans. In brown algae, alginates have a high affinity for divalent cations and sulfated polysaccharides give account of the capture of trivalent cations. Besides, the reuse of seaweed biomass after the remediation process in order to obtain products with commercial value is an approach that makes attractive the biosorption process at an industrial scale. The chemical composition of brown algae biomass makes it suitable for the production of different by-products such as biofuels, after the biomass has been demineralized. Nevertheless, much more efforts must be done in order to generate quantitative data regarding the performance and the operational costs for biosorption processes using dead seaweed biomass at an industrial level.

Author details

Claudia Ortiz-Calderon*, Héctor Cid Silva and Daniel Barros Vásquez

*Address all correspondence to: claudia.ortiz@usach.cl

Faculty of Chemistry and Biology, Department of Biology, University of Santiago of Chile, Santiago, Chile

References

[1] Atkinson BW, Bux F, Kasan HC. Considerations for application of biosorption technology to remediate metal-contaminated industrial effluents. Water SA. 1998;24(2):129–135.

[2] Davis TA, Volesky B, Mucci A. A review of the biochemistry of heavy metal biosorption by brown algae. Water Res. 2003;37(18):4311–4330.

[3] Raize O, Argaman Y, Yannai S. Mechanisms of biosorption of different heavy metals by brown marine macroalgae. Biotechnol Bioeng. 2004;87(4):451–458.

[4] Jalali R, Ghafourian H, Asef Y, Davarpanah SJ, Sepehr S. Removal and recovery of lead using nonliving biomass of marine algae. J Hazard Mater. 2002;92(3):253–262.

[5] Romera E, González F, Ballester A, Blázquez ML, Muñoz JA. Comparative study of biosorption of heavy metals using different types of algae. Bioresour Technol. 2007;98(17):3344–3353.

[6] Deniaud-Bouët E, Kervarec N, Michel G, Tonon T, Kloareg B, Hervé C. Chemical and enzymatic fractionation of cell walls from Fucales: insights into the structure of the extracellular matrix of brown algae. Ann Bot. 2014;114(6):1203–1216.

[7] Vieira RHSF, Volesky B. Biosorption: a solution to pollution? Int Microbiol. 2000;3(1):17–24.

[8] Murphy V, Hughes H, McLoughlin P. Cu(II) binding by dried biomass of red, green and brown macroalgae. Water Res. 2007;41(4):731–740.

[9] Murphy V, Hughes H, McLoughlin P. Comparative study of chromium biosorption by red, green and brown seaweed biomass. Chemosphere. 2008;70(6):1128–1134.

[10] Rindi F, Soler-Vila A, Guiry MD. Taxonomy of marine macroalgae used as sources of bioactive compounds. In: Hayes M, editor. Marine bioactive compounds: sources, characterization and applications [Internet]. Boston, MA: Springer US; 2012. p. 1–53. doi:10.1007/978-1-4614-1247-2_1

[11] Balboa EM, Conde E, Moure A, Falqué E, Domínguez H. *In vitro* antioxidant properties of crude extracts and compounds from brown algae. Food Chem [Internet]. 2013;138(2–3):1764–1785. doi:10.1016/j.foodchem.2012.11.026

[12] Cock JM, Sterck L, Rouzé P, Scornet D, Allen AE, Amoutzias G, et al. The Ectocarpus genome and the independent evolution of multicellularity in brown algae. Nature. 2010;465(7298):617–621.

[13] Chen JP. Decontamination of heavy metals: processes, mechanisms, and applications. Vol. 97, CRC Press/Taylor and Francis Group; 2012. III–IV.

[14] Holan ZR, Volesky B. Biosorption of lead and nickel by biomass of marine algae. Biotechnol Bioeng. 1994;43(11):1001–1009.

[15] Mata YN, Blázquez ML, Ballester A, González F, Muñoz JA. Biosorption of cadmium, lead and copper with calcium alginate xerogels and immobilized *Fucus vesiculosus*. J Hazard Mater. 2009;163(2–3):555–562.

[16] Rahman MS, Sathasivam KV. Heavy metal adsorption onto *kappaphycus* sp. from aqueous solutions: the use of error functions for validation of isotherm and kinetics models. Biomed Res Int [Internet]. 2015;2015:1–13. Available from: http://www.hindawi.com/journals/bmri/2015/126298/.

[17] Pavasant P, Apiratikul R, Sungkhum V. Biosorption of Cu^{2+}, Cd^{2+}, Pb^{2+} and Zn^{2+} using dried marine green macroalga *Caulerpa lentillifera*. Bioresour Technol. 2006;97:2321–2329.

[18] Sheng PX, Ting YP, Chen JP, Hong L. Sorption of lead, copper, cadmium, zinc, and nickel by marine algal biomass: characterization of biosorptive capacity and investigation of mechanisms. J Colloid Interface Sci. 2004;275(1):131–141.

[19] Guiry MD, Guiry G. AlgaeBase. World-wide electronic publication, National University of Ireland, Galway. [Internet]. 2016 [cited 2016 Jul 8]. Available from: http://www.algaebase.org.

[20] Hamdaoui O, Naffrechoux E. Modeling of adsorption isotherms of phenol and chlorophenols onto granular activated carbon. Part I. Two-parameter models and equations allowing determination of thermodynamic parameters. J Hazard Mater. 2007;147(1–2):381–394.

[21] Michel G, Tonon T, Scornet D, Cock JM, Kloareg B. The cell wall polysaccharide metabolism of the brown alga *Ectocarpus siliculosus*. Insights into the evolution of extracellular matrix polysaccharides in Eukaryotes. New Phytol. 2010;188(1):82–97.

[22] Holan ZR, Volesky B. Biosorption of heavy metals: review. Biotechnol Prog. 1995;11:235–250.

[23] Vierira R, Volesky B. Biosorption: a solution to pollution? Int Microbiol. 2000;3:17–24.

[24] Figueira MM, Volesky B, Ciminelli VST, Roddick FA. Biosorption of metals in brown seaweed biomass. Water Res. 2000;34(1):196–204.

[25] Rowbotham JS, Dyer PW, Greenwell HC, Selby D, Theodorou MK. Copper(II)-mediated thermolysis of alginates: a model kinetic study on the influence of metal ions in the thermochemical processing of macroalgae. Interface Focus. 2013;3:1–16.

[26] Pohl P, Schimmack W. Adsorption of radionuclides (134Cs, 85Sr, 226Ra, 241Am) by extracted biomasses of cyanobacteria (*Nostoc carneum, N. insulare, Oscillatoria geminata* and *Spirulina laxis-sima*) and Phaeophyceae (*Laminaria digitata* and *L. japonica*; waste products from alginat. J Appl Phycol. 2006;18(2):135–143.

[27] Mata YN, Torres E, Blázquez ML, Ballester A, González F, Muñoz JA. Gold(III) biosorption and bioreduction with the brown alga *Fucus vesiculosus*. J Hazard Mater. 2009;166(2–3):612–618.

[28] Mata YN, Blázquez ML, Ballester A, González F, Muñoz JA. Characterization of the biosorption of cadmium, lead and copper with the brown alga *Fucus vesiculosus*. J Hazard Mater. 2008;158(2–3):316–323.

[29] Michalak I, Chojnacka K. Interactions of metal cations with anionic groups on the cell wall of the macroalga *Vaucheria* sp. Eng Life Sci. 2010;10(3):209–217.

[30] Gupta VK, Rastogi A. Biosorption of lead from aqueous solutions by green algae *Spirogyra* species: kinetics and equilibrium studies. J Hazard Mater. 2008;152(1):407–414.

[31] Mehta SK, Gaur JP. Use of algae for removing heavy metal ions from wastewater: progress and prospects. Crit Rev Biotechnol. 2005;25(3):113–152.

[32] Romera E, González F, Ballester A, Blázquez ML, Muñoz JA. Biosorption with algae: a statistical review. Crit Rev Biotechnol. 2006;26(4):223–235.

[33] Yang F, Liu H, Qu J, Paul Chen J. Preparation and characterization of chitosan encapsulated *Sargassum* sp. biosorbent for nickel ions sorption. Bioresour Technol [Internet]. 2011;102(3):2821–2828. doi: 10.1016/j.biortech.2010.10.038

[34] Zheng Y-M, Liu T, Jiang J, Yang L, Fan Y, Wee ATS, et al. Characterization of hexavalent chromium interaction with Sargassum by X-ray absorption fine structure spectroscopy, X-ray photoelectron spectroscopy, and quantum chemistry calculation. J Colloid Interface Sci [Internet]. 2011;356(2):741–748. doi:10.1016/j.jcis.2010.12.070

[35] Fourest E, Volesky B. Contribution of sulfonate groups and alginate to heavy metal biosorption by the dry biomass of *Sargassum fluitans*. Environ Sci Technol. 1996;30(1):277–282.

[36] Cid H, Ortiz C, Pizarro J, Barros D, Castillo X, Giraldo L, et al. Characterization of copper (II) biosorption by brown algae *Durvillaea antarctica* dead biomass. Adsorption. 2015;21(8):645–658.

[37] Ahmady-Asbchin S, Andrès Y, Gérente C, Cloirec PL. Biosorption of Cu(II) from aqueous solution by *Fucus serratus*: surface characterization and sorption mechanisms. Bioresour Technol. 2008;99(14):6150–6155.

[38] Karthikeyan S, Balasubramanian R, Iyer CSP. Evaluation of the marine algae *Ulva fasciata* and *Sargassum* sp. for the biosorption of Cu(II) from aqueous solutions. Bioresour Technol. 2007;98(2):452–455.

[39] Davis TA, Volesky B, Vieira RHSF. Sargassum seaweed as biosorbent for heavy metals. Water Res. 2000;34(17):4270–7278.

[40] Kleinübing SJ, da Silva EA, da Silva MGC, Guibal E. Equilibrium of Cu(II) and Ni(II) biosorption by marine alga *Sargassum filipendula* in a dynamic system: competitiveness and selectivity. Bioresour Technol. 2011;102(7):4610–4617.

[41] López Errasquín E, Vázquez C. Tolerance and uptake of heavy metals by Trichoderma atroviride isolated from sludge. Chemosphere. 2003;50(1):137–143.

[42] Bishnoi NR, Kumar R, Kumar S, Rani S. Biosorption of Cr(III) from aqueous solution using algal biomass Spirogyra spp. J Hazard Mater. 2007;145(1–2):142–147.

[43] Ebrahimi B, Shojaosadati SA, Ranaie SO, Mousavi SM. Optimization and evaluation of acetylcholine esterase immobilization on ceramic packing using response surface methodology. Process Biochem. 2010;45(1):81–87.

[44] Fosso-Kankeu E, Mulaba-Bafubiandi AF. Review of challenges in the escalation of metal-biosorbing processes for wastewater treatment: applied and commercialized technologies. Afr J Biotechnol. 2014;13(17):1756–1771.

[45] Schiewer S, Volesky B. Ionic strength and electrostatic effects in biosorption of protons. Environ Sci Technol. 1997;31(7):1863–1871.

[46] Chu KH, Hashim MA. Copper biosorption on immobilized seaweed biomass: column breakthrough characteristics. J Environ Sci. 2007;19(8):928–932.

[47] Omar HA, Abdel-Razek AS, Sayed MS. Biosorption of Cesium-134 from aqueous solutions using immobilized marine algae: equilibrium and kinetics. Nat Sci. 2010;8(11):140–147.

[48] Holan ZR, Volesky B, Prasetyo I. Biosorption of cadmium by biomass of marine algae. Biotechnol Bioeng. 1993;41(8):819–825.

[49] Aravindhan R, Madhan B, Rao JR, Nair BU. Recovery and reuse of chromium from tannery wastewaters using *Turbinaria ornata* seaweed. J Chem Technol Biotechnol. 2004;79(11):1251–1258.

[50] Cossich ES, Da Silva EA, Tavares CRG, Filho LC, Ravagnani TMK. Biosorption of chromium(III) by biomass of seaweed *Sargassum* sp. in a fixed-bed column. Adsorption. 2004;10(2):129–138.

[51] Borba CE, Guirardello R, Silva EA, Veit MT, Tavares CRG. Removal of nickel(II) ions from aqueous solution by biosorption in a fixed bed column: experimental and theoretical breakthrough curves. Biochem Eng J. 2006;30(2):184–191.

[52] Tsui MTK, Cheung KC, Tam NFY, Wong MH. A comparative study on metal sorption by brown seaweed. Chemosphere. 2006;65(1):51–57.

[53] Leusch A, Holan ZR, Volesky B. Biosorption of heavy metals (Cd, Cu, Ni, Pb, Zn) by chemically-reinforced biomass of marine algae. J Chem Technol Biotechnol. 1995;62:279–288.

[54] Park D, Yun Y-S, Young Cho H, Park JM. Chromium biosorption by thermally treated biomass of the brown seaweed, *Ecklonia* sp. Ind Eng Chem Res. 2004;43(26):8226–8232.

[55] Park D, Yun YS, Jong MP. Studies on hexavalent chromium biosorption by chemically-treated biomass of *Ecklonia* sp. Chemosphere. 2005;60(10):1356–1364.

[56] Vijayaraghavan K, Balasubramanian R. Is biosorption suitable for decontamination of metal-bearing wastewaters? A critical review on the state-of-the-art of biosorption processes and future directions. J Environ Manage. 2015;160:283–296.

[57] Vijayaraghavan K, Palanivelu K, Velan M. Treatment of nickel containing electroplating effluents with *Sargassum wightii* biomass. Process Biochem. 2006;41(4):853–859.

[58] Patrón-Prado M, Acosta-Vargas B, Serviere-Zaragoza E, Méndez-Rodríguez LC. Copper and cadmium biosorption by dried seaweed *Sargassum sinicola* in saline wastewater. Water Air Soil Pollut. 2010;210(1–4):197–202.

[59] Zhou JL, Huang PL, Lin RG. Sorption and desorption of Cu and Cd by macroalgae and microalgae. Environ Pollut. 1998;101(1):67–75.

[60] Sheng PX, Ting YP, Chen JP. Biosorption of heavy metal ions (Pb, Cu, and Cd) from aqueous solutions by the Marine Alga *Sargassum* sp. in single- and multiple-metal systems. Industrial and Engineering Chemistry Research. 2007;46(8):2438–2444.

[61] Volesky B. Biosorption and me. Water Res. 2007;41(18):4017–4029.

[62] Lim SF, Zheng YM, Zou SW, Chen JP. Characterization of copper adsorption onto an alginate encapsulated magnetic sorbent by a combined FT-IR, XPS, and mathematical modeling study. Environ Sci Technol. 2008;42(7):2551–2556.

[63] He J, Chen JP. A comprehensive review on biosorption of heavy metals by algal biomass: materials, performances, chemistry, and modeling simulation tools. Bioresour Technol. 2014;160:67–78.

[64] Song D, Park S, Kang HW, Park S Bin, Han J. Recovery of Lithium (I), Strontium (II), and Lanthanum (III) using Ca-alginate beads. J Chem Eng. 2013;58:2455–2464.

[65] Banks C. Scavenging trace concentrations of metals. In: Wase J, Forster C, editors. Biosorbents for metal ions. London, United Kingdom: Taylor & Francis Ltd; 1997. p. 115–140.

[66] Tsezos M. Biosorption of metals. The experience accumulated and the outlook for technology development. Hydrometallurgy. 2001;59(2–3):241–243.

[67] Pohl P. Method of production of adsorption material. US patent number 5648313. 1997.

[68] Volesky B, Kuyucak N. Biosorbent for gold. US patent number 4769223. 1988.

[69] Brierley JA, Brieley CL, Goyak GM. AMT-BIOCLAIM: A new wastewater treatment and metal recovery technology. In: Lawrence RW, Branion RMR, Ebner HG, editors. Fundamental and applied biohydrometallurgy. Boca Raton: CRC Press; 1986. p. 291–308.

[70] Brielly J. Production and application of a Bacillus-based product for use in metals biosorption. In: Volesky B, editor. Biosorption of heavy metals. Boca Raton: CRC Press; 1990. p. 305–311.

[71] Chojnacka K. Biosorption and bioaccumulation—the prospects for practical applications. Environ Int [Internet]. 2010;36(3):299–307. doi: 10.1016/j.envint.2009.12.001

[72] Volesky B. Biosorption of heavy metals. In: Volesky B, editor. CRC Press; 1990. 3–5 p.

[73] Bv-sorbex. BV Sorbex Technology Description www.bvsorbex.net/sxProcess.pdf (revised on July 2016).

[74] Vijayaraghavan K, Won SW, Mao J, Yun YS. Chemical modification of *Corynebacterium glutamicum* to improve methylene blue biosorption. Chem Eng J. 2008;145(1):1–6.

[75] Park D, Yun Y-S, Park JM. The past, present, and future trends of biosorption. Biotechnol Bioprocess Eng. 2010;15(1):86–102.

[76] BV SORBEX Inc. BV SORBEX Inc. The company. http://www.bvsorbex.net/sx.htm (revised on July 2016).

[77] Fomina M, Gadd GM. Biosorption: Current perspectives on concept, definition and application. Bioresour Technol [Internet]. 2014;160:3–14. doi:10.1016/j.biortech.2013.12.102

[78] Gadd GM. Biosorption: Critical review of scientific rationale, environmental importance and significance for pollution treatment. J Chem Technol Biotechnol. 2009;84(1):13–28.

[79] Volesky B. Detoxification of metal-bearing effluents: biosorption for the next century. Hydrometallurgy. 2001;59(2–3):203–216.

[80] Aderhold D, Williams CJ, Edyvean RGJ. The removal of heavy-metal ions by seaweeds and their derivatives. Bioresour Technol. 1996;58(1):1–6.

[81] Rubio J, Souza ML, Smith RW. Overview of flotation as a wastewater treatment technique. Miner Eng. 2002;15(3):139 155.

[82] Qin JJ, Wai MN, Oo MH, Wong FS. A feasibility study on the treatment and recycling of a wastewater from metal plating. J Memb Sci. 2002;208(1–2):213–221.

[83] Madaeni SS, Mansourpanah Y. COD removal from concentrated wastewater using membranes. Filtr Sep [Internet]. 2003;40(6):40–46.

[84] Michalak I, Chojnacka K, Witek-Krowiak A. State of the art for the biosorption process—a review. Appl Biochem Biotechnol. 2013;170(6):1389–1416.

[85] Saeid A, Chojnacka K, Korczyński M, Korniewicz D, Dobrzański Z. Biomass of Spirulina maxima enriched by biosorption process as a new feed supplement for swine. J Appl Phycol. 2013;25(2):667–675.

[86] Michalak I, Chojnacka K. Edible macroalga Ulva prolifera as microelemental feed supplement for livestock: the fundamental assumptions of the production method. World J Microbiol Biotechnol. 2009;25(6):997–1005.

[87] Michalak I, Chojnacka K. The application of macroalga *Pithophora varia* Wille enriched with microelements by biosorption as biological feed supplement for livestock. J Sci Food Agric. 2008;88(7):1178–1186.

[88] Michalak I, Chojnacka K. The new application of biosorption properties of *Enteromorpha prolifera*. Appl Biochem Biotechnol. 2010;160(5):1540–1556.

[89] Nkemka VN, Murto M. Evaluation of biogas production from seaweed in batch tests and in UASB reactors combined with the removal of heavy metals. J Environ Manage. 2010;91(7):1573–1579.

[90] Eyras MC, Defossé GE, Dellatorre FG. Seaweed compost as an amendment for horticultural soils in patagonia, Argentina. Compost Sci Util. 2008;16(2):119–124.

[91] Diaz-Vazquez LM, Rojas-Perez A, Fuentes-Caraballo M, Robles IV, Jena U, Das KC. Demineralization of *Sargassum* spp. macroalgae biomass: selective hydrothermal liquefaction process for bio-oil production. Front Energy Res. 2015;3(February):1–11.

[92] Ross AB, Jones JM, Kubacki ML, Bridgeman T. Classification of macroalgae as fuel and its thermochemical behaviour. Bioresour Technol. 2008;99(14):6494–6504.

[93] Bach QV, Sillero MV, Tran KQ, Skjermo J. Fast hydrothermal liquefaction of a Norwegian macro-alga: screening tests. Algal Res. 2014;6(PB):271–276.

[94] Neveux N, Yuen AKL, Jazrawi C, Magnusson M, Haynes BS, Masters AF, et al. Biocrude yield and productivity from the hydrothermal liquefaction of marine and freshwater green macroalgae. Bioresour Technol. 2014;155:334–341.

Chapter 17

Valorisation of Lignocellulosic Biomass Wastes for the Removal of Metal Ions from Aqueous Streams: A Review

Carlos Escudero-Oñate, Núria Fiol, Jordi Poch and
Isabel Villaescusa

Additional information is available at the end of the chapter

http://dx.doi.org/10.5772/65958

Abstract

Heavy metal pollution derived from anthropogenic activities is a relevant environmental threat nowadays due to their toxic nature, persistence and accumulation potential in the food chain. A wide variety of lignocellulosic-based biomaterials have been thoroughly assessed by the scientific community as sorbents for the removal of metals from aqueous streams. This kind of biomaterials, mainly constituted by lignin and cellulose, bear functional groups such as alcohol, ketone and carboxylates that provide active sorption points for the effective removal of heavy metals. The role of lignin in the sorption process is especially relevant, since this substance provides polyhydroxy and polyphenol functional groups—especially effective in the coordination of metals—and that provide ion exchange functionality to the material. Depending on their nature, these materials can be used either in their raw form or chemically modified form so as to enhance their sorption capacity and/or to achieve improved mechanical and mass transfer properties.

Keywords: lignocellulosic wastes, immobilisation, heavy metals, sorption, desorption, kinetics, equilibrium, modelling, recovery

1. Introduction

1.1. Heavy metal pollution

Heavy metals occur as natural constituents of the earth crust and some of them considered as persistent environmental pollutants. They may vary in oxidation state and hydrochemical speciation, but cannot be degraded or destroyed. The problems related to heavy metal pollution are transversal; water, air and soil components are susceptible of being severely polluted [1].

To a large degree, industrial activities are mainly responsible for environmental discharges and pollution. There are many industrial sources of pollution including manufacturing processes such as smelting and refining, electricity generation and nuclear power, agricultural fertilisation, wastewater treatment, fuel combustion, production of batteries, alloys manufacturing, electroplating, waste incineration, ceramics production and glass colouring [2].

Human exposure and intake of hazardous levels of heavy metals may occur through food, air and water [1]. It should be highlighted as well that some metals such as barium, cadmium, chromium, mercury, lead and hazardous metalloids such as arsenic can be bioaccumulated in different organs and may exert their toxic effects after long time of periodic ingestion of even low levels.

1.2. Structure of lignocellulosic biomass

Lignocellulosic biomass has been pointed out as a valuable source of chemicals and materials for different applications and a very relevant actor in the design of strategies to reduce the social and economic reliance on fossil resources. These materials are naturally produced from incorporation of CO_2 and water (driven by solar power) through the photosynthesis process. In this context, lignocellulosic biomass is the most abundant and biorenewable biomass on earth [3].

The major components of these materials are cellulose, hemicellulose and lignin, and cellulose and hemicellulose are considered to be polysaccharides. The structure of cellulose is based on the molecular formula $(C_6H_{10}O_5)_n$ and is one of the most relevant polysaccharides occurring in the plant cell wall. Hundreds of glucose units are linked through glucosidic linkage (**Figure 1a**) and individual chains usually interact with one another through hydrogen bonds. This material has found an extensive use in the paper industry. Hemicellulose is another important polysaccharide usually contained in lignocellulosic biomass with a more intricate structure and linkages than cellulose. Hemicellulose is composed of heteropolymers and may contain xyloglucan, xylan, glucomannans and galactoglucomannans in variable ratios depending on the type of biomass.

When it comes to heavy metal sorption by raw, unmodified lignocellulosic biomass, lignin has been found to play a key role. Lignin is a three-dimensional structure made of phenolic polymers that consists of three types of phenylpropanoid units: *p*-coumaryl alcohol, coniferyl alcohol and sinapyl alcohol (**Figure 1b**). A short section of a lignin polymer is presented in **Figure 1(c)**. This substance acts as cellular glue, joining the individual fibres and conferring strength to the plant tissue. The richness in electron-donor active sites that are provided in the polyhydroxy and polyphenol functional groups of lignin offers a unique frame for the interaction and binding of cationic heavy metals [6].

http://dx.doi.org/10.5772/65958

(a)

(b)

(c)

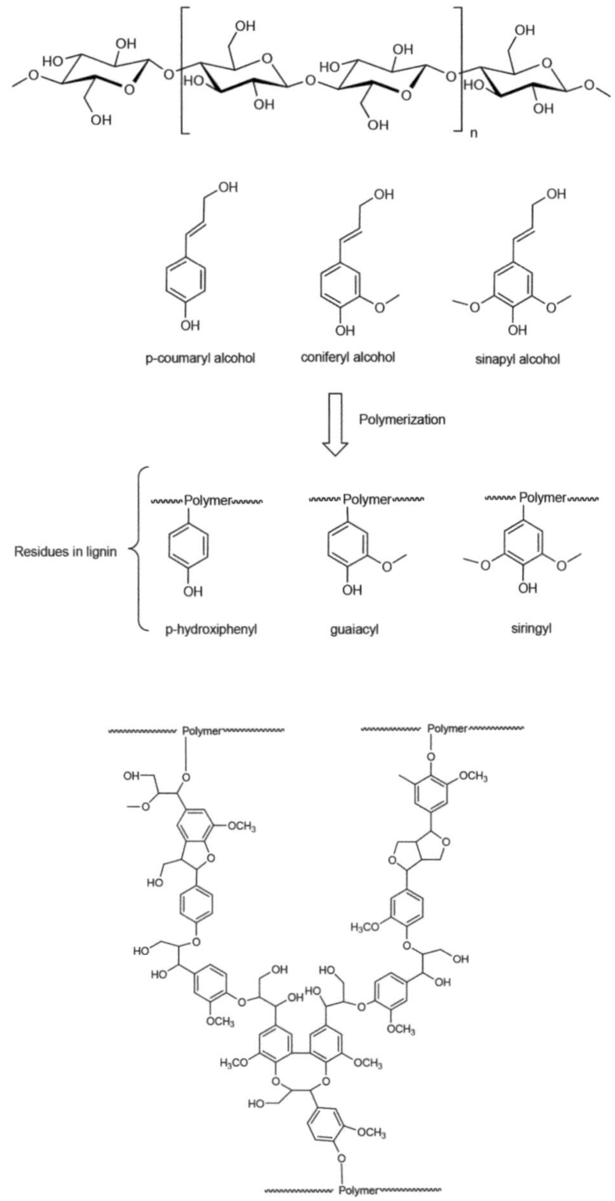

Figure 1. Structure of (a) cellulose, (b) main phenylpropane units and their related residues found in lignin and (c) structure of a lignin template [4, 5].

Abdolali et al. reported the chemical composition of some common lignocellulosic materials studied in metal sorption processes [6]. In general terms, lignocellulosic sorbents are composed of cellulose (30–35%), hemicellulose (20–40%), lignin (15–25%) and small amounts of water, ash, cyclic hydrocarbons and extractives. The chemical compositions in lignocellulosic materials are in different percentages depending on plant and also on the part of plant: leaves contain lower percentage of cellulose (15–25%) and lignin (5–10%) but higher percentage of hemicellulose (70–80%), while in stones and nuts, the lignin content is higher than cellulose and hemicellulose (30–40%). Pujol et al. reported a composition of about 25% total lignin and 23% polysaccharides in exhausted coffee wastes [7]. They also reported a relative composition depending on the particle size when characterising the different components of grape stalks (GS) [8].

The main constituents of lignocellulosic materials contain a variety of functional groups that play an important role in metal sorption. It was reported that acetamido groups, carbonyl, phenolic, structural polysaccharides, amino, amido, sulphydryl, carboxyl groups, alcohol and esters, present in lignocellulosic materials have affinity for metal complexation [9]. Ion exchange between positive cations and sodium, potassium, calcium and magnesium present in the adsorbent was identified as an important mechanism in several studies. These two mechanisms, together with chelation, are the main mechanisms known for metal sorption in lignocellulosic sorbents [9].

2. Lignocellulosic-based materials for the removal of heavy metals from aqueous effluents

Lignocellulosic materials from agricultural or plant wastes have been widely studied due to their renewable nature, large production and great local availability. A review of the recent literature reveals that hundreds of lignocellulosic materials around the world were tested to be used as low-cost sorbents for heavy metals and most of them are considered efficient and promising sorbents. The only negative aspects related to the use of raw lignocellulosic materials are connected to the lower sorption capacity compared to activated carbons and commercial ion exchange resins, as well as the potential release of organic matter that might cause a secondary pollution in the water treated.

Recently, Malik et al. reviewed low-cost adsorbents from different plant parts: husks, shells, straws, stems and woods, leaves, barks, grasses, stalks, seeds and hulls, fibres, fruit peels and pulps, bagasse and other lignocellulosic-based materials such as corn cob, oaks or fruit stones to be used as biosorbents [10]. The particular physicochemical properties of each material determine its properties as a sorbent.

Metal sorption by these heterogeneous materials is a complex process affected not only by the main mechanism, but also by other secondary mechanisms and combinations of different phenomena including chemisorption, adsorption on surface pores, adsorption by physical forces, entrapment in inter- and intra-fibrillar capillaries and spaces of the structural polysaccharides network, diffusion through cell walls and membranes, surface precipitation and metal reduction [9, 11]. The main mechanisms involved in the interaction between a specific sorbent material and metals could be predicted and verified by combination of different spectroscopic techniques and conventional techniques such as titration, chemical blocking of functional groups and related release of cations from sorbent [6].

The efficiency of each sorbent extremely depends on the operating conditions such as temperature, sorbent particle size, pH, contact time or initial metal concentration. For this reason, a thorough study to find out the optimal sorption conditions is required for each couple sorbent metal. Hundreds of these studies can be found in the literature, where the optimal conditions for a specific pair of lignocellulosic sorbent metal were reported. The main results in this field can be found in reviews published in prestigious journals, helping the researchers to update the information about potential biosorbents suitable for a given application [6, 9, 11–14]. In these reviews, the performance of different types of lignocellulosic sorbents is usually expressed as maximum sorption capacity (q_{max}) and the efficiency comparison between materials used to be based in these values. Nevertheless, additional information is needed to compare sorption efficiency between sorbents for a specific metal due to differences in particle sorbent sizes, sorbent dosages and other experimental conditions used in each work.

A large share of current research on metal sorption by lignocellulosic materials has focused on the removal of heavy metal cations, such as Pb(II), Cd(II), Zn(II) and Cu(II) [4], but other toxic metals such as Ni(II), Al(III), Hg(II), Fe(II), chromium and metalloids (arsenic, selenium, molybdenum, and vanadium) have been also studied as target pollutant in biosorption processes [15].

Nowadays however, there is a growing interest in biosorption processes that use low-cost materials as sorbents for the recovery of rare earths or precious metals [16, 17]. The recovery of valuable metals by biosorption opens a challenging and exciting new scenario in sorption studies far beyond the regular water treatment, since the recovered material is expected to have an intrinsic economic value.

Despite metal sorption using biomass has been regarded as an environmental-friendly technique and provides a set of potential advantages against traditional technologies; most of the studies have been focused on synthetic solutions. To the best of our knowledge, studies reporting the use of biosorption-based technologies for metal removal from real scenarios of contaminated wastewaters are scarce. Few studies have so far explored the use of lignocellulosic materials in the treatment of real polluted wastewater. Fruit shell of gulmohar and olive stones were investigated for the removal of Cr(VI) from an electroplating wastewater [18, 19], and rice agro-wastes, coconut shell, neem leaves and hyacinth roots were used for the removal of Pb(II) from wastewater in a battery industry [20], and recently Liu et al. [21] published a complete method for chromium electroplating wastewater treatment based on biosorption using exhausted coffee.

3. Structural characterisation of lignocellulosic wastes and their interactions with heavy metals

3.1. Infrared spectroscopy

Infrared spectroscopy (IR) is a technique based on the vibrations of the atoms on a given molecule and is a result of the molecular vibration mechanism, which refers to energy-matter interaction [22]. Usually, an infrared spectrum is obtained by passing infrared radiation through a sample and quantifying the fraction of incident radiation of each frequency that has been absorbed. The energy at which a peak appears in an absorption (or transmission) spectra corresponds to the characteristic energy of the vibration in a part of the molecule. The selection rule for IR spectroscopy is that an electric dipole moment in the molecule has to change during the vibration.

Techniques based on IR spectroscopy have become a powerful tool for determining the functional groups and the mechanisms involved in the removal of heavy metals by different type of biomass. A magnification in the region 800–1800 cm^{-1} of a raw lignocellulosic material (grape stalk wastes) and exposed to either Cr(III) or Cr(VI) solutions are presented in **Figure 2**. In the spectra, the positions of the main bands that have suffered modifications in their frequencies have been indicated with arrows.

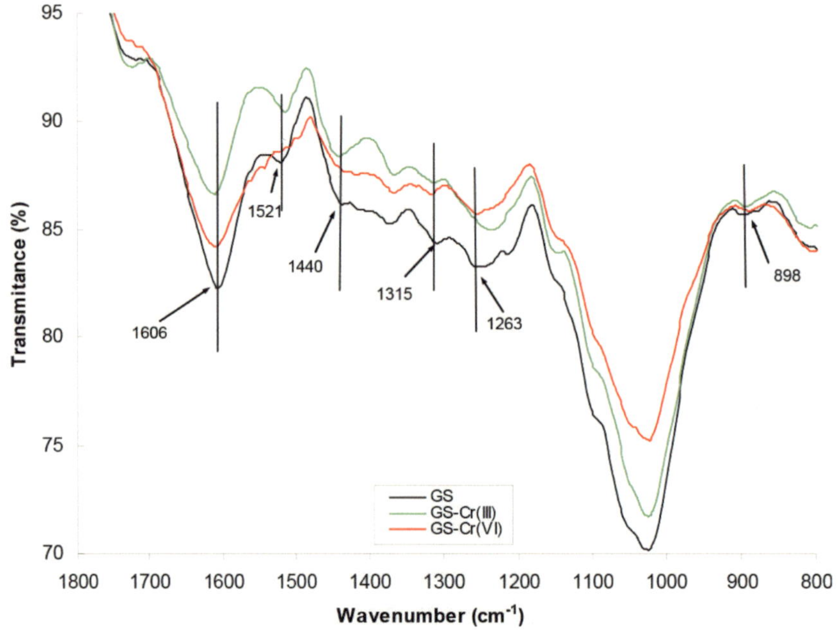

Figure 2. Grape stalks (GS) FTIR-ATR spectra before and after treatment with Cr(III) and Cr(VI) solutions. Initial metal concentration: 800 mg·L^{-1}. Agitation time: 24 h, pH$_0$ = 3.

The initial and final positions of the modified bands have been summarised in **Table 1**, including the assignment of the band to the functional group/s involved in Cr(III) and Cr(VI) sorption.

Grape stalks treated with Cr(III) and Cr(VI) solutions show modifications in the characteristic bands of syringyl and guaiacyl moieties (1263 and 1315 cm^{-1}, respectively), the methoxy deformation (1440 cm^{-1}), the aromatic skeleton vibration (1521 cm^{-1}) and the aromatic ring vibration (1606 cm^{-1}). All these bands, characteristic of the lignin macromolecule, indicate that both Cr(III) and Cr(VI) are adsorbed onto this component of the sorbent, making the cellulose almost unaltered during the sorption process. The important role of lignin in sorption of copper and nickel onto the cork had been previously noticed in results of ^{13}CP-MAS-NMR on the solid phase of the sorbent [25].

Frequency (cm^{-1})				
GS	GS-Cr(III)	GS-Cr(VI)	Assignment	Ref.
898	897	898	Carbohydrates (unaltered)	[23]
1263	1238	1247	Guayacyl/C–O phenolic	[24]
1315	1319	1321	Syringyl	[23]
1440	1444	1421	Methoxy deformation	[23]
1521	1515	1511	Aromatic skeleton vibration	[23]
1606	1614	1612	C=C Aromatic vibration	[23]

Table 1. Observed frequencies and assignment.

3.2. Scanning electron microscopy-energy dispersive X-ray analysis

Scanning electron microscopy (SEM) coupled to energy dispersive X-ray analysis (EDX) is a powerful instrumental combination to assess the mechanisms governing heavy metal removal by lignocellulosic biomass. SEM provides a great magnification of the surfaces and allows gathering high-quality images, from which the morphology and topography of the materials can be assessed. The additional use of detection of backscattered electrons (BSE) helps finding target regions where metals may have been selectively accumulated in lignocellulosic biomaterials. EDX is widely used coupled to SEM and allows obtaining local elemental analysis of the surfaces under the observation. The technique becomes especially relevant in the assessment of sorption processes where local microprecipitation is likely to occur.

Metal microprecipitation takes place when the solubility of the sorbate reaches its limit. Microprecipitation in metal-removal processes—despite not being directly related to sorption—positively contributes to the overall detoxification of the effluent, since the metal microprecipitate remains immobilised in the surface and remains separated from the solution.

This may occur even due to local conditions, e.g. on or inside the sorbent, and not necessarily in the bulk of the solution. These favourable conditions for microprecipitation may be created by local deviations in physical conditions such as pH or by the presence of materials released from the sorbent itself. Escudero et al. [26] using SEM-BSE-EDX techniques observed microprecipitation phenomena when exploring the removal of hexavalent chromium by grape stalk wastes (GS) entrapped into calcium alginate gel beads. Micrographs in the SEM mode (a, d), BSE mode (b) and characteristic EDX spectra (c) are presented in **Figure 3**.

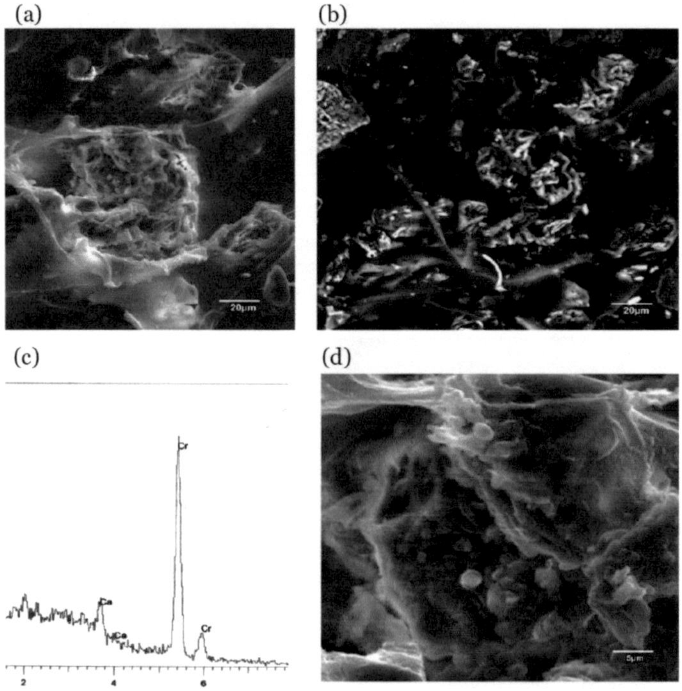

Figure 3. Electronic micrograph of grape stalks entrapped into a calcium alginate gel matrix exposed to a Cr(VI) solution. (a) Scanning mode, (b) backscattered electrons mode, (c) EDX local analysis and (d) focus on the grape stalk region in the SEM mode [26].

The observations indicated that chromium is mostly accumulated in the grape stalk surface, as it may be evidenced by the bright colour observed in the BSE picture (b). While magnifying the region, small nodules appear onto the surface of the grape stalks, indicating the formation of microprecipitates (d). The evidence of chromium accumulation in these nodules was further demonstrated through the EDX local analysis (c). Using further electron spin resonance analysis of the solids, the authors evidenced the formation of Cr(III). It was hypothesised then that Cr(VI) undergoes a reduction process that partially converts hexavalent into trivalent chromium with an associated large proton consumption (**Scheme 1a**). In this local alkalinised environment, Cr(III) precipitates onto the surface of the lignocellulosic material as hydroxide through reaction 1 (b) and remains isolated by the bulk of the solution by the surrounding calcium alginate gel.

(a) $$HCrO_4^- + 7H^+ + 3e^- \xrightarrow{\text{Raw GS} \rightarrow \text{Oxidized GS}} Cr^{3+} + 4H_2O$$

(b) $$Cr^{3+} + 3OH^- \longrightarrow Cr(OH)_3 \ (\downarrow)$$

Scheme 1. Hexavalent chromium removal through a combined reduction-precipitation process promoted by grape stalks [26].

4. Immobilisation techniques: the use of hydrogels as high-permeable immobilising materials

Native biomass shows low density and poor mechanical strength and rigidity [27]. These properties make its application difficult in biosorption processes either in batch or in column. In the batch mode, mechanical stirring and shear forces can break fragile particles of the biosorbent resulting in particle attrition (breakage, fragmentation, fines formation) and then, a continuous change on the original particle size distribution. Particle size is a key parameter in the performance of sorption processes; it is desirable that size distribution remains homogeneous throughout the entire process. It is known that, to maximise mass transfer and sorption efficiency, the sorbent particle size of the sorbent must be as small as possible (maximising surface-to-mass ratio). Limitations in the use of small particle sizes are (i) sorbent should be easily handled and (ii) clogging in filters, columns, valves and pipes must be prevented.

In the case of continuous bed up-flow sorption processes (columns), additional problems due to the native biomass physical properties must be considered. Biomass needs to be wetted to allow free swelling of the particles prior to column filling and to remove the finest particles by natural flotation. In addition to this, the non-uniform shape of the native biomass hinders the estimation of a form factor (sphere, cylinder, ellipsoid) of the particles. This parameter is needed when formulating a model to describe sorption processes. Most of the aforementioned drawbacks and limitations can be overcome, however, by immobilisation of the biomass in a water-permeable polymeric matrix. Immobilisation of biomass provides several advantages over native biomass: (i) increase of mechanical resistance, (ii) increase of density, (iii) possibility of enhancement of effective surface area due to the use of biomass powder, (iv) easy handling and (v) achievement of precision in form factor, since immobilisation can yield quasi-spherical beads or granules.

Biomass immobilisation consists of the attachment or entrapment of biomass on a support. Immobilisation can be carried out by four different techniques: adsorption, covalent binding, entrapment and membrane confinement. Entrapment in a gel matrix or gel encapsulation is among the most widely studied methods for immobilisation of enzymes, microbial biomass and animal and plant cells, being calcium alginate one of the regularly used matrixes [28]. The support selection is of crucial importance and must be chosen according to the target application of the immobilised material. Applications of immobilisation techniques are found in different scopes of several fields such as biotechnology and pharmaceutical, environment, food and biosensor industries [29].

For the treatment of wastewater, support materials need to meet (among others) the following criteria: insoluble, non-biodegradable, non-toxic, high mechanical and chemical stability, high diffusivity and ease of immobilisation procedure [30]. Natural polymer derivatives of algal polysaccharides (alginate, carrageenan, agar and agarose) and chitosan (an amino polysaccharide derived from chitin) have been experimentally used. Crini [31] has recently reviewed the most important features of polysaccharide-based materials used as adsorbents.

One of the most used immobilisation supports is alginate [32]. Alginate is a common term used for natural polymers composed of linear, unbranched chains of varying lengths, proportions and sequences of 1,4 linked residues of β-D-mannuronic acid and α-L-guluronic acid residues [33]. As the composition of alginates depends on the source from which they are extracted, they can exhibit different physical and chemical properties. The main source of alginates are different species of brown seaweeds *Laminaria digitata* [34] and *Macrocystis pyrifera* [35]. One of the properties of alginates is that they can cross-link with divalent metal ions such as calcium, barium, copper, zinc and lead [36, 37] to form M^{2+}-alginate hydrogels. Alginate gelation takes place when divalent metals interact with blocks of acid residues resulting in the formation of a 3D network, which is usually described by an 'egg-box' model [38]. Calcium alginate is one of the most utilised entrapment matrixes in environmental studies. Heavy metals adsorption takes place via ion exchange between Ca^{2+} ions from the hydrogel and metal ions. Pandey et al. [39] studied Cr(VI, Pb(II) and Cu(II) by using calcium alginate beads and Jodra and Mijangos [40] determined the ion exchange selectivity of calcium alginate gels for heavy metals and found the selectivity order: Pb > Cu > Cd > Ni > Zn > Co. During the last years different types of biomass used for heavy metals sorption have been entrapped in calcium alginate with the aim of enhancing their sorption performance. Studies on fungi [41, 42], algae [43, 44], bacteria [45] and yeast [46] encapsulated in calcium alginate have been recently carried out. Nevertheless, studies on metal sorption by using vegetable wastes and specifically lignocellulosic materials encapsulated in calcium alginate are scarce. The most probable reason is the high capacity of calcium alginate itself to sorb heavy metal ions. The only example of vegetable waste can be found in the work of Ansari et al. [47]. The authors encapsulated rose waste biomass in calcium alginate beads to sorb Pb(II) and found that the immobilised biomass yielded to higher metal ion sorption. Unfortunately, the authors did not report the amount of lead sorbed by pure calcium alginate beads.

Grape stalks, a lignocellulosic waste produced in wine production, was entrapped in calcium alginate and used for Cr(VI) removal at pH 3 in batch mode [48–50] and in column [51] from single metal solutions and from binary mixtures containing Cr(VI) and Cr(III) [52]. A scheme of a typical arrangement to obtain gel beads by the dropping technique is presented in **Figure 4**.

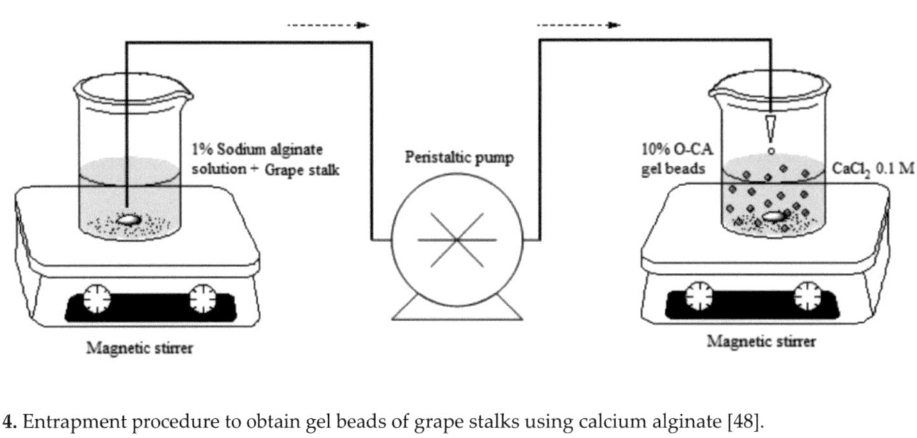

Figure 4. Entrapment procedure to obtain gel beads of grape stalks using calcium alginate [48].

Grape stalks were reported to be a very effective reducing agent to reduce Cr(VI) to Cr(III) [53]. Results of the above-mentioned studies with grape stalk demonstrated that both sorption and reduction processes were enhanced when the sorbent was immobilised into calcium alginate. Furthermore, the Cr(III) formed as a consequence of the reduction reaction was sorbed onto calcium alginate beads via ion exchange with calcium ions. It must be remarked that at pH 3 major chemical species of Cr(VI) is $HCrO_4^-$ that is hardly sorbed by calcium alginate beads. Therefore, the joint action of both sorbents under appropriate conditions can lead to the total elimination of Cr(VI) [26].

Treatment phase			
Electrochemical treatment		**Biosorption-based treatment**	
1. Electrochemical reduction	**Subtotal**	**1. Batch reactor**	**Subtotal**
Electricity	1.560	Stirring	780
2. Reagent addition	**Subtotal**	**2. Biosorbent addition**	**Subtotal**
Sodium hydroxide (NaOH)	500	Sodium hydroxide (NaOH)	500
Hydrochloric acid (HCl)	100	Hydrochloric acid (HCl)	100
Coagulant	2000	Grape stalks	21
3. Control sensors and separation unit	**Subtotal**	**3. Control sensors and separation unit**	**Subtotal**
Electrodes (pH and redox)	1000	Electrodes (pH and redox)	1000
Filter clothes	100	Filter clothes	100
Energy	200	Energy	200
4. Waste management	**Subtotal**	**4. Waste management**	**Subtotal**
Waste collection	240	Waste collection	240
post-treatment phase			
Electrochemical treatment		**Biosorption-based treatment**	
5. Treatment with activated carbon	**Subtotal**	**5. Treatment with biosorbent**	**Subtotal**
Activated carbon	134.4	Grape stalks	1.2
		Sodium alginate	339.4
Other expenses			
Electrochemical treatment		**Biosorption-based treatment**	
Internal labour costs	3000	Internal labour costs	1500
External labour costs	1000	External labour costs	750
Pumps and valves	1000	Pumps and valves	1000
Total	**10,834€**	**Total**	**6532€**

Table 2. Comparison between electrolytic and biosorption-based schemes for the treatment of 300 m^3/year of wastewater from an electroplating water industry. Costs are expressed in €.

The authors of the present chapter carried out a thorough techno-economical study concerning implementation of biosorption-based technology for the treatment of wastewater from an electroplating industry (unpublished data). A brief economic summary showing the different technological processes involved and their related expenses is presented in **Table 2**. The table compares the costs related to the use of regular technology (based on electroreduction) with the ones associated with a water treatment based on biosorption/bioreduction by grape stalks (GS) and precipitation/refining using GS entrapped into calcium alginate (CA).

In a regular wastewater treatment, Cr(VI) is reduced to Cr(III) using electricity. In acidic media, 3 mols of electrons are consumed in this process per mol of Cr(VI) reduced. In the case of the GS sorption/reduction process, there is no electric current applied to the solution and the only energetic requirements are related to the mixing of the material with the Cr(VI) solution. Avoiding electrochemical reduction in the treatment phase, decreases the costs related to energy consumption to half (from 1560 to 780€). In terms of reagents, NaOH and HCl are reagents used in both treatment schemes. In the secondary phase (post-treatment or refining), the costs related to the use of activated carbon are about 134€. Using GS and CA for the removal of residual Cr(VI) and the formed Cr(III) involve about 341€. In terms of other expenses, there is both an internal and external labour costs reduction when biosorption-based technologies are used. This saving is due to the frequent cleaning and, in the last term, replacement of the electrodes in the electrochemical cells. In the case of a biosorption unit, repairs are less frequent and maintenance less expensive.

Overall, the use of GS and CA in the detoxification of Cr(VI) polluted effluents involves a saving close to 40% if compared to the electrochemical reduction process. These data are just an example that clearly shows the viability of lignocellulosic materials for real wastewater treatment. There is a clear opportunity to go beyond the laboratory studies and implement properly scaled biosorption processes to the treatment of real industrial effluents.

5. Equilibrium and kinetics modelling

5.1. Equilibrium models in single component solutions

A precise mathematical description of the equilibrium isotherms is of paramount importance for the effective design of sorption systems. The most widely used adsorption isotherms found in the literature to describe the amount of solute adsorbed as a function of the equilibrium concentration in solution are summarised in **Table 3**.

From these models, by far the most widely employed are Langmuir and Freundlich. While Langmuir isotherm model relies on the adsorption theory and assumes the formation of a sorbate monolayer, Freundlich isotherm is an empirical model to correlate the concentration of a sorbate on the solid phase and the concentration of the sorbate in the fluid.

Isotherm model	Equation	Linear expression	Ref.
Langmuir	$q_e = \frac{Q_m b C_e}{1+bC_e}$	$\frac{C_e}{q_e} = \frac{1}{bQ_m} + \frac{C_e}{Q_m}$	[54–56]
Freundlich	$q_e = k_F C_e^{\frac{1}{n_F}}$	$\ln q_e = \ln k_F + \frac{1}{n}\ln C_e$	[54–56]
Redlich-Peterson	$q_e = \frac{K_{RP} C_e}{1+\alpha_{RP} C_e^{\beta}}$	$\ln\left(\frac{K_{RP} C_e}{q_e} - 1\right) = \ln\alpha_{RP} + \beta \ln C_e$	[55, 56]
Sips	$q_e = \frac{Q_m b C_e^{n_S}}{1+bC_e^{n_S}}$		[54–56]
Temkin	$q_e = \frac{RT}{b_T}\ln\left(K_T C_e\right)$	$q_e = \frac{RT}{b_T}\ln K_T + \frac{RT}{b_T}\ln C_e$	[54–56]
Dubinin-Radushkevich	$q_e = Q_m \exp\left(-B_D \varepsilon^2\right)$ $\varepsilon = RT ln\left(1+\frac{1}{C_e}\right)$	$\ln q_e = \ln Q_m - 2B_D R^2 T^2 \ln\left(1+\frac{1}{C_e}\right)$	[55, 56]
Toth	$q_e = \frac{Q_m b C_e}{\left(1+\left(bC_e\right)^t\right)^{\frac{1}{t}}}$		[54, 56]

Table 3. Mathematical equations of single component isotherm models.

Obtaining isotherm parameters from experimental data is a key aspect to consider. Using linear regression to determine the sorption isotherm parameters and decide which model provides the best fit was not only the easiest way at the time this approach was proposed, but also has become a 'custom' extended to our days. Obviously, the linearisation of the equations involves a transformation of the data in a process that alters the structure of errors, violates the error variance assumptions and the normality of the least squares method [57–59]. This set of error sources explains that the linear parameters obtained through Freundlich model produces isotherms that best fit data at low concentrations, while the linear parameters of the Langmuir tend to better match isotherm data at higher concentrations [57, 60]. Taking into account the aforementioned drawbacks of the linear procedures, direct calculation of the isotherm parameters through non-linear optimisation methods is strongly recommended. The non-linear optimisation is a more complex method than the linearisation approach and requires the proper choice of an error function to accurately evaluate the fit of the experimental results to the chosen isotherm, since the choice of the error function may affect the derived parameters. The most widely used error functions to fit sorption isotherm data are summarised in **Table 4**.

Error function	Expression	Ref.
The sum of the squares of the errors (SSR)	$\sum_{i=1}^{N}\left(q_{i,\exp}-q_{i,cal}\right)^2$	[54, 57, 61]
Hybrid fractional error function (HYBRD)	$\frac{100}{N-p}\sum_{i=1}^{N}\frac{\left(q_{i,\exp}-q_{i,cal}\right)^2}{q_{i,\exp}}$	[54, 57, 61]
Marquardt's percent standard deviation (MPSD)	$100\sqrt{\frac{1}{N-p}\sum_{i=1}^{N}\left(\frac{q_{i,\exp}-q_{i,cal}}{q_{i,\exp}}\right)^2}$	[54, 57, 61]
Average relative error (ARE)	$\frac{100}{N}\sum_{i=1}^{N}\left\lvert\frac{q_{i,\exp}-q_{i,cal}}{q_{i,\exp}}\right\rvert$	[54, 57, 61]
Sum of the absolute errors (EABS)	$\sum_{i=1}^{N}\left\lvert q_{i,\exp}-q_{i,cal}\right\rvert$	[54, 57, 61]

Table 4. Error functions.

The sum of the squares of the errors (SSR) is the most widely used error function, but it shows an important drawback; the isotherm parameters obtained provides a better fit in the case of high concentrations due to the fact that errors and therefore their squares increase, when it does the concentration.

Hybrid fractional error function (HYBRD) is an error function developed with the attempt of improving the adjustment of the SSR at low concentrations divided by the measured value and includes the degrees of freedom of the system.

Marquardt's percent standard deviation (MPSD) is similar in some aspect to a geometric mean error distribution modified according to the number of degrees of freedom of the system.

Average relative error function (ARE) attempts to minimise the fractional error distribution across the entire concentration range.

Sum of the absolute errors (EABS) uses a similar approach to that used in the SSR. Isotherm parameters determined using this error function provide a better fit as the magnitude of the error increases, biasing the fit towards the high concentration data [61].

In terms of least-squares regression analysis, one of the central premises is that the independent variable x remains fixed, i.e. there is no measured error in x. Therefore, there should be no error in the measurement of the equilibrium concentration in solution (C_e). However, the determination of the remaining concentration in the solution is not free from error, despite careful calibration and repetition of sample analysis is performed. Hence, the application of orthogonal distance regression (ODR) may be appropriate [59, 62–65].

The ODR takes into account the errors in both variables (C_e and q). The most widely used error functions are summarised in **Table 5**.

Error function	Expression	Ref.
Theoretical orthogonal distance regression	$\sum_{i=1}^{N}\left[\left(\frac{q_{i,\exp}-q_{i,cal}}{\sigma_{\varepsilon i}}\right)^2+\left(\frac{C_{ei,\exp}-C_{ei,cal}}{\sigma_{\delta i}^{e}}\right)^2\right]$	[59, 63]
Orthogonal distance regression	$\sum_{i=1}^{N}\left[\left(\frac{q_{i,\exp}-q_{i,cal}}{q_{i,\exp}}\right)^2+\left(\frac{C_{ei,\exp}-C_{ei,cal}}{C_{ei,\exp}}\right)^2\right]$	[62, 63, 65, 66]
Experimental weighted orthogonal distance regression	$\sum_{i=1}^{N}\left[\left(\frac{q_{i,\exp}-q_{i,cal}}{sdy_i}\right)^2+\left(\frac{C_{ei,\exp}-C_{ei,cal}}{sdx_i}\right)^2\right]$	[63]
Triplicate orthogonal distance regression	$\sum_{i=1}^{N}\left[\left(\frac{q_{i,\exp}-q_{i,cal}}{q_{i,\exp}}\right)^2+\left(\frac{C_{ei,\exp}-C_{ei,cal}}{C_{ei,\exp}}\right)^2\right]$	[63]

Table 5. Error functions.

5.2. Equilibrium models in multicomponent solutions

Since the interaction of a component with other components in a mixture can be synergistic, antagonistic or non-interactive, the biosorption results cannot be predicted on the basis of studies derived from single component solutions. The behaviour of each species in a multi-component system depends largely on the physical and chemical properties of both, sorbent and sorbate. However, most of the isotherms of a single component can be extended to describe a multi-component adsorption system. The most frequently used extensions are summarised in **Table 6**.

In the scientific literature, there are few models to describe synergistic effects in sorption processes. A straightforward and simple way to develop a synergistic model is based on the use of monocomponent models and introduction of correction factors, for example, by combining a single isotherm (Langmuir, Freundlich, etc.) with a form factor $1+F_i(C)$ [72, 73].

When it comes to obtaining the parameters of these models, the same comments stated previously for monocomponent sorption isotherms applies. The error function most widely used is based on the sum of the SSR for each component although the most suitable method would be the use of ODR.

Isotherm model	Equation	Ref.
Extended Langmuir:	$q_{e,i} = \dfrac{q_{0,i} b_i C_{e,i}}{1+\sum_j b_j C_{e,j}}$	[67–69]
Modified competitive Langmuir:	$q_{e,i} = \dfrac{q_{0,i} b_i \dfrac{C_{e,i}}{\eta_i}}{1+\sum_j b_j \dfrac{C_{e,j}}{\eta_j}}$	[67, 70, 71]
Extended Freundlich:	$q_{e,i} = \dfrac{K_{F,i} C_{e,i}^{\frac{1}{n_i}+x_i}}{C_{e,i}^{x_i} + y_i C_{e,j}^{z_j}}$	[67, 68]
Competitive non modified Redlich Peterson model:	$q_{e,i} = \dfrac{K_{RP,i} C_{e,i}}{1+\sum_j a_{RP,j} C_{e,j}^{\beta_j}}$	[67, 68]
Competitive modified Redlich Peterson model:	$q_{e,i} = \dfrac{K_{RP,i} \dfrac{C_{e,i}}{\eta_i}}{1+\sum_j a_{RP,j} \left(\dfrac{C_{e,j}}{\eta_j}\right)^{\beta_j}}$	[68]

Table 6. Multicomponent equilibrium models.

5.3. Kinetics models

A summary on the most widely used models to describe the time-course profile of metal removal in sorption processes is presented in **Table 7**.

The scientific literature to date shows that, in most cases, the obtention of the characteristic parameters of the aforementioned models is performed through linearisation. As previously discussed, this process involves a transformation of the data set and the subsequent alteration of the error structure and therefore a given bias in the so obtained parameters. To avoid this, non-linear regression should be used using the SSR as target error function. In this case, the use of ODR is not strictly needed since it can be considered that there is no error in the accurate determination of the time variable.

As a final remark, it should be highlighted that a good fitting of an experimental data set to a given kinetic or equilibrium model does not necessarily imply that the sorption process is governed by the mechanisms on which the model relies [74].

Kinetic model	Equation	Linear expression	Ref.
Pseudo-first-order or Lagergren equation	$\frac{dq_t}{dt} = K_1(q_e - q_t)$ $q_t = q_e(1-\exp(-K_1 t))$	$\ln(q_e - q_t) = \ln q_e - K_1 t$	[56, 74]
Pseudo-second-order	$\frac{dq_t}{dt} = K_2(q_e - q_t)^2$ $q_t = \frac{K_2 q_e^2 t}{1+K_2 q_e t}$	$\frac{t}{q_t} = \frac{1}{K_2 q_e^2} + \frac{t}{q_e}$	[56, 74]
Weber-Morris or intra-particle diffusion	$q_t = K_p t^{\frac{1}{2}}$		[56, 74]
Elovich equation	$\frac{dq_t}{dt} = a_E \exp(-\alpha_E q_t)$ $q_t = \frac{1}{\alpha_E}\ln(t + t_0) - \frac{1}{\alpha_E}\ln(t_0)$ $t_0 = \frac{1}{a_E \alpha_E}$		[74]

Table 7. Kinetic models.

6. Metal desorption and regeneration of lignocellulosic wastes

Metal laden biomass can either be directly disposed or incinerated, delivering the ashes to a regular hazardous waste landfill. Incineration reduces waste volume, but enhances the metal content per mass unit and may cause environmental issues due to potential toxic metal leaching. Disposal or incineration of metal-laden biomass is recommended when biomass is abundant and cheap and metal is not worth recovering. Another alternative is regeneration of the biomass by elution of the loaded metal using desorbing agents. The selection criteria that guides to choose the appropriate desorbing agent are: (i) small volume of eluent should yield high metal concentration in the resulting solution, (ii) the structural integrity of the biomass must not be severely affected, and (iii) the eluent should be economic and environmentally friendly.

Most of the desorbing agents used for sorbents regeneration and metal recovery are based on strong mineral acids (HCl, HNO_3, H_2SO_4), short-chain organic acids (HCOOH, CH_3COOH), bases (NaOH, $NaHCO_3$, Na_2CO_3, KOH, K_2CO_3), salts (NaCl, KCl, $CaCl_2$, KNO_3), chelating agents (ethylenediaminetetraacetic (EDTA), diethylenediaminepentacetic (DTPA), nitrilotri-acetic (NTA)) or buffer solutions (phosphate, bicarbonate). Recently, the most widely used desorbing agents have been reviewed [75].

The most used desorption agents to remove metal-laden lignocellulosic biomass are acids. An acidic solution (pH 2.0) was found to be the best desorbing solution for Cu(II) and Cr(VI) loaded onto commercial coffee wastes [76]. A 0.1 M HCl solution resulted to be effective to desorb Cd(II) (83.9%) from rice husk in both batch and column modes [77]; Cu(II) (99.4%), Cd(II) (98.5%) and Zn(II) (99.3%) from papaya wood [78]; As(V) from coconut coir pith [79]; Cu(II) from cork and yohimbe bark [80]; and 0.2 and 0.5 M HCl for chromium recovery from avena by-products and Ga(III) desorption from coir [71]. A 0.1 M HNO_3 solution was used by Gupta and Nayak [81] to desorb 98.2% Cd(II) loaded onto orange peel powder with Fe_3O_4. Higher concentrations than 0.01 M of HNO_3 and HCl were needed to effectively desorb Cd(II) loaded onto coffee beans [82]. *Ficus religiosa* lead(II) laden was regenerated by using 0.05 M HNO_3 [83]. HCl and EDTA solutions were used to desorb U(VI) from citrus waste material. The best desorption yields were obtained by 0.1 EDTA (94.7%) followed by 0.1 HCl (89.71%) [84]. HCl and EDTA were also tested by Martinez et al. to recover Pb, Cu, Cd and Ni from grape stalk and olive stones wastes, the former being the most effective [85, 86].

Alkaline solutions were also successfully tested as desorbing agents for metal ions from biomass. Note that 0.5 M sodium citrate was able to remove lead(II) laden on hop by-products [87]. Elution of arsenic-laden rice polish and rice husk packed columns was achieved by passing through the column of 10% NaOH [88] and 1 M KOH [89] solutions, respectively.

The main conclusion retrieved from the literature survey is that recovery of metals from exhausted lignocellulosic materials and other low-cost sorbent materials and the regeneration of the sorbent is not nowadays the focus of researchers. It may be foreseen however that in the future, when some metals become scarce, further research will be conducted to find out selective metal desorbing agents that meet the most of the aforementioned criteria.

7. Conclusion

Lignocellulosic biomass has a huge potential as a low cost, renewable source and environmentally friendly alternative to conventional methods for the removal of heavy metals from aqueous polluted streams. This technology is of special interest to treat large volumes of effluent containing low metal concentration to produce a final effluent that does not pose environmental hazards. A thorough screening and selection of the most effective low-cost sorbents with sufficiently high metal-binding capacity and selectivity for heavy metal ions are prerequisites for full-scale implementation in industrial processes. Despite further research, efforts towards full understanding of sorption mechanisms and development of more accurate mathematical models might be required, the technology can be considered mature enough as to face scale-up scenarios to large scale. Industrial stakeholders, policymakers and regulators have nowadays a challenging and exciting opportunity to take a step forward towards environmental sustainability considering sorption onto biomaterials on their wastewater treatment schemes.

http://dx.doi.org/10.5772/65958

Annex 1: Notation

q_e	Amount of sorbate adsorbed per unit mass of sorbent at equilibrium
Q_m	Maximum uptake of sorbate per unit mass of sorbent
b	Langmuir constant related to sorbent affinity
C_e	Concentration of sorbate in solution at equilibrium
k_F	Freundlich constant related to the amount of sorbate adsorbed
n_F	Freundlich constant related to sorption intensity
K_{RP}	Redlich-Peterson constant related to the amount of sorbate adsorbed
α_{RP}	Redlich-Peterson constant
β	Redlich-Peterson exponent $0 \leq \beta \leq 1$
n_S	Sips constant related to sorption intensity
R	Universal gas constant
T	Temperature
b_T	Temkin constant
K_T	Temkin constant related to binding
B_D	Dubinin-Radushkevich constant related to sorption energy
ε	Polanyi potential
τ	Toth model exponent
$q_{i,exp}$	Experimental data values
$q_{i,cal}$	Data values calculated from the model
N	Number of data
p	Number of model parameters
$\sigma_{\varepsilon i}$	Population standard deviation of measurement error in dependent variable
$\sigma_{\delta i}^{e}$	Population standard deviation of measurement error in equilibrium concentration
sdy_i	Estimates of population standard deviation for dependent variable
sdx_i	Estimates of population standard deviation for independent variable
η	Dimensionless interaction factor
q_t	Amount of sorbate adsorbed per unit mass of sorbent at time t
K_1	First-order reaction rate equilibrium constant
K_2	Second-order reaction rate equilibrium constant
t	Time
K_p	Intra-particle diffusion rate constant
a_E	Elovich equation constant
α_E	Elovich equation constant

Author details

Carlos Escudero-Oñate[1*], Núria Fiol[2], Jordi Poch[3] and Isabel Villaescusa[2]

*Address all correspondence to: carlos.escudero@niva.no

1 Norwegian Institute for Water Research (NIVA), Gaustadalléen, Oslo, Norway

2 Chemical Engineering Department, EPS, University of Girona, Girona, Spain

3 Applied Mathematics Department, EPS, University of Girona, Girona, Spain

References

[1] Järup L. Hazards of heavy metals contamination. British Medical Bulletin, 2003. 68; 167–182. DOI: 10.1093/bmb/ldg032.

[2] Wang J, Chen C. Biosorbents for heavy metals removal and their future. Biotechnology Advances. 2009. 27; 195–226. DOI: 10.1016/j.biotechadv.2008.11.002.

[3] Zhou CH, Xia X, Lin CX, Tong DS, Beltramini J. Catalytic conversion of lignocellulosic biomass to fine chemicals and fuels. Chemical Society Reviews. 2011. 40(11); 5588–5617. DOI: 10.1039/C1CS15124J.

[4] Vanholme R, Demedts B, Morreel K, Ralph J, Boerjan W. Plant Physiology, 2010. 153(3); 895–905. DOI: 10.1104/pp.110.155119.

[5] Walker GM. Bioethanol: Science and technology of fuel alcohol. Ventus Publishing ApS. Frederiksberg, Denmark, 2010. ISBN 978-87-7681-681-0.

[6] Abdolali A, Guo WS, Ngo HH, Chen SS, Nguyen NC, Tung KL. Typical lignocellulosic wastes and by-products for biosorption process in water and wastewater treatment: A critical review. Bioresource Technology. 2014. 160; 57–66. DOI: 10.1016/j.biortech. 2013.12.037.

[7] Pujol D, Liu C, Gominho J, Olivella MA, Fiol N, Villaescusa I, Pereira H. The chemical composition of exhausted coffee waste. Industrial Crops and Products. 2013. 50; 423–429. DOI: 10.1016/j.indcrop.2013.07.056.

[8] Pujol D, Liu C, Fiol N, Olivella MA, Gominho J, Villaescusa I, Pereira H. Chemical characterization of different granulometric fractions of grape stalks waste. Industrial Crops and Products. 2013. 50; 494–400. DOI: 10.1016/j.indcrop.2013.07.051.

[9] Sud D, Mahajan G, Kaur MP. Agricultural waste material as potential adsorbent for sequestering heavy metal ions from aqueous solutions – A review. Bioresource Technology. 2008. 14(99); 6017–6027. DOI: 10.1016/j.biortech.2007.11.064.

[10] Malik DS, Jain CK, Yadav AK. Removal of heavy metals from emerging cellulosic low-cost adsorbents: A review. Applied Water Science. 2016. 1–24. DOI: 10.1007/s13201-016-0401-8.

[11] Miretzky P, Fernandez Cirelli A. Cr(VI) and Cr(III) removal from aqueous solution by raw and modified lignocellulosic materials: A review. Journal of Hazardous Materials. 2010. 180(1–3); 1–19. DOI: 10.1016/j.jhazmat.2010.04.060.

[12] Demirbas A. Heavy metal adsorption onto agro-based waste materials: A review. Journal of Hazardous Materials. 2008. 2–3; 220–229. DOI: 10.1016/j.jhazmat.2008.01.024.

[13] Hubbe MA, Hasan SH, Ducoste JJ. Cellulosic substrates for removal of pollutants from aqueous systems: A review. Metals BioResources. 2011. 1. 2161–2287 ISSN: 1930–2126.

[14] Michalak I, Chojnacka K, Witek-Krowiak A. State of the art for the biosorption process—A review. Applied Biochemistry and Biotechnology. 2013. 170; 1389. DOI: 10.1007/s12010-013-0269-0.

[15] Volesky B. Sorption and biosorption. BV Sorbex, Inc, Montreal, Canada. 2004.

[16] Das N. Recovery of precious metals through biosorption—A review. Hydrometallurgy. 2010. 103(1–4); 180–189. DOI: 10.1016/j.hydromet.2010.03.016.

[17] Das N, Das D. Recovery of rare earth metals through biosorption: An overview. Journal of Rare Earths. 2013. 31(10); 933–943. DOI: 10.1016/S1002-0721(13)60009-5.

[18] Prasad AGD, Abdullah MA. Biosorption of Cr (VI) from synthetic wastewater using the fruit shell of gulmohar (*Delonix regia*): Application to electroplating wastewater. BioResources. 2010. 5(2); 838–853. ISSN: 1930–2126.

[19] Martín-Lara MA, Blázquez G, Trujillo MC, Pérez A, Calero M. New treatment of real electroplating wastewater containing heavy metal ions by adsorption onto olive stone. Journal of Cleaner Production. 2014. 81; 120–129. DOI: 10.1016/j.jclepro.2014.06.036.

[20] Singha B, Das SK. Removal of Pb(II) ions from aqueous solution and industrial effluent using natural biosorbents. Environmental Science and Pollution Research. 2012. 19; 2212. DOI: 10.1007/s11356-011-0725-8.

[21] Liu C, Fiol N, Poch J, Villaescusa I. A new technology for the treatment of chromium electroplating wastewater based on biosorption. Journal of Water Process Engineering. 2016. 11. 143–151 DOI: 10.1016/j.jwpe.2016.05.002.

[22] Chipara M, Hamilton J, Chipara AC, George T, Chipara DM, Ibrahim EE, Lozano K, Sellmyer DJ. Fourier transform infrared spectroscopy and wide-angle X-ray scattering: Investigations on polypropylene–vapor-grown carbon nanofiber composites. Journal of Applied Polymer Science. 2012. 125(1); 353–360. DOI: 10.1002/app.35576.

[23] Dupont L, Guillon E. Removal of hexavalent chromium with a lignocellulosic substrate extracted from wheat bran. Environmental Science and Technology. 2003. 15(18); 4235–4241. DOI: 10.1021/es0342345.

[24] Twardowska I, Kyziol J. Sorption of metals onto natural organic matter as a function of complexation and adsorbent-adsorbate contact mode. Environment International. 2003. 8; 783–791. DOI: 10.1016/S0160-4120(02)00106-X.

[25] Villaescusa I, Fiol N, Cristiani F, Floris C, Lai S, Nurchi VM. Copper(II) and nickel(II) uptake from aqueous solutions by cork wastes: A NMR and potentiometric study. Polyhedron. 2002. 21(14); 1363–1367. DOI: 10.1016/S0277-5387(02)00957-9.

[26] Escudero C, Fiol N, Villaescusa I, Bollinger JC. Effect of chromium speciation on its sorption mechanism onto grape stalks entrapped into alginate beads. Arabian Journal of Chemistry. 2013. DOI: 10.1016/j.arabjc.2013.03.011.

[27] Misra SP. Adsorption-desorption of heavy metal ions. Current Science. 2014. 107(4); 601–612.

[28] Chaplin MF, Bucke C. The preparation and kinetics of immobilised enzymes in enzyme technology. Cambridge University Press. Cambridge, United Kingdom. 1990. Chapter 3. pp. 81–89.

[29] Peinado RA, Moreno JJ, Maestre O, Mauricio JC. Use of a novel immobilization yeast system for winemaking. Biotechnology Letters. 2005. 27; 1421–1424. DOI: 10.1007/s10529-005-0939-2.

[30] Martins SCS, Martins CM, Fiúza LMCG, Santaella ST. Immobilization of microbial cells: A promising tool for treatment of toxic pollutants in industrial wastewater. African Journal of Biotechnology. 2013. 12(28); 4412–4418. DOI: 10.5897/AJB12.2677.

[31] Crini G. Recent developments in polysaccharide-based materials used as adsorbents in wastewater treatment. Progress in Polymer Science. 2005. 30; 38–70. DOI: 10.1016/j.progpolymsci.2004.11.002.

[32] Eroglu E, Smith SM, Raston CL. Application of various immobilization techniques for algal bioprocesses. Biofuel and Biorefinery Technologies. 2015. 2; 19–44. DOI: 10.1007/978-3-319-16640-7_2.

[33] Feng L, Cao Y, Su D, You S, Han F. Influence of sodium alginate pretreated by ultrasound on papain properties: Activity, structure, conformation and molecular weight and distribution. Ultrasonic Sonochemistry. 2016. 32; 224–230. DOI: 10.1016/j.ultsonch.2016.03.015.

[34] Vauchel P, Kaas R, Arhaliass A, Baron R, Legrand J. A new process for extracting alginates from Laminaria digitata: Reactive extrusion. Food and Bioprocess Technology. 2008. 1(3); 297–300. DOI: 10.1007/s11947-008-0082-x.

[35] Hernández-Carmona G, McHugh DJ, Arvizu-Higuera DL, Rodríguez-Montesinos E. Pilot plant scale extraction of alginate from *Macrocystis pyrifera*. 1. Effect of pre-extraction treatments on yield and quality of alginate. Journal of Applied Phycology. 1999. 10; 507–513. DOI: 10.1023/A:1008004311876.

[36] Bajpai SK, Sharma S. Investigation of swelling degradation behaviour of alginate beads crosslinked with Ca^{2+} and Ba^{2+} ions. Reactive & Functional Polymers. 2004. 59; 129–140. DOI: 10.1016/j.reactfunctpolym.2004.01.002.

[37] Hassan R, Tirkistani F, Zaafarany I, Fawzy A, Khairy M, Iqbal S. Polymeric biomaterial hydrogels. I. Behavior of some ionotropic cross-linked metal-alginate hydrogels especially copper-alginate membranes in some organic solvents and buffer solutions. Advances in Bioscience and Biotechnology. 2012. 3; 845–854. DOI: 10.4236/abb.2012.37105.

[38] Grant GT, Morris ER, Rees DA, Smith PJC, Thom D. Biological interactions between polysaccharides and divalent cations: The egg-box model. FEBS Letters. 1973. 32(1); 195–198. DOI: 10.1016/0014-5793(73)80770-7.

[39] Pandey A, Bera D, Shukla A, Ray L. Studies on Cr(VI), Pb(II) and Cu(II) adsorption-desorption using calcium alginate as biopolymer. Chemical Speciation & Bioavailability. 2007. 19(1); 17–24. DOI: 10.3184/095422907X198031.

[40] Jodra Y, Mijangos F. Ion exchange selectivities of calcium alginate gels for heavy metals. Waters Science & Technology. 2001. 43(2); 237–244.

[41] Bishnoi NR, Kumar R, Bishnoi K. Biosorption of Cr(VI) with *Trichoderma viride* immobilised fungal biomass and cell free Ca-alginate beads. Indian Journal of Experimental Biology. 2007. 45(7); 657–664.

[42] Wuyep PA, Chuma AG, Awodi S, Nok AJ. Biosorption of Cr, Mn, Fe, Ni, Cu and Pb metals from petroleum refinery effluent by calcium alginate immobilized mycelia of *Polyporus squamosus*. Scientific Research & Essay. 2007. 2(7); 217–221.

[43] Mata YN, Blázquez ML, Ballester A, González F, Muños JA. Biosorption of cadmium, lead and copper with calcium algiante xerogels and immobilized *Fucus vesiculosus*. Journal of Hazardous Materials. 2009. 163(2–3); 555–562. DOI: 10.1016/j.jhazmat.2008.07.015.

[44] Olusola B, Aransiola MN. Biosorption of zinc (Zn^{2+}) and Iron (Fe^{2+}) from wastewater using *Botrydium granulatum* and *Euglena texta*. Bioscience and Bioengineering. 2015. 1(2); 17–21.

[45] Samuel J, Pulimi M, Paul ML, Maurya A, Chandrasekaran N, Mukherjee A. Batch and continuous flow studies of adsorptive removal of Cr(VI) by adapted bacterial consortia immobilized in alginate beads. Bioresource Technology. 2013. 128; 423–430. DOI: 10.1016/j.biortech.2012.10.116.

[46] Gad A, Attia M, Ahmed HA. Heavy metals bio-remediation by immobilizaed *Saccharomyces cerevisiae* and *Opuntia ficus indica*. Journal of American Science. 2010. 6(8); 79–87.

[47] Ansari TM, Hanif MA, Mahmood A, Ijaz U, Khan MA, Nadeem R, Ali M. Immobilization of rose waste biomass for uptake of Pb(II) from aqueous

solutions. Biotechnology Research International. 2011. 2011; 1–9. DOI: 10.4061/2011/685023.

[48] Fiol N, Poch J, Villaescusa I. Chromium(VI) uptake by grape stalks wastes encapsulated in calcium alginate beads: Equilibrium and kinetics studies. Chemical Speciation & Bioavailability. 2004. 16(1/2). 25–33 DOI: 10.3184/095422904782775153.

[49] Fiol N, Poch J, Villaescusa I. Grape stalks wastes encapsulated in calcium alginate beads for Cr(VI) removal from aqueous solutions. Separation Science and Technology. 2005. 40; 1013–1028. DOI: 10.1081/SS-200048041.

[50] Sillerová H, Komárek M, Liu C, Poch J, Villaescusa I. Biosorbent encapsulation in calcium alginate: Effects of process variables on Cr(VI) removal from solutions. International Journal of Biological Macromolecules. 2015. 80; 260–270. DOI: 10.1016/j.ijbiomac.2015.06.032.

[51] Fiol N, Escudero C, Poch J, Villaescusa I. Preliminary studies on Cr(VI) removal from aqueous solution using grape stalk wastes encapsulated in calcium alginate beas in a packed bed up-flow column. Reactive & Functional Polymers. 2006. 66; 795–807. DOI: 10.1016/j.reactfunctpolym.2005.11.006.

[52] Escudero C, Fiol N, Villaescusa I. Chromium sorption on grape stalks encapsulated in calcium alginate beads. Environmental Chemistry Letters. 2006. 4; 239–242. DOI: 10.1007/s10311-006-0055-0.

[53] Fiol N, Escudero C, Villaescusa I. Chromium sorption and Cr(VI) reduction to Cr(III) by grape stalks and yohimbe bark. Bioresource Technology. 2008. 99(11); 5030–5036. DOI: 10.1016/j.biortech.2007.09.007.

[54] Anirudhan TS, Radhakrishnan PG. Kinetic and equilibrium modelling of cadmium(II) ions sorption onto polymerized tamarind fruit shell. Desalination. 2009. 249(3); 1298–1307. DOI: 10.1016/j.desal.2009.06.028.

[55] Rangabhashiyam S, Anu S, Giri Nandagopal MS, Selvaraju N. Relevance of isotherm models in biosorption of pollutants by agricultural by-products. Journal of Environmental Chemical Engineering. 2014. 2(1); 398–414. DOI: 10.1016/j.jece.2014.01.014.

[56] Abdolali A, Ngo HH, Guo WS, Lee DJ, Tung KL, Wang XC. Development and evaluation of a new multi-metal binding biosorbent. Bioresource Technology. 2014. 160; 98–106. DOI: 10.1016/j.biortech.2013.12.038.

[57] Allen SJ, Mckay G, Porter JF. Adsorption isotherm models for basic dye adsorption by peat in single and binary component systems. Journal of Colloid and Interface Science. 2004. 280(2); 322–333. DOI: 10.1016/j.jcis.2004.08.078.

[58] Badertscher M, Pretsch E. Bad results from good data. Trends in Analytical Chemistry. 2006. 25(11); 1131–1138. DOI: 10.1016/j.trac.2006.09.003.

[59] El-Khaiary MI. Least-squares regression of adsorption equilibrium data: Comparing the options. Journal of Hazardous Materials. 2008. 158(1); 73–87. DOI: 10.1016/j.jhazmat.2008.01.052.

[60] Ayoob S, Gupta AK. Insights into isotherm making in the sorptive removal of fluoride from drinking water. Journal of Hazardous Materials. 2008. 152; 976–985. DOI: 10.1016/j.jhazmat.2007.07.072.

[61] Foo KY, Hameed BH. Insights into the modeling of adsorption isotherm systems. Chemical Engineering Journal. 2010. 156; 2–10. DOI: 10.1016/j.cej.2009.09.013.

[62] Poch J, Villaescusa I. Orthogonal distance regression: A good alternative to least squares for modeling sorption data. Journal of Chemical & Engineering Data. 2012. 57(2); 490–499. DOI: 10.1021/je201070u.

[63] Marković DD, Lekić BM, Rajaković-Ognjanović VN, Onjia AE, Rajaković LV. A new approach in regression analysis for modeling adsorption isotherms. The Scientific World Journal. 2014. 1–17 DOI: 10.1155/2014/930879.

[64] Yan F, Chu Y, Zhang K, Zhang F, Bhandari N, Ruan G, Dai Z, Liu Y, Zhang Z, Kan AT, Tomson MB. Determination of adsorption isotherm parameters with correlated errors by measurement error models. Chemical Engineering Journal. 2015. 281(1); 921–930. DOI: 10.1016/j.cej.2015.07.021.

[65] Millar GJ, Miller GL, Couperthwaite SJ, Papworth S. Factors influencing kinetic and equilibrium behaviour of sodium ion exchange with strong acid cation resin. Separation and Purification Technology. 2016. 163; 79–91. DOI: 10.1016/j.seppur.2016.02.045.

[66] Beltrán JL, Pignatello JJ, Teixidó M. ISOT_Calc: A versatile tool for parameter estimation in sorption isotherms. Computers & Geosciences. 2016. 94; 11–17. DOI: 10.1016/j.cageo.2016.04.008.

[67] Gupta A, Balomajumder C. Simultaneous adsorption of Cr(VI) and phenol onto tea waste biomass from binary mixture: Multicomponent adsorption, thermodynamic and kinetic study. Journal of Environmental Chemical Engineering. 2015. 3; 785–796. DOI: 10.1016/j.jece.2015.03.003.

[68] Chan OS, Cheung WH, McKay G. Single and multicomponent acid dye adsorption equilibrium studies on tyre demineralised activated carbon. Chemical Engineering Journal. 2012. 191; 162–170. DOI: 10.1016/j.cej.2012.02.089.

[69] Villaescusa I, Fiol N, Martínez M, Miralles N, Poch J, Serarols J. Removal of copper and nickel ions from aqueous solutions by grape stalks wastes. Water Research. 2004. 38(4); 992–1002. DOI: 10.1016/j.watres.2003.10.040.

[70] Escudero C, Poch J, Villaescusa I. Modelling of breakthrough curves of single and binary mixtures of Cu(II), Cd(II), Ni(II) and Pb(II) sorption onto grape stalks waste. Chemical Engineering Journal. 2013. 217(1); 129–138. DOI: 10.1016/j.cej.2012.11.096.

[71] Suryavanshi US, Shukla SR. Adsorption of Ga(III) on oxidized coir. Industrial and Engineering Chemical Research. 2009. 48; 870–876. DOI: 10.1021/ie801259c.

[72] Aksu Z, Alper Isoglu I. Use of dried sugar beet pulp for binary biosorption of Gemazol Turquoise Blue-G reactive dye and copper(II) ions: Equilibrium modeling. Chemical Engineering Journal. 2007. 127; 177–188. DOI: 10.1016/j.cej.2006.09.014.

[73] Pujol D, Bartrolí M, Fiol N, de la Torre F, Villaescusa I, Poch J. Modelling synergistic sorption of Cr(VI), Cu(II) and Ni(II) onto exhausted coffee wastes from binary mixtures Cr(VI)–Cu(II) and Cr(VI)–Ni(II). Chemical Engineering Journal. 2013. 230; 396–405. DOI: 10.1016/j.cej.2013.06.033.

[74] Liu Y, Liu YJ. Biosorption isotherms, kinetics and thermodynamics. Separation and Purification Technology. 2008. 61(3); 229–242. ISSN 1383-5866. DOI: 10.1016/j.seppur.2008.07.017.

[75] Lata S, Singh PK, Samadder SR. Regeneration of adsorbents and recovery of heavy metals. International Journal of Environmental Science and Technology. 2015. 12; 1461–1478. DOI: 10.1007/s13762-014-0714-9.

[76] Kyzas GZ. Commercial coffee wastes as material for adsorption of heavy metals from aqueous solutions. Materials. 2012. 5; 1826–1840. DOI: 10.3390/ma5101826.

[77] Ajmal M, Rao RAK, Anwar S, Ahmad J, Ahmad R. Adsorption studies on rice husk: Removal and recovery of Cd(II) from wastewater. Bioresource Technology. 2003. 86(2); 147–149. DOI: 10.1016/S0960-8524(02)00159-1.

[78] Saeed A, Akhter MW, Iqbal M. Removal and recovery of heavy metals from aqueous solution using papaya wood as a new biosorbent. Separation & Purification Technology. 2005. 45; 25–31. DOI: 10.1016/j.seppur.2005.02.004.

[79] Anirudhan TS, Unnithan MR. Arsenic (V) removal from aqueous solutions using an anion exchanger derived from coconut coir pith and its recovery. Chemosphere. 2007. 66(1); 60–66. DOI: 10.1016/j.chemosphere.2006.05.031.

[80] Villaescusa I, Martínez M, Miralles N. Heavy metal uptake from aqueous solution by cork and yohimbe bark. Journal of Chemical Technology & Biotechnology. 2000. 75; 812–816. DOI: 10.1002/1097-4660(200009)75:9.

[81] Gupta VK, Nayak A. Cadmium removal and recovery from aqueous solutions by novel adsorbents prepared from orange peel and Fe_2O_3 nanoparticles. Chemical Engineering. 2012. 180; 81–90. DOI: 10.1016/j.cej.2011.11.006.

[82] Kaikake K, Hoaki K, Sunada H, Dhakal RP, Baba Y. Removal characteristics of metal ions using degreased coffee beans: Adsorption equilibrium of cadmium(II). Bioresource Technology. 2007. 98(15); 2787–2791. DOI: 10.1016/j.biortech.2006.02.040.

[83] Qaiser S, Saleemi AR, Umar M. Biosorption of lead from aqueous solution by *Ficus religiosa* leaves: Batch and column study. Journal of Hazardous Materials. 2009. 166(2–3); 998–1005. DOI: 10.1016/j.jhazmat.2008.12.003.

[84] Saleem N, Bhatti HM. Adsorptive removal and recovery of U(VI) by citrus waste biomass. BioResources. 2011. 6(3); 2522–2538. ISSN: 1930–2126.

[85] Martínez M, Miralles N, Hidalgo S, Fiol N, Villaescusa I, Poch J. Removal of lead(II) and cadmium(II) from aqueous solutions using grape stalk waste. Journal of Hazardous Materials. 2006. B133; 203–211. DOI: 10.1016/j.jhazmat.2005.10.030.

[86] Fiol N, Villaescusa I, Martínez M, Miralles N, Poch J, Serarols J. Sorption of Pb(II), Ni(II), Cu(II) and Cd(II) from aqueous solution by olive stone waste. Separation & Purification Technology. 2006. 50(1); 132–140. DOI: 10.1016/j.seppur.2005.11.016.

[87] Gardea-Torresdey J, Hejazi M, Tiemann K, Parsons JG, Duarte-Gardea M, Henning J. Use of hop (*Humulus lupulus*) agricultural by-products for the reduction of aqueous lead(II) environmental health hazards. Journal of Hazardous Materials. 2002. B91; 95–112. DOI: 10.1016/S0304-3894(01)00363-6.

[88] Ranjan D, Talat M, Hasan SH. Rice polish: An alternative to conventional adsorbents for treating arsenic. Industrial and Engineering Chemical Research. 2009. 48; 10180–10185. DOI: 10.1021/ie900877p.

[89] Amin MN, Kaneco S, Kitagawa T, Begum A, Katsumata H, Suzuki T, Ohta K. Removal of arsenic in aqueous solutions by adsorption onto waste rice husk. Industrial and Engineering Chemical Research. 2006. 45(24); 8105–8110. DOI: 10.1021/ie060344j.

Chapter 18

Progress Towards Engineering Microbial Surfaces to Degrade Biomass

Grace L. Huang and Robert T. Clubb

Additional information is available at the end of the chapter

http://dx.doi.org/10.5772/65509

Abstract

Lignocellulosic biomass is a promising feedstock to sustainably produce useful biocommodities. However, its recalcitrance to hydrolysis limits its commercial utility. One attractive strategy to overcome this problem is to use consolidated bioprocessing (CBP) microbes to directly convert biomass into chemicals and biofuels. Several industrially useful microbes possess desirable consolidated bioprocessing characteristics, yet they lack the ability to degrade biomass. Engineering these microbes' surfaces to display cellulases and cellulosome-like structures could endow them with potent cellulolytic activity, enabling them to be used in CBP. In this chapter, we discuss recent progress in engineering the surfaces of *Saccharomyces cerevisiae, Escherichia coli, Bacillus subtilis, Corynebacterium glutamicum,* and lactic acid bacteria. We discuss the techniques used to display cellulases on their surfaces, their recombinantly achieved cellulolytic activities, and current obstacles that limit their utility.

Keywords: lignocellulose, consolidated bioprocessing, cellulase, minicellulosome, cell surface display

1. Introduction

As concerns over limited petroleum supplies rise, the momentum to produce renewable fuels, chemicals, and other materials from biomass has increased [1–3]. Second-generation biofuels derived from sustainable feedstocks such as lignocellulosic biomass are attractive, as plant biomass is cheap and highly abundant; over 1 billion tons of lignocellulosic biomass are produced annually in the United States, while an estimated 10–50 billion tons of waste lignocellulose are produced worldwide [4–6]. However, the resistance of lignocellulose to hydrolysis limits its use in biofuel production and has driven the search for new technologies to cost-effectively exploit this valuable resource [7, 8]. To produce fermentable sugars to generate cellulosic ethanol, conventional industrial processes utilize a multistep procedure to degrade lignocellulose (**Figure 1A**) [7, 9]. Typically, the biomass is first pretreated with strong acids and high temperatures to remove lignin and to partially degrade its cellulose and hemicellulose components [7]. It is then exposed to purified cellulase enzymes that hydrolyze the remaining polysaccharides into shorter polysaccharides and monosaccharides that can be fermented into ethanol by yeast. Significant effort has been put forth to optimize these steps, including the development of a range of new pretreatment approaches and enzyme cocktails [9–11]. It is generally believed that the cost of converting biomass into useful biocommodities could be greatly reduced by using a consolidated bioprocessing (CBP) microbe, a single microorganism that produces all of the necessary enzymes to degrade lignocellulose and then utilizes the resulting sugars to produce high levels of the biocommodity (**Figure 1B**) [12–14]. A CBP microbe would decrease costs by reducing the number of processing steps required to generate the final product and avoid the use of costly purified cellulase enzymes that are estimated to contribute $0.68–1.47 to the per gallon cost of cellulosic ethanol [15]. Given their great potential, a number of research groups have sought to develop a lignocellulose-utilizing CBP microbe using native and recombinant strategies [12]. In the native strategy, product production pathways are engineered into naturally cellulolytic microbes, while in the recombinant strategy, genetically well-studied microorganisms that may already be capable of producing a desired product are engineered to possess cellulolytic activity. Here, we describe progress towards creating recombinant cellulolytic microbes to convert biomass into useful commodities by engineering their surfaces to display cellulase enzymes.

Lignocellulose is recalcitrant to degradation and requires the action of many different types of enzymes to break it down [15]. It is composed of varying amounts of cellulose (25–55%), hemicellulose (8–30%), and lignin (18–35%) [7, 16]. Cellulose is a polymer of β-1,4-linked glucose molecules that can hydrogen-bond with other cellulose polymers to form both amorphous and crystalline regions [17]. It is synergistically degraded by three types of cellulases: endoglucanases, exoglucanases, and β-glucosidases [7]. Endoglucanases attack within a cellulose strand to hydrolyze the β-1,4-glucosidic bonds, producing new reducing and nonreducing ends that can be further broken down by exoglucanases [18]. The shorter cellodextrin chains that are produced by these enzymes, including the disaccharide cellobiose, are then degraded into glucose monomers by β-glucosidases [18]. Hemicellulose is a sugar polymer that is composed of a number of different types of pentose and hexose sugars [16]. Xylan is its main component and is the second most abundant polysaccharide in lignocellulose. As compared to cellulose, hemicellulose is more accessible to degradation by a range of enzymes with different substrate specificities, including among others: xylanases, arabinases, and mannanases. Finally, lignin surrounds and blocks enzyme access to cellulose and hemicellulose and is a complex polymer containing a mixture of phenolic compounds linked through radical coupling reactions [19]. A large number of enzymes are needed to degrade it, including peroxidases and laccases [20].

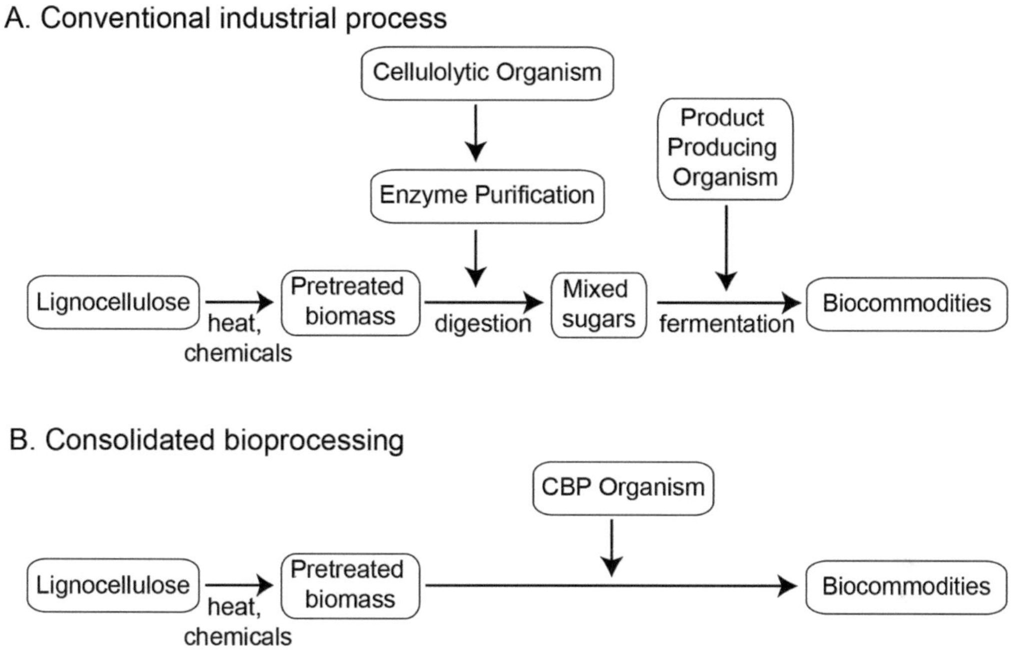

Figure 1. Conventional and consolidated bioprocessing (CBP) methods to convert biomass into biocommodities. (A) Many conventional industrial methods first pretreat biomass using heat and chemicals to remove lignin and to partially digest cellulose and hemicellulose. Complex enzyme mixtures are then added to the pretreated biomass to degrade its cellulose more fully. The resultant sugars are then fermented by a microorganism to generate the final product (currently only cellulosic ethanol). (B) A consolidated bioprocessing (CBP) microorganism could directly convert either pretreated or untreated lignocellulose into a biocommodity. It would bypass multiple steps in the conventional process, including the use of purified enzymes to degrade the biomass.

One promising recombinant strategy to create a useful CBP microbe is to engineer it to display a range of enzymes that degrade lignocellulose, thereby allowing it to produce sugars that can be further metabolized by the microorganism into useful chemicals [21–26]. In this approach, the goal was to engineer the microbial surface to effectively mimic the activity of naturally cellulolytic anaerobic bacteria. These microbes have the impressive capacity to adhere to, and degrade, untreated biomass and are typified by the cellulolytic anaerobic bacterium *Clostridium thermocellum* [27]. This eubacterium efficiently degrades biomass using a surface displayed complex called a cellulosome (**Figure 2A**) [27]. The cellulosome contains a central scaffolding protein that coordinates the binding of different enzymes. The primary scaffoldin, CipA, contains nine type-I cohesin modules that bind to cellulases harboring type-I dockerin modules [28]. CipA also contains a carbohydrate-binding module (CBM) that tethers the cellulosome complex to its cellulose substrate and a C-terminal type-II dockerin module [29]. The type-II dockerin module anchors the cellulosome complex to cell surface proteins by interacting with their type-II cohesin domains. As these surface proteins can contain multiple cohesin domains (1, 2, or 7 domains), large polycellulosomal structures can be displayed. Even more complex polycellulosomal structures exist in other species of bacteria and can contain over 100 enzymes [28].

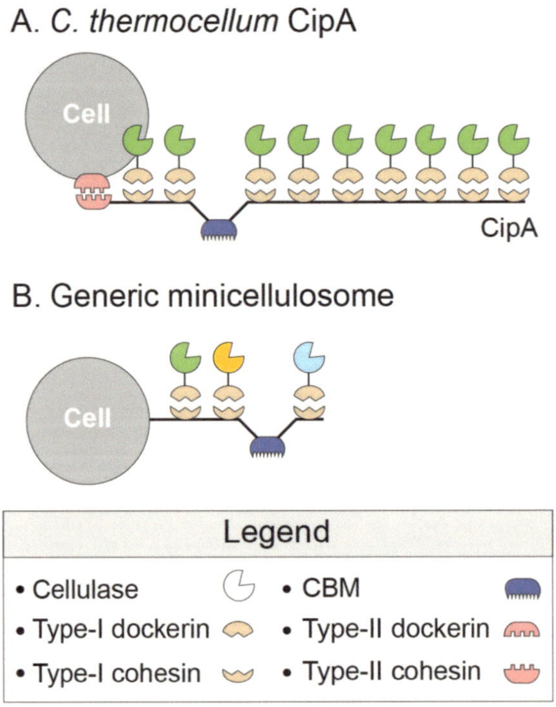

Figure 2. Natural cellulosomes and recombinant minicellulosomes. (A) The prototypical cellulosome from *C. thermocellum*. It contains the CipA scaffoldin that is capable of coordinating the binding of nine cellulases. CipA contains nine type-I cohesin domains that bind to cellulases containing type-I dockerin domains. It also contains a carbohydrate-binding module (CBM) that can bind to cellulose, holding the enzymes and the microorganism in close proximity to the substrate. CipA is anchored to the cell surface through its type-II dockerin domain that interacts with type-II cohesin modules present in the cell wall bound SdbA, Orf2, or OlpB proteins. (B) Basic structure of a minicellulosome that is displayed on a recombinant microbe. It is similar to the cellulosome, but contains fewer binding sites for enzymes. Color and symbol code: grey circles, cells; pink units, type-II cohesin-dockerin pairs; tan units, type-I cohesin-dockerin pairs; dark blue comb, CBM; and partial circular units: cellulases.

Microbes displaying cellulosomes are believed to degrade lignocellulose much more efficiently than microbes that degrade biomass by secreting cellulases [30]. There are three main reasons why increased efficiencies may be obtained. First, secreting enzymes presumably imposes a higher metabolic burden upon the microorganism as compared to displaying the enzymes on the cell surface. This is because the secreted enzymes can be lost to the environment with no guarantee that the sugars that they will produce will ultimately be accessible to the microbe for use as nutrients. As a result, larger quantities of the enzymes must be produced if they are to be secreted. For example, aerobic fungi (such as *Trichoderma reesei*, from which many purified industrial enzyme cocktails are derived) secrete 1–10 g of enzymes per liter of culture to degrade biomass, while cellulolytic anaerobic bacteria (such as *C. thermocellum*) need to produce only ~0.1 g of enzymes per liter of culture [31]. Second, colocalization of enzymes with different substrate preferences within a cellulosome promotes synergistic enzyme-enzyme and enzyme-proximity interactions, where the cellulolytic activity of the complexed enzymes is greater than that of individual enzymes due to their complementary activities and

http://dx.doi.org/10.5772/65509

optimal enzyme spacing [32]. The presence of both hemicellulases and cellulases in a cellulosome complex also enables hemicellulose and cellulose fibers to be removed simultaneously, thereby overcoming potential physical hindrances. The benefits of the *C. thermocellum* cellulosomal system have been quantified: its specific activity against crystalline cellulose is 15-fold higher than the secreted enzyme system from *T. reesei* [33]. Finally, the placement of the cellulosome on the microbial surface increases the rate of hydrolysis by promoting cellulose–enzyme–microbe (CEM) synergy [34]. In this process, sugar uptake by the microbe presumably becomes more efficient by promoting import of the products into cells and removing potential enzyme inhibitors such as glucose and cellobiose from the environment [35].

Microbe-based CBP technologies are not currently used industrially to produce biocommodities from lignocellulosic biomass. However, a major step towards CBP of starch into ethanol has recently been demonstrated by Lallemand and Mascoma. These companies created TransFerm®, a genetically modified yeast strain that secretes a glucoamylase and that is optimized to produce higher yields of ethanol. Although the cells do not fully degrade starch on their own, they reduce by 20–45% the amount of exogenous enzymes that needs to be added to process starch. At present, microbe-based CBP of lignocellulose is not being performed industrially, and ongoing research is primarily focused on constructing and identifying microorganisms with optimal cellulolytic and biocommodity production capabilities. Furthermore, detailed cost analyses of CBP versus conventional pretreatment and saccharification approaches have not been reported [36]. This is because the specific costs associated with CBP will vary based on the biocommodity produced, and the microbe and biomass source that is employed. However, the greatest cost savings associated with CBP will likely be obtained by reducing the costs of saccharification. A detailed cost analysis has been performed for cellulosic ethanol production from corn stover using dilute acid pretreatment, enzymatic saccharification, and cofermentation [37]. In this analysis, on-site production of fungal enzymes was estimated to contribute $0.34 per gallon of ethanol (assuming enzyme loadings of 20 mg enzyme per gram of biomass), which could in principle be eliminated using a CBP microbe. Another analysis by Johnson explored the potential cost savings associated with altering the source of enzyme production from off-site to on-site cultivation, specifically on biomass as a primary substrate [38]. The estimated full cost of producing cellulosic ethanol was reduced by 19% if the enzymes were produced on-site because it eliminated the need for enzyme purification, formulation of the enzyme mixture to preserve stability, and transport. Similar substantial gains could be obtained using microbe-based CBP.

In this chapter, we review progress towards engineering microbes to display "minicellulosomes," smaller cellulosome-like complexes that can degrade biomass (**Figure 2B**). A list of the microorganisms engineered to display minicellulosomes discussed in this chapter is presented in **Table 1**. We discuss recent developments in displaying these structures on industrially useful microorganisms, including *Saccharomyces cerevisiae, Escherichia coli, Bacillus subtilis, Corynebacterium glutamicum,* and lactic acid bacteria. The mechanisms used to display enzymes, their cellulolytic activities, and current obstacles that limit their utility are discussed.

Anchor	Assembly (# enzymes/ complex)	Enzymes displayed		References
S. cerevisiae				
Aga2	*Ex vivo* (3)	*C. cellulolyticum*:	Exoglucanase (CelE)	[73]
		C. thermocellum	Endoglucanase (CelA)	
		and		
		C. cellulolyticum	Endoglucanase (CelG)	
		or		
		C. thermocellum:	β-Glucosidase (BglA)	
Aga2	*Ex vivo* consortium (3)	*C. thermocellum*	Endoglucanase (CelA)	[74]
		T. aurantiacus	β-Glucosidase (Bgl1)	
		and		
		C. cellulolyticum	Exoglucanase (CelE)	
		or		
		T. reesei:	Cellobiohydrolase (CBHII)	
α-Agglutinin	*Ex vivo* consortium (3)	*C. thermocellum*: *T. aurantiacus*: *T. reesei*:	Endoglucanase (CelA) β-Glucosidase (Bgl1) Cellobiohydrolase (CBHII)	[75]
Aga2	*Ex vivo* consortium (2)	*C. thermocellum*: *T. reesei*:	Endoglucanase (CelA) Exoglucanase (CBHII)	[76]
Aga2	*Ex vivo* adaptive assembly (4)	*C. cellulolyticum*: *C. thermocellum*:	Endoglucanase (CelG) β-Glucosidase (BglA)	[77]
Aga2	*In vivo* (3)	*A. aculeatus*: *T. reesei*:	β-Glucosidase 1 (BGL1) Cellobiohydrolase II (CBHII) Endoglucanase II (EGII)	[32]
Aga2	*In vivo* (5)	*A. aculeatus*: *T. reesei*: *H. insolens*: *T. aurantiacus*:	β-Glucosidase 1 (BGL1) Cellobiohydrolase II (CBHII) Endoglucanase II (EGII) Cellobiose dehydrogenase (CDH) Lytic polysaccharide monooxygenase (LPMO)	[78]
Aga2	*In vivo* adaptive assembly (12 max)	*C. cellulolyticum*:	Endoglucanase (CelCCA) Cellobiohydrolase (CelCCE) β-Glucosidase (Ccel_2454)	[79]
α-Agglutinin	*Ex vivo* (3)	*A. awamori*: *A. niger*: *T. lanuginosus*:	Acetylxylan esterase (AwAXEf) β-xylosidase (XlnDt) Endoxylanase (XynAc)	[87]
Aga2	*In vivo* (3)	*A. niger*:	Arabinofuranosidase (AbfB)	[88]

http://dx.doi.org/10.5772/65509

Anchor	Assembly (# enzymes/ complex)	Enzymes displayed		References
S. cerevisiae				
		T. reesei:	β-Xylosidase (XlnD)	
			Endoxylanase (XynII)	
α-Agglutinin	*In vivo* (4)	*A. aculeatus*:	β-Glucosidase 1 (BGL1)	[89]
		T. reesei:	Endoglucanase II (EGII)	
Aga2	*Ex vivo* (N/A)	*C. cellulolyticum*:	Endoglucanase (CelA)	[90]
B. subtilis				
LysM	*Ex vivo* (3)	*B. subtilis*:	Endoglucanase (Cel5)	[31]
		C. phytofermentans:	Cellobiohydrolase (Cel48)	
		C. thermocellum:	Endoglucanase (Cel9)	
LPXTG from *S. aureus*	*Ex vivo* (3)	*C. cellulolyticum*:	Exoglucanase (Cel9E)	[118]
		C. thermocellum:	Endoglucanase (Cel9G)	
			Endoglucanase (Cel8A)	
Lactic acid bacteria				
LPXTG from *S. pyogenes*	*Ex vivo* (1-2)	*E. coli*:	β-Glucuronidase (UidA)	[127]
LPXTG from *S. pyogenes*	*Ex vivo* (2)	*E. coli*:	β-Glucuronidase (UidA)	[128]
			β-Galactosidase (LacZ)	
LPXTG from *L. plantarum*	*Ex vivo* consortium (2)	*T. fusca*:	Endoglucanase (Cel6A)	[129]
			Xylanase (Xyn11A)	
C. glutamicum				
MscCG	*In vivo* (2)	*C. thermocellum*:	Endoglucanase (CelE)	[133]
			β-Glucosidase (BglA)	

Table 1. Multi-enzyme display in minicellulosomes.

2. Displaying enzymes on *Saccharomyces cerevisiae*

Significant effort has been put forth to display cellulolytic enzymes on the surface of *S. cerevisiae* because of its established role in producing bioethanol [39]. The yeast cell surface is comprised of β-glucans, mannoproteins, and small amounts of chitin [40]. As β-glucan is the major constituent, displayed proteins are typically anchored to this fibrous scaffold. The most widely used approach to display cellulases and minicellulosomes employs a glycosylphosphatidylinositol (GPI)-anchoring system (**Figure 3A**). The protein of interest is expressed as a fusion with a polypeptide GPI attachment signal that is typically derived from the α-agglutinin protein or other cell wall proteins such as Sed1 and Cwp2 [41, 42]. After protein synthesis, GPI is added to the ω-site amino acid in the anchor signal sequence by the GPI transamidase complex in the endoplasmic reticulum [43, 44]. The GPI-linked protein is then directed towards the lipid bilayer and subsequently covalently linked to the cell wall β-1,6-glucan by the putative Dfg5 and Dcw1 cross-linkers [45–48]. Cellulase can also be displayed indirectly. In this approach, the enzyme is expressed as a fusion protein with the a-agglutinin Aga2 subunit, which in turn, forms disulfide bonds to the Aga1 subunit that is covalently linked to the cell wall via a GPI anchor. An estimated 1×10^4 to 1×10^5 proteins per cell are displayed using the Aga2 subunit [49]. The number of proteins displayed on the surface can be increased by genetically modifying the yeast strain. For example, deletion of a major endogenous GPI-anchored cell wall protein, SED1, can greatly improve heterologous protein display by reducing competition for cell wall-anchoring sites [50]. Display levels can also be improved by lowering the mannan content of the cell surface [51] and by employing the SED1 signal peptide, promoter and anchor [52, 53]. Reducing protein glycosylation also improves minicellulosome display [54]. The general strategies used to enhance enzyme display in yeast may also be useful in optimizing the other recombinant microbes that are discussed below.

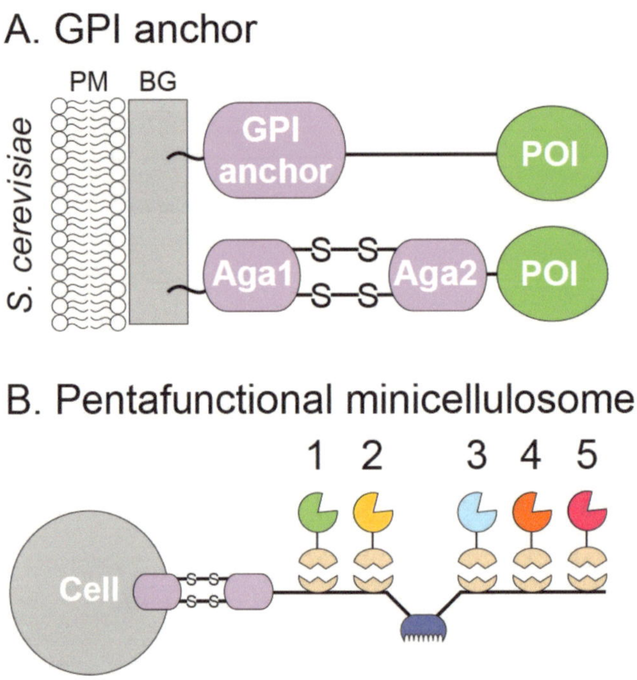

Figure 3. Cellulase and minicellulosome display in *S. cerevisiae*. (A) Methods used to display cellulases and their complexes on the cell surface. Proteins of interest (POI) are attached to the cell surface by a GPI anchor that is covalently linked to the β-glucan of the cell wall. The GPI anchor is attached to the displayed protein by expressing it as a fusion protein that contains the GPI attachment signal from either the α-agglutinin, SED1, or Cwp2 proteins. Proteins can also be displayed indirectly by expressing them as a fusion with the Aga2 protein that in turn interacts with the GPI anchored a-agglutinin Aga1 located on the cell surface. (B) One of the most active cellulolytic minicellulosomes that has been displayed on the surface of *S. cerevisiae*. It contains five enzyme functionalities and is displayed through fusion of the miniscaffoldin to the Aga2 protein. Symbols have been described in **Figure 2**. Enzymes 1–5 correspond to *T. reesei* endoglucanase (EGII), *T. reesei* cellobiohydrolase (CBHII), *A. aculeatus* β-glucosidase (BGL1), *T. aurantiacus* GH61a lytic polysaccharide monooxygenase (LPMO), and *H. insolens* cellobiose dehydrogenase (CDH), respectively. Abbreviations: POI, protein of interest; PM, plasma membrane; BG, β-glucan.

2.1. Individual cellulases

Individual cellulases were originally displayed on the cell surface by the Tanaka group [55]. They engineered cells to codisplay individual carboxymethylcellulase (CMCase) and β-glucosidase (BGL1) enzymes derived from *Aspergillus aculeatus* by fusing them to a C-terminal GPI anchor sequence from α-agglutinin. The cells could degrade cellodextrins, soluble glucose polymers that contain up to six glucose residues. Subsequently, the Kondo laboratory significantly advanced this technology by developing *S. cerevisiae* strains that display additional enzymatic activities. Initially, they constructed strains that codisplayed the *A. aculeatus* BGL1 and *T. reesei* endoglucanase II (EGII) enzymes, which when cultured in nutrient-rich media fermented β-glucan into ethanol [56]. Later, the inclusion of the *T. reesei* cellobiohydrolase II (CBHII) enzyme allowed the strains to ferment more complex amorphous phosphoric acid swollen cellulose (PASC) into ethanol [57]. Separately, the Kondo group also demonstrated that cellulose degradation can be further improved using a variety of approaches, including engineering a displayed cellulase to contain multiple cellulose-binding domains, integrating multiple copies of the cellulases genes into the yeast chromosome, and by strain diploidization [58–60]. Increasing the number and density of the displayed cellulases significantly improves the ability of *S. cerevisiae* to degrade cellulose. For example, a strain displaying the BGL1, EGII, and CBHII enzymes produces more ethanol from PASC than a strain that displays only BGL1 and secretes EGII and CBHII [61, 62]. Enzyme proximity on the cell surface also appears to have a major effect on activity, as the cellulolytic properties of cells displaying one or more cellulases improved as the enzyme density increased [63]. Interestingly, in these studies, good activity was observed when the enzymes were estimated to be separated on the cell surface by 10–100 nm, an enzyme spacing that is presumably similar to the spacing found in certain types of bacterial cellulosomes [64].

Yeast displaying cellulases may be industrially useful. For example, strains displaying the BGL1, CBHII, and EGII enzymes produced 43.1 g/L ethanol from 200 g/L liquid hot water pretreated rice straw in 72 h [65]. While supplementation with a purified cellulase cocktail at 10 filter paper units (FPU)/g-biomass was necessary to achieve this high ethanol yield, a control strain that did not display the enzymes required 10-fold more added cellulase to produce similar quantities of ethanol. Attractively, the cells displaying the enzymes could be reused in five fermentation cycles without significantly losing their activity [66]. Displaying enzymes also reduces the amount of purified cellulase lost through irreversible adsorption onto crystalline cellulose, facilitating more efficient cellulose degradation and higher ethanol yields [67]. Recently, studies using cellulase displaying cells have further reduced the amount of purified cellulase cocktail that needs to be added to convert biomass into ethanol [68]. These newer generation cells require 44% less commercial enzyme supplementation to degrade pretreated biomass by displaying four enzymes using the Sed1 anchor: *A. aculeatus* BGL1, *T. reesei* EGII, *Talaromyces emersonii* cellobiohydrolase I (CBHI), and *Chrysosporium lucknowense* cellobiohydrolase II (CBH2). When supplemented with 1 FPU of commercial enzyme cocktails, the cells yielded 18 g/L ethanol from 100 g/L of pretreated and milled rice straw in 96 h, obtaining 80% of the theoretical yield. In contrast, without enzymatic supplementation, only 7% of the theoretical yield was obtained using the same cells. Therefore, while the use of cellulase displaying yeast cells can significantly reduce the amount of commercial enzyme that needs to be added to degrade pretreated biomass, current generation cells still require enzyme supplementation.

Towards the goal of improving their cellulolytic activity, several studies have engineered cellulase displaying yeast cells to also produce complementary enzymes and transporters that improve cellulose solubilization and utilization. The Kondo group constructed cells that co-expressed three displayed cellulases, as well as the *Neurospora crassa* cellodextrin transporter (CTDI) and intracellular BGL1 enzyme [69]. This strain produces 1.7-fold more ethanol from PASC than a strain that only displays the cellulases, presumably by reducing the build-up of enzyme inhibitory products on the cell surface. Additionally, strains displaying cellulases as well as proteins that disrupt the structure of cellulose show improvements in ethanol production; cells codisplaying three cellulases and an *Aspergillus oryzae* expansin-like protein (AoelpI) produce 1.4-fold more ethanol from PASC than a control strain that only displayed the cellulases [70]. Finally, the Ueda group demonstrated that sequential exposure of biomass to engineered yeast cells could be beneficial. In this work, pretreatment of hydrothermally processed rice straw with yeast displaying the *Trametes* sp. Ha1. laccase I enzyme enabled 1.21-fold more ethanol to be produced from the biomass after it was subsequently exposed to yeast displaying three cellulases [71]. These studies highlight interesting enzyme functionalities that should be considered for inclusion in other recombinant microorganisms.

2.2. *Ex vivo* assembled minicellulosomes

Several research groups engineered *S. cerevisiae* to display minicellulosomes that are smaller than natural cellulosomes (**Figure 2B**). In the minicellulosome, a cell-surface displayed "miniscaffoldin" protein that contains cohesin domains noncovalently binds cellulases via their dockerin modules. These recombinant miniscaffoldins are either attached directly to the cell wall by fusing them to a GPI anchor sequence (from α-agglutinin or Cwp2) or indirectly by fusing them to the a-agglutinin Aga2 subunit. Two general methods are used to display minicellulosomes, an *ex vivo* approach in which cellulases need to be added to cells that display a miniscaffoldin (**Figure 4A**, **4B**), or an *in vivo* approach in which the microbe produces all of the protein components that are necessary to assemble and display the minicellulosome (**Figure 4C**).

In 2009, two groups demonstrated *ex vivo* minicellulosome assembly by creating cells that display a miniscaffoldin and then adding to the cells cellulases that contained a dockerin domain. The Volschenk group displayed a miniscaffoldin by fusing two cohesin modules and a CBM to the Cwp2 GPI anchor. They visually demonstrated binding of this yeast strain to filter paper via the CBM and constructed a minicellulosome by adding a purified endoglucanase–dockerin enzyme to the cells [72]. Concurrently, the Chen laboratory constructed an Aga2-fused recombinant minicellulosome that contained three cohesin modules derived from three different bacterial species [73]. This more elaborate minicellulosome used species-specific cohesin–dockerin interactions to target β-glucosidase, exoglucanase, and endoglucanase enzymes to specific sites within the complex. Quantitatively, the researchers demonstrated that incorporation into the minicellulosome enabled the enzymes to function synergistically to

http://dx.doi.org/10.5772/65509

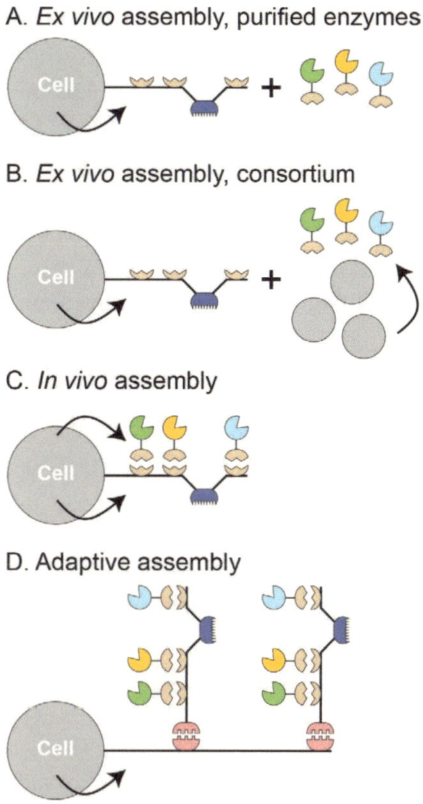

Figure 4. Summary of strategies used to assemble minicellulosomes on microbial surfaces. (A) *Ex vivo* assembly. Exogenous cellulase-dockerin fusion proteins (either purified or derived from cell lysates) are added to cells engineered to display a miniscaffoldin that binds to the cellulases. (B) *Ex vivo* assembly using a microbial consortium. Similar to (A), but miniscaffoldin displaying cells are cultured with a consortium of microbes that have been engineered to secrete cellulase-dockerin fusion proteins. (C) *In vivo* assembly. A single microorganism is engineered to express all of the components of the minicellulosome, enabling its spontaneous assembly on the cell surface. (D) Adaptive assembly method used to construct larger minicellulosomes. The miniscaffoldin contains two basic parts, a non-catalytic anchor-anchoring scaffoldin that attaches the structure to the cell wall and a primary scaffoldin that binds to the cellulase-dockerin fusion proteins. Color and symbol code as described in **Figure 2**.

produce ethanol from PASC. Later, the Chen laboratory bypassed the need to add purified enzymes or cell lysates to cells to produce the minicellulosome by using a consortium of yeast strains to produce the cellulases [74]. This was accomplished using four strains: one that displayed the miniscaffoldin and three strains that secreted each of the enzymes needed to form the complex. After determining the optimal ratio of strains, they demonstrated that ~1.9 g/L of ethanol could be produced from 10 g/L of PASC in 48 h. Additionally, they demonstrated that fusing the miniscaffoldin directly to the cell wall via an attached α-agglutinin GPI anchor instead of indirectly through the Aga2 subunit improved minicellulosome display [75]. Similarly, the Hahn laboratory utilized a consortium approach to build an *ex vivo* assembled minicellulosome that contained three identical cohesin–dockerin pairs and randomly incorporated enzymes [76]. Cells displaying this structure also obtained similar ethanol yields, 1.8 g/L of ethanol from 10 g/L of PASC in 94 h.

In order to build larger surface structures with higher enzyme densities, the Chen laboratory used an adaptive assembly approach [77]. In this procedure, the minicellulosome is built using two separate scaffoldin pieces: a primary scaffoldin that binds to the catalytic components and an anchoring scaffoldin that is attached to the cell surface and only binds to the primary scaffoldin (**Figure 4D**). They produced a yeast strain that displayed an Aga2-fused anchoring scaffoldin that contained two cohesin domains from *Acetivibrio cellulolyticus* and *Bacteroides cellulosolvens*. Two *E. coli* strains were then used to produce two primary scaffoldins that each contained the *C. thermocellum* and *R. flavefaciens* cohesins, and either the dockerin module from *A. cellulolyticus* or *B. cellulosolvens*. Thus, the final complex contained a total of four enzymes, with two copies each of the *C. cellulolyticum* endoglucanase (CelG) and *C. thermocellum* β-glucosidase (BglA) enzymes. This strain produced only 1.9 g/L of ethanol from 10 g/L of PASC in 72 h, presumably because it lacked exoglucanase activity.

2.3. *In vivo* assembled minicellulosome

The need to add enzymes to cells displaying a miniscaffoldin may make the use of *ex vivo* assembled minicellulosomes industrially impractical. To overcome this problem, two research groups engineered *S. cerevisiae* to express all of the components of the minicellulosome such that it could form spontaneously (*in vivo* assembly). The Zhao laboratory constructed a yeast strain that produced an Aga2-fused miniscaffoldin, *T. reesei* EGII and CBHII, and *A. aculeatus* BGL1 enzymes [32]. By comparing the display of a single enzyme to the simultaneous display of two or three noncomplexed enzymes, they demonstrated enzyme-enzyme synergy, which improved activity by 5.5-fold. They also demonstrated enzyme-proximity synergy, as the trifunctional minicellulosome degraded biomass better than cells containing three independently displayed enzymes (1.6-fold improvement). Quantitative measurements indicated that each cell displays 3×10^4 unifunctional miniscaffoldins, whereas 1.8×10^4 tri-functional minicellulosomes are displayed per cell. *In vivo* construction of increasingly more elaborate minicellulosomes decreased protein expression, likely due to metabolic burden. Compared to unifunctional minicellulosomes, the production of the miniscaffoldin and two enzymes caused the expression of all components to decrease slightly, and overexpression of an additional third enzyme caused protein expression levels of CBHII to drop significantly. This suggests that it will be increasingly more challenging to display larger, more complex enzyme structures on the cell surface. This is unfortunate, because adding additional enzyme functionalities is advantageous to cellulose solubilization. Although difficult to achieve, for example, the tri-functional displayed minicellulosome constructed by the Zhao group produces 1.8 g/L of ethanol from 10 g/L of PASC in 70 h. However, pentafunctional minicellulosomes are more effective, as they produce 2.7 g/L of ethanol from 10 g/L of PASC in 96 h and 1.8 g/L of ethanol from 10 g/L of Avicel in 96 h [78]. This larger pentafunctional complex contains two additional enzymes with new functionalities, the *Thermoascus aurantiacus* GH61a lytic polysaccharide monooxygenase (LPMO) and *Humicola insolens* cellobiose dehydrogenase (CDH) (**Figure 3B**). Even though fermentative growth was reduced because of the presence of oxygen needed for monooxygenase function, greater ethanol yields were still obtained.

The largest *in vivo* assembled minicellulosome reported thus far was constructed by the Tan group using an adaptive assembly approach [79]. This structure could display up to 12 enzymes and was formed using a cell surface-associated Aga2-fused anchoring scaffoldin that contained type-II cohesin domains that could bind four primary miniscaffoldins harboring type-II dockerin domains. Each primary scaffoldin also contained a CBM and three species-specific type-I cohesin modules for controlled cellulase binding. Combined, this strain displays in a single complex four copies each of the *C. cellulolyticum* endoglucanase, cellobiohydrolase, and β-glucosidase enzymes. However, the investigators found that as longer anchoring scaffoldins were used, a smaller percentage of cells displayed the scaffoldin. For this reason, they demonstrated the production of 1.4 g/L ethanol from 10 g/L Avicel in 96 h from a smaller six enzyme displaying strain.

2.4. Hemicellulases and hemicellulosomes

In order to develop xylose-fermenting strains of yeast, similar strategies have been employed to display hemicellulases and hemicellulosomes on the cell surface. The Kondo laboratory displayed an individual *T. reesei* endoxylanase II (XYNII) enzyme using the α-agglutinin anchor and demonstrated that the cells degraded birchwood xylan into xylobiose, xylotriose, and xylotetraose [80]. Subsequently, the cells were further engineered to produce ethanol from birchwood xylan by adding the *A. oryzae* β-xylosidase (XylA) to the cell surface and by expressing intracellular xylose utilization proteins (the *Pichia stipitis* xylose reductase (XYL1) and xylitol dehydrogenase (XYL2), as well as the *S. cerevisiae* xylulokinase (XKS1)) [81]. In a separate set of studies, the investigators later demonstrated cellodextrin degradation and xylose assimilation by yeast displaying *A. aculeatus* BGL1 enzyme, which enabled the microbe to produce ethanol from the sulfuric acid hydrolysate of wood chips [82]. Interestingly, the ability of this strain to co-utilize cellobiose and xylose avoided carbon catabolite repression [83]. Additional engineering created cells that displayed the XylA, XYNII, and BGL1 enzymes, enabling the xylose-utilizing strain to produce 8.2 g/L of ethanol from 80% (v/v) rice straw hydrolysate [84]. When *P. stipitis* xylose reductase (XR) was produced intracellularly in place of the three aforementioned xylose utilization proteins, xylitol accumulated, which is a starting material for chemical production of some pharmaceuticals [85]. Using an alternative pathway to utilize xylose, the Ueda group produced α-agglutinin anchored *Clostridium cellulovorans* xylose isomerase (XI) which converted xylose to xylulose extracellularly before fermenting it to ethanol [86].

Clustering hemicellulases within surface displayed complexes leads to improved enzymatic activity. The Silva group developed an *ex vivo* assembled xylanosome using the α-agglutinin anchor [87]. The xylanosome contained the *Thermomyces lanuginosus* endoxylanase (XynAC), *Aspergillus niger* β-xylosidase (XlnDt), *Aspergillus awamori* acetylxylan esterase (AwAXEf) enzymes, as well as the xylose-binding domain (XBD) from *Thermotoga maritima*. The xylanases in the complex functioned synergistically, with enzyme-enzyme synergy improving xylan hydrolysis by 1.6-fold, and enzyme-substrate synergy improving hydrolysis by 3.3-fold as compared to the free enzymes. Proximity of the endoxylanase to the XBD improved xylan hydrolysis by 2.5-fold, suggesting that placement of substrate binding domains can contribute significantly to hydrolysis. The Zhao laboratory created an *in vivo* assembled minihemicellulosome that is structurally related to their minicellulosome complex described above [88]. The minihemicellulosome contained the *A. niger* arabinofuranosidase (AbfB), β-xylosidase (XlnD), and *T. reesei* endoxylanase (XynII) enzymes. The yeast strain also possessed *P. stipitis* xylose utilization enzymes and could hydrolyze arabinoxylan better than the uni- and bifunctional yeast strains. Surprisingly, against birchwood xylan, a bifunctional minihemicellulosome containing XynII and XlnD exhibited a higher hydrolysis rate than the tri-functional complex, producing 0.95 g/L of ethanol from 10 g/L of birchwood xylan in 80 h.

2.5. Artificial cellulosome structures

In order to better control enzyme placement and increase the number of displayed enzymes on the cell surface, several groups have created artificial cellulosomes that are structurally distinct from naturally occurring cellulosomes. These structures use unique protein-protein interaction domains to tether the enzymes to the scaffoldin instead of naturally occurring cohesin-dockerin interactions. The Kondo laboratory created an *in vivo* assembled minicellulosome using a miniscaffoldin that contained two cohesin domains and two Z domains derived from *Staphylococcus aureus* protein A [89]. This enabled codisplay of the *A. aculeatus* BGL1—dockerin fusion protein and the *T. reesei* EGII protein fused to the Fc domain of human IgG. The group suggested that this artificial construct enables tighter regulation of the display ratio of the enzymes. Similarly, the use of type-I and type-II cohesin-dockerin pairs, or cohesin-dockerin pairs from different species, allows control over enzyme incorporation. The advantages of this noncellulosome-like structure were not extensively investigated, but activity against β-glucan was demonstrated. The Su group has displayed enzymes using a novel *ex vivo* assembled complex that contains an amyloid-like scaffoldin [90]. A multistep process was required to assemble the complex. Initially, yeast expressed an Aga2 anchored GFP-specific antibody fragment. Then, cells were incubated with a GFP-dockerin fusion protein to prime the cells to bind the protein scaffold. The protein scaffold itself was then created through the fibrillation of a recombinant-purified protein that contained a cohesin and a hydrophilic linker region that was fused to the N-terminus of Ure2, an amyloid-like yeast protein. In this way, multiple cohesin domains were incorporated into a large protein scaffold. As compared to cells displaying a single cohesin-dockerin pair harboring the *C. cellulolyticum* endoglucanase (CelA), cells displaying the amyloid-like scaffoldin had 8.5-fold greater activity against CMC. However, the process of assembly is complex, and certain cohesin domains may be rendered nonfunctional during fibrillation. The use of extremely large supramolecular scaffoldins to coordinate enzyme binding may also reduce the potential benefits of CEM synergy. Whether the enzymatic activities of these artificial cellulosomes are superior to complexes that more closely resemble natural cellulosomes remains to be determined, as their ability to degrade complex cellulose substrates to produce ethanol was not reported.

3. *Escherichia coli* surface display of individual cellulases

The model Gram-negative bacterium *E. coli* has several features that are attractive for CBP, including a robust genetic system that enables surface and metabolic engineering, as well as the microbe's innate ability to utilize the main components of lignocellulose, glucose, arabinose, and xylose [91, 92]. Its cell envelope is comprised of an inner membrane, a peptidoglycan layer, and an outer membrane. Only individual, noncomplexed cellulases and xylanases have been displayed on the surface of *E. coli* by fusing them to lipoproteins or integral membrane proteins (**Figure 5**). Lipoprotein anchors that have been used to display cellulases include the *E. coli*-derived Lpp-OmpA fusion protein [93, 94], the *P. syringae* ice nucleation protein (INP) [95–97], and the *E. coli* bacterial lipocalin (Blc) protein [98–100]. Cellulases have also been displayed by fusing them to outer membrane proteins such as the: *B. subtilis* poly-ɣ-glutamate synthase A (PgsA) [101, 102], *E. coli* outer membrane protein C (OmpC) [103], and the *E. coli* outer membrane protein X (OmpX) [104]. Cellulases can also be displayed through a unique mechanism in which they are fused to the *E. coli* autotransporter protein (AIDA-I) [105]. Other less commonly used approaches to display cellulases include fusing them to the *E. coli* inner membrane protein HdeD [100], or to the *B. anthracis* BclA exosporal protein whose mechanism of display in *E. coli* is not well understood [106]. In most cases, except for the work reported by the Kondo and the Karim groups, only single enzymes have been displayed on the surface, leading to cells that have limited cellulolytic activity [100, 101].

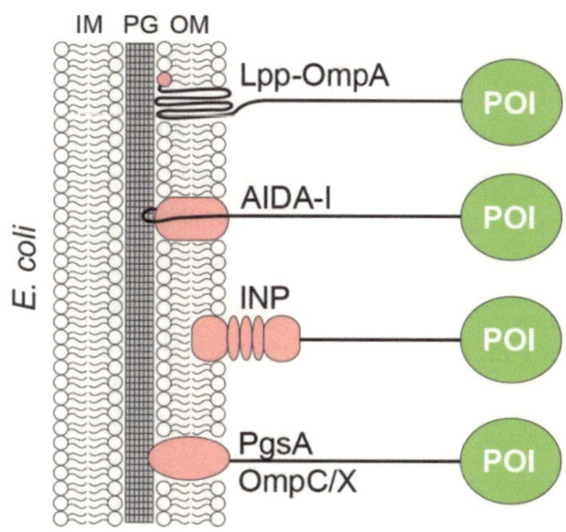

Figure 5. Methods used to display cellulases in *E. coli*. Proteins of interest are displayed as a result of fusion to lipoproteins (*E. coli* Lpp-OmpA, *P. syringae* INP, *E. coli* Blc), fusion to outer membrane proteins (*B. subtilis* PgsA, *E. coli* OmpC, *E. coli* OmpX), or fusion to the *E. coli* autotransporter AIDA-I. Not all anchoring methods are depicted here. At present, only individual enzymes have been displayed on the cell surface. Abbreviations: protein of interest (POI), inner membrane (IM), peptidoglycan (PG),and outer membrane (OM).

Pilot studies have shown that *E. coli* displaying cellulases can be used to produce ethanol or isopropanol from cellulosic substrates. Using the *E. coli* AIDA-I anchor, the display of the *T. fusca* β-glucosidase enzyme yielded cells that produced approximately 17–17.9 g/L of ethanol from 40 g/L of cellobiose in 72 h [105]. By utilizing the *E. coli* Blc anchor to display the *T. fusca* β-glucosidase, cells could produce 69.0 ± 11.6 mM of isopropanol in a 21 h fermentation containing 50 g/L of cellobiose [99]. The low yield of isopropanol, as compared to when glucose was supplied, suggested that the β-glucosidase activity was too low to efficiently produce glucose from cellobiose. The most complex *E. coli* display system thus far achieved displayed three enzymes using the *B. subtilis* PgsA anchor [101]. This strain displayed endocellulase, exocellulase, and β-glucosidase enzymes derived from *C. cellulolyticum*. From 10 g/L of PASC, it produced 3.59 ± 0.15 g/L of ethanol in 60 h. It could also make ethanol from dilute sulfuric acid pretreated corn stover hydrolysate (2 g/L), producing 0.71 ± 0.12 g/L of ethanol in 48 h. One notable study performed by Bokinsky et al. demonstrated that *E. coli* strains that secreted cellulases could produce advanced biofuels from ionic liquid pretreated biomass [107]. A consortium of two *E. coli* strains was used to produce either the *Bacillus* sp. D04 endocellulase and the *Cellvibrio japonicus* β-glucosidase, or the *Clostridium stercorarium* endoxylanase and the *C. japonicus* xylobiosidase. The consortia could grow on ionic liquid (IL)-treated biomass, albeit to half the cell density of strains grown on glucose. Additional engineering enabled cells to produce fatty-acid ethyl esters, butanol, and pinene from IL-pretreated switchgrass. However, only 5 and 6% of the available cellulose and hemicellulose were digested, respectively. Improved cellulolytic activity could be achieved by displaying cellulase containing complexes, but *E. coli* cells that display minicellulosomes have yet to be constructed.

4. Displaying cellulases and minicellulosomes in Gram-positive bacteria

Many species of Gram-positive bacteria are used industrially to produce biocommodities and have great promise as agents to produce second-generation biofuels. However, they are not naturally cellulolytic, prompting efforts designed to decorate their surfaces with cellulases. Below, we discuss progress towards creating cellulolytic strains of *B. subtilis*, *C. glutamicum*, and several industrially useful species of lactic acid bacteria. A common feature of these microbes is the absence of an extensive outer membrane, as they are surrounded by a thick peptidoglycan layer that in some instances is further surrounded by additional protective layers. Thus, the mechanisms used to display proteins on their surfaces are distinct from those employed to decorate Gram-negative bacteria.

4.1. *Bacillus subtilis*

B. subtilis is a model Gram-positive bacterium that is used industrially to produce commercial enzymes. It has several desirable traits that could enable its use in consolidated bioprocessing including established genetic tools to manipulate its genome as well as its generally recognized as safe status (GRAS) [108, 109]. It is also tolerant to high concentrations of salts and solvents and has the ability to utilize both pentose and hexose sugars produced from lignocellulose [110]. Finally, *B. subtilis* naturally secretes large quantities of extracellular enzymes (20–25 g of protein per liter of growth culture), suggesting that it should be capable of robustly exporting the enzymes needed to build cellulase containing complexes [111, 112]. The Gram-positive cell wall offers many sites for both covalent and noncovalent protein anchoring. Three mechanisms have been used to display cellulase on vegetative *B. subtilis* cells: membrane association via a lipoprotein anchor, noncovalent cell wall interactions using the LysM domain, and covalent attachment to the cell wall using sortase enzymes (**Figure 6A**). Covalent attachment occurs through sortase enzyme processing of a C-terminal cell wall sorting signal (CWSS) that contains a LPXTG motif and has been estimated to display ~2.4×10^5 proteins per cell [113, 114]. Noncovalent methods enable ~1.2×10^7 proteins to be displayed per cell when the protein of interest is fused to the LysM domain, a binding module that interacts with the N-acetyl-muramic acid and N-acetyl-D-glucosamine components of the peptidoglycan [115, 116]. Cellulases have also been displayed by expressing them as a fusion protein with the membrane-associated lipoprotein, PrsA [117]. However, this display mechanism requires lysozyme treatment of the cell to remove the peptidoglycan.

Two groups have displayed *ex vivo* assembled minicellulosomes on the surface of *B. subtilis*. The Zhang laboratory displayed a minicellulosome with a miniscaffoldin that contained three cohesin domains, a CMB, and three LysM domains [31]. An estimated 2×10^4 miniscaffoldins bound to the surface of each cell. Minicellulosomes were assembled by incubating the cells with purified *B. subtilis* endoglucanase, *C. thermocellum* endoglucanase, and *Clostridium phytofermentans* cellobiohydrolase enzymes. Against regenerated amorphous cellulose (RAC) and crystalline cellulose, respectively, cell-tethered minicellulosomes degraded substrate 2.3- and 4.5-fold better than free minicellulosomes. As compared to commercial fungal enzymes dosed at the same protein concentrations, minicellulosome-displaying cells degraded RAC to a similar extent after 72 h, but exhibited 30% greater hydrolytic activity on Avicel. The Clubb group also demonstrated *ex vivo* assembly of a minicellulosome complex using a LPXTG anchor that can be processed by a sortase enzyme [118]. Purified *C. cellulolyticum* endoglucanase and exoglucanase, and *C. thermocellum* endoglucanase enzymes associated with a covalently cell wall attached miniscaffoldin. A major challenge specific to *B. subtilis* is the large amount of proteases that this microbe produces that can degrade heterologous surface-exposed proteins [119]. The Zhang group addressed this problem by displaying the cellulosome in a *B. subtilis* strain in which six extracellular proteases had been deleted, while the Clubb group improved protein display by deleting WprA, a cell wall-associated protease. Display of an *in vivo* assembled minicellulosome has yet to be achieved. Notably, an *in vivo* assembled minicellulosome was reported that could degrade untreated biomass, but this work was later retracted.

B. subtilis sporulates to produce a highly stress-resistant, dormant spore cell that can be decorated with cellulases. During the process of spore formation within the mother cell, genomic DNA is encapsulated within multiple protective layers including a cortex, coat, and crust [120]. Attractively, proteins do not need to be translocated across the cytoplasmic membrane in order to be displayed on the spore. There has been a significant amount of research performed to optimize protein display on *B. subtilis* spores [121]. While spore coat proteins are typically used as carriers to display proteins, a recent study found that native, unmodified proteins could be overexpressed in the mother cell and absorbed to the spore surface for display (**Figure 6B**) [122]. Using this approach, monomeric *B. subtilis* carboxymethylcellulase (CelB) and multimeric *E. coli* β-galactosidase (LacZ) were successfully displayed on the spore surface, and vigorous physiochemical treatment was shown to be needed to

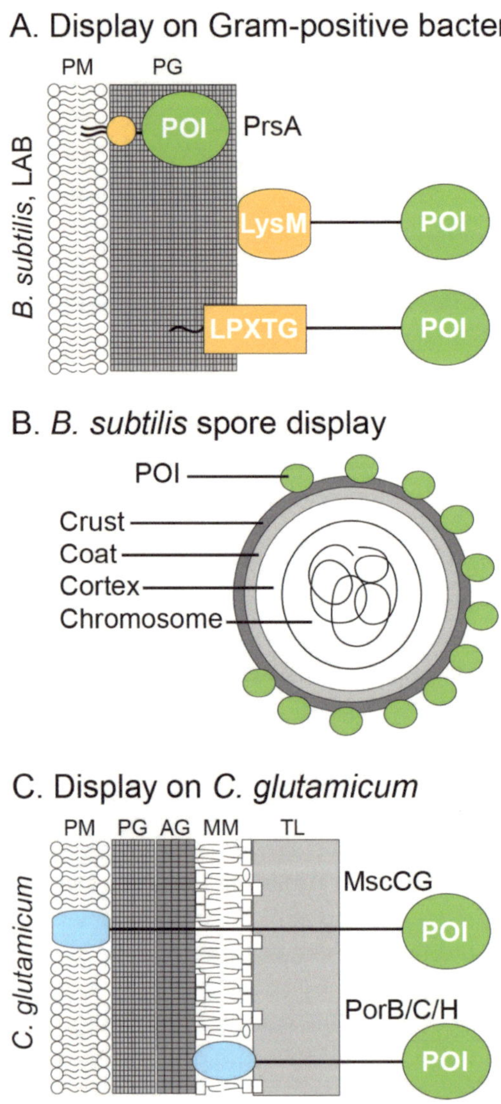

Figure 6. Cellulase and minicellulosome display in Gram-positive bacteria. (A) Methods used to display cellulases and their complexes. Proteins can be anchored to the plasma membrane through fusion to PrsA, but lysozyme treatment is necessary to expose this cellulase to the external environment. More conventionally, proteins are cell surface displayed either through noncovalent interactions with cell wall peptidoglycan by protein fusion to the LysM domain, or through covalent attachment to cell wall peptidoglycan by protein fusion to a cell wall sorting signal containing the LPXTG motif, which is processed by sortase transpeptidase. (B) Proteins can be displayed on the surface of *B. subtilis* spores. Native proteins, without fusion to anchor proteins, can be absorbed by the spore surface during spore formation. (C) Methods used to display cellulases and their complexes on the cell surface of *C. glutamicum*. Proteins of interest are displayed through fusion to membrane protein MscCG or porin proteins PorB, PorC, and PorH. Abbreviations: Protein of interest (POI), plasma membrane (PM), peptidoglycan (PG), arabinogalactan (AG), mycomembrane (MM), and top layer (TL).

remove the enzymes. Interestingly, proteins can also be directly adsorbed on the spore surface by incubating them with spores, resulting in 7.75×10^3 to 1.55×10^4 individual proteins being displayed per particle [123]. Thus, cellulases that cannot be secreted or expressed as fusion proteins can be readily displayed on spores using this method.

4.2. Lactic acid bacteria (LAB)

Gram-positive lactic acid bacteria (LAB) are widely used in the food industry to ferment sugars into lactic acid [124]. They have potential in biomass processing, as they can utilize pentose and hexose sugars, and some members of this group, namely *Lactobacillus plantarum*, are tolerant to low pH and ethanol concentrations up to 13% [26, 125]. Similar methods are used to display proteins in LAB and *B. subtilis*, as their cell wall envelopes are structurally related (**Figure 6A**). In particular, proteins can be displayed noncovalently using LysM-binding modules, or covalently using LPXTG anchors that are processed by sortase enzymes. Comparison studies using *Lactococcus lactis* have shown that covalent attachment using sortase enzymes leads to the largest number of functionally displayed proteins on this microbe's surface, but the relative efficiencies of the sortase and LysM display systems in other species of LAB have not been determined [126].

Two groups have used sortases to assemble minicellulosomes *ex vivo* in LAB. The Martin group anchored a variety of nuclease A-fused miniscaffoldin constructs onto the surface of a single protease-deficient strain of *L. lactis* [127]. These constructs either contained one cohesin, two cohesins, a cohesin and a CBM, or only a CBM. Using a purified *E. coli* β-glucuronidase (UidA)-dockerin fusion protein, they estimated that ~10^4 complexes are displayed per cell. Interestingly, unlike work in yeast, scaffoldin size was not a limiting factor for display. Instead, it was suggested that protein secretion might be limiting, as scaffoldins containing a CBM showed improved secretion, presumably due to more rapid folding facilitated by the CBM. Using similar approaches, this group later produced a bifunctional minicellulosome using specifically associating type-I and type-II cohesin-dockerin pairs [128]. They demonstrated that the order of enzyme docking affected the activities of the *E. coli* UidA and LacZ enzymes, due to steric factors that influenced enzyme binding.

In separate studies, the Mizrahi group engineered *L. plantarum* to display a sortase attached miniscaffoldin and then used a consortium of cellulase secreting strains to assemble the minicellulosome *ex vivo* [129]. In these studies, recombinant *T. fusca* endoglucanase (Cel6A) and xylanase (Xyn11A) activities were studied in their free form, individually bound to cells, or bound to surface displayed minicellulosomes. Against hypochlorite pretreated wheat straw, secreted enzymes had the best initial activity, followed by the minicellulosome. However, the sugar production rate of the surface displayed minicellulosome slowly increased over time, while the rate for the secreted enzymes decreased. Eventually, cellulosomal activity overtook that of the secreted enzymes, presumably because the cell-associated enzymes are more stable. Interestingly, cells displaying individually anchored enzymes had minimal activity. This is consistent with results obtained in yeast that have shown that enzymes with different substrate specificities are most active against cellulose when they are densely clustered with one another on the cell surface to promote synergistic interactions [63].

4.3. *Corynebacterium glutamicum*

C. glutamicum is an industrially important microbe that produces several tons of glutamate and lysine annually. Although it is a Gram-positive bacterium, its peptidoglycan layer is covered by an arabinogalactan layer, a mycolic acid bilayer, and a top layer composed of polysaccharides, glycolipids, and proteins [130]. The presence of the outer mycolic acid layer confers Gram-negative characteristics to the bacterium and, along with the cytoplasmic membrane, is the primary point to which cellulases are attached (**Figure 6C**). The Kondo laboratory demonstrated the feasibility of displaying heterologous proteins by attaching them to the mycolic acid layer [131]. This was achieved by expressing the cellulase as a fusion protein in which it was joined to the PorC porin. *C. glutamicum* displaying a *Saccharophagus degradans* β-glucosidase-PorC fusion produced 1.08 g/L of lysine from 20 g/L of cellobiose in 96 h [132]. The Han group also displayed an *in vivo* assembled bifunctional minicellulosome that associated with the cytoplasmic membrane. This was accomplished by expressing the scaffoldin as a fusion protein with the mechanosensitive channel (MscCG) [133]. The complex contained the *C. thermocellum* endoglucanase E (CelE) and β-glucosidase A (BglA) enzymes and could release sugars from pretreated rice straw, miscanthus, and rape stem. In the future, engineering these cells to more efficiently degrade complex lignocellulose may lead to more cost-effective ways to produce lysine and glutamate from inexpensive biomass.

5. Conclusions

Towards the goal of cost effectively converting biomass into useful biocommodities, several research groups have developed creative ways to display cellulases on microbes to endow them with cellulolytic activity. Comparing the biomass degradation efficiencies of different types of recombinant microorganisms is difficult. This is because investigators have measured their activities using a variety of cellulosic substrates that can vary dramatically in their resistance to enzymatic degradation as they differ in their solubility, enzyme accessibility, crystallinity, degree of polymerization, fraction of reducing ends, and the presence of hemicellulose and lignin [134]. Moreover, different methods are frequently used to pretreat biomass and to measure the extent of degradation, which can be reported as enzymatic activity, sugar released, biomass remaining, or product produced [17, 135, 136]. However, despite these qualifiers, the results of studies reported to date enable several major conclusions to be drawn. In particular, they provide convincing evidence that clustering enzymes on the surface within minicellulosomes leads to improved microbial cellulolytic activity by promoting synergistic interactions, and in some instances, by improving enzyme stability [32]. Interestingly, synergistic enzyme interactions can also be obtained by displaying different types of individual enzymes, as long as they are densely clustered and have complementary activities [63]. Because complexed enzyme systems require smaller amounts of enzymes to be produced than secreted enzyme systems to achieve efficient degradation of lignocellulose, the cellulosomal system may be optimal for CBP microorganisms, since conserved energy may be directed towards product production.

http://dx.doi.org/10.5772/65509

Displaying large, enzymatically diverse recombinant minicellulosomes remains a challenging problem, but progress has been made in *S. cerevisiae, B. subtilis, C. glutamicum,* and lactic acid bacteria (summarized in **Table 1**). Surface engineering in *S. cerevisiae* is the most advanced, with several groups developing strains that can assemble minicellulosomes *in vivo*, leading to significant improvements in biomass utilization [32, 78, 79, 89]. Complexes containing up to 12 enzymes have been displayed, but great challenges remain as expressing large structures with more enzymes reduces cellular display levels [79]. A variety of promising approaches may help to overcome this limitation, including adaptive assembly and host engineering methods. Furthermore, greater cellulolytic activity through the incorporation of new synergistic enzyme functionalities may offset the decreases associated with the production of larger proteins, as seen by the increased activity of a pentafunctional minicellulosome [78]. Approaches to display minicellulosome structures on eubacteria that have developed genetic systems are less advanced. Thus far, only *ex vivo* assembled complexes have been displayed in Gram-positive *B. subtilis* and LAB, and only individual enzymes have been displayed on the Gram-negative bacterium *E. coli*. The potential of displaying complexes on the cell surface of *E. coli* would seem to be bleak, as the presence of a second outer membrane in this microbe hampers protein secretion. However, studies of *C. glutamicum* are promising, as it is the only bacterium that has been engineered to display an *in vivo* assembled minicellulosome, although this complex contained only two enzymes. In *B. subtilis*, exogenous proteases appear to be a limiting factor, while in LAB protein expression levels are suboptimal [119, 137]. However, it would seem likely that these limitations can be overcome by optimizing display using genetic engineering and directed evolution approaches, as well as the use of cell consortiums to construct complexes *ex vivo*.

A great variety of surface engineering approaches have been developed to create ever more impressive recombinant cellulolytic organisms. However, it is clear that the lignocellulose hydrolysis rates of these recombinant microorganisms needs to be improved if they are to be used in CBP. Future studies may improve their cellulolytic activity by using directed evolution to enhance complex display and stability, by judiciously displaying cellulases that exhibit maximal enzyme synergy, and by devising new methods to stably attach proteins to the cell surface. Genetic engineering of the host will also be critical, enhancing expression, secretion, and display levels, and by eliminating proteins or factors affecting the stability and retention of anchored protein complexes over time. Displayed complexes will also have to be optimized to be maximally active against different types of biomass that have different sugar compositions and structures [138–140]. When these cells are further engineered to produce useful chemicals, their ability to cost-effectively produce biocommodities from biomass will be an important step towards reducing the world's dependency on oil [141–144].

Acknowledgements

This material is based on work supported by the U.S. Department of Energy Office of Science, Office of Biological and Environmental Research program under Award Number DE-FC02-

02ER63421. GL Huang was supported by a Ruth L. Kirschstein National Research Service Award GM007185.

Nomenclature

consolidated bioprocessing	CBP
cellulose–enzyme–microbe	CEM
phosphoric acid swollen cellulose	PASC
regenerated amorphous cellulose	RAC
ionic liquid	IL
glycosylphosphatidylinositol	GPI
cell wall sorting signal	CWS
cellulose-binding module	CBM

Author details

Grace L. Huang[1,2] and Robert T. Clubb[1,3*]

*Address all correspondence to: rclubb@mbi.ucla.edu

1 Department of Chemistry and Biochemistry, University of California, Los Angeles, Los Angeles, CA, USA

2 UCLA-DOE Institute of Genomics and Proteomics, University of California, Los Angeles, Los Angeles, CA, USA

3 Molecular Biology Institute, University of California, Los Angeles, Los Angeles, CA, USA

References

[1] Kerr, R.A., Energy. World oil crunch looming? Science, 2008. 322(5905): p. 1178–1179.

[2] Robertson, G.P., et al., Agriculture. Sustainable biofuels redux. Science, 2008. 322(5898): p. 49–50.

[3] Chu, S. and A. Majumdar, Opportunities and challenges for a sustainable energy future. Nature, 2012. 488(7411): p. 294–303.

[4] Perlack, R.D., et al. Biomass as feedstock for a bioenergy and bioproducts industry the technical feasibility of a billion-ton annual supply. 2005; Available from: http://www.ornl.gov/~webworks/cppr/y2001/rpt/123021.pdf.

[5] Lynd, L.R., et al., Fuel ethanol from cellulosic biomass. Science, 1991. 251(4999): p. 1318–1323.

[6] Taha, M., et al., Commercial feasibility of lignocellulose biodegradation: possibilities and challenges. Curr Opin Biol, 2016. 38: p. 190–197.

[7] Himmel, M.E., et al., Biomass recalcitrance: engineering plants and enzymes for biofuels production. Science, 2007. 315(5813): p. 804–807.

[8] Chundawat, S.P.S., et al., Deconstruction of lignocellulosic biomass to fuels and chemicals. Annu Rev Chem Biomol Eng, 2011. 2: p. 121–145.

[9] Menon, V. and M. Rao, Trends in bioconversion of lignocellulose: biofuels, platform chemicals & biorefinery concept. Prog Energy Combust Sci, 2012. 38(4): p. 522–550.

[10] Zhao, X., K. Cheng, and D. Liu, Organosolv pretreatment of lignocellulosic biomass for enzymatic hydrolysis. Appl Microbiol Biotechnol, 2009. 82(5): p. 815–27.

[11] Hendriks, A.T. and G. Zeeman, Pretreatments to enhance the digestibility of lignocellulosic biomass. Bioresour Technol, 2009. 100(1): p. 10–8.

[12] Olson, D.G., et al., Recent progress in consolidated bioprocessing. Curr Opin Biotechnol, 2012. 23(3): p. 396–405.

[13] la Grange, D.C., R. den Haan, and W.H. van Zyl, Engineering cellulolytic ability into bioprocessing organisms. Appl Microbiol Biotechnol, 2010. 87(4): p. 1195–1208.

[14] Lynd, L.R., et al., Consolidated bioprocessing of cellulosic biomass: an update. Curr Opin Biotechnol, 2005. 16(5): p. 577–583.

[15] Klein-Marcuschamer, D., et al., The challenge of enzyme cost in the production of lignocellulosic biofuels. Biotechnol Bioeng, 2012. 109(4): p. 1083–1087.

[16] Zhao, X., L. Zhang, and D. Liu, Biomass recalcitrance. Part I: the chemical compositions and physical structures affecting the enzymatic hydrolysis of lignocellulose. Biofuels Bioprod Biorefin, 2012. 6(4): p. 465–482.

[17] Zhao, X.B., L.H. Zhang, and D.H. Liu, Biomass recalcitrance. Part II: fundamentals of different pre-treatments to increase the enzymatic digestibility of lignocellulose. Biofuels Bioprod Bioref, 2012. 6(5): p. 561–579.

[18] Ghose, T., Cellulase biosynthesis and hydrolysis of cellulosic substances, in Advances in Biochemical Engineering, T. Ghose, Editor. 1977, Springer Berlin/Heidelberg.

[19] Boerjan, W., J. Ralph, and M. Baucher, Lignin biosynthesis. Annu Rev Plant Biol, 2003. 54: p. 519–546.

[20] Bugg, T.D., et al., Pathways for degradation of lignin in bacteria and fungi. Nat Prod Rep, 2011. 28(12): p. 1883–1896.

[21] Wieczorek, A.S., D. Biot-Pelletier, and V. J.J, Recombinant Cellulase and Cellulosome Systems, in Cellulose - Biomass Conversion, J. Kadla, Editor. 2013: InTech.

[22] Huang, G.L., T.D. Anderson, and R.T. Clubb, Engineering microbial surfaces to degrade lignocellulosic biomass. Bioengineered, 2014. 5(2): p. 96–106.

[23] Tanaka, T. and A. Kondo, Cell surface engineering of industrial microorganisms for biorefining applications. Biotechnol Adv, 2015. 33(7): p. 1403–1411.

[24] Liu, Z., et al., Recent advances in yeast cell-surface display technologies for waste biorefineries. Bioresour Technol, 2016. 215: p. 324–333.

[25] Munoz-Gutierrez, I. and A. Martinez, Polysaccharide hydrolysis with engineered Escherichia coli for the production of biocommodities. J Ind Microbiol Biotechnol, 2013. 40(5): p. 401–410.

[26] Michon, C., et al., Display of recombinant proteins at the surface of lactic acid bacteria: strategies and applications. Microbial Cell Factor, 2016. 15.

[27] Bayer, E.A., et al., Cellulosomes-structure and ultrastructure. J Struct Biol, 1998. 124(2–3): p. 221–234.

[28] Fontes, C.M. and H.J. Gilbert, Cellulosomes: highly efficient nanomachines designed to deconstruct plant cell wall complex carbohydrates. Annu Rev Biochem, 2010. 79: p. 655–681.

[29] Leibovitz, E., et al., Characterization and subcellular localization of the Clostridium thermocellum scaffoldin dockerin binding protein SdbA. J Bacteriol, 1997. 179(8): p. 2519–2523.

[30] Ding, S.Y., et al., A biophysical perspective on the cellulosome: new opportunities for biomass conversion. Curr Opin Biotechnol, 2008. 19(3): p. 218–227.

[31] You, C., et al., Enhanced microbial utilization of recalcitrant cellulose by an ex vivo cellulosome-microbe complex. Appl Environ Microbiol, 2012. 78(5): p. 1437–1444.

[32] Wen, F., J. Sun, and H. Zhao, Yeast surface display of trifunctional minicellulosomes for simultaneous saccharification and fermentation of cellulose to ethanol. Appl Environ Microbiol, 2010. 76(4): p. 1251–1260.

[33] Lynd, L.R., et al., Microbial cellulose utilization: fundamentals and biotechnology. Microbiol Mol Biol Rev, 2002. 66(3): p. 506–577, table of contents.

[34] Lu, Y., Y.H. Zhang, and L.R. Lynd, Enzyme-microbe synergy during cellulose hydrolysis by Clostridium thermocellum. Proc Natl Acad Sci U S A, 2006. 103(44): p. 16165–16169.

[35] Demain, A.L., M. Newcomb, and J.H. Wu, Cellulase, clostridia, and ethanol. Microbiol Mol Biol Rev, 2005. 69(1): p. 124–154.

[36] Klein-Marcuschamer, D. and H.W. Blanch, Renewable fuels from biomass: technical hurdles and economic assessment of biological routes. AIChE J, 2015. 61(9): p. 2689–2701.

[37] Humbird, D., National Renewable Energy Laboratory (U.S.), and Harris Group Inc., Process design and economics for biochemical conversion of lignocellulosic biomass to ethanol dilute-acid pretreatment and enzymatic hydrolysis of corn stover, in Nrel/Tp 5100-47764. 2011, National Renewable Energy Laboratory: Golden, CO. p. 1 online resource (ix, 136 p.) ill.

[38] Johnson, E., Integrated enzyme production lowers the cost of cellulosic ethanol. Biofuels Bioprod Bioref, 2016. 10(2): p. 164–174.

[39] Tanaka, T. and A. Kondo, Cell-surface display of enzymes by the yeast Saccharomyces cerevisiae for synthetic biology. FEMS Yeast Res, 2014.

[40] Lipke, P.N. and R. Ovalle, Cell wall architecture in yeast: new structure and new challenges. J Bacteriol, 1998. 180(15): p. 37353740.

[41] Van der Vaart, J.M., et al., Comparison of cell wall proteins of Saccharomyces cerevisiae as anchors for cell surface expression of heterologous proteins. Appl Environ Microbiol, 1997. 63(2): p. 615–620.

[42] van der Vaart, J.M., et al., The retention mechanism of cell wall proteins in Saccharomyces cerevisiae. Wall-bound Cwp2p is beta-1,6-glucosylated. Biochim Biophys Acta, 1996. 1291(3): p. 206–214.

[43] Fujita, M. and T. Kinoshita, Structural remodeling of GPI anchors during biosynthesis and after attachment to proteins. FEBS Lett, 2010. 584(9): p. 1670–1677.

[44] Ohishi, K., N. Inoue, and T. Kinoshita, PIG-S and PIG-T, essential for GPI anchor attachment to proteins, form a complex with GAA1 and GPI8. EMBO J, 2001. 20(15): p. 4088–4098.

[45] Orlean, P. and A.K. Menon, Thematic review series: lipid posttranslational modifications. GPI anchoring of protein in yeast and mammalian cells, or: how we learned to stop worrying and love glycophospholipids. J Lipid Res, 2007. 48(5): p. 993–1011.

[46] Lu, C.F., J. Kurjan, and P.N. Lipke, A pathway for cell wall anchorage of Saccharomyces cerevisiae alpha-agglutinin. Mol Cell Biol, 1994. 14(7): p. 4825–4833.

[47] Lu, C.F., et al., Glycosyl phosphatidylinositol-dependent cross-linking of alpha-agglutinin and beta 1,6-glucan in the Saccharomyces cerevisiae cell wall. J Cell Biol, 1995. 128(3): p. 333–340.

[48] Orlean, P., Architecture and biosynthesis of the Saccharomyces cerevisiae cell wall. Genetics, 2012. 192(3): p. 775–818.

[49] Chao, G., et al., Isolating and engineering human antibodies using yeast surface display. Nat Protoc, 2006. 1(2): p. 755–768.

[50] Kotaka, A., et al., Enhancement of beta-glucosidase activity on the cell-surface of sake yeast by disruption of SED1. J Biosci Bioeng, 2010. 109(5): p. 442–446.

[51] Matsuoka, H., et al., Cell wall structure suitable for surface display of proteins in Saccharomyces cerevisiae. Yeast, 2014. 31(2): p. 67-76.

[52] Inokuma, K., T. Hasunuma, and A. Kondo, Efficient yeast cell-surface display of exo— and endo-cellulase using the SED1 anchoring region and its original promoter. Biotechnol Biofuels, 2014. 7.

[53] Inokuma, K., et al., Enhanced cell-surface display and secretory production of cellulolytic enzymes with Saccharomyces cerevisiae Sed1 signal peptide. Biotechnol Bioeng, 2016.

[54] Suzuki, H., et al., Deglycosylation of cellulosomal enzyme enhances cellulosome assembly in Saccharomyces cerevisiae. J Biotechnol, 2012. 157(1): p. 64–70.

[55] Murai, T., et al., Assimilation of cellooligosaccharides by a cell surface-engineered yeast expressing beta-glucosidase and carboxymethylcellulase from aspergillus aculeatus. Appl Environ Microbiol, 1998. 64(12): p. 4857–4861.

[56] Fujita, Y., et al., Direct and efficient production of ethanol from cellulosic material with a yeast strain displaying cellulolytic enzymes. Appl Environ Microbiol, 2002. 68(10): p. 5136–5141.

[57] Fujita, Y., et al., Synergistic saccharification, and direct fermentation to ethanol, of amorphous cellulose by use of an engineered yeast strain codisplaying three types of cellulolytic enzyme. Appl Environ Microbiol, 2004. 70(2): p. 1207–1212.

[58] Ito, J., et al., Improvement of cellulose-degrading ability of a yeast strain displaying Trichoderma reesei endoglucanase II by recombination of cellulose-binding domains. Biotechnol Prog, 2004. 20(3): p. 688–691.

[59] Yamada, R., et al., Cocktail delta-integration: a novel method to construct cellulolytic enzyme expression ratio-optimized yeast strains. Microb Cell Fact, 2010. 9: p. 32.

[60] Yamada, R., et al., Direct ethanol production from cellulosic materials using a diploid strain of Saccharomyces cerevisiae with optimized cellulase expression. Biotechnol Biofuels, 2011. 4: p. 8.

[61] Yanase, S., et al., Ethanol production from cellulosic materials using cellulase-expressing yeast. Biotechnol J, 2010. 5(5): p. 449–455.

[62] Liu, Z., et al., Combined cell-surface display- and secretion-based strategies for production of cellulosic ethanol with Saccharomyces cerevisiae. Biotechnol Biofuels, 2015. 8.

[63] Bae, J., K. Kuroda, and M. Ueda, Proximity effect among cellulose-degrading enzymes displayed on the Saccharomyces cerevisiae cell surface. Appl Environ Microbiol, 2015. 81(1): p. 59–66.

[64] Vazana, Y., et al., A synthetic biology approach for evaluating the functional contribution of designer cellulosome components to deconstruction of cellulosic substrates. Biotechnol Biofuels, 2013. 6(1): p. 182.

[65] Matano, Y., T. Hasunuma, and A. Kondo, Display of cellulases on the cell surface of Saccharomyces cerevisiae for high yield ethanol production from high-solid lignocellulosic biomass. Bioresour Technol, 2012. 108: p. 128–33.

[66] Matano, Y., T. Hasunuma, and A. Kondo, Cell recycle batch fermentation of high-solid lignocellulose using a recombinant cellulase-displaying yeast strain for high yield ethanol production in consolidated bioprocessing. Bioresour Technol, 2013. 135: p. 403–9.

[67] Matano, Y., T. Hasunuma, and A. Kondo, Simultaneous improvement of saccharification and ethanol production from crystalline cellulose by alleviation of irreversible adsorption of cellulase with a cell surface-engineered yeast strain. Appl Microbiol Biotechnol, 2013. 97(5): p. 2231–2237.

[68] Liu, Z., et al., Engineering of a novel cellulose adherent cellulolytic Saccharomyces cerevisiae for cellulosic biofuel production. Sci Rep, 2016. 6: p. 24550.

[69] Yamada, R., et al., Efficient direct ethanol production from cellulose by cellulase—and cellodextrin transporter-co-expressing Saccharomyces cerevisiae. AMB Express, 2013. 3(1): p. 34.

[70] Nakatani, Y., et al., Synergetic effect of yeast cell-surface expression of cellulase and expansin-like protein on direct ethanol production from cellulose. Microb Cell Factor, 2013. 12.

[71] Nakanishi, A., et al., Effect of pretreatment of hydrothermally processed rice straw with laccase-displaying yeast on ethanol fermentation. Appl Microbiol Biotechnol, 2012. 94(4): p. 939–948.

[72] Lilly, M., et al., Heterologous expression of a Clostridium minicellulosome in Saccharomyces cerevisiae. FEMS Yeast Res, 2009. 9(8): p. 1236–1249.

[73] Tsai, S.L., et al., Functional assembly of minicellulosomes on the Saccharomyces cerevisiae cell surface for cellulose hydrolysis and ethanol production. Appl Environ Microbiol, 2009. 75(19): p. 6087–6093.

[74] Tsai, S.L., G. Goyal, and W. Chen, Surface display of a functional minicellulosome by intracellular complementation using a synthetic yeast consortium and its application to cellulose hydrolysis and ethanol production. Appl Environ Microbiol, 2010. 76(22): p. 7514–7520.

[75] Goyal, G., et al., Simultaneous cell growth and ethanol production from cellulose by an engineered yeast consortium displaying a functional mini-cellulosome. Microb Cell Fact, 2011. 10: p. 89.

[76] Kim, S., et al., Cellulosic ethanol production using a yeast consortium displaying a minicellulosome and beta-glucosidase. Microb Cell Fact, 2013. 12: p. 14.

[77] Tsai, S.L., N.A. DaSilva, and W. Chen, Functional display of complex cellulosomes on the yeast surface via adaptive assembly. ACS Synth Biol, 2013. 2(1): p. 14–21.

[78] Liang, Y., et al., Engineered pentafunctional minicellulosome for simultaneous saccharification and ethanol fermentation in Saccharomyces cerevisiae. Appl Environ Microbiol, 2014. 80(21): p. 6677–6684.

[79] Fan, L.H., et al., Self-surface assembly of cellulosomes with two miniscaffoldins on Saccharomyces cerevisiae for cellulosic ethanol production. Proc Natl Acad Sci U S A, 2012. 109(33): p. 13260–13265.

[80] Fujita, Y., et al., Construction of whole-cell biocatalyst for xylan degradation through cell-surface xylanase display in Saccharomyces cerevisiae. J Mol Catal B Enzym, 2002. 17(3–5): p. 189–195.

[81] Katahira, S., et al., Construction of a xylan-fermenting yeast strain through codisplay of xylanolytic enzymes on the surface of xylose-utilizing Saccharomyces cerevisiae cells. Appl Environ Microbiol, 2004. 70(9): p. 5407–5414.

[82] Katahira, S., et al., Ethanol fermentation from lignocellulosic hydrolysate by a recombinant xylose- and cellooligosaccharide-assimilating yeast strain. Appl Microbiol Biotechnol, 2006. 72(6): p. 1136–1143.

[83] Nakamura, N., et al., Effective xylose/cellobiose co-fermentation and ethanol production by xylose-assimilating S-cerevisiae via expression of beta-glucosidase on its cell surface. Enzym Microb Technol, 2008. 43(3): p. 233–236.

[84] Sakamoto, T., et al., Direct ethanol production from hemicellulosic materials of rice straw by use of an engineered yeast strain codisplaying three types of hemicellulolytic enzymes on the surface of xylose-utilizing Saccharomyces cerevisiae cells. J Biotechnol, 2012. 158(4): p. 203–210.

[85] Guirimand, G., et al., Cell surface engineering of Saccharomyces cerevisiae combined with membrane separation technology for xylitol production from rice straw hydrolysate. Appl Microbiol Biotechnol, 2016. 100(8): p. 3477–3487.

[86] Ota, M., et al., Display of Clostridium cellulovorans xylose isomerase on the cell surface of Saccharomyces cerevisiae and its direct application to xylose fermentation. Biotechnol Prog, 2013. 29(2): p. 346–51.

[87] Srikrishnan, S., W. Chen, and N.A. Da Silva, Functional assembly and characterization of a modular xylanosome for hemicellulose hydrolysis in yeast. Biotechnol Bioeng, 2013. 110(1): p. 275–85.

[88] Sun, J., et al., Direct conversion of xylan to ethanol by recombinant Saccharomyces cerevisiae strains displaying an engineered minihemicellulosome. Appl Environ Microbiol, 2012. 78(11): p. 3837–3845.

[89] Ito, J., et al., Regulation of the display ratio of enzymes on the saccharomyces cerevisiae cell surface by the immunoglobulin g and cellulosomal enzyme binding domains. Appl Environ Microbiol, 2009. 75(12): p. 4149–4154.

[90] Han, Z.L., et al., self-assembled amyloid-like oligomeric-cohesin scaffoldin for augmented protein display on the saccharomyces cerevisiae cell surface. Appl Environ Microbiol, 2012. 78(9): p. 3249–3255.

[91] van Bloois, E., et al., Decorating microbes: surface display of proteins on Escherichia coli. Trends Biotechnol, 2011. 29(2): p. 79–86.

[92] Clomburg, J.M. and R. Gonzalez, Biofuel production in Escherichia coli: the role of metabolic engineering and synthetic biology. Appl Microbiol Biotechnol, 2010. 86(2): p. 419–434.

[93] Francisco, J.A., et al., Specific adhesion and hydrolysis of cellulose by intact Escherichia coli expressing surface anchored cellulase or cellulose binding domains. Biotechnology (NY), 1993. 11(4): p. 491–495.

[94] Qu, W., Y. Xue, and Q. Ding, Display of fungi xylanase on Escherichia coli cell surface and use of the enzyme in xylan biodegradation. Curr Microbiol, 2015. 70(6): p. 779–785.

[95] Jung, H.C., et al., Expression of carboxymethylcellulase on the surface of Escherichia coli using Pseudomonas syringae ice nucleation protein. Enzym Microb Technol, 1998. 22(5): p. 348–354.

[96] Kim, Y.S., H.C. Jung, and J.G. Pan, Bacterial cell surface display of an enzyme library for selective screening of improved cellulase variants. Appl Environ Microbiol, 2000. 66(2): p. 788–793.

[97] Liu, W., et al., Engineering of Clostridium phytofermentans endoglucanase Cel5A for improved thermostability. Appl Environ Microbiol, 2010. 76(14): p. 4914–4917.

[98] Tanaka, T., et al., Creation of a cellooligosaccharide-assimilating Escherichia coli strain by displaying active beta-glucosidase on the cell surface via a novel anchor protein. Appl Environ Microbiol, 2011. 77(17): p. 6265–6270.

[99] Soma, Y., et al., Direct isopropanol production from cellobiose by engineered Escherichia coli using a synthetic pathway and a cell surface display system. J Biosci Bioeng, 2012. 114(1): p. 80–85.

[100] Tanaka, T., et al., Creation of cellobiose and xylooligosaccharides-coutilizing Escherichia coli displaying both beta-glucosidase and beta-xylosidase on its cell surface. ACS Synth Biol, 2014. 3(7): p. 446–453.

[101] Ryu, S. and M.N. Karim, A whole cell biocatalyst for cellulosic ethanol production from dilute acid-pretreated corn stover hydrolyzates. Appl Microbiol Biotechnol, 2011. 91(3): p. 529–542.

[102] Chen, Y.P., et al., Enhancing the stability of xylanase from Cellulomonas fimi by cell-surface display on Escherichia coli. J Appl Microbiol, 2012. 112(3): p. 455–463.

[103] Ko, K.C., et al., Bacterial cell surface display of a multifunctional cellulolytic enzyme screened from a bovine rumen metagenomic resource. J Microbiol Biotechnol, 2015. 25(11): p. 1835–1841.

[104] Yim, S.S., et al., Isolation of a potential anchoring motif based on proteome analysis of Escherichia coli and its use for cell surface display. Appl Biochem Biotechnol, 2013. 170(4): p. 787–804.

[105] Munoz-Gutierrez, I., et al., Cell surface display of a beta-glucosidase employing the type V secretion system on ethanologenic Escherichia coli for the fermentation of cellobiose to ethanol. J Ind Microbiol Biotechnol, 2012. 39(8): p. 1141–1152.

[106] Park, T.J., et al., Surface display of recombinant proteins on Escherichia coli by BclA exosporium of Bacillus anthracis. Microb Cell Fact, 2013. 12: p. 81.

[107] Bokinsky, G., et al., Synthesis of three advanced biofuels from ionic liquid-pretreated switchgrass using engineered Escherichia coli. Proc Natl Acad Sci U S A, 2011. 108(50): p. 19949–19954.

[108] Schallmey, M., A. Singh, and O.P. Ward, Developments in the use of Bacillus species for industrial production. Can J Microbiol, 2004. 50(1): p. 1–17.

[109] Liu, L., et al., Developing Bacillus spp. as a cell factory for production of microbial enzymes and industrially important biochemicals in the context of systems and synthetic biology. Appl Biochem Biotechnol, 2013. 97(14): p. 6113–6127.

[110] Zhang, X.-Z. and Y.-H.P. Zhang, One-step production of biocommodities from lignocellulosic biomass by recombinant cellulolytic Bacillus subtilis: Opportunities and challenges. Eng Life Sci, 2010. 10(5): p. 398–406.

[111] van Dijl, J.M. and M. Hecker, Bacillus subtilis: from soil bacterium to super-secreting cell factory. Microbial Cell Factories, 2013. 12(3).

[112] Kang, Z., et al., Molecular engineering of secretory machinery components for high-level secretion of proteins in Bacillus species. J Ind Microbiol Biotechnol, 2014. 41(11): p. 1599–1607.

[113] Spirig, T., E.M. Weiner, and R.T. Clubb, Sortase enzymes in Gram-positive bacteria. Mol Microbiol, 2011. 82(5): p. 1044–1059.

http://dx.doi.org/10.5772/65509

[114] Nguyen, H.D. and W. Schumann, Establishment of an experimental system allowing immobilization of proteins on the surface of Bacillus subtilis cells. J Biotechnol, 2006. 122(4): p. 473–482.

[115] Buist, G., et al., LysM, a widely distributed protein motif for binding to (peptido)glycans. Mol Microbiol, 2008. 68(4): p. 838–847.

[116] Chen, C.L., et al., Development of a LytE-based high-density surface display system in Bacillus subtilis. Microb Biotechnol, 2008. 1(2): p. 177–90.

[117] Kim, J.H., I.S. Park, and B.G. Kim, Development and characterization of membrane surface display system using molecular chaperon, prsA, of Bacillus subtilis. Biochem Biophys Res Commun, 2005. 334(4): p. 1248–1253.

[118] Anderson, T.D., et al., Assembly of minicellulosomes on the surface of Bacillus subtilis. Appl Environ Microbiol, 2011. 77(14): p. 4849–4858.

[119] Westers, L., H. Westers, and W.J. Quax, Bacillus subtilis as cell factory for pharmaceutical proteins: a biotechnological approach to optimize the host organism. Biochim Biophys Acta, 2004. 1694(1–3): p. 299–310.

[120] McKenney, P.T., A. Driks, and P. Eichenberger, The Bacillus subtilis endospore: assembly and functions of the multilayered coat. Nat Rev Microbiol, 2013. 11(1): p. 33–44.

[121] Isticato, R. and E. Ricca, Spore Surface Display. Microbiol Spectr, 2014. 2(5).

[122] Pan, J.G., et al., Display of native proteins on Bacillus subtilis spores. FEMS Microbiol Lett, 2014. 358(2): p. 209–217.

[123] Sirec, T., et al., Adsorption of beta-galactosidase of Alicyclobacillus acidocaldarius on wild type and mutants spores of Bacillus subtilis. Microbial Cell Factories, 2012. 11(100).

[124] Chapot-Chartier, M.P. and S. Kulakauskas, Cell wall structure and function in lactic acid bacteria. Microbial Cell Factories, 2014. 13(Suppl 1: S9).

[125] G-Alegria, E., et al., High tolerance of wild Lactobacillus plantarum and Oenococcus oeni strains to lyophilisation and stress environmental conditions of acid pH and ethanol. FEMS Microbiol Lett, 2004. 230(1): p. 53–61.

[126] Kyla-Nikkila, K., U. Alakuijala, and P.E.J. Saris, Immobilization of Lactococcus lactis to cellulosic material by cellulose-binding domain of Cellvibrio japonicus. J Appl Microbiol, 2010. 109(4): p. 1274–1283.

[127] Wieczorek, A.S. and V.J. Martin, Engineering the cell surface display of cohesins for assembly of cellulosome-inspired enzyme complexes on Lactococcus lactis. Microb Cell Fact, 2010. 9: p. 69.

[128] Wieczorek, A.S. and V.J.J. Martin, Effects of synthetic cohesin-containing scaffold protein architecture on binding dockerin-enzyme fusions on the surface of Lactococcus lactis. Microbial Cell Factories, 2012. 11(160).

[129] Morais, S., et al., A combined cell-consortium approach for lignocellulose degradation by specialized Lactobacillus plantarum cells. Biotechnology for Biofuels, 2014. 7(112).

[130] Burkovski, A., Cell envelope of corynebacteria: structure and influence on pathogenicity. ISRN Microbiol, 2013. 2013: p. 935736.

[131] Tateno, T., et al., Development of novel cell surface display in Corynebacterium glutamicum using porin. Appl Microbiol Biotechnol, 2009. 84(4): p. 733–739.

[132] Adachi, N., et al., Direct L-lysine production from cellobiose by Corynebacterium glutamicum displaying beta-glucosidase on its cell surface. Appl Microbiol Biotechnol, 2013. 97(16): p. 7165–7172.

[133] Kim, S.J., et al., Bi-functional cellulases complexes displayed on the cell surface of Corynebacterium glutamicum increase hydrolysis of lignocelluloses at elevated temperature. Enzym Microb Technol, 2014. 66: p. 67–73.

[134] Zhang, P.Y.H., M.E. Himmel, and J.R. Mielenz, Outlook for cellulase improvement: screening and selection strategies. Biotechnol Adv, 2006. 24(5): p. 452–481.

[135] Yang, B. and C.E. Wyman, Pretreatment: the key to unlocking low-cost cellulosic ethanol. Biofuels Bioprod Bioref, 2008. 2(1): p. 26–40.

[136] Dashtban, M., et al., Cellulase activities in biomass conversion: measurement methods and comparison. Crit Rev Biotechnol, 2010. 30(4): p. 302–309.

[137] Le Loir, Y., et al., Protein secretion in Lactococcus lactis: an efficient way to increase the overall heterologous protein production. Microbial Cell Factories, 2005. 4(2).

[138] Banerjee, G., et al., Rapid optimization of enzyme mixtures for deconstruction of diverse pretreatment/biomass feedstock combinations. Biotechnology for Biofuels, 2010. 3(22).

[139] Engel, P., et al., Rational approach to optimize cellulase mixtures for hydrolysis of regenerated cellulose containing residual ionic liquid. Bioresour Technol, 2012. 115: p. 27–34.

[140] Kim, I.J., et al., Customized optimization of cellulase mixtures for differently pretreated rice straw. Bioprocess Biosyst Eng, 2015. 38(5): p. 929–937.

[141] Dellomonaco, C., F. Fava, and R. Gonzalez, The path to next generation biofuels: successes and challenges in the era of synthetic biology. Microb Cell Fact, 2010. 9: p. 3.

[142] Jang, Y.S., et al., Engineering of microorganisms for the production of biofuels and perspectives based on systems metabolic engineering approaches. Biotechnol Adv, 2012. 30(5): p. 989–1000.

[143] Rabinovitch-Deere, C.A., et al., Synthetic biology and metabolic engineering approaches to produce biofuels. Chem Rev, 2013. 113(7): p. 4611–4632.

[144] Li, H., A.F. Cann, and J.C. Liao, Biofuels: biomolecular engineering fundamentals and advances. Annu Rev Chem Biomol Eng, 2010. 1: p. 19–36.

Chapter 19

Determination of the Biomass Content of End-of-Life Tyres

Leticia Saiz Rodríguez, José M. Bermejo Muñoz,
Adrien Zambon and Jean P. Faure

Additional information is available at the end of the chapter

http://dx.doi.org/10.5772/65830

Abstract

Studies have been conducted in France and Spain for (1) the validation of sampling methods to achieve representative samples of end-of-life tyre (ELT) materials and (2) the comparison and validation of test methods to quantify their biomass content. Both studies conclude that the ^{14}C techniques are the most reliable techniques for determining the biomass content of end-of-life tyres. Indeed, thermogravimetry and pyrolysis-GC/MS do not lead to results consistent with the theoretical content of biogenic materials present in tyres, and results in both cases differ considerably from the known natural rubber content of the reference samples studied using thermogravimetric analysis. Furthermore, in the two last techniques, natural isoprene cannot be distinguished from synthetic isoprene. Results obtained with radiocarbon analysis based on ^{14}C contents could be used as reference values of the biomass content of the ELTs: in the ranges of 18–22% for passenger car tyres and 29–34% for truck tyres, in line with actual natural rubber and other components content. Additionally, the presence of textile fibres and stearic acid, which are known sources of biomass in the tyre, cannot be evaluated by thermogravimetry and pyrolysis-GC/MS techniques.

Keywords: end-of-life tyres, biomass, natural rubber, reference values, alternative fuel, pyrolysis-GC-MS, thermogravimetry, radiocarbon analysis, isoprene

1. Introduction

End-of-life tyre (ELT) is any pneumatic tyre removed from any vehicle and not selected to be mounted on a vehicle again [1]. Because the end-of-life tyre is a non-reusable tyre in its original form, it enters a waste management system based on product/material recycling and energy recovery. For that, it is important to know the composition of the tyres and their properties in order to recycle them or recover their energy.

Tyres could be classified according to the type of vehicle in which they are suitable to be mounted, so the main groups established by the industry are passenger car, truck and bus, and agro tyres, all of them included in this chapter.

Tyres are composite materials, essentially made of rubber, metal wire and textile fibres, in Ref. [1]. The rubber component is based on a vulcanized mixture of elastomers with some different chemicals (see **Table 1**), such as carbon black, silica and extender oils which also constitute a high percentage of the tyres.

Their compositions and combustion properties are similar to, or even better than those of coal (see **Table 2**). Due to their high carbon content (60–70%), they have become an interesting alternative fuel with a net calorific value in the range of 26.4–30.2 MJ/kg [2].

The use of secondary fuels is progressively increasing, not only because of its economic benefits, but also because of the environmental advantages of using solid recovered fuels [3]. These include natural resource savings, the preservation of fossil fuels such as petroleum coke, and above all the reduction in net emissions of CO_2 due to the biogenic origin of some components of the tyre, mainly natural rubber. Indeed, according to Directive 2003/87/EC [4], emissions associated with biomass fraction are considered to be neutral with regard to the greenhouse effect.

It also leads to the reduction in other pollutants [5] such as SO_x, mainly because the sulphur content in tyres (1–2%), used for the vulcanization process, is in any case lower than the quantity in most fossil fuels (see **Table 2**).

ELTs contain a fraction of biogenic carbon that mainly comes from their natural rubber content. This is not, however, the only source of biogenic carbon. Most tyre formulations also include stearic acid in small quantities, used as activator of the vulcanization reaction, and

Material	Passenger car tyre (%)	Trucks tyre (%)
Rubber/elastomers	43	42
Carbon black and silica	28	24
Metal	13	25
Textile	5	–
Zinc oxide	2	2
Sulphur	1	1
Accelerators/antidegradants	8	–
Stearic acid	1	n.a
Oils	7	n.a

Table 1. Main components of the passenger car tyre and truck tyre.

Content in % mass				
Major elements (>1%)				
	Passenger car	**Truck**	**Coal**	**Petroleum coke**
Carbon (C)	68 to 70%	60 to 63%	63.9%	84 to 97%
Iron (Fe)	11 to 12%	25 to 27%		0 to 0.2%
Hydrogen (H)	6 to 6.3%	5.3 to 5.6%	3.6%	0 to 5%
Oxygen (O)	3.3 to 3.8%	1.5 to 2.2%		
Silica (Si)	1.5 to 1.9%	0.3 to 0.5%	2%	Nd
Zinc (Zn)	1.3 to 1.5%	1.3 to 1.8%		
NCV (MJ/kg)	30.2	26.4	26	32
Sulphur (S)	1 to 1.5%	1.2 to 1.6%	1.3%	0.2 to 6%
Biomass fraction (% mass)	18.3%	29.1%	0%	0%

Table 2. Elemental analysis of tyres.

also smaller quantities of rayon, a natural fibre used as a reinforcement material in the manufacturing of some tyre carcasses. Nowadays, cotton could only be found in carcasses of older tyres [6, 7].

Despite the range of variation in the formulation of tyres, in practice their composition hardly varies, and thus they are one of the more dependable fuels, coming from wastes.

The greatest variation in biomass content is found among tyres of different types: passenger car tyres, truck tyres or agro vehicle tyres [1].

Each tyre particles obtained from treatment by shredding is quite heterogeneous in terms of biomass content. This intrinsic heterogeneity of tyres particles at the microscopic level is related to their composition. For example, elastomer mixtures are not the same in each part of the tyre (see **Figure 1**).

Although the heterogeneity does not appear at the industrial scale (consumption of around one ton per hour), the microscopic heterogeneity is of importance when it is necessary to take representative laboratory samples and to prepare them for analysis, in order to prevent different results.

In order to quantify the biomass content of an ELT, three analytical techniques have been identified. The first one is a method that determines the biogenic carbon content by measuring the activity of the ^{14}C isotope, a technique employed in archaeology to date organic materials [8–10]. Another widely used technique for determining the composition of vulcanized elastomers is thermogravimetric analysis (TGA) [11]. This method is based on the measurement of the weight variation of a sample when it is submitted to a progressive increase in temperature in a controlled atmosphere. The third method, pyrolysis-gas chromatography/

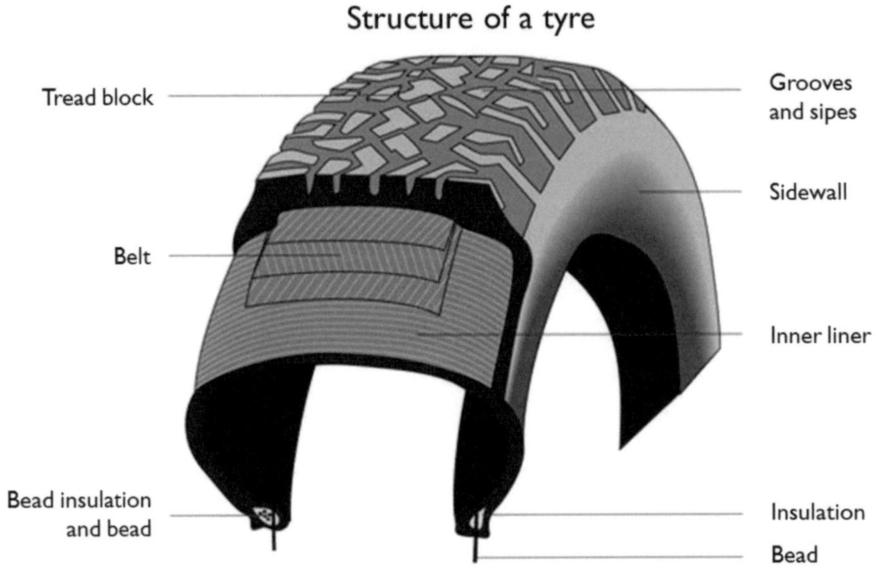

Figure 1. Different layers of a tyre.

mass spectrometry (Py-GC/MS), has been used extensively for qualitative and quantitative identification of polymer blends [12].

The results obtained under sampling, testing and analysis conditions respecting good practices with regards to heterogeneous materials show a remarkable stability in the measured parameters.

This chapter provides first the appropriate methodology to make a selection of test samples from tyres, and secondly, novel information on the use of different technologies for the determination of their biomass content.

Two cases studies conducted independently by ALIAPUR and SIGNUS, in France and Spain, respectively, have probed which techniques are appropriate to measure biomass content in tyres and which are not. Also in both studies reference values of biogenic content have been established, and results are quite close to theoretical values.

This chapter then offers an analysis of the differences between the results of the three techniques as well as the advantages, disadvantages and problems.

2. Sampling procedure for the estimation of biogenic fraction of end-of-life tyres

The management of ELTs to be used as secondary fuel mainly consists of the shredding of relevant quantities of different types of tyres coming from a diverse range of origins.

http://dx.doi.org/10.5772/65830

Shredded material can be stocked in piles, in which each particle contains parts of different layers of the tyre, each with a particular composition. So, the first problem to solve is how to estimate the biomass content taking random portions of materials. If this process is not performed carefully, there is an important risk not to be representative enough of the total stock. This is especially critical if we take into account that any of the used analytical techniques hardly needs a few milligrams of material.

One of the key points for this calculation of biomass content is then the design of a sampling plan which is representative of the big sized lots. In the case of samples taken on the shredding site, the samples will be representative of one to several days of production. Depending on the size of the facility, the sample could represent tens or even hundreds of tons of material.

In the case of samples taken during loading or unloading processes at the storage site, the samples will be representative of several weeks of production; the stocks in this case could even be of thousands of tons of material.

2.1. General criteria for the definition of sampling procedure

The minimum mass to produce a representative laboratory sample from the lot should be determined by the following formula [13]:

$$m_m = \frac{\pi}{6} \cdot d_{95}^3 \cdot s \cdot \lambda \cdot g \cdot \frac{(1-p)}{(Cv)^2 \cdot p} \tag{1}$$

where,

- m_m is the mass of the minimum sample size, in grams as received.
- d_{95} is the nominal top size of a particle (a mass fraction of 95% of the particles are smaller than d_{95}), in mm. This value is measured by means sieves following the method described within CEN/TS 15415:2006 [14].
- s is the shape factor, in mm^3/mm^3; reference value of 1.0 in the case of granular materials with nominal size smaller than 50 mm [13].
- λ is the average particle density of the particles in the solid recovered fuel, in g/mm^3 as received [15].
- g is the correction factor for distribution in the particles size. Its value is related to the superior nominal size d_{95} and the minimum size of the particle d_{05}.
- p is the fraction of the particles with a specific characteristic (such as a specific contaminant), in g/g, and is equal to 0.1 [13].
- Cv is the coefficient of variation. Its value is 0.1 [13].

2.2. General criteria for the preparation of laboratory samples

Samples were prepared using a riffle splitter with 14 slots of 27 mm in width, until the mass of the sample is greater than the minimum size of the laboratory sample necessary to guarantee total representativeness for few milligrams. This must be calculated according to the third-power law [16], which for granular materials is expressed as:

$$m > \alpha \cdot d_{95}^{\ 3} \tag{2}$$

where m is the mass retained after each sample division step in grams, d_{95} is the nominal top size in millimetres, and α is a constant over the whole sample preparation procedure for a particular material in g/mm³.

Depending on the size of the laboratory sample for each method, a particular number of tests are needed to guarantee the representativeness of the sample.

2.3. Methodology for taking representative samples from ELTs

The preparation of the sample consists in a reduction made in several stages of fragmentation/quartering until the different subpopulations (rubbers, metal wires, textile fibres) are obtained (see **Figure 2**). The standard means and procedures for reducing test samples are carried out in a laboratory with the appropriate equipment (sample division, cryogenic mill…). **Figure 3** shows the flow diagram of the whole process for collection and preparation of the representative sample to obtain a test portion.

Taking into account the size of the average production lots, it was estimated that 1.5 t of whole tyres is the minimum necessary quantity of ELTs. For the Spanish case, four different samples were taken in order to estimate the biomass content by type of tyre.

Starting with 1.5 t of tyres, the first step is the shredding of tyres to reduce the size of the sample and to obtain a mix of particles from different tyres. This first-size reduction can be carried out in a primary shredder to obtain pieces within the interval of 35–200 mm.

The sample must be taken at the production platform, using a tool of the open rectangular shovel type by completely cutting the flow of falling material. Samples are considered valid if a total quantity of at least 25 kg is taken per increment to represent a production of 1.5 t.

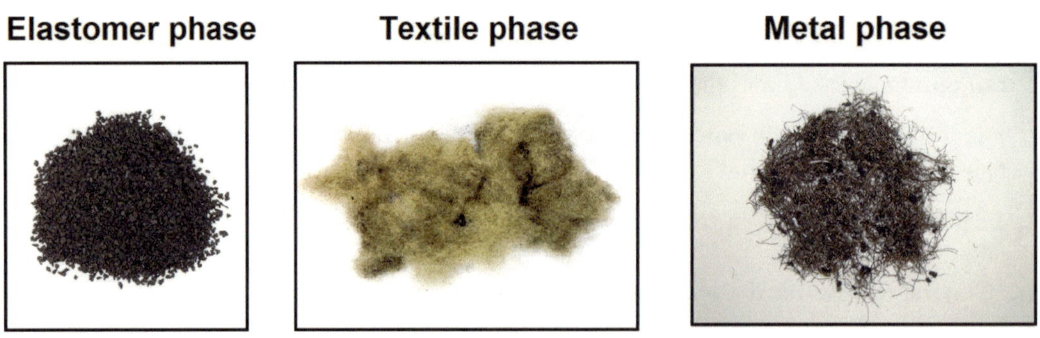

Figure 2. Different fractions of the tyre.

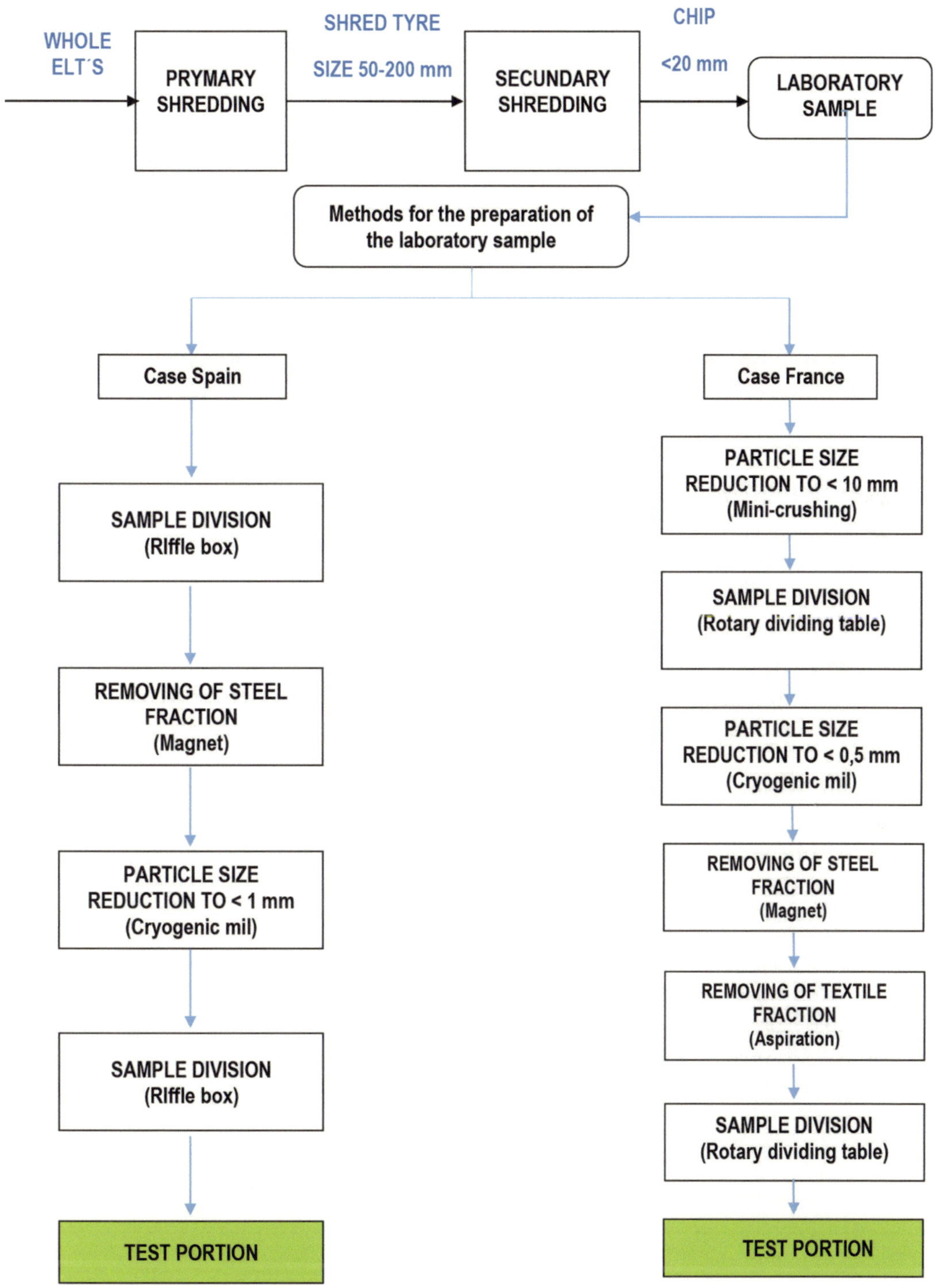

Figure 3. Complete scheme for test sampling preparation.

In a second step, those particles are reduced under a maximum size of 20 mm. The samples produced by this secondary shredder should be taken in smaller portions, at constant intervals of time, called increments.

No less than 24 increments are recommended to achieve a representative sample.

The obtained sample must be quartered to obtain subsamples of a quantity between 1.5 and 3 kg. That is bigger than the minimum quantity necessary for this type of material (around 1.0 kg).

Then the steel fraction should be removed from the sample using a magnet, taking special care in leaving the rest of material (rubber and textile fibres) that should be reduced in particles sized under 1 mm. Different methods could be used for this purpose, especially cryogenics. The obtained product should then be quartered to obtain a test portion.

For the French case, the procedure is similar to the above described (see **Figure 3**).

3. Test methods for the determination of biomass content in tyres

Different test methods were identified in the literature for the determination of natural rubber content in blends that can thus be indirectly used for the quantification of biomass content in tyres [17].

Some of those methods are based on the determination of content in elastomers and are particularly able to distinguish the presence of isoprene, the main component of the natural rubber, among other different elastomers of the rubber blend. This natural rubber has been identified as the main source of biogenic carbon in a tyre; nonetheless, this component is not the only bio-based one. Thus, some of the test methods identified for this chapter are not really conclusive about the total content of biomass in tyres. On the other hand, there are significant differences in the time to analyse the samples (see **Table 3**).

3.1. Pyrolysis-GC/MS

This method is based on the degradation of a sample in an electric furnace at 500–600°C and keeping the sample within this temperature range [18]. This temperature range is recommended to obtain rapid pyrolysis without excessive degradation or carbonization of the rubber sample. However, a temperature of 550°C is advised to obtain the maximum quantity of pyrolysate for NR, IR, BR, SBR, IIR, BIIR and CIIR that are the major elastomeric components of a tyre.

	TGA	**Py-GC/MS**	**^{14}C**	
			^{14}C (LSC)	**^{14}C (BI)**
Test duration	A few hour	A few days	2 months	A few days

Table 3. Test duration of the different methods.

http://dx.doi.org/10.5772/65830

This pyrolysis must be performed passing a stream of nitrogen through the pyrolysis reactor. Nitrogen serves to displace air, prevents oxidation and facilitates the transfer of the pyrolysis products to the gas chromatographer.

The gas chromatographer is equipped with 30-m-long capillary chromatographic column in a fused non-polar-type silica.

The gas chromatographer is coupled to a mass spectrometer operating in scan mode. It detects and registers certain decomposition substances between 35 and 550 atomic mass units (amu).

The pyrolysis-GC/MS carried out in one of the studies is based on ISO standard 7270-2 and requires a calibration curve by pyrolysing the samples with known styrene/butadiene/isoprene ratios.

The approach of this method is to evaluate the natural rubber content in a sample of tyres: It is possible to calculate the total concentration of elastomers in samples and also the concentration in natural elastomers by reporting the result on the previously produced calibration curve.

The authors of this study observed several problems during the application of this method; the main one being related to the non-possibility to distinguish natural isoprene from synthetic isoprene.

One major drawback is that determining the content comparing results with a curve made with different ratio of known samples of styrene/butadiene/isoprene rubbers only gives relative values inside the elastomeric fraction and not in the whole sample. An unrealistic composition on biomass could then be reported with this method.

Another problem derived from the use of PY-GC/MS is the non-detection of other biogenic components of the rubber. A variability of results with pyrolysis temperature and the extraction time in solvents before pyrolysis is also reported. Finally, the presence of brominated butyl could also disturb the results.

Taking into account all these issues, Pyrolysis-GC/MS is then not considered as a valid technique for the evaluation of biogenic content.

3.2. Thermogravimetric analysis (TGA)

This method is based on the continuous measurement of the weight loss of a sample submitted to a ramp of temperature in a controlled atmosphere [11, 19].

In the case of TGA, each type of elastomer has a particular temperature at which the loss of mass occurs. When a sample of vulcanized rubber is tested, some particular peaks appear at specific temperatures.

At lower temperatures, below 300°C, moisture, volatile components derived from plasticizers and other simple chemicals of the rubber blend, volatize.

In the range from 300 to 525°C most of the elastomers in a tyre rubber blend are degraded by the heat. The first thermal decomposition corresponds to natural rubber NR, and the maximum weight-loss rate occurs in the 300–400°C interval. Styrene-butadiene SBR maximum weight-loss rate occurs between 420 and 550°C.

One of the studies tried to produce a reference calibration curve based on different binary NR/SBR rubber samples of known composition and the intensity—or height—of the peaks of the DTG curves. Each measurement produces a typical graph of weight loss as a function of the increase in temperature. **Figure 4** shows an example of a weight loss curve for a (NR 75%, SBR 25%) blend and its derivative weight loss curves. This curve shows one minimum per elastomer, H_{NR} and H_{SBR}.

The calibration curve in **Figure 5** represents the relationship between the peak height ratio H_{NR}/H_{SBR} and the NR content of each of the reference samples. The resulting values are adjusted to a second degree polynomial equation [19]:

$$r = \left(\frac{NR}{NR + SBR}\right)\% = a\left(\frac{H_{NR}}{H_{SBR}}\right)^2 + b\left(\frac{H_{NR}}{H_{SBR}}\right) + c \tag{3}$$

$$a = -22.331;\ b = 84.503;\ c = 3.0782$$

where r is the percentage of NR in the elastomeric fraction (NR+SBR), H_{NR} is the maximum rate of weight loss in the area where NR decomposes, and H_{SBR} is the maximum rate where SBR decomposes.

Using this calibration curve, the value of r of an unknown sample can be then determined, based on the height of the peaks for NR and SBR.

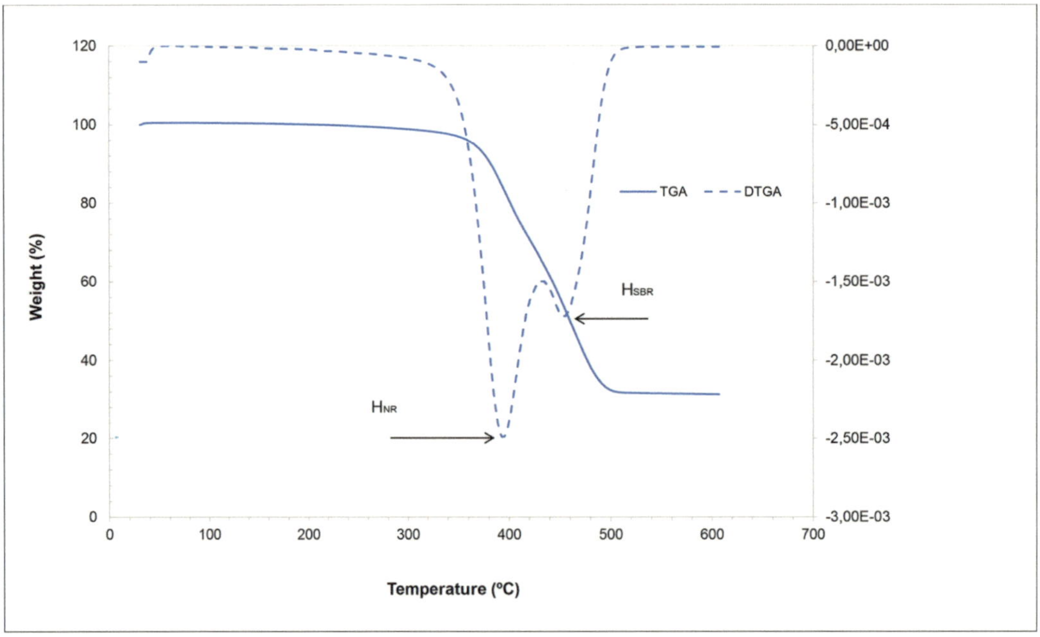

Figure 4. Weight loss and derivative weight loss curve for a polymer blend (NR75%/SBR25%).

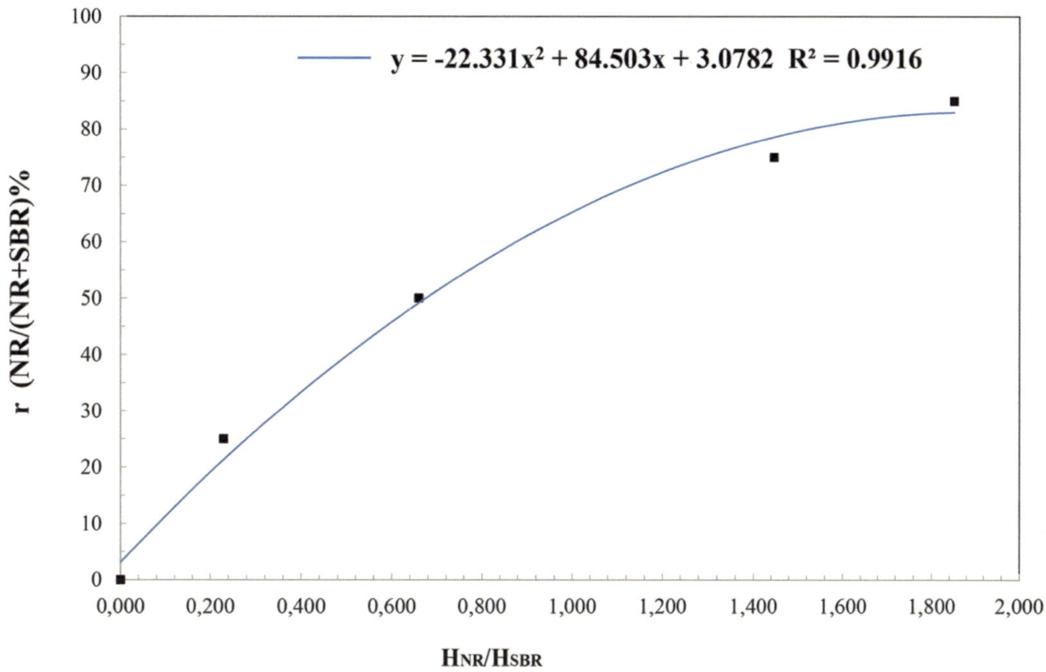

Figure 5. Plot of r (NR/(NR+SBR)%) vs. the peak heights ratio (H_{NR}/H_{SBR}) to determine the biomass content in the tyres studied.

However, it has been reported by authors of this chapter [17] that the correlation of the result for NR obtained by TG analysis and the actual content of NR in samples with known quantities of this elastomer is very bad. Like in the case of PY-GC/MS, this analysis technique is only valid for an estimation of the natural rubber content, independently of the accuracy of the results. Indeed, there is not any possibility to distinguish natural isoprene from synthetic isoprene.

Furthermore, problems derived from the use of this technique start when there is a combination of more than two elastomers in the sample; in such case, the identification and quantification of the elastomers are very difficult because of the overlapping of peaks. Finally, other biogenic components of the rubber, such as cotton or stearic acid, are not detectable by the use of this technique.

Taking into account all these issues, thermogravimetric analysis is then not considered as a valid technique for the evaluation of biogenic content.

3.3. Radiocarbon analysis: ^{14}C methods

The determination of biomass content in different materials using ^{14}C methods is based on analytical procedures used for the determination of the age of carbon contained in materials [8, 9].

Three well-known methods for the determination of ^{14}C content are described in the literature, two of which have been used in the studies made by the authors of this chapter. Those methods are commonly accepted for the determination of the age of objects, especially employed in archaeology, to date organic materials. The percentage of recent carbon in samples gives an accurate idea of non-fossil carbon content.

This technique is based on the principle that all the carbon atoms in organic materials have either a contemporary origin, proceeding directly or indirectly from the fixation of contemporary atmospheric CO_2 by means of photosynthesis, or a fossil origin and were fixed millions of years ago.

Every living organism contains a quantity of ^{14}C proportional to the relative abundance of ^{14}C in the atmosphere. Thus, the percentage of biomass in a material is directly proportional to its ^{14}C content. Fossil fuels, however, do not contain ^{14}C, as its half-life is 5,700 years [20, 21].

3.3.1. $^{14}C/^{12}C$ determination by beta-ionization (BI)

One of the studies used the analysis of the carbon concentration of bio-based origin (Beta ionization method) focusing on the biomass ^{12}C assay, considered to be more accurate.

The test has been developed according to standard ASTM D6866-08 and has determined the biogenic carbon content specifically for this purpose.

Particularly, in this case, this test method has been adapted and used for measuring ^{14}C, in elastomeric fraction and textile fraction.

3.3.2. Liquid scintillation spectrometer (LSC) ^{14}C determination

Another alternative is the ^{14}C determination by liquid scintillation spectrometer LSC using butyl-PBD, as scintillation agent, added to benzene (C_6H_6) samples, previously prepared by the following chemical reactions:

$$2CO_2 + 10Li \rightarrow Li_2C_2 + 4Li_2O \tag{4}$$

$$Li_2C_2 + 2H_2O \rightarrow C_2H_2 + 2LiOH \tag{5}$$

$$3C_2H_2 \rightarrow C_6H_6 \tag{6}$$

CO_2 is then produced in a combustion chamber by burning an appropriate sample of rubber coming from tyre. It was ensured that the reacting CO_2 was only coming from the sample.

The ^{14}C activity was corrected by the isotopic fractionation according to the directives of the ASTM D6866-05 standard test methods for the determination of biomass content [10]. To do this, the $^{12}C/^{13}C$ ratio was established in the stable isotopes laboratory of reference.

4. Discussion of results of the biomass content in tyres

Two different studies have been conducted recently, one in France and the other Spain for the quantification of the biomass content in tyres.

	Spanish case (number of tests)		French case (number of tests)	
Description of the sample	TGA	^{14}C (method C)	Py-GC/MS	^{14}C (method B)
Passenger car tyres	8	4	8	8
Truck and bus	4	2	4	4
Agro tyres	4	2	–	–
Total number of tests	12	8	12	12

Table 4. Tests are conducted for each method.

In both cases, the purpose of the studies was not only the quantification of the biomass content in tyres but also the validation or not of different techniques for this purpose.

In paragraph 3 of this chapter, three techniques have been compared: PY-GC/MS and TGA were discarded, and techniques related with ^{14}C determination have been accepted and are highly recommended. **Table 4** shows the number of tests conducted for each method.

Table 5 indicates the content of each sample of tyres in the recycling plant selected to conduct this study in Spain, taking into account the Spanish shared market, in terms of sizes. Lot coded PT140611-1 represents a sample of motorcycle, passenger car, SUV and vans tyres representative in percentage of the market in Spain. The other sample PA140611-2 was randomly taken from a stocked pile of those types of tyres. In addition, two other samples, one corresponding to truck tyres, and other sample with agro tyres were prepared.

Tyres were shredded and reduced to granules following the procedure described in paragraph 2 of this chapter. Samples were appropriately divided using a riffle box with the appropriate number of slots.

After reducing the size of the particles below 20 mm, the steel wires were removed from samples. This process was carried out with a magnet and the percentage of steel in samples is given in **Table 6**. Some rubber particles remain attached to the steel samples, so a calcination of these samples was performed to calculate the actual content of metal wires in the representative samples.

In the same way, the rubber granulates samples were reduced to fine powder, under 1 mm.

Sample code	Description of the sample	Number of tyres	Total weight in tons
PT140611-1	Passenger car tyres of a representative sample of the market	200	1.62
PA140611-2	Random sample of passenger car tyres	190–210	1.58
MC190711-3	Truck and bus	30	1.67
MA190711-4	Agro tyres	30	1.76

Table 5. Samples of different types of tyres (Spanish case).

	% in weight of the steel fraction	% percentage of steel after calcination	% real content per sample
PT140611-1	23	70.25	15.9
PA140611-2	24	62.22	15.1
MC190711-3	36	77.35	27.9
MA190711-4	6	90.15	5.2

Table 6. Percentage of steel fraction (Spanish case).

The French study used a specific laboratory equipment that permits a complete separation of the three different fractions of materials with no contamination between them, starting from shredding stage. So, the phase of calcination of steel fraction was not used. Summary of content of different materials are listed in **Tables 7** and **8**. Then the rubber and textile fractions were also reduced to particles under 1 mm.

In both studies, a cryogenic laboratory mill was used to achieve this particle size total reduction in the samples.

4.1. Results of the ^{14}C analysis

According to the sampling plan, the sample for laboratory test is representative of the lot if quantity is over 66 mg. In the case of liquid scintillation spectrometry, the mass of the sample should be between 10 and 12 g and then, only one single test per sample is necessary. Nevertheless, all the samples have been tested twice to ensure the repeatability of the results.

Results of ^{14}C per sample of rubber and textile using this LSC are shown in **Table 9**.

The percentage of modern carbon is calculated with the following expression.

Sample	Rubber	Metal	Textile
VL1	80.90	12.16	6.93
VL2	82.54	10.05	7.41
VL3	82.30	13.15	4.55
VL4	80.30	13.57	6.13
VL5	83.59	10.77	5.64
VL6	83.08	11.25	5.68
VL7	84.54	12.58	2.89
VL8	84.03	12.28	3.68
Average reference values	82.7	12.0	5.4

Table 7. Summary of content of different fractions per sample in French case (passenger car tyres).

Sample	Rubber	Metal	Textile
PL1	78.74	21.26	0.00
PL2	79.67	20.33	0.00
PL3	73.66	25.50	0.84
PL4	72.55	27.45	0.00
Average reference values	73.1	26.5	0.42

Table 8. Summary of content of different fractions per sample in French case (truck and bus tyres).

$$pMC = \frac{A_{SN}}{A_{ON}} \tag{7}$$

where A_{SN} is the isotopic activity of the sample, standardized for isotopic division, A_{ON} is the activity of the oxalic acid reference, also standardized for isotopic division and pMC is the percentage of modern carbon. Using this Eq. (7), the results of the biomass content in the Spanish study are summarized in **Table 10**.

In general terms, duplicated samples show similar values, all of them are under the limits of tolerance established for this analytical technique.

Table 10 shows the average value of the percentage of biomass in the rubber-textile fraction for each sample. The final result of biomass in the tyres sample, taking into account the content of steel in each one, is written in the **Table 11**.

As a conclusion, the biomass content in the sample corresponding to the passenger car tyres both random and representative to the Spanish share market are exactly the same with an average value of 22.2%.

Sample code	Weight of the sample burned (g)	Yield		Weight of C_6H_6 obtained (g)	Weight of C_6H_6 analyzed (g)	Isotopic division ($^0/_{00}$)
		C_2H_2%	C_6H_6%			
PT140611-1.1	12,1	99,2%	94,5%	7,4	6,188	-26,6 ± 0,2
PT140611-1.2	9,7	99,9%	97,5%	7,8	6,077	-26,6 ± 0,2
PT140611-2.1	11,2	99,6%	97,8%	8,1	6,123	-27,2 ± 0,2
PT140611-2.2	10,1	99,5%	97,9%	7,6	6,121	-27,7 ± 0,2
MC190711-3.1	10,8	99,3%	97,4%	7,7	6,112	-26,7 ± 0,2
MC190711-3.2	10	99,6%	97,6%	7,7	6,100	-26,7 ± 0,2
MA190711-4.1	9,7	97,9%	98,4%	7,8	6,103	-26,0 ± 0,2
MA190711-4.2	9,9	96,0%	97,5%	7,6	6,127	-26,0 ± 0,2

Table 9. Summary of synthetized samples for ^{14}C analysis (Spanish case).

Sample	Rubber and textile fraction	
	pMC (%)	Biomass (%)
PT140611-1.1	29,2 ± 0,7	27,2 ± 0,7
PT140611-1.2	27,5 ± 0,7	25,6 ± 0,7
PA140611-2.1	27,8 ± 0,7	25,8 ± 0,7
PA140611-2.2	28,5 ± 0,7	26,5 ± 0,7
MC190711-3.1	50,2 ± 0,7	46,7 ± 0,7
MC190711-3.2	50,9 ± 0,7	47,3 ± 0,7
MA190711-4.1	29,7 ± 0,7	27,6 ± 0,7
MA190711-4.2	30,1 ± 0,7	28,0 ± 0,7

Table 10. Results of ^{14}C analysis for the biomass content in tyres (Spanish case).

Sample	% biomass of the rubber-textile fraction	% biomass of the total sample	Range of biomass content in tyres (%)
Passenger car tyres of a representative sample of the market	26.4	22.2	21.5–22.9
Random sample of passenger car tyres	26.15	22.2	21.9–22.5
Truck and bus	47.0	33.9	33.7–34.1
Agro tyres	27.8	26.4	26.2–26.5

Table 11. Results of biomass content in tyres samples (Spanish case).

Biomass values obtained for tyres samples corresponding to truck and bus tyres resulted in an average percentage of 33.9% by weight, and the average value for biomass content of agro tyres resulted in a 26.4% by weight.

On the other hand, the study carried out in France on several tyre samples, both passenger car, truck and bus tyres, for the determination of biomass content have been tested by the beta-ionization technique.

Results obtained for different types of samples are shown in **Tables 12** and **13**. The value of biomass for passenger car tyre samples is in average 18.3 and 29.1% for the truck and bus samples.

In any case, the magnitude order of the biomass value obtained in the French case are in accordance with approximate theoretical values of natural rubber and other chemical substances with bio-based content in tyres. More specifically, the different values of natural rubber content in car and truck tyre have been obtained and are in line with theoretical values.

http://dx.doi.org/10.5772/65830

Sample	Biomass content (%)	Standard deviation
VL1	17	1.69
VL2	18.9	0.36
VL3	17.0	1.69
VL4	18.4	0.01
VL5	20.3	4.00
VL6	18.1	0.04
VL7	17.3	1.00
VL8	19.4	1.21
Average reference values	18.3	

Table 12. Biomass content of passenger car tyres measured by BI (French case).

Sample	Biomass content (%)	Standard deviation
PL1	30.7	0.36
PL2	31.4	1.69
PL3	29.7	0.16
PL4	28.6	2.25
Average reference values	29.1	

Table 13. Biomass content of truck and bus tyres measured by BI (French case).

5. Conclusions

A tyre is a complex product with a biomass content heterogeneously distributed inside it. Furthermore, there is also a huge variability in composition inside the market, by brands and types, each one having a different biomass content. Therefore, to carry out a proper study of biomass content of a batch or a sample representing a significant number of tyres, it is absolutely necessary to make a good selection of the laboratory sample by means of statistical methods to ensure representativeness of the lot. In this chapter, the procedure for managing tyres samples to obtain representative samples by continuous size reduction and quartering has been described.

Furthermore, it should be taken into account that necessary quantities for the biomass content determination are in the order of a few grams or even milligrams, so the sampling step is crucial to obtain proper results.

Several methods have been explored for determining the biomass fraction of tyres, but some of them are not completely reliable and conclusive, mainly for the following reasons:

The results of the thermogravimetric method differed considerably from the known natural rubber content of the reference samples as well as results obtained from the ^{14}C technique that is because the synthetic isoprene cannot be distinguished from natural isoprene. The pyrolysis-GC-FID method is neither considered as a reliable method, mainly for the same reason and because results are affected by temperature and extraction time.

In addition, the presence of textile fibres and stearic acid in the tyre, a well-known source of biomass, cannot be evaluated for both techniques (pyrolysis-GC-MS and TGA),

French and Spanish studies conclude that the ^{14}C techniques are the most reliable for determining the biomass content of end-of-life tyres. Both methods, beta-ionization (BI) and liquid scintillation spectrometry (LSC), lead to results close to the actual biomass content in tyres.

Finally, the reference values of the biomass content of the end-of-life tyres have been established. Average content for passenger car tyres is 18.3% for the French market and 22% for the Spanish market.

In the case of truck and bus tyres, content in both cases are in the range from 29.1% in the French case to 34% in the Spanish case. Apart from errors of different analytical techniques and laboratories, differences found in both studies could be related to different market share in term of sizes and brands in both countries.

In the Spanish case, a reference value has also been established for agro tyres with an average biomass content of 26.4%.

Author details

Leticia Saiz Rodríguez[1]*, José M. Bermejo Muñoz[1], Adrien Zambon[2] and Jean P. Faure[2]

*Address all correspondence to: lsaiz@signus.es

1 Signus Ecovalor S.L, Madrid, Spain

2 Aliapur, Lyon, France

References

[1] R&D ALIAPUR. Life cycle assessment of 9 recovery methods for end-of-life tyres [Internet]. 2010. Available from: http://www.etrma.org/uploads/Modules/Documentsmanager/aliapur_lca-reference-document-june-2010.pdf [Accessed: 2016-03-10]

[2] R&D ALIAPUR. Using used tyres as an alternative source of fuel [Internet]. 2009. Available from: https://www.aliapur.fr/sites/default/files/files/RetD/Extract_-_Used_tyres_as_an_alternative_source_of_fuel_July_09.pdf [Accessed: 2016-03-10]

[3] W. Staber, S. Flamme, J. Fellner: Methods for determining the biomass content of waste. Waste Management & Research. 2008; 26: 78–87. doi:10.1177/0734242X07087313

[4] European Parliament and of the Council. Directive 2003/87 of the European Parliament and Council, Establishing a scheme for greenhouse gas emission allowance trading within the community and amending Council Directive 96/61/EC [Internet]. 2003. Available from: http://eur-lex.europa.eu/legal-content/EN/TXT/PDF/?uri=CELEX:02003L0087-20140430&from=EN [Accessed: 2016-03-10]

[5] S. Singh, W. Nimmo, B.M. Gibbs, P.T. Williams: Waste tyre rubber as a secondary fuel for power plants. Fuel. 2009; 88: 2473–2480. doi:10.1016/j.fuel.2009.02.026

[6] M.N. Chukwu, I. C. Madufor et al: Effect of stearic acid level on the physical properties of natural rubber vulcanisate. The Pacific Journal of Science and Technology. 2011; 12: 344–350. url:www.akamaiuniversity.us/PJST12_1_344.pdf

[7] B. Lah, D.Klinar, B. Likozar: Pyrolysis of natural, butadiene, styrene–butadiene rubber and tyre components: modelling kinetics and transport phenomena at different heating rates and formulations. Chemical Engineering Science. 2013; 87: 1–13. doi:10.1016/j.ces.2012.10.003

[8] CEN (European Committee for Standardization). CEN/TS 15747–SRF, "^{14}C-based methods for the determination of the biomass content" [Internet]. 2008. Available from: http://standards.globalspec.com/standards/detail?docid=1166222&familyid=EUPDNCAAAAAAAAAA [Accessed: 2016-03-10]

[9] CEN (European Committee for Standardization). EN 14440–SRF, "Methods for the determination of biomass content" [Internet]. 2011. Available from: http://standards.globalspec.com/std/1465300/cen-en-15440 [Accessed: 2016-03-10]

[10] ASTM International Standard. D 6866-05. Determining the biobased content of natural range materials using radiocarbon and isotope ratio mass spectrometry analysis. Book of Standards. 2005; 0803. doi:10.1520/D6866-16

[11] S. R. Shield, G. N. Ghebremeskel and C. Hendrix: Pyrolysis-GC/MS and TGA as tools for characterizing blends of SBR and NBR. Rubber Chemistry and Technology. 2001; 74: 803–813. doi:10.5254/1.3547654

[12] Y.S. Lee, W. Lee, S. Cho, I. Kim, C. Ha, J.: Quantitative analysis of unknown compositions in ternary polymer blends: a model study on NR/SBR/BR system. Journal of Analaytical and Applied Pyrolysis. 2007; 78: 1, (85–94). doi:10.1016/j.jaap.2006.05.001

[13] CEN (European Committee for Standardization). EN 15413:2011–"Solid recovered fuels. Methods for the preparation of the test sample from the laboratory sample". [Internet]. 2011. Available from: http://www.aenor.es/aenor/normas/normas/fichanorma.asp?tipo=N&codigo=N0049240#.V9ktiPmLSJA [Accessed: 2016-03-10]

[14] CEN (European Committee for Standardization). EN 15415-1:2011, "Solid recovered fuels. Determination of particle size distribution. Screen method for small dimension particles". [Internet]. 2011. Available from: http://shop.bsigroup.com/ProductDetail/?pid=000000000030219013 [Accessed: 2016-03-10]

[15] CEN (European Committee for Standardization). CEN/TS 15401:2010, "Solid recovered fuels. Determination of bulk density". [Internet]. 2010. Available from: http://shop.bsigroup.com/ProductDetail/?pid=000000000030209997 [Accessed: 2016-03-10]

[16] CEN (European Committee for Standardization). EN 15443:2011–"Solid recovered fuels. Methods for the preparation of the Laboratory sample". [Internet]. 2011. Available from: http://www.aenor.es/aenor/normas/normas/fichanorma.asp?tipo=N&codigo=N0048493#.V9kuy_mLSJA [Accessed: 2016-03-10]

[17] L.Saiz, J.M. Bermejo et al: Comparative analysis of ^{14}C and TGA techniques for the quantification of the biomass content of end-of-life tires. Rubber Chemistry and Technology. 2014; 87: 4 (664–678). doi:10.5254/rct.14.85993

[18] ISO 7270-2_2012 "Rubber—Analysis by pyrolytic gas-chromatographic methods—part 2: determination of styrene/butadiene/isoprene ratio". [Internet]. 2012. Available from: http://www.iso.org/iso/catalogue_detail.htm?csnumber=53025 [Accessed: 2016-03-10]

[19] M.J. Fernández-Berridi, N. González, A. Mugica, C. Bernicot: Pyrolysis-FTIR and TGA techniques as tools in the characterization of blends of natural rubber and SBR. Thermochimica Acta. 2006; 444: 65–70. doi:10.1016/j.tca.2006.02.027

[20] J. Mohn, S. Szidat, J. Fellner, H. Rechberger, R. Quartier, B. Buchmann, L. Emmenegger: Determination of biogenic and fossil CO_2 emitted by waste incineration based on $^{14}CO_2$ and mass balances. Bioresource Technology. 2008; 99: 6471–6479. doi:10.1016/j.biortech.2007.11.042

[21] S.W.L. Palstra, H.A.J. Meijer: Carbon-14 based determination of the biogenic fraction of industrial CO_2 emissions—application and validation. Bioresource Technology. 2010; 101: 3702–3710. doi:10.1016/j.biortech.2009.12.004

Chapter 20

Reaction Behaviors of Bagasse Modified with Phthalic Anhydride in 1-Allyl-3-Methylimidazolium Chloride with Catalyst 4-Dimethylaminopyridine

Hui-Hui Wang, Xue-Qin Zhang, Yi Wei and Chuan-Fu Liu

Additional information is available at the end of the chapter

http://dx.doi.org/10.5772/65508

Abstract

The modification of lignocellulose with cyclic anhydrides could confer stronger hydrophilic properties to lignocellulose, which could be used in many industrial fields. To elucidate the modification mechanism of lignocellulose, bagasse was phthalated comparatively with its three main components in 1-allyl-3-methylimidazolium chloride (AmimCl) using 4-dimethylaminopyridine as catalyst and phthalic anhydride as acylation reagent in the present study. From FT-IR and 2D HSQC analyses, the skeleton of bagasse and the fractions were not significantly changed during phthalation in AmimCl. 2D HSQC results suggested that the reactive hydroxyls in bagasse were partially phthalated, and the reactivity of the hydroxyls in anhydroglucose units followed the order C-6 > C-2 > C-3. Similarly, the reactivity order of hydroxyls in anhydroxylose units was C-2 > C-3. For lignin, the predominant diesterification occurred during the homogeneous modification, and both aliphatic and aromatic hydroxyls were phthalated. The reactivity order of phenolic hydroxyls was S-OH > G-OH > H-OH, which was distinct from that without catalyst. In addition, it was found that the thermal stability of phthalated bagasse was affected by the disruption of cellulose crystallinity and the degradation of components. The thermal stability of the phthalated bagasse decreased upon chemical modification and regeneration.

Keywords: bagasse, phthalic anhydride, esterification, reaction behavior, 2D HSQC NMR, ^{31}P NMR, AmimCl

1. Introduction

Nowadays, the growing environmental concerns resulting from petroleum-based plastic materials and excessive use of fossil fuel have driven a strong global interest in renewable bio-based polymers and composites derived from natural resources [1, 2]. Biomass resource is one of the most abundant, inexpensive, and currently underutilized products from agricultural and biorefinery industries [3]. Biomass resources, especially agricultural residues with great renewability, biocompatibility, and biodegradability, are directing the development of next generation of chemicals, materials, energy, and processes [4]. Therefore, it could be particularly advisable to shift society's dependence away from fossil resources to biomass resources.

Bagasse, sugar industry residues, is mainly composed of lignocellulose, which represents one of the most abundant renewable resources on earth. Chemical modification represents an efficient process to introduce new functionalities into lignocellulose. Because of the poor dissolubility of lignocellulose in common molecular solvent, the heterogeneous modification has been studied for decades. However, heterogeneous modification is always accomplished at low efficiency and nonuniformity, and accompanied with the occurrence of side reactions [5, 6]. Even worse, the solvent or reaction medium for heterogeneous modification could not be recovered after modification [7], which caused environmental pollution and wastage of resources. It is a never-ending endeavor to explore homogeneous reaction systems to compensate the deficiency of heterogeneous systems.

Fortunately, it has been found that several dual solvent systems and ionic liquids (ILs) could dissolve lignocellulose [8]. Compared with other solvent systems, ILs exhibit negligible vapor pressure, theoretically making infinite recycling possible. In addition, the high electrochemical thermal stability, nonflammability, and designability made ILs widely used in diverse fields, such as solvent for synthesis and catalysis, separation, analytical chemistry, and biomass refinery [9–11]. The bridge between lignocellulose and ILs were built by Swaltloshi et al., who reported the nonderivatizating dissolution of cellulose in BmimCl [12]. In 2007, it was reported that woods with different hardness could be readily dissolved in various imidazium-based ILs under mild conditions [13, 14], which opened a new window of opportunities in homogeneous functionalization of lignocellulose. Since then, the chemical modification of lignocellulose in ILs has been expanding fast. Due to the existence of abundant hydroxyls in lignocellulose, the modification of lignocellulose with carboxylic anhydride, acid chloride, and isocyanides could be achieved using ILs as reaction media [15].

The chemical modification of lignocellulose with linear carboxylic acid anhydrides or acid chloride can produce the corresponding carboxylic acid or HCl as a by-product. However, the modification of lignocellulose with dicarboxylic acid anhydride is an effective process and can attach carboxyl groups onto lignocellulose. The attachment of carboxyl group can greatly improve the compatibility between modified lignocellulose and other polymer materials by producing new chemical linkages and the effect of hydrogen bonds [16]. Besides, the introduction of carbonyl group could also increase the hydrophilicity [17], achieve ion exchange [18], and remove heavy mental ion [19]. Our previous works showed that the homogeneous chemical modification of the isolated cellulose from bagasse with succinic anhydride and phthalic anhydride without catalyst could be achieved, and the optimal reaction conditions (reaction temperature 90°C and reaction time 90 min) for preparing modified cellulose with maximum degree of substitution (DS) were also screened [20, 21]. In 2010, Li et al. reported chemical modification of cellulose in BmimCl with catalyst DMAP could remarkably improve the DS of modified products [22]. However, the structural parameters of lignocellulosic derivatives were

http://dx.doi.org/10.5772/65508

complex and difficult to control due to the complicated mixture of cell wall components. Therefore, it is very essential to elucidate the modification mechanism of lignocellulose to produce tunable materials directly from lignocellulose, although some simple investigations have been published. Especially, the application of catalyst DMAP could obviously increase the degree of substitution of lignocellulose modified with cyclic anhydride, which is beneficial to produce polymer composites based on esterified lignocellulose. As far as the authors are aware, there have been no reports of the detailed structural changes of lignocellulose during the homogeneous phthalation with catalyst DMAP. We therefore investigated the reaction behaviors of lignocellulose during the homogenous phthalation with catalyst DMAP.

In the present study, bagasse was modified with phthalic anhydride with catalyst DMAP in AmimCl. Three main components cellulose, hemicelluloses, and lignin were isolated from bagasse, and were phthalated under the same conditions to elucidate the modification mechanism. The physicochemical properties of phthalated samples were characterized by FT-IR, CP/MAS ^{13}C NMR, ^{1}H NMR, liquid-state ^{13}C NMR, and 2D HSQC NMR. The thermal stability of the unmodified, regenerated, and modified materials was also investigated using TG.

2. Material and methods

2.1. Materials

Bagasse was obtained from a local factory (Jiangmen, China). It was dried in sunlight and then cut into small pieces. The cut bagasse was ground and screened to prepare 40–60 mesh size particles (450–900 μm). The dried ground samples were dewaxed with toluene-ethanol (2:1 v/v) and then dried in a cabinet oven with air circulation at 50°C for 24 h [15]. Ionic liquid AmimCl was purchased from the Shanghai Cheng Jie Chemical Co., Ltd. and used as received. All other chemicals used were of analytical grade and purchased from Guangzhou Chemical Reagent Factory (Guangzhou, China).

2.2. Isolation of cellulose, hemicelluloses and lignin from bagasse

Cellulose and hemicelluloses were isolated according to the previous literatures [21, 23]. Briefly, the extractive-free bagasse was delignified at 75°C for 2 h with sodium chlorite at pH 3.8–4.0, followed by the extraction with 10% NaOH with the ratio of solid to liquid at 1:20 g/mL for 10 h at room temperature (four times). The solid residues were filtered out, washed thoroughly with distilled water, then washed with ethanol and dried in an oven with air circulation at 50°C for 24 h to obtain cellulose. The filtrate from NaOH extraction (the first time) was neutralized, concentrated, and transferred into three volumes of 95% ethanol with agitation. The precipitates were filtered out, washed with 70% ethanol, and freeze-dried to obtain hemicelluloses.

Lignin was isolated from the extractive-free bagasse according to the previous literatures [24, 25]. The obtained crude lignin with high content of carbohydrates was purified as follows: dried crude lignin (1 g) was dissolved in 2 mL of acetone/water (9:1, v/v) in a beaker. The obtained lignin solution was added dropwise into 200 mL of distilled water with stirring, and the resulted suspension was further centrifuged. The obtained solid residues were washed with distilled water (thrice, total 90 mL), and then freeze-dried. The dried residues were dissolved into 2 mL of 1,2-dichloroethane/ethanol (2:1, v/v). The obtained solution was added dropwise into 200 mL of anhydrous ether in a beaker with stirring at room temperature, and the resulted suspension was further agitated for 30 min. The obtained suspension was centrifuged, and the solid residues was washed with anhydrous ether (thrice, total 90 mL) and freeze-dried.

2.3. Homogeneous phthalation of bagasse and the isolated fractions (cellulose, hemicelluloses and lignin)

The homogeneous phthalation of bagasse and its isolated fractions was performed according to the previous literature [26]. The extractive-free ground bagasse (12 g) was first finely ball-milled for 4 h in a planetary mill (Grinder BM4, China, Beijing) at 400 rpm using two 500 mL ZrO2 jars (500 g, 1 cm diameter). Cellulose isolated from bagasse was also ball-milled for 4 h under the same conditions, while hemicelluloses and lignin were directly used in the present study.

Typically, the prepared materials (0.2 g) was added to 10 g of AmimCl at room temperature under nitrogen atmosphere for 5 min and heated at 90°C under stirring for 4 h to obtain a clear solution. Phthalic anhydride (the weight ratio of phthalic anhydride to material, 4:1) and DMAP (the weight ratio of DMAP to phthalic anhydride, 5%) were added to the solution. The flask was continuously purged with nitrogen atmosphere for 5 min. The reaction was performed at 90°C for 90 min with agitation. Subsequently, the resulted solution was precipitated with ethanol (99 wt%, 200 mL) under agitation. The suspension was further stirred for 24 h. The solid residues were filtered out, washed with ethanol (three times, total 600 mL) to remove unreacted phthalic anhydride, catalyst DMAP, and AmimCl, and free-dried for further characterization.

2.4. Determination of WPG and DS

WPG [15] was calculated according to Eq. (1):

$$WPG = \frac{M_1 - M_0}{M_0} \times 100 \tag{1}$$

where M_0 and M_1 are the vacuum-dried weights of the bagasse or fractions before and after chemical modification, respectively.

The phthalation degree of phthalated samples was evaluated by DS, according to Eq. (2):

$$DS = \frac{n_{OH'}}{n_{OH}} \times 100 \tag{2}$$

where n_{OH}' is the substituted hydroxyls contents, and n_{OH} is the hydroxyl content of unmodified sample.

The substituted hydroxyl content of the phthalated samples was determined based on the equivalent volume of known molarity NaOH and HCl by back-titration method [21]. A known weight of the sample was dissolved in 10 mL 0.1 mol/L NaOH. The excess of NaOH was back-titrated with 0.025 mol/L HCl by using automatic potentiometric titrator, and the final pH of end titration point was set as 7.0. The titration was repeated three times, and the average value of HCl volume was used as the calculations of the substituted hydroxyl content according to Eq. (3):

$$n_{OH'} = \frac{c_1V_1 - c_2V_2}{2} \times \frac{1000}{m} \tag{3}$$

where n_{OH}' is the substituted hydroxyls contents, m (g) is the dry weight of sample analyzed, c_1 (mol/L) is the molarity of NaOH, V_1 (mL) is the volume of NaOH, c_2 (mol/L) is the molarity of HCl, and V_2 (mL) is the equivalent volume of known molarity HCl.

Based on the assumption that the cellulose and hemicelluloses are composed of AGU and AXU, respectively, the theoretical hydroxyl content of unmodified cellulose and hemicelluloses were calculated from their macromolecular structure according to Eqs. (4) and (5), respectively. The total hydroxyl content of unmodified lignin was 6.31 mmol/g by ^{31}P NMR analysis. The hydroxyl content of unmodified bagasse was calculated from Eq. (6), according to the contents of cellulose, hemicelluloses, and lignin in the extractive-free bagasse:

$$n_C = \frac{1000}{162} \times 3 \tag{4}$$

where n_C (mmol/g) is the theoretical hydroxyl content of unmodified cellulose, 162 g/mol is the molar mass of AGU, and 3 is the number of hydroxyl groups on each AGU, and

$$n_H = \frac{1000}{132} \times 2 \tag{5}$$

where n_H (mmol/g) is the theoretical hydroxyl content of unmodified hemicelluloses, 132 g/mol is the molar mass of AXU, and 2 is the number of hydroxyl groups on each AXU, and

$$n_B = n_C \times 44.85 + n_H \times 33.13 + n_L \times 19.24 \tag{6}$$

where n_B and n_L (mmol/g) are the hydroxyl contents of unmodified bagasse and lignin, respectively, and 44.85%, 33.13%, and 19.14% are the contents of cellulose, hemicelluloses, and lignin, respectively, in the extractive-free bagasse according to the standard NREL methods.

2.5. Characterization

FT-IR spectra were collected on an FT-IR spectrophotometer (Nicolet 510) using a KBr disk containing approximately 1% finely ground samples. Thirty-two scans were taken for each sample with a resolution of 2 cm^{-1} in transmittance mode in the range of 4000–400 cm^{-1}.

^{1}H NMR, ^{13}C NMR, ^{31}P NMR, and 2D HSQC spectra were recorded on Bruker AV-III HD 600 spectrometer (Germany). The ^{1}H NMR spectra were recorded at 298.0 K from the dried samples (40 mg) dissolved in 0.5 mL DMSO-d_6. The detailed collecting and processing parameters for ^{1}H NMR analysis were as follows: number of scans, 16; receiver gain, 31; acquisition time, 2.7263 s; relaxation delay, 1.0 s; pulse width, 11.0 s; spectra frequency, 600.17 Hz; and spectra width, 12019.2 Hz.

For CP/MAS ^{13}C NMR analysis, dried samples (200 mg) were tested at 303.0 K. The detailed collecting and processing parameters were as follows: number of scans, 2000; receiver gain, 567; acquisition time, 0.0344 s; relaxation delay, 1.5 s; pulse width, 4.55 s; spectra frequency, 100.61Hz; and spectra width, 29761.9 Hz.

For liquid-state ^{13}C NMR analysis, dried samples (40 mg) was dissolved in 0.5 mL DMSO-d_6 at 298.0 K, the detailed collecting and processing parameters were as follows: number of scans, 15,000; receiver gain, 1290; acquisition time, 1.2583 s; relaxation delay, 1.5 s; pulse width, 12.00 s; spectra frequency, 100.61 Hz; and spectra width, 26041.7 Hz.

For 2D HSQC NMR analysis, the spectra were acquired using 40 mg samples in the 0.5 mL DMSO-d_6 at 299.9 K. The detailed collecting and processing parameters were as follows: number of scans, 16; receiver gain, 187; acquisition time, 0.1420 s; relaxation delay, 2.0 s; pulse width, 11.0 s; spectra frequency, 600.17/150.91 Hz; and spectra width, 7211.51/24875.6 Hz.

The hydroxyls contents of lignin samples were determined by ^{31}P NMR spectroscopy, according to the reported method [25]. The detailed collecting and processing parameters were as follows: number of scans, 300; receiver gain, 187; acquisition time, 0.3408 s; relaxation delay, 2.0 s; pulse width, 12.0 s; spectra frequency, 242.95 Hz; and spectra width, 96,153.8 Hz.

TG of samples was performed on a simultaneous thermal analyzer (SDT Q600, TA Instrument). The sample weighted between 8 and 10 mg. The scans were first run from room temperature to 100°C, were equilibrated at 100°C for 5 min, and then were cooled to 40°C at the cooling rate of 20°C/min. The scans were again heated from 40°C to 700°C at the heating rate of 10°C/min under nitrogen flow.

3. Results and discussion

3.1. Homogeneous phthalation of bagasse and its fractions

The reaction behaviors of bagasse during homogeneous phthalation in AmimCl were investigated comparatively with the isolated fractions under the same conditions. The contents of cellulose, hemicelluloses, and lignin in the extractive-free bagasse were determined as 44.85%, 33.13%, and 19.14%, respectively, according to the standard NREL methods [27].

As shown in **Table 1**, the DS of phthalated cellulose, hemicelluloses, and lignin were 19.88%, 13.67%, and 27.10%, respectively, indicating the reactivity of lignin was the highest among the three main components during bagasse homogeneous modification in AmimCl. The DS of phthalated bagasse (10.80%) was lower than phthalated fractions, which was probably due to the complicated linkages between bagasse fractions. Besides, the WPG of phthalated cellulose, hemicelluloses, and lignin were 17.04%, 5.98%, and -49.70%, respectively, suggesting that the

	Phthalation conditions				n_{OH}' (mmol/g)	DS (%)	WPG (%)
Sample no	Molar ratio[a]	DMAP[b]	Time (min)	Temperature (°C)			
B_s	4:1	5%	90	60	1.57	10.80	-10.51%
C_s	4:1	5%	90	60	3.68	19.88	17.04%
H_s	4:1	5%	90	60	2.07	13.67	5.98%
L_s	4:1	5%	90	60	1.71	27.10	-49.70%

[a]Represents the weight ratio (w/w) of phthalic anhydride/materials.
[b]Represents the dosage of catalyst dimethylaminopyridine (DMAP) by weight ratio (w/w).
B_s, C_s, H_s, and L_s represent the phthalated bagasse, cellulose, hemicelluloses, and lignin, respectively.

Table 1. The substituted hydroxyl content, weight percent gain, and degree of substitution of samples.

significant degradation of lignin occurred during homogeneous phthalation. The WPG of phthalated bagasse was -10.51%, probably resulting from the degradation of lignin during bagasse homogeneous modification.

3.2. ^{31}P NMR analysis

To investigate the detailed information of lignin hydroxyls during homogeneous modification, ^{31}P NMR spectroscopy was applied in the present study to quantify the hydroxyl contents of lignin samples [28, 29]. The ^{31}P NMR spectra of unmodified lignin (L_o, spectrum a) and phthalated lignin (L_s, spectrum b) are shown in **Figure 1**. According to the reported literatures [30, 31], the assignments of signals in ^{31}P NMR spectra were as followed: international standard (IS), 145.50–144.80 ppm; aliphatic OH, 149.50–145.50 ppm; phenolic S-OH, 143.50–142.50 ppm; phenolic G-OH, 140.00–138.50 ppm; phenolic H-OH 138.00–137.50 ppm; COOH, 135.50–134.50 ppm.

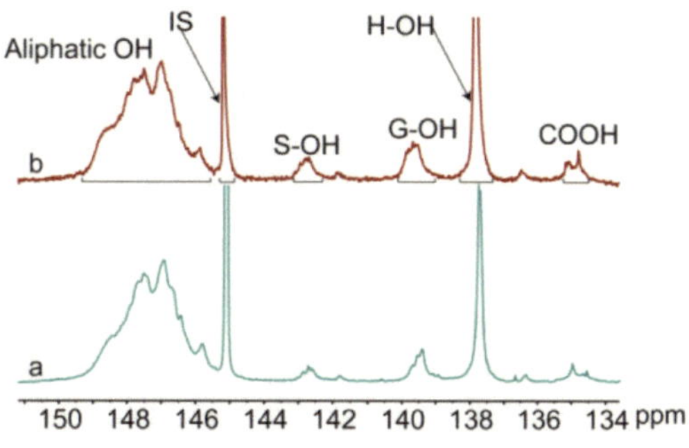

Figure 1. ^{31}P NMR spectra of unmodified lignin (L_o, spectrum a) and phthalated lignin (L_s, spectrum b).

The hydroxyl contents of unmodified and phthalated lignin are present in **Table 2**, and the decreased percentages of lignin hydroxyls are shown in **Figure 2**. After phthalation, the aliphatic hydroxyl content decreased from 4.71 mmol/g in unmodified lignin to 3.33 mmol/g in phthalated lignin, and the total phenolic hydroxyl content was also reduced from 1.60 (L_o) to 1.27 mmol/g (L_s). These results indicated that both aromatic and phenolic hydroxyls were phthalated during lignin modification in AmimCl. Besides, the total hydroxyl content of lignin decreased from 6.31 (L_O) to 4.60 mmol/g (L_S), while the carboxyl content slightly increased from 0.08 (L_O) to 0.09 (L_S) mmol/g. The result suggested that the predominant diester form [32] during lignin homogeneous modification in AmimCl. For lignin phenolic hydroxyls, the content of phenolic S-OH decreased from 0.16 to 0.09 mmol/g after phthalation; the content of phenolic G-OH decreased from 0.40 (L_O) to 0.25 (L_S) mmol/g; the decrease in the content of phenolic H-OH from 1.04 (L_r) to 0.93 (L_S) mmol/g was also observed. These results indicated phenolic S-, G-, and H-OH were phthalated during lignin modification in AmimCl.

Sample	L_o	L_s
Aliphatic OH (mmol/g)	4.71	3.33
Phenolic OH (mmol/g)	1.60	1.27
S-OH (mmol/g)	0.16	0.09
G-OH (mmol/g)	0.40	0.25
H-OH (mmol/g)	1.04	0.93
COOH (mmol/g)	0.08	0.09
S/G ratio (mmol/g)	0.40	0.38

Table 2. The hydroxyl contents of lignin samples.

As shown in **Figure 2(a)**, the decreased percentage of aromatic hydroxyls was higher than that of phenolic hydroxyls, indicating the higher reactivity of aliphatic hydroxyls during lignin homogeneous phthalation. For phenolic hydroxyls, the decreased percentage of phenolic S-OH, G-OH, and H-OH were present (**Figure 2b**). Therefore, the phthalation reactivity of phenolic hydroxyls followed the order of S-OH > G-OH > H-OH.

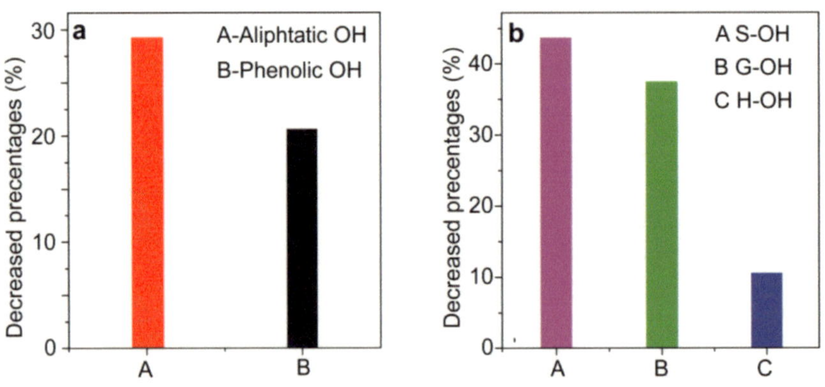

Figure 2. The decreased percentages of lignin hydroxyls.

3.3. FT-IR

The FT-IR spectra of unmodified cellulose (C_o, spectrum a), regenerated cellulose (C_r, spectrum b), and phthalated cellulose (C_S, spectrum c) are shown in **Figure 3**. The absorbance at 3433,

2895, 1646, 1376, 1165, and 1057 cm^{-1} were associated with unmodified cellulose [20]. These peaks present in regenerated and phthalated cellulose indicated that the skeleton of cellulose was stable during cellulose dissolution and modification in AmimCl. Compared with the spectrum of unmodified cellulose, the spectrum of the phthalated cellulose (spectrum c) provides evidence of phthalation by showing the presence of three new bands at 1718, 1580, and 749 cm^{-1} [21], corresponding to the stretching of carbonyl group, antisymmetric stretching of carboxylic anions, and the out-of-plane C-H bending of ortho disubstituted benzene, respectively. In addition, the intensity of peak at 1057 cm^{-1} for C-O-C stretching in spectrum c slightly decreased indicating the degradation of cellulose macromolecules during dissolution and modification. This degradation was also reported in the previous literature [21].

Figure 4 illustrates the FT-IR spectra of unmodified hemicelluloses (spectrum a), regenerated hemicelluloses (spectrum b), and phthalated hemicelluloses (spectrum c). Similarly, the

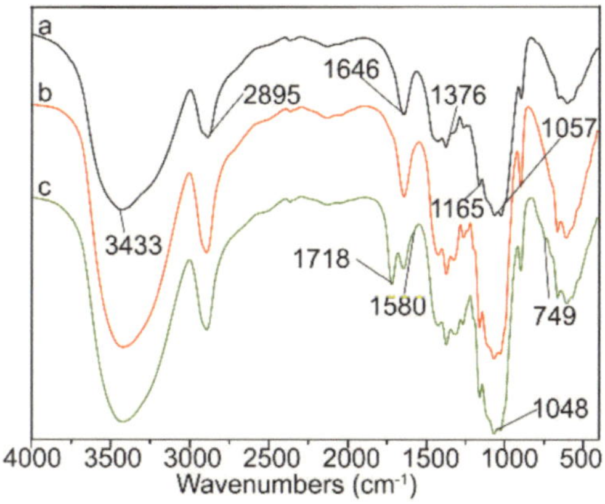

Figure 3. FT-IR spectra of unmodified cellulose (C_o, spectrum a), regenerated cellulose (C_r, spectrum b), and phthalated cellulose (C_s, spectrum c).

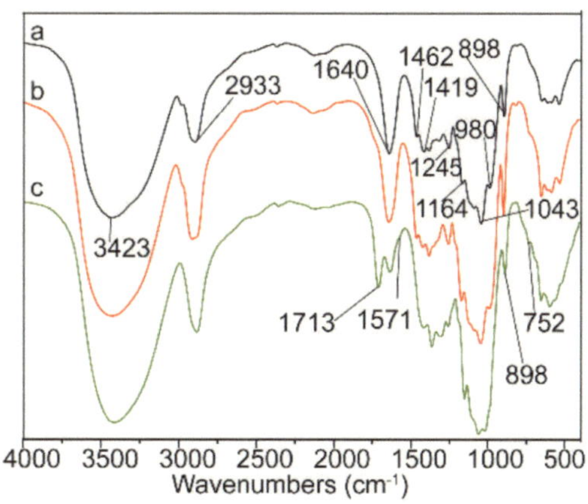

Figure 4. FT-IR spectra of unmodified hemicelluloses (H_o, spectrum a), regenerated hemicelluloses (H_r, spectrum b), and phthalated hemicelluloses (H_s, spectrum c).

skeleton of hemicelluloses remained unchanged after dissolution and modification in AmimCl. Compared with unmodified hemicelluloses, the new bands at 1713, 1571, and 752 cm^{-1} in phthalated hemicelluloses indicated the phthalation of hemicelluloses [33].

The FT-IR spectra of unmodified lignin (spectrum a), regenerated lignin (spectrum b), and phthalated lignin (spectrum c) are present in **Figure 5**. The bands at 3418, 2930, 1692, 1504, 1595, 1267, 1118, 1031, 824, and 747 cm^{-1} are related to unmodified lignin [34, 35]. Similar to cellulose and hemicelluloses, the skeleton of lignin was also stable after dissolution and modification in AmimCl. Compared with unmodified lignin, the intensity of bands at 1702 and 748 cm^{-1} in phthalated lignin obviously increased, suggesting the occurrence of esterification between lignin and phthalic anhydride [36].

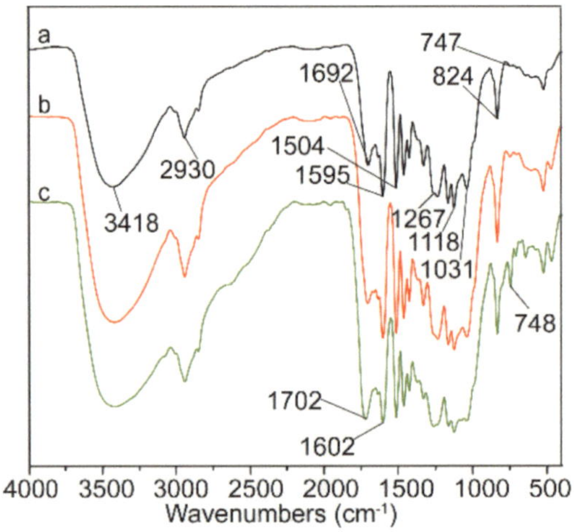

Figure 5. FT-IR spectra of unmodified lignin (L_o, spectrum a), regenerated lignin (L_r, spectrum b), and phthalated lignin (L_s, spectrum c).

FT-IR spectra of unmodified bagasse (B_o, spectrum a), regenerated bagasse (B_r, spectrum b), and phthalated bagasse (B_s, spectrum c) are shown in **Figure 6**. The bands were assigned according to the previous literature [37]. The peaks at 1605, 1515, 1375, and 897 cm^{-1} remained predominant in modified bagasse, suggesting the skeleton of bagasse fractions were unchanged during bagasse homogeneous modification in ionic liquid AmimCl [15]. The intensities of the peak at 1725 and 744 cm^{-1} increased after modification, providing the direct evidence of esterification between phthalic anhydride and bagasse. In conclusion, the phthalation of bagasse and its fractions occurred during homogeneous modification according to FT-IR analyses.

3.4. NMR analysis

3.4.1. Solid-state ^{13}C NMR of bagasse

The poor dissolubility of bagasse samples in DMSO-d_6 made high-resolution liquid-state 2D HSQC NMR spectroscopy impossible, and CP/MAS ^{13}C NMR spectroscopy was applied for bagasse samples analysis, as shown in **Figure 7**. The signals in CP/MAS ^{13}C NMR spectra were assigned based on the reported literatures [15, 21]. The signals at 191.2, 170.5, and 130.5 ppm

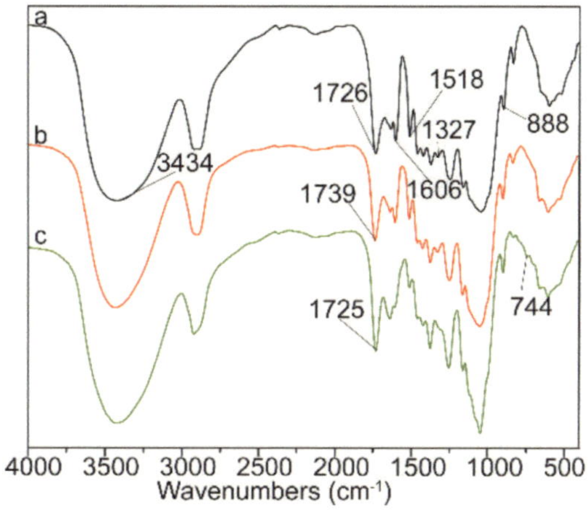

Figure 6. FT-IR spectra of unmodified bagasse (B_o, spectrum a), regenerated bagasse (B_r, spectrum b), and phthalated bagasse (B_s, spectrum c).

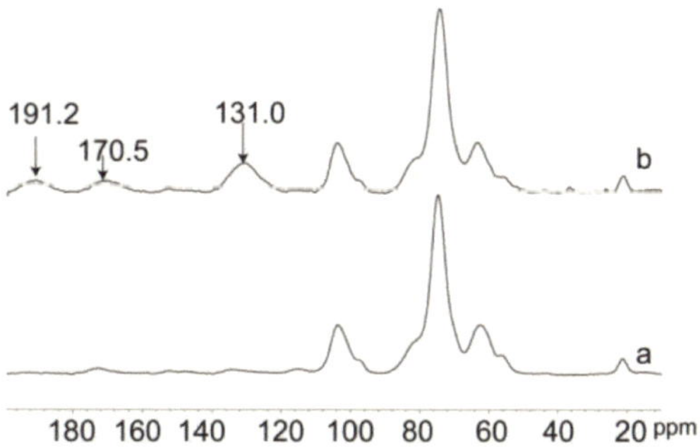

Figure 7. Solid-state ^{13}C NMR spectra of unmodified bagasse (B_o, spectrum a) and phthalated bagasse (B_s, spectrum b).

correspond to benzene ring, carbonyl group, and carboxylic groups, respectively, in phthalated bagasse. Compared with unmodified bagasse, the intensities of the three signals in phthalated bagasse remarkably increased, indicating the attachment of phthaloyl group onto bagasse.

The signals between 60 and 105 ppm relate to the carbons of carbohydrates. The chemical shifts and intensities of signals in this region remained basically unchanged, indicating the carbon skeleton of bagasse polysaccharides (cellulose and hemicelluloses) was stable during homogeneous modification, which was consistent with the results from FT-IR. Due to the low resolution of CP/MAS ^{13}C NMR spectroscopy, the detailed information of bagasse fractions during homogeneous phthalation need to be further revealed with 2D HSQC NMR spectroscopy.

3.4.2. Detailed reaction behaviors of cellulose during homogeneous phthalation

The HSQC spectra of unmodified cellulose (spectrum a) and phthalated cellulose (spectrum b) are present in **Figure 8**. The primary correlations of cellulose were well assigned, as reported in the previous literatures [20, 21]. The cross-peaks of cellulose were clearly observed in **Figure 8(a)** and **(b)** at 73.49/3.06 [C-C_2 (C_2/H_2)], 75.33/3.36 [C-C_3 (C_3/H_3)], 80.85/3.33 [C-C_4 (C_4/H_4)], 77.14/

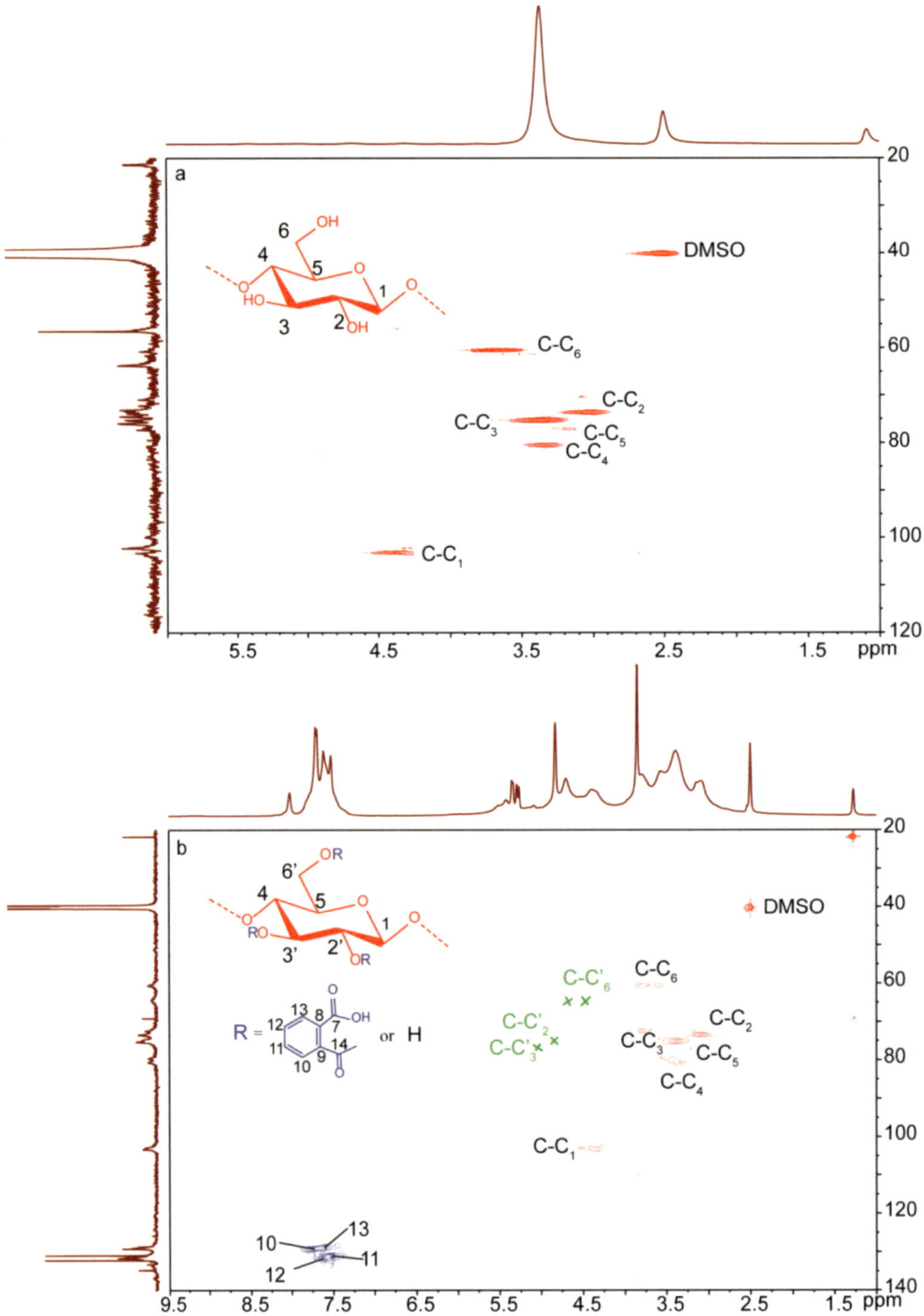

Figure 8. 2D HSQC spectra of unmodified cellulose (C_o, spectrum a) and phthalated cellulose (C_s, spectrum b).

3.18 [C-C_5 (C_5/H_5)], and 103.44/4.33 [C-C_1 (C_1/H_1)] ppm; the two C-C_6 (C_6/H_6) peaks were also distinctively located at 60.77/3.79 and 60.77/3.58 ppm, respectively. These results confirmed that the skeleton of cellulose was stable during homogeneous phthalation in AmimCl, consistent with the FT-IR analysis.

The cross-peaks at 131.49/7.62, 131.11/7.52, 129.43/7.78, and 128.94/7.61 ppm were associated with C_{10}/H_{10}, C_{11}/H_{11}, C_{12}/H_{12}, and C_{13}/H_{13} in the aryl group of phthalated cellulose, respectively. The presence of these peaks in phthalated cellulose confirmed the attachment of phthaloyl group onto cellulose. More importantly, two peaks from substituted C-6 in AGU C-C_6' (C_6/H_6) appeared at 64.39/4.39 and 64.50/4.67 ppm, and peaks from substituted C-2 [C-C_2' (C_2/H_2)] and C-3 [C-C_3' (C_3/H_3)] in AGU were located at 74.32/4.60 and 75.30/4.83 ppm, respectively. These results suggested the successful phthalation of hydroxyls at C-6, C-2, and C-3 positions in AGU. The DS of hydroxyls on different positions could be evaluated upon the integral area of the characteristic substituted correlations. By integration, the DS of C_6-OH, C_2-OH, and C_3-OH in AGU were 12.05, 2.54, and 1.84%, respectively. Therefore, the DS of hydroxyls in AGU followed the order C-6 > C-2 > C-3. The phthalated cellulose was previously prepared under the similar experimental conditions without catalyst in our laboratory, and the DS of hydroxyls at C-6, C-2, and C-3 positions were 6.30, 2.01, and 0% in AGU, respectively. By contrast, it could be concluded that the addition of catalyst DMAP could improve the uniformity and DS of phthalated cellulose.

3.4.3. *Detailed reaction behaviors of hemicelluloses during homogeneous phthalation*

The 2D HSQC spectra of unmodified hemicelluloses (spectrum a) and phthalated hemicelluloses (spectrum b) are present in **Figure 9**. According to the previous literatures [38, 39], the correlations of xylan were at 102.12/4.25 [X-C_1 (C_1/H_1)], 73.39/3.03 [X-C_2 (C_2/H_2)], 74.60/3.23 [X-C_3 (C_3/H_3)], and 75.84/3.4 [X-C_4 (C_4/H_4)] ppm, respectively, and the cross-peaks from X-C_5 (C_5/H_5) appeared at 63.59/3.16 and 63.59/3.86 ppm in both unmodified and phthalated hemicelluloses. Besides, the cross-peaks of arabinose were also well resolved at 107.76/5.24 [A-C_1 (C_1/H_1)], 86.62/3.89[A-C_4 (C_4/H_4)], 81.32/3.76[A-C_2 (C_2/H_2)], 78.32/3.57 [A-C_3 (C_3/H_3)], and 62.13/3.40 [A-C_5 (C_5/H_5)] ppm, respectively. These results indicated that the hemicelluloses isolated from bagasse were mainly composed of arabinose and xylan, consistent with the previous literatures [25, 27]. Compared with unmodified hemicelluloses, the cross-peaks at 128.81/7.66, 129.34/7.76, 130.99/7.51, and 131.31/7.63 ppm were for C_{12}/H_{12}, C_9/H_9, C_{10}/H_{10}, and C_{11}/H_{11}, respectively, in the phthaloyl group of phthalated hemicelluloses. These results further confirmed the occurrence of the esterification between hemicelluloses and phthalic anhydride during homogeneous modification in AmimCl. The cross-peaks at 74.95/4.70 and 76.58/5.02 ppm relate to the phthalated hydroxyls at C-2 and C-3 positions in AXU. The DS of hydroxyls at C-2 and C-3 positions in AXU were 3.56 and 1.54% by integral evaluation, indicating the higher reactivity of C_2-OH in AXU during hemicelluloses homogeneous phthalation.

3.4.4. *Detailed reaction behaviors of lignin during homogeneous phthalation*

The 2D HSQC NMR spectra of unmodified lignin (spectrum a) and phthalated lignin (spectrum b) are illustrated in **Figure 10**, and the structure of identified lignin subunits with

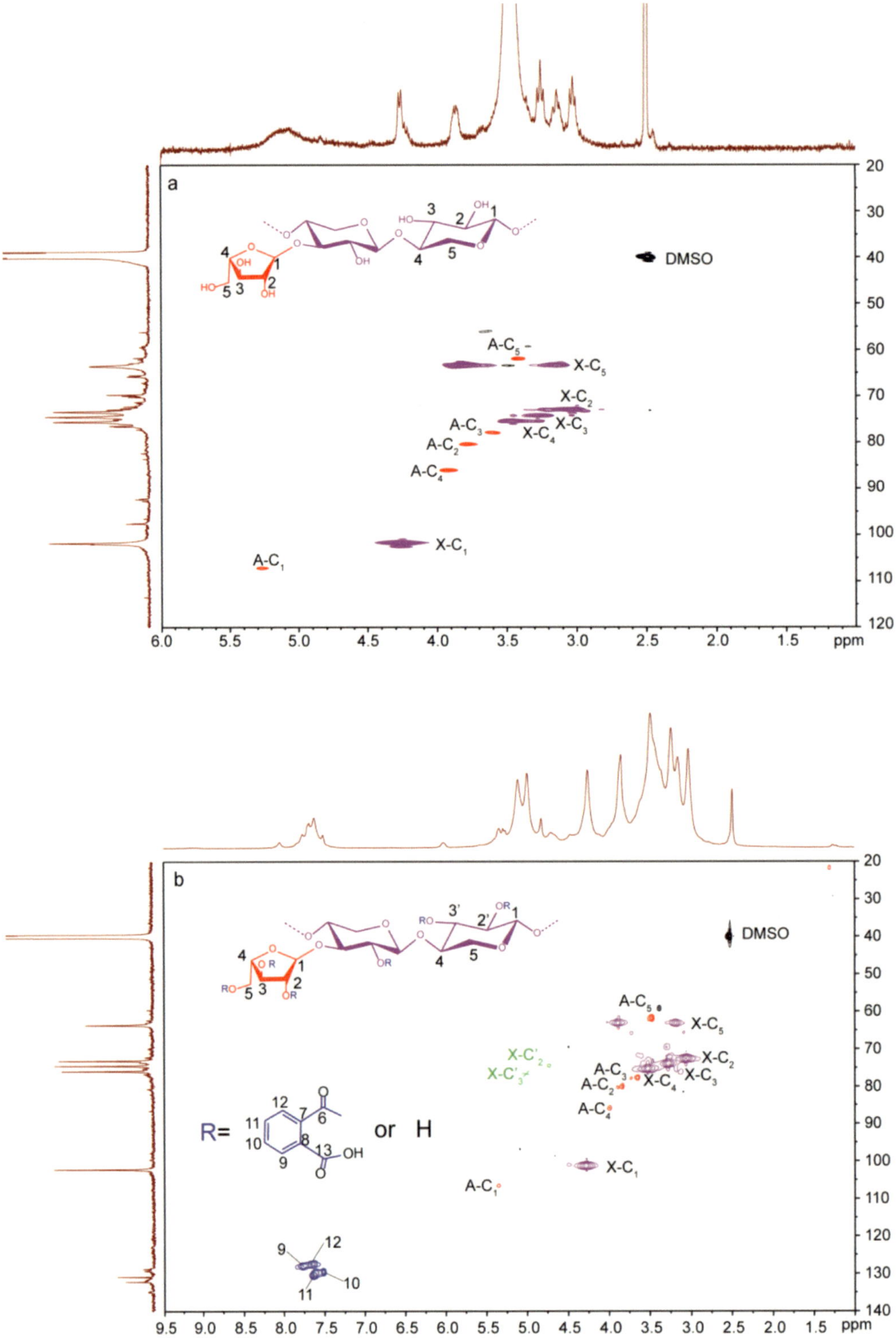

Figure 9. 2D HSQC spectra of unmodified hemicelluloses (H_o, spectrum a) and phthalated hemicelluloses (H_s, spectrum b).

http://dx.doi.org/10.5772/65508

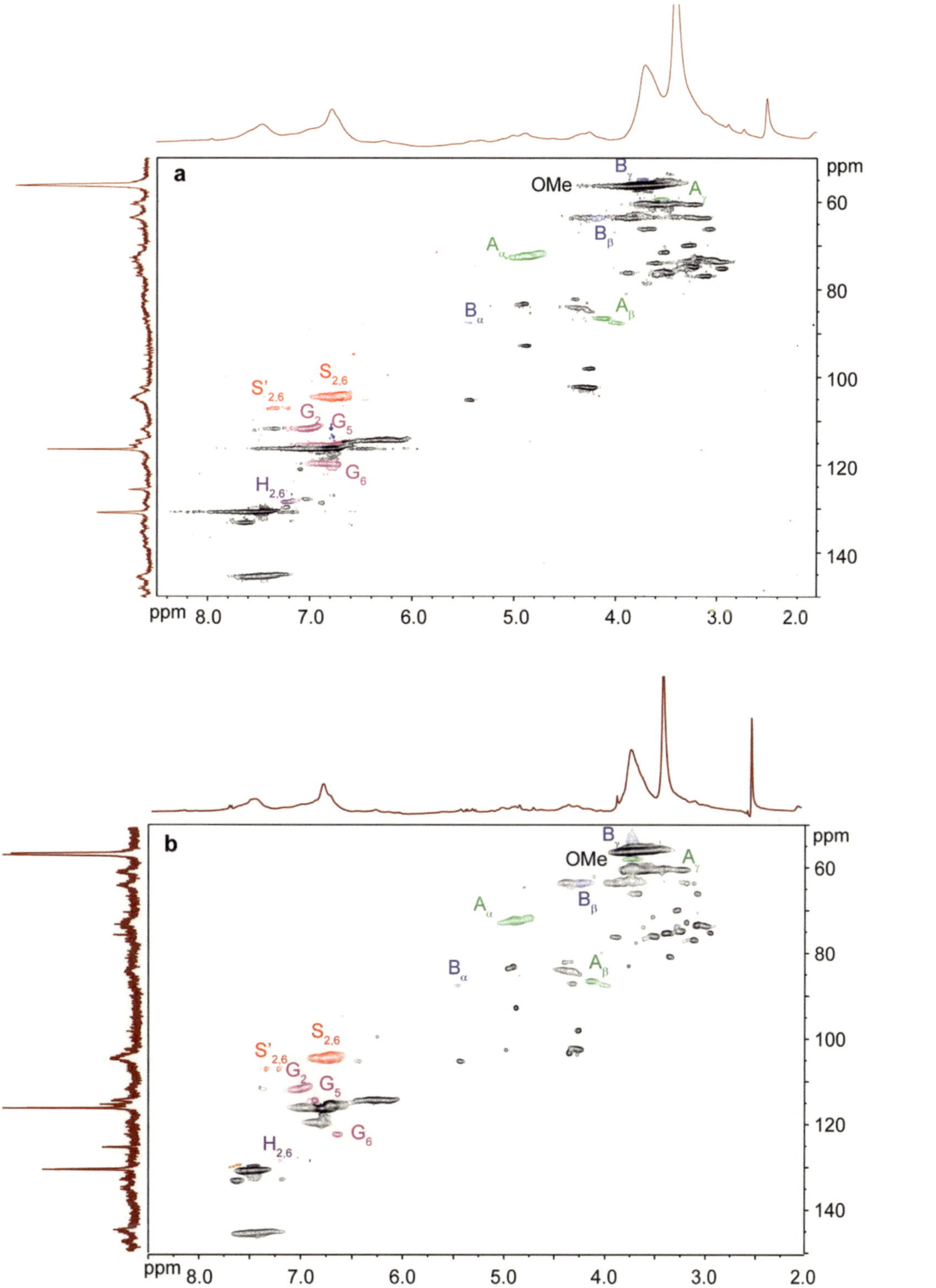

Figure 10. 2D HSQC spectra of unmodified lignin (L_O, spectrum a) and phthalated lignin (L_S, spectrum b).

hydroxyls are shown in **Figure 11**, including aryl ether (β-O-4′, A), phenylcoumaran (β-5′, B), guaiacyl unit (G-unit, C), syringyl unit (S-unit, S), and *p*-hydrophenyl unit (H-unit, H). According to the previous publications [24, 25], the correlations of these signals were well assigned. The cross-peaks from S-units $C_{2,6}/H_{2,6}$ was at 104.68/6.72 ppm, and the signals for C_α-oxidized S-units (S′) appeared at δ_C/δ_H 107.15/7.34 ppm. The cross-peaks originated from G-units were well distinguished: C_2/H_2 (δ_C/δ_H 111.69/7.00 ppm), C_5/H_5 (δ_C/δ_H 115.33/6.71 ppm), and C_6/H_6 (δ_C/δ_H 119.75/6.80 ppm). The $C_{2,6}/H_{2,6}$ correlations in H-units at δ_C/δ_H 118.66/7.20 ppm was also observed. For lignin side-chains, lignin substructures with hydroxyls were also well recognized. The C_α/H_α, C_β/H_β (linked to G/H units), C_β/H_β (linked to S units), and C_γ/H_γ correlations in β-*O*-4′ linkages were at δ_C/δ_H 72.43/4.84, 84.16/4.35, 86.69/4.09, and 60.35/3.48 ppm, respectively. In addition, phenylcoumaran (β-5′, substructure B) appeared in noticeable amounts as indicated by the C_α/H_α, C_β/H_β, and C_γ/H_γ correlations at δ_C/δ_H 87.67/5.43, 55.36/3.70, and 63.50/4.21 ppm, respectively.

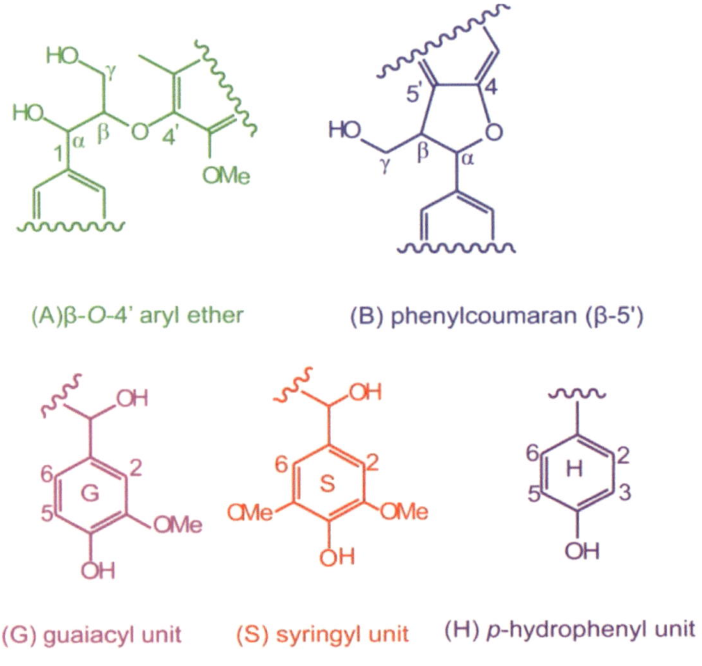

Figure 11. The structure of lignin subunits.

The relative quantities of the primary substructures were calculated according to the previous literatures [40, 41]. The aromatic lignin S-, G-, and H-units were expressed as a fraction of 100%, and the relative molar quantities of aryl ether, phenylcoumaran were expressed as a percentage of the total aromatic S-, G-, and H-units. The detailed information of quantitative lignin substructures calculated from HSQC spectra are listed in **Table 3**.

After phthalation, the content of aryl ether increased from 57.80/100Ar (L_o) to 58.33/100Ar (L_s), and the content of phenylcoumaran also increased from 2.76/100Ar (L_o) to 3.13/100Ar (L_s). These increases were probably due to the decreased intensity of $C_{2,6}/H_{2,6}$ correlations from

Sample	L_o	L_s
Aliphatic percentages of (S+G+H)		
Aryl ether (A)	57.80/100Ar	58.33/100Ar
Phenylcoumaran (B)	2.76/100Ar	3.13/100Ar
Aromatic molar percentages (S+G+H)		
Syringyl (S)	49.54%	57.81%
Guaiacyl (G)	45.78%	37.50%
p-hydroxyphenyl (H)	4.59%	4.69%
S/G	1.08	1.54

Table 3. Quantitative information of unmodified lignin (L_o) and phthalated lignin (L_s) determined with 2D HSQC NMR.

aromatic units resulting from the condensation of lignin [41] during the homogeneous phthalation. The relative percentages of G-units decreased from 45.78% to 37.50%, while the increases in the relative percentage of S- and H-units were observed. This result suggested that the G-units were easily degraded during lignin homogeneous modification [42, 43]. Besides, the increase in the S/G ratio from 1.08 (L_O) to 1.54 (L_S) further confirmed the degradation of G-units during lignin homogeneous modification.

3.5. Thermal analysis

In the present study, the effect of dissolution, regeneration, and modification in AmimCl on the thermal stability of samples was investigated by comparing unmodified, regenerated, and phthalated samples. **Figure 12** illustrates the TG curves of cellulose (A), hemicelluloses (B), lignin (C), and bagasse (D), respectively.

The onset and midpoint degradation temperature of regenerated cellulose were higher than those of unmodified cellulose, indicating the improved thermal stability after dissolution and regeneration in IL. This was probably because cellulose fractions with low molecular weight could not be regenerated from IL [15]. As shown in **Table 4**, the onset and midpoint degradation temperature of phthalated cellulose were lower than those of regenerated cellulose, suggesting the thermal stability of phthalated cellulose decreased after homogeneous modification. The decreased thermal stability of phthalated cellulose was probably due to partial hydrolysis, degradation of cellulose, and decrystallization of crystalline cellulose macromolecules [21].

Similar to cellulose, the thermal stability of regenerated hemicelluloses were slightly higher than that of unmodified hemicelluloses, probably resulting from the loss of hemicellulosic fractions with low molecular weights during dissolution and regeneration in AmimCl. Compared with regenerated hemicelluloses, the onset and midpoint degradation temperature of phthalated hemicelluloses decreased to 212 and 265°C, respectively, indicating the decreased thermal stability of phthalated hemicelluloses after homogeneous modification. This decrease was also possibly resulted from the degradation of hemicelluloses during homogeneous modification [44].

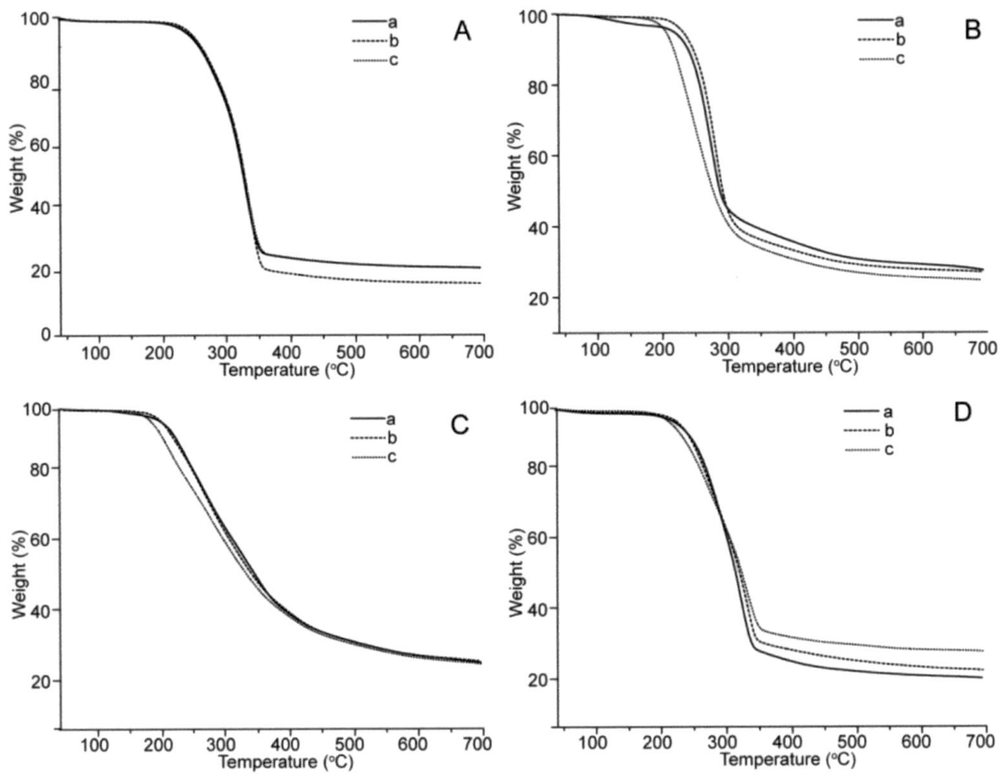

Figure 12. TG curves of cellulose (A), hemicelluloses (B), lignin (C), and bagasse (D), respectively. The curves a, b, and c represent unmodified, regenerated, and phthalated samples, respectively.

Sample	Onset T (°C)	Midpoint T (°C)
C_o	279	315
C_r	292	320
C_s	282	319
H_o	240	281
H_r	243	282
H_s	212	265
L_o	200	358
L_r	196	345
L_s	174	345
B_o	256	305
B_r	248	302
B_s	244	304

Table 4. The onset and midpoint degradation temperature of samples.

http://dx.doi.org/10.5772/65508

The onset and midpoint degradation temperature of regenerated lignin were both lower than those of unmodified lignin, as listed in **Table 4**, indicating the decreased thermal stability of lignin after dissolution and regenerating. Compared with regenerated lignin, the thermal stability of phthalated lignin after homogeneous modification was also decreased. This decrease was probably resulted from the degradation of lignin during dissolution, regeneration, and modification in AmimCl.

For bagasse, the thermal stability of bagasse samples decreased after dissolution and regeneration in ionic liquid AmimCl. This was probably resulted from the degradation of lignin during dissolution and regeneration. Compared with regenerated bagasse, the onset degradation temperature of phthalated bagasse decreased, while the midpoint degradation temperature of phthalated bagasse increased. These results suggested the thermal stability of lignocellulose was affected by the loss of fractions with low molecular weights and the degradation of phthalated samples. The former led to the increase in the thermal stability, while the later resulted in the decreased thermal stability of phthalated lignocellulose.

4. Conclusions

The phthalation of bagasse and its fractions occurred during homogeneous modification in AmimCl, and the reactivity of the three main components followed the order: lignin > cellulose > hemicelluloses. The FT-IR and NMR analyses suggested the skeleton of bagasse remained stable during dissolution, regeneration, and modification; the reactivity of hydroxyls in AGU followed the order: C-6 > C-2 > C-3; and the reactivity order in AXU was as follows: C-2 > C-3. For lignin, both aromatic and phenolic hydroxyls were phthalated; the reactivity of aromatic hydroxyls was higher than phenolic hydroxyls; and the reactivity order of lignin phenolic hydroxyls was as follows: S-OH > G-OH > H-OH. The thermal stability of modified lignocellulose was affected by the disruption of cellulose crystallinity and the degradation reaction of the modified components.

Acknowledgments

This work was financially supported by the National Natural Science Foundation of China (31170550, 31170555), the Fundamental Research Funds for the Central Universities (2014ZG0046), and the National Program for Support of Top-notch Young Professionals.

Nomenclature

AmimCl	1-allyl-3-methylimidazolium chloride
DMAP	4-dimethylaminopyridine
ILs	ionic liquids

DS	degree of substitution
FT-IR	Fourier transform infrared resonance
CP/MAS	solid-state cross polarization/magnetic angle spinning
NMR	nuclear magnetic resonance
2D HSQC	two dimensional heteronuclear single quantum correlation
AGU	anhydroglucose units
AXU	anhydroxylose units
S	syringyl
G	guaiacyl
H	*p*-hydroxyphenyl
DMSO-d_6	perdeutero-dimethylsulfoxide
BmimCl	1-butyl-3-methylimicazium chloride
TG	thermogravimetric analysis
WPG	weight percentage gain

Author details

Hui-Hui Wang, Xue-Qin Zhang, Yi Wei and Chuan-Fu Liu*

*Address all correspondence to: chfliu@scut.edu.cn

State Key Laboratory of Pulp and Paper Engineering, South China University of Technology, Guangzhou, PR China

References

[1] Nagarajan V, Mohanty AK, Misra M. Sustainable green composites: Value addition to agricultural residues and perennial grasses. ACS Sustainable Chemistry & Engineering. 2013;**1**:325–333. DOI: 10.1021/sc300084z

[2] Ragauskas AJ, Williams CK, Davison BH, Britovsek G, Cairney J, Eckert CA, Frederick WJ, Hallett JP, Leak DJ, Liotta CL. The path forward for biofuels and biomaterials. Science. 2006;**311**:484–489. DOI: 10.1126/science.1114736

[3] Liam B, Philip O. Biofuels from microalgae—A review of technologies for production, processing, and extractions of biofuels and co-products. Renewable and Sustainable Energy Reviews. 2010;**14**:557–577. DOI: 10.1016/j.rser.2009.10.009

[4] Minjares-Fuentes R, Femenia A, Garau MC, Candelas-Cadillo MG, Simal S, Rosselló C. Ultrasound-assisted extraction of hemicelluloses from grape pomace using response surface methodology. Carbohydrate Polymers. 2016;**138:**180–191. DOI: 10.1016/j.carbpol.2015.11.045

[5] Massimo B, Giovanna F, Mariastella S, Antonino L. Surface chemical modification of natural cellulose fibers. Journal of Applied Polymer Science. 2002;**83:**38–45. DOI: 10.1002/app.2229

[6] Freire CSR, Silvestre AJD, Neto CP, Belgacem MN, Gandini A. Controlled heterogeneous modification of cellulose fibers with fatty acids: Effect of reaction conditions on the extent of esterification and fiber properties. Journal of Applied Polymer Science. 2006;**100:**1093–1102. DOI: 10.1002/app.23454

[7] Viera RGP, Rodrigues Filho G, de Assunção RMN, Meireles CdS, Vieira JG, de Oliveira GS. Synthesis and characterization of methylcellulose from sugar cane bagasse cellulose. Carbohydrate Polymers. 2007;**67:**182–189. DOI: 10.1016/j.carbpol.2006.05.007

[8] Fasching M, Schröder P, Wollboldt RP, Weber HK, Sixta H. A new and facile method for isolation of lignin from wood based on complete wood dissolution. Holzforschung. 2008;**62:**15–23. DOI: 10.1515/HF.2008.003

[9] Sheldon RA. Green solvents for sustainable organic synthesis: State of the art. Green Chemistry. 2005;**7:**267–278. DOI: 10.1039/b418069k

[10] Olivier-Bourbigou H, Magna L, Morvan D. Ionic liquids and catalysis: Recent progress from knowledge to applications. Applied Catalysis A: General. 2010;**373:**1–56. DOI: 10.1016/j.apcata.2009.10.008

[11] Zakrzewska ME, Bogel-Łukasik E, Bogel-Łukasik R. Ionic liquid-mediated formation of 5-hydroxymethylfurfural? A promising biomass-derived building block. Chemical Reviews. 2010;**111:**397–417. DOI: 10.1021/cr100171a

[12] Swatloski RP, Spear SK, Holbrey JD, Rogers RD. Dissolution of cellose with ionic liquids. Journal of the American Chemical Society. 2002;**124:**4974–4975. DOI: 10.1021/ja025790m

[13] Kilpeläinen I, Xie HB, King AS, Granstrom M, Heikkinen S, Argyropoulos DS. Dissolution of wood in ionic liquids. Journal of Agricultural and Food Chemistry. 2007;**55:**9142–9148. DOI: 10.1021/jf071692e

[14] Fort DA, Remsing RC, Swatloski RP, Moyna P, Moyna G, Rogers RD. Can ionic liquids dissolve wood? Processing and analysis of lignocellulosic materials with 1-n-butyl-3-methylimidazolium chloride. Green Chemistry. 2007;**9:**63–69. DOI: 10.1039/b607614a

[15] Chen MJ, Chen CY, Liu CF, Sun RC. Homogeneous modification of sugarcane bagasse with maleic anhydride in 1-butyl-3-methylimidazolium chloride without any catalysts. Industrial Crops and Products. 2013;**46:**380–385. DOI: 10.1016/j.indcrop.2013.02.023

[16] Ibrahim MM, Dufresne A, El-Zawawy WK, Agblevor FA. Banana fibers and microfibrils as lignocellulosic reinforcements in polymer composites. Carbohydrate Polymers. 2010;**81:**811–819. DOI: 10.1016/j.carbpol.2010.03.057

[17] Xie HB, King A, Kilpelainen I, Granstrom M, Argyropoulos DS. Thorough chemical modification of wood-based lignocellulosic materials in ionic liquids. Biomacromolecules. 2007;**8:**3740–3748. DOI: 10.1021/bm700679s

[18] Nada AA, Hassan ML. Ion exchange properties of carboxylated bagasse. Journal of Applied Polymer Science. 2006;**102:**1399–1404. DOI: 10.1002/app.24255

[19] Anirudhan TS, Radhakrishnan PG. Chromium (III) removal from water and wastewater using a carboxylate-functionalized cation exchanger prepared from a lignocellulosic residue. Journal of Colloid and Interface Science. 2007;**316:**268–276. DOI: 10.1016/j.jcis.2007.08.051

[20] Liu CF, Sun RC, Zhang AP, Ren JL, Wang XA, Qin MH, Chao ZN, Luo W. Homogeneous modification of sugarcane bagasse cellulose with succinic anhydride using a ionic liquid as reaction medium. Carbohydrate Research. 2007;**342:**919–926. DOI: 10.1016/j.carres.2007.02.006

[21] Liu CF, Sun RC, Zhang AP, Qin MH, Ren JL, Wang XA. Preparation and characterization of phthalated cellulose derivatives in room-temperature ionic liquid without catalysts. Journal of Agricultural and Food Chemistry. 2007;**55:**2399–2406. DOI: 10.1021/jf062876g

[22] Li WY, Lan W, Chen D, Liu CF, Sun RC. DMAP-catalyzed phthalylation of cellulose with phthalic anhydride in [bmim] Cl. BioResources. 2011;**6:**2375–2385.

[23] Peng F, Ren JL, Xu F, Bian J, Peng P, Sun RC. Fractional study of alkali-soluble hemicelluloses obtained by graded ethanol precipitation from sugar cane bagasse. Journal of Agricultural and Food Chemistry. 2009;**58:**1768–1776. DOI: 10.1021/jf9033255

[24] Moghaddam L, Zhang Z, Wellard RM, Bartley JP, O'Hara IM, Doherty WOS. Characterisation of lignins isolated from sugarcane bagasse pretreated with acidified ethylene glycol and ionic liquids. Biomass and Bioenergy. 2014;**70:**498–512. DOI: 10.1016/j.biombioe.2014.07.030

[25] Zhang AP, Liu CF, Sun RC, Xie J. Extraction, purification, and characterization of lignin fractions from sugarcane bagasse. BioResources. 2013;**8:**1604–1614.

[26] Chen MJ, Shi QS. Transforming sugarcane bagasse into bioplastics via homogeneous modification with phthalic anhydride in ionic liquid. ACS Sustainable Chemistry & Engineering. 2015;**3:**2510–2515. DOI: 10.1021/acssuschemeng.5b00685

[27] Lan W, Liu CF, Sun RC. Fractionation of bagasse into cellulose, hemicelluloses, and lignin with ionic liquid treatment followed by alkaline extraction. Journal of Agricultural and Food Chemistry. 2011;**59:**8691–8701. DOI: 10.1021/jf201508g

[28] Argyropoulos DS. Quantitative phosphorus-31 NMR analysis of lignins, a new tool for the lignin chemist. Journal of Wood Chemistry and Technology. 1994;**14:**45–63.

[29] Crestini C, Argyropoulos DS. Structural analysis of wheat straw lignin by quantitative 31P and 2D NMR spectroscopy. The occurrence of ester bonds and α-O-4 substructures. Journal of Agricultural and Food Chemistry. 1997;**45:**1212–1219. DOI: 10.1021/jf960568k

[30] Pu YQ, Cao SL, Ragauskas AJ. Application of quantitative 31P NMR in biomass lignin and biofuel precursors characterization. Energy & Environmental Science. 2011;**4:**3154–3166. DOI: 10.1039/c1ee01201k

[31] Jasiukaityte E, Kunaver M, Crestini C. Lignin behaviour during wood liquefaction—Characterization by quantitative ^{31}P, ^{13}C NMR and size-exclusion chromatography. Catalysis Today. 2010;**156:**23–30. DOI: 10.1016/j.cattod.2010.02.001

[32] Vaidya AA, Gaugler M, Smith DA. Green route to modification of wood waste, cellulose and hemicellulose using reactive extrusion. Carbohydrate Polymers. 2016;**136:**1238–1250. DOI: 10.1016/j.carbpol.2015.10.033

[33] Peng XW, Ren JL, Sun RC. Homogeneous esterification of xylan-rich hemicelluloses with maleic anhydride in ionic liquid. Biomacromolecules. 2010;**11:**3519–3524. DOI: 10.1021/bm1010118

[34] Sun YC, Xu JK, Xu F, Sun RC. Efficient separation and physico-chemical characterization of lignin from eucalyptus using ionic liquid–organic solvent and alkaline ethanol solvent. Industrial Crops and Products. 2013;**47:**277–285. DOI: 10.1016/j.indcrop.2013.03.025

[35] Wang Y, Song H, Hou JP, Jia CM, Yao S. Systematic isolation and utilization of lignocellulosic components from sugarcane bagasse. Separation Science and Technology. 2013;**48:**2217–2224. DOI: 10.1080/01496395.2013.791855

[36] Chen MJ, Zhang XQ, Liu CF, Sun RC, Lu FC. Approach to renewable lignocellulosic biomass film directly from bagasse. ACS Sustainable Chemistry & Engineering. 2014;**2:**1164–1168. DOI: 10.1021/sc400555v

[37] Liu CF, Sun RC, Qin MH, Zhang AP, Ren JL, Ye J, Luo W, Cao ZN. Succinoylation of sugarcane bagasse under ultrasound irradiation. Bioresource Technology. 2008;**99:**1465–1473. DOI: 10.1016/j.biortech.2007.01.062

[38] Lin Y, King JY, Karlen SD, Ralph J. Using 2D NMR spectroscopy to assess effects of UV radiation on cell wall chemistry during litter decomposition. Biogeochemistry. 2015;**125:**427–436. DOI: 10.1007/s10533-015-0132-1

[39] Guan Y, Zhang B, Qi XM, Peng F, Yao CL, Sun RC. Fractionation of bamboo hemicelluloses by graded saturated ammonium sulphate. Carbohydrate Polymers. 2015;**129:**201–207. DOI: 10.1016/j.carbpol.2015.04.042

[40] Sette M, Wechselberger R, Crestini C. Elucidation of lignin structure by quantitative 2D NMR. Chemistry–A European Journal. 2011;**17:**9529–9535. DOI: 10.1002/chem.201003045

[41] Wen JL, Yuan TQ, Sun SL, Xu F, Sun RC. Understanding the chemical transformations of lignin during ionic liquid pretreatment. Green Chemistry. 2014;**16:**181–190. DOI: 10.1039/c3gc41752b

[42] Kim JY, Shin EJ, Eom IY, Won K, Kim YH, Choi D, Choi IG, Choi JW. Structural features of lignin macromolecules extracted with ionic liquid from poplar wood. Bioresource Technology. 2011;**102:**9020–9025. DOI: 10.1016/j.biortech.2011.07.081

[43] Varanasi P, Singh P, Arora R, Adams PD, Auer M, Simmons BA, Singh S. Understanding changes in lignin of *Panicum virgatum* and *Eucalyptus globulus* as a function of ionic liquid pretreatment. Bioresource Technology. 2012;**126:**156–161. DOI: 10.1016/j.biortech.2012.08.070

[44] Wang HT, Yuan TQ, Meng LJ, She D, Geng ZC, Sun RC. Structural and thermal characterization of lauroylated hemicelluloses synthesized in an ionic liquid. Polymer Degradation and Stability. 2012;**97:**2323–2330. DOI: 10.1016/j.polymdegradstab.2012.07.033

Chapter 21

Review of Biomass Thermal Gasification

Mohammed Abed Fattah Hamad,
Aly Moustafa Radwan and Ashraf Amin

Additional information is available at the end of the chapter

http://dx.doi.org/10.5772/66362

Abstract

Gasification of biomass is one of the most attractive methods for producing hydrogen rich gas. Syngas production from biomass is an attractive solution for energy crisis. The production of energy from biomass reduces the dependence of developing countries on fossil fuels, as ample biomass is available in the developing countries and is renewable. Downdraft gasifiers are fixed bed gasifiers where the gasifying agent and biomass are flowing downwards, developed for high-volatile fuels such as wood or biomass gasification. Cocurrent flow regime throughout the oxidation and reduction zones reduces the tars and particulates in syngas, which will reduce the necessity of complicated cleaning methods compared to updraft gasifiers especially if the gas is used as a burnable gas in a small community. It is important to ensure homogenous distribution of gasifying agent at the downdraft gasifier throat. This chapter presents latest trends in gasification of biomass using downdraft gasification.

Keywords: gasification, hydrogen, agriculture waste, catalysts, downdraft gasifier

1. Introduction

Gasification of biomass is one of the most attractive methods for producing hydrogen rich gas. Syngas production from biomass is an attractive solution for energy crisis. The production of energy from biomass reduces the dependence of developing countries on fossil fuels; as ample biomass is available in the developing countries and is renewable. Downdraft gasifiers are fixed bed gasifiers where the gasifying agent and biomass are flowing downwards, developed for high volatile fuels like wood or biomass gasification. Cocurrent flow regime throughout the oxidation and reduction zones reduces the tars and particulates in syngas, which will reduce the necessity of complicated cleaning methods compared to updraft gasifiers especially if the gas is used as a burnable gas in a small community. It is important to ensure homogenous distribution of gasifying agent at the downdraft gasifier throat [1–3].

Gasification is a process under development to utilize the energy conserved in biomass. Gasification can be used as a source of energy in rural and off-grid areas to fill the power gaps. The limited supply and the increasing demands of fossil fuels have led the world to investigate alternative energy sources. Renewable energy sources have been studied widely, and biomass appears as the most promising renewable energy source. Biomass can be used to overcome the depletion of fossil fuels and to reduce the environmental impact of the conventional fuels such as greenhouse gas emissions using one of these four technologies: direct combustion, thermochemical processes, biochemical processes, and agrochemical processes. Biomass is the third energy source after coal and oil. Biomass covers 35% of the energy demand of the developing countries corresponding to 13% of the world energy demand. Biomass is widely available in quantities enough to meet the world energy demand [1–7].

The oldest way to generate energy is to burn biomass. Due to environmental and technical difficulties associated with burning biomass, innovative processes should be developed to utilize biomass [5, 8–10]. Developing more effective techniques to utilize biomass will reduce the disposal problem and create profits. Hydrolysis, pyrolysis, gasification, and hydrogenation are the principal processes for biomass conversion in the literature [7, 11]. Gasification represents efficient and environmentally friendly method for producing the syngas as a biofuel from different sources of biomass [12–14], and to produce second-generation biofuels such as methanol, ethanol, and hydrogen [8, 10, 12, 15]. Gasification can be defined as the partial (incomplete) combustion of biomass, and gasification could extract up to 60–90% of the energy stored in biomass [16, 17]. To develop second-generation biofuels, economical and Feasible clean technologies of syngas are required. [15]. However, economical gasification of biomass may produce burnable gases, which can be used to provide heat requirements instead of LPG [12]. Gasifiers were developed to replace biomass burners. Gasifiers will prevent the necessity of on-site power generation [18, 19]. Gasification is the conversion of biomass into a combustible gas and charcoal by partial oxidation of biomass at temperature range of 800–900°C [6, 19, 20].

The charcoal is finally reduced to H_2, CO, CO_2, O_2, N_2, and CH_4 [6, 8–10, 21]. Char gasification starts at temperatures above 350°C [7]. The products of gasification consist of the following components: ash, volatile alkali metals, tars, and syngas. Tars represent a challenge for the commercialization of gasification product as an alternative fuel [22]. Frequently using tar may result in complete shutdown and repair of the industrial unit [18, 22]. Tars set and amount vary considerably based on reaction conditions and gasifier type [18]. Gas produced from gasifier can be cleaned by removing tars either physically or chemically [18]. Physical removal can be achieved using bag filters or wet scrubbers. Chemical removal methods depend on converting tars to lighter hydrocarbons either using thermal conversion or catalytic conversion processes [18, 22].

Gasification of such material may help in reducing the gap between electricity requirements and available energy sources. Decentralized power regeneration units will help to fill power gap in rural and off-grid locations [4]. Yet, it is still difficult to develop a decentralized power generation unit based on biomass energy which can be used to fill the gap in energy needs in rural areas and farms [4]. Technical difficulties prevent further commercialization of gasification units in accordance to lower conversion efficiency [23, 24]. Leung et al. [25] proposed a governmental support to accomplish faster steps toward gasification units commercialization. However, all over the world, biomass energy has been widely incorporated in the power generation system; U.S. started partial and full conversion of conventional power plants to biomass [24]. Throughout this chapter, we will discuss the latest trends in agricultural waste gasification. Our goal is to provide a full description of the process starting from basic understanding and ending by design of a gasification unit.

2. Chemistry of gasification

The reactions taking place in the gasifier can be summarized as indicated below [3, 4, 21]:

Partial oxidation:

$$C+\tfrac{1}{2}O_2 \leftrightarrow CO \quad \Delta H = -268 \text{ kJ/mol} \tag{1}$$

Complete oxidation:

$$C+O_2 \leftrightarrow CO_2 \quad \Delta H = -406 \text{ kJ/mol} \tag{2}$$

Water gas phase reaction:

$$C+H_2O \leftrightarrow CO+H_2 \quad \Delta H = 118 \text{ kJ/mol} \tag{3}$$

Boudouard reaction:

$$C+CO_2 \leftrightarrow 2CO \quad \Delta H = 170.7 \text{ kJ/mol} \tag{4}$$

The heat required for water gas phase and Boudouard reactions is provided by complete and partial oxidation reactions, and complete oxidation provides around 60% of the heat requirements during gasification [3, 17]. In addition to the previous reactions that are common in combustion and gasification, hydrogen, steam, and carbon monoxide undergo further reactions as shown below [3, 24]:

Water gas shift reaction:

$$CO+H_2O \leftrightarrow CO_2+H_2 \quad \Delta H = -42 \text{ kJ/mol} \tag{5}$$

Methane formation:

$$CO+3H_2 \leftrightarrow CH_4+H_2O \quad \Delta H = -88 \text{ kJ/mol} \tag{6}$$

The water gas shift and methane formation reactions are in equilibrium and the governing parameters are: pressure, temperature, and concentration of reaction species.

3. Gasifier design

The unit design is a very important factor in determining the syngas quality and heating value [15]. The gasifier will hold two processes: conversion of biomass to charcoal and then conversion of charcoal to hydrogen and carbon monoxide. The mixture of hydrogen and carbon monoxide can be used for direct heating in rural areas [16]. Leung et al. [25] identified four types of gasifiers: updraft, open core, downdraft, and circulating fluidized bed (CFB) gasifiers. The maximum efficiency of the updraft, downdraft, and CFB gasifiers may reach to 75%, the maximum energy output is 10E6, 4E6, and 40E6 kJ/h, respectively. According to Chopra and Jain [13], the fixed bed gasifiers can be further divided into: updraft, Imbert downdraft, throatless downdraft, cross draft, and two-stage gasifiers. The fixed-bed gasifier is suitable for producing low heating value gas for small and medium applications [13, 26]. The downdraft gasifier is distinguished by a simple design, high carbon conversion, long residence time, low cost, low pressure, relatively clean gas, and low gas velocity. The downdraft gasifier is suitable for producing low heating value burnable gas or for generating electricity of small-scale systems in the range of 10 kW up to 1 MW [12, 26–28].

3.1. Design of downdraft gasifiers

Downdraft gasifiers are fixed bed gasifiers where the gasifying agent and biomass are flowing downwards, developed for high volatile fuels like wood or biomass gasification. Cocurrent flow regime throughout the oxidation and reduction zones reduces the tars and particulates in syngas, which will reduce the necessity of complicated cleaning methods compared to updraft gasifiers especially if the gas is used as a burnable gas in a small community [12, 17]. It is important to ensure homogenous distribution of gasifying agent at the downdraft gasifier throat.

Bhavanam and Sastry [24] provided design procedures for different types of downdraft gasifiers. The gasification reaction in a downdraft gasifier undergoes several steps, starting with drying step at 100°C, followed by pyrolysis step between 200 and 300°C resulting in release of around 70% of biomass weight as volatile matter and tars [16, 24]. After pyrolysis, the remaining biomass and volatile matter react with the incoming oxygen in the combustion step. Finally, various reactions take place in the reduction zone including carbon and steam reaction to produce CO and hydrogen, water-gas shift reaction, and CO and steam to form methane and carbon dioxide [24]. The four gasification reaction steps are illustrated in **Figure 1**. However, a limited experience has been gained in the field of biomass gasification while it represents an attractive renewable energy route [16]. **Table 1** illustrates the design specifications for two types of downdraft gasifiers: Imbert and stratified downdraft gasifiers. **Table 1** is developed based on extensive discussion in Bhavanam and Sastry [24].

Imbert downdraft gasifier is a cylindrical chamber of varying inner diameter across chamber length. The upper part of the cylindrical chamber is loaded with biomass according to

http://dx.doi.org/10.5772/66362

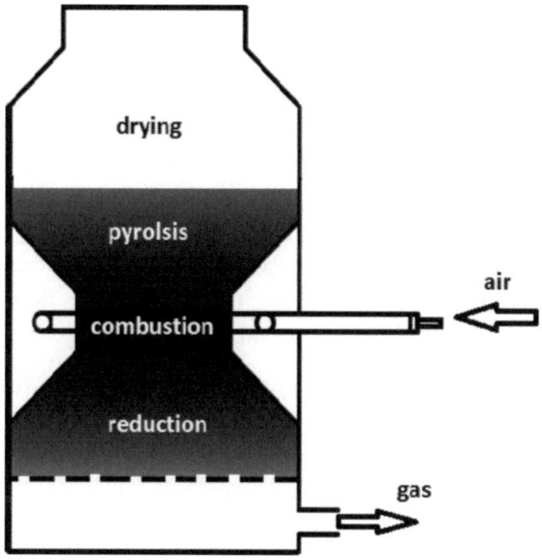

Figure 1. Different reaction zones in downdraft gasifier.

Design considerations		Imbert	Stratified
Biomass	Material	Uniform woody	Small size
	Moisture content	<20%	<20%
	Ash content	<5%	–
Reactor type		Packed bed supported on a throat	No-throat cylindrical packed bed with open top
Biomass feeding		Hopper	Open top
Gas feeding		Nozzle in the combustion zone	Enters from top mixed with biomass
Produced gas		Tar oils <1% T = 700°C	Less tar
Maximum capacity		500 kW	Easy to scale up

Table 1. Design consideration for Imbert and stratified downdraft gasifiers.

requirement. Air nozzles, attached to distribution manifold, permit air to be drawn into biomass to improve mixing of gasifying agent and biomass. A charcoal balance is established around the nozzles. Below the air nozzles, a classical Imbert hearth forms the reduction part. Insulating the reduction hearth reduces the amount of tars in the produced syngas and increases gasification efficiency. The hot gases are forced to go through the hot zone due to hearth constriction. The char bed on the grate removes the dust, which should be cleaned eventually to prevent clogging, and dropping in airflow or channeling [17].

Stratified or open-top downdraft gasifier is a uniform diameter gasifier, usually made of a cylindrical vessel with a hearth near the bottom. The stratified gasifier is an improved, easy to design gasifier compared to Imbert downdraft gasifier. The open-top helps in maintaining uniform access of gasifying agent to the pyrolysis zone, which prevents localized heating. Biomass is added through the open-top to the top layer of gasifier. The length of the gasifier can be divided into four reaction zones: unreacted biomass zone at which air enters, the flaming pyrolysis zone at which air reacts with biomass, adiabatic char gasification zone at which gases from flaming pyrolysis zone reacts with charcoal, and finally the unreacted charcoal zone that is located just above the grate which acts as a buffer for ash and charcoal. The stratified downdraft gasifier can be mathematically modeled easily as a plug flow reactor at which air and biomass are uniformly mixed. With such simple design, it is expected that stratified downdraft gasifier will replace the Imbert downdraft gasifier in commercial applications [17].

Wander et al. [29] illustrated the design of 12 kg/h downdraft stratified gasifier for sawmill dust gasification. The reactor is a cylindrical body of 270 mm internal diameter and 1100 mm of height made of SAE 1020 steel. Internal rods are used to mix the sawdust in the reduction zone. Ash box is used to reduce the ash content of the produced syngas. The reactor is insulated using 50 mm of rock wall. Air was introduced from the open top as a gasifying agent and a secondary air was used to provide air required for internal burner. A gas chromatography was used to analyze the gas samples and three water condensers in an ice bath were used to measure tars and humidity content.

Zainal et al. [30] developed a downdraft gasifier for the gasification using around 50 kg/h of wood chips. The temperature in the combustion zone may reach 1000°C, which reduces the tar content of produced syngas. The gasifier is made of mild steel pipe with a diameter of 0.6 m and a height of 2.5 m. A cone structure is used inside the gasifier above the combustion zone with an inclination of 60° to facilitate the wood material movement. The air supply was accomplished using a 40 mm diameter stainless-steel pipe with eight 10 mm diameter nozzles. The air is preheated by positioning the supply tube inside the gasifier. The heating value of the produced syngas was in the range of 4.65–5.62 MJ/Nm^3 depending on operating conditions.

Panwar et al. [31] developed an open-top downdraft gasifier for wood gasification to provide the heat requirement for the food processing industry. The downdraft gasifier was lined with ceramic and designed for a wood input of 60 kg/h equivalent to 180 kW. The gasifier body is made of mild steel. The air distribution system consists of six air tuyeres of 20 mm in diameter. A cyclone was used to remove solid particulates from produced syngas. The complete combustion of the syngas is achieved in a premixed burner to provide heat needed for the food processing industry unit. Note that 30 kg of charcoal and 10 kg of wood were used to provide hear required for gasifier start up.

Sheth and Babu [1] showed a design of an Imbert downdraft gasifier for wood waste gasification with a total height of 1.1 m. The diameter of pyrolysis zone and reduction zone are 0.31 and 0.15 m, respectively. The gasifier has throated combustion zone, a bed of char supported by a grate follows the combustion zone. The air is supplied through two nozzles in the oxidation zone. The high temperature in the combustion zone ensures cracking of tars into volatiles and water. The diameter of pyrolysis, reduction, and oxidation zones is 310, 150, and 53 mm, respectively. The grate is movable to unclog it for removing ash.

Vervaeke et al. [32] illustrated the design of a small-scale pilot plant downdraft gasifier equivalent to 100 kW of electricity generation. The downdraft unit used in this study is a pilot scale of the Xylowatt gasifier. The downdraft gasifier is a batch gasifier with a capacity of 90–105 kg. The gasification system consists of downdraft gasifier and inside it are ash collection container, cyclone, filter, and a scrubber.

Lv et al. [33] developed a downdraft gasifier to produce hydrogen from biomass using air and oxygen/steam as gasifying agents. Total 5 kg of char were supported on the grate to reduce tar content and to act as a catalyst to upgrade syngas. Biomass is pine wood blocks used in cubes cut into 3 cm × 3 cm × 3 cm. The gasifier height is 1.3 m and the diameter is 35 cm. The gasifying agent is preheated in a chamber inside the gasifier. Gas is cleaned using triple-stage spray shower filled with steel wire rings. The internal diameter of the gasifier is calculated according to the power output. The height is calculated based on batch operation time. The internal diameter is calculated in meters using the following equation [26]:

$$D = \left(\frac{1.27 * \mathrm{FCR}}{\mathrm{SGR}}\right)^{0.5} \tag{7}$$

where FCR is the fuel consumption rate (kg/h) and SGR is the specific gasification rate (kg/h/m^2).

The height can be determined in meter using the following equation [26]:

$$H = \frac{\mathrm{SGR} \times t}{\rho} \tag{8}$$

where t is the operation time (h) and ρ is the feedstock bulk density (kg/m^3).

The power output P_0 can be calculated in kW from the following equation [26]:

$$P_0 = \frac{\mathrm{FCR} \times \mathrm{HHV} \times \eta}{3.6} \tag{9}$$

where HHV is the higher heating value of the feedstock in MJ/kg and η is the efficiency of the gasifier usually around 0.7. The amount of air needed during operation can be calculated in Nm^3/h from the following equation [26]:

$$\mathrm{AFR} = \frac{\varepsilon \times \mathrm{FCR} \times \mathrm{SA}}{\rho_a} \tag{10}$$

where ε is the equivalence ratio, FCR is the fuel consumption rate, SA is the stoichiometric amount of air required for chemical reaction, and ρ_a is the density of air (1.18 kg/m^3). Finally, the size of the air nozzle, which is required for uniform air distribution, can be calculated in mm^2 from the following equation [26]:

$$A = \frac{\mathrm{AFR} \times 10^3}{v \times 3.6} \tag{11}$$

where v is the inlet velocity of air (m/s).

4. Factors affecting on the gasification process

Zhou et al. [34] discussed the ongoing gasification projects taking place in China. The biomass gasification units were divided based on scale: small-, medium-, and large-scale biomass gasification and power generation units. Pretreatment of biomass includes size reduction, size screening, separation of magnetic materials, and storing as wet biomass. Then prior to gasification, drying and storing as dry material are accomplished to reduce the moisture content to 10–15% [35]. Feedstock type and feedstock preparation are important factors affecting the yield and quality of produced syngas. Shredding and drying are two processes conducted to prepare the biomass raw material for gasification process [14]. The main parameters affecting the gasification are clarified below:

Equivalence ratio (ER): The equivalence ratio is the air/biomass ratio divided by the theoretical air/biomass ratio. Increasing ER will decrease the heating value of the produced gas due to decreasing H_2 and CO concentration and increasing CO_2 concentration. Higher ER helps in reducing tars and provides more O_2 to react with volatiles. Typical values of ER ranges between 0.2 and 0.4 [24]. Guo et al. [36] reported that increasing ER decreases the concentration of combustible gases (H_2, CO, CH_4, and C_nH_m). The heating value was higher than 4 MJ/Nm^3 when ER is kept lower than 0.4. Increasing ER improves the reaction temperature and carbon conversion, and reduces the tar yield. For a downdraft stratified gasifier, Wander et al. [29] suggested an equivalence ratio of 0.3:0.35 kg-O_2/kg-wood. A higher ratio is required when higher heat loss is expected, an equivalence ratio of 2:2.4 kg-air/kg-wood is optimum for producing a syngas with low heating value of 4–6 MJ/Nm^3. For woody material in a downdraft gasifier, Zainal et al. [30] suggested an equivalence ratio of 0.268–0.43 with 0.38 showed optimum value (corresponding to a heating value of 5.62 MJ/Nm^3). While Sheth and Babu [1] defined that the optimum equivalence ratio for wood gasification in Imbert downdraft gasifier is 0.205.

Effect of biomass characteristics: Biomass characteristic is a major factor affecting produced syngas quality. The physical properties that may have major effect are: absolute and bulk density, and particulate size. The chemical composition parameters that are of major importance to define the syngas [17] quality including volatile matter, moisture content, fixed carbon, ash content, and gross calorific value and the ultimate analysis comprises the carbon, oxygen, nitrogen, and sulfur of the dry biomass on a weight% [19].

Moisture content: The moisture content can be determined by complete drying of biomass sample. The moisture content is calculated by subtracting the sample weight after drying from fresh sample weight. Maximum allowable moisture content in downdraft gasifier is 40% on dry weight basis. Updraft gasifier can handle biomass with higher moisture content. The higher moisture content in biomass will increase the consumed energy for drying, and will reduce the pyrolysis of biomass. As a general rule, increasing moisture content decreases the conversion [1, 24].

Superficial velocity: The superficial velocity is the ratio of the syngas production rate at normal conditions and the narrowest cross-sectional area of gasifier. Lower superficial velocity is linked with high yield of char, and large quantities of unburned tars, which may deactivate catalyst, plug lines, and destroy compressors. On the other hand, higher superficial velocity results in reduced amount of char and low overall process efficiency [24].

Operating temperature: Operating temperature affects conversion, tar content, gas composition, gas heating value, and char conversion. To select the optimum temperature, gasifier type, and biomass source should be considered. Usually, temperature higher than 800°C should be used to obtain high conversion and low tar content in the produced syngas [24]. Low temperature is associated with low tar content, low H_2 and CO content in the produced syngas [12]. Increasing temperature will increase gas yield, hydrogen, heating value, and ash agglomeration. To overcome the ash agglomeration problem, practical temperature does not exceed 750°C [24].

Gasifying agent: Gasifying agents in use are air, steam, steam/oxygen mixture, and CO_2. Gasifying agent affects the heating value of the produced syngas. The heating value increases with increasing steam content of the gasifying agent, while heating value decreases as air increases in the gasifying agent [24]. The steam/oxygen mixture represents a zero nitrogen-gasifying agent which increase heating value and allow liquefying the produced gas after proper treatment [37]. Using almond shells, the lower heating value was 5.9–6.7, 6.3–8.4, and 10.9–11.7 MJ/Nm^3 using the gasifying agent: 35 wt.% O_2 enriched air, 50 wt.% O_2 enriched air, and steam/oxygen mixture, respectively. Campoy et al. [38] reported a heat value of syngas produced from gasification to have an average value of 4–6 and 9–13 MJ/Nm^3 using air and oxygen/steam mixture, respectively. In addition to lower efficiency compared to air/steam mixture, enriched oxygen-air requires high capital cost for oxygen [38]. The addition of steam will shift toward the reforming reaction and heterogeneous gasification reactions.

Residence time: Residence time has a remarkable impact on the composition and produced tars. Increasing residence time decreases the fraction of oxygen-containing compounds, decreases yield of one and two atomic ring compounds, and increases three and four ring compounds [24].

Pressure: Atmospheric and higher pressures are commonly used in gasification process. Selecting the optimum pressure depends on the application of the produced syngas. If the syngas is used for producing methanol or synthetic auto-fuels, higher pressures are preferred to improve the process yield and to reduce tar content. For generating burnable gases, atmospheric pressure should be used [12]. High pressure applications are recommended for large-scale gasification, while atmospheric pressure is recommended for small-scale gasification [35]. High pressure gasification is still not well developed and further research is needed to further commercialize such process [39].

Catalyst: Catalyst type is a very important factor affecting gasification quality and produced syngas. Catalyst affects the composition of the syngas by manipulating the percentage volume of hydrogen, carbon dioxide, methane, and carbon monoxide. Optimum catalyst should play a role in minimizing the gas content of carbon dioxide and maximizing the gas content of hydrogen, carbon monoxide, and methane [40]. The catalyst type and loading on the gasification of cotton stalks and saw dust were studied. The catalysts selected are USY zeolite, dolomite, CaO, granulated slag, red brick clay, olivine, and cement kiln dust. The results demonstrate that the cement kiln dust and calcium hydroxide are more effective for increasing the gas yield and decreasing the char yield [8, 10].

Effect of biomass/steam ratio: Biomass/steam ratio affects hydrogen content in the produced syngas. Contradictory reports are found in literature, while Lv et al. [41] reported a positive effect on hydrogen content when biomass/steam ratio increases. Lv et al. [33] reported a negative effect of biomass/steam ratio increase on syngas hydrogen content. This variation can be understood by considering that biomass/steam ratio effect is altered according to the entire system configuration.

Lower values of biomass/steam ratio shift the reaction to produce more solid carbon and methane, since number of moles of steam increases in the feed. While at higher values of biomass/steam ratio, Co and H_2 are increased in the syngas as carbon and methane produced are decreased consequently.

5. Cost of biomass gasification process

The cost of any industrial process is governed by the capital cost and the running cost. Selection of best gasifier type depends on cost of fabrication, ease of manufacture, tar content, lower heating value, feedstock elasticity, and application of syngas [26]. The fixed bed gasifiers are more suitable for small- and medium-scale applications, while fluidized bed gasifiers are suitable for large-scale applications (equivalent to >15 MW) [25]. For example in China rice hulls, fluidized bed gasifiers are used in a production scale equivalent to 1–1.2 MW while downdraft gasifiers are used in a production scale equivalent to 60–200 kW [42]. The capital cost of the gasifier is divided into three items: gasifier and gas cleaning system cost, fuel gas utilization equipment cost, and fitting and system construction cost [25]. Cleaning systems and removing tars will add a significant cost to the produced syngas, which reduce the feasibility of using syngas in internal combustion engines [12]. Optimizing tar content can be achieved by varying the operating conditions and feedstock [43].

Upgrading using catalytic treatment represents the most economical and efficient method for syngas upgrading since it provides a way for removing tars and other particulates and converting tars to hydrocarbons [12]. Downdraft gasifier represents a reasonable cost production method for generating syngas with low tar content [29]. Especially small gasifiers that has proven economic feasibility [27].Wu et al. [44] recommended implementing biomass gasification depending on the low biomass price. By comparing different technologies to generate electricity based on 1 MW scale, Wu et al. [44] mentioned that the capital cost of fluidized bed gasifier system for biomass gasification-power generation system is 60–70% of the capital cost of coal power station and much lower compared to the capital cost of conventional power station. For producing combustible gases, Bridgwater et al. [45] reported that for syngas produced from fluidized bed, updraft, and downdraft gasifiers: hydrogen volume percentage is 9, 11, and 17%, respectively; CO volume percentage is 14, 24, and 21%, respectively; and a heating value of 5.4, 5.5, and 5.7 MJ/Nm^3, respectively. The downdraft represents the ideal solution to produce combustible (burnable) gases for household uses.

Biomass gasification economics are very sensitive to the scale of produced MW [44]. Leung et al. [25] mentioned two disadvantages of small- and medium-size gasifiers: capital cost limitation that may prevent incorporating important processes like tar removal, and the environmental demands imposed by new regulations which is difficult to be met by different biomass gasification technologies. Wu et al. [44] identified 160 kW as a critical scale of biomass gasification unit, less than 160 kW biomass gasification units loses the economical attraction. Note that 1–5 MW was recommended as the most competitive size for biomass gasification unit. Lower than a unit capacity of 160 kW, the price of kWh increases sharply from 0.4 to 1.8 Yuan RMB/kWh for very small capacities. For unit with capacities higher than 160 Wh, the price will decrease gradually as the unit size increases. At the 600 kW capacity, the price will be around 0.3 Yuan RMB/kWh; while the price may reach 0.25 for a unit capacity of 1000 kW [44].

It is recommended to conduct gasification at pilot plant scale to mimic large scale to figure out the approximate industrial process scale economics [38]. The steam enhances the reforming and heterogeneous gasification reactions, the temperature inside the gasifier should be kept enough to support such reactions [38]. Combining gasification unit with heat and power generation systems will improve the economics of the process [3]. Gasification units combined with heat and power generation systems are expected to have an overall efficiency of 85% compared to a maximum efficiency of 35–55% for conventional power station, in addition to a substantial saving in carbon emissions. Total 1000 kg/year of carbon are saved for each MW when gasification units hybrid with heat and power generation systems [3].

Downdraft gasifiers are economically competitive even to conventional LPG heating unit. Panwar et al. [31] found that replacing LPG heating system with a downdraft wood gasification system could save $13,850 US for 3000 h of operation. The payback period of the gasification system was only 1100 h. According to the extensive study of literature, the recommended gasification process consists of the following steps [13, 17, 19]:

1. Straw collection and preparation (milling and pelletization of straws).
2. Belt conveyor for feeding of the gasifier.
3. Downdraft gasifier.
4. Blower for suction of air and gas produced.
5. Gas cleaning and separation of tars.
6. Gas holder for storage of gas.
7. Gas distribution net.
8. Gas application devices.
9. Gas metering devices.

6. Preliminary techno-economic studies of downdraft gasifier

The aim of this chapter is to illustrate a detailed design of biomass gasification system to generate syngas for household applications. The stratified gasifier is selected based on the

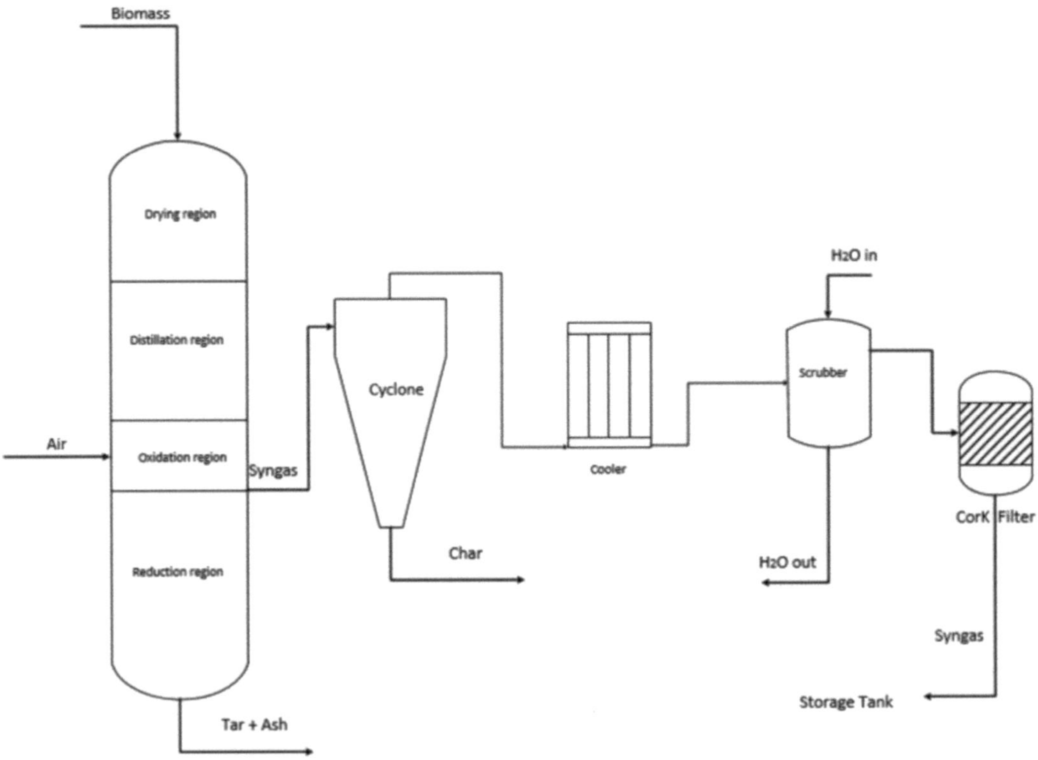

Figure 2. Gasification system for producing burnable gas.

following parameters: easiness in design and scaling up, and production of syngas with tar content lower than that of Imbert gasifier [16–18, 26]. The gasification system comprises of a downdraft burnable syngas gasifier followed by a gas cleaning and distributing system. Throughout this chapter, the design specification of the downdraft gasifier is presented. This system can be used to convert solid agricultural waste to a syngas that is a burnable gas used to provide energy requirements for small communities, as shown in **Figure 2**.

The energy (household) requirements: for 50 families.

The gas demand per day is: 500 m^3/day.

The syngas gas will be produced on two batches: Morning and afternoon (each one will last for 250 m^3/batch).

The first batch will take place from 7 to 10 am.

The second batch will take place from 2 to 5 pm.

The storage unit will hold around 200 m^3 gas and accordingly will provide heating requirements during the period of nonoperation.

http://dx.doi.org/10.5772/66362

The system consists of the following units: biomass shredding, grinding unit, gasification unit, air controlling system, air heating and gas precooling unit, cyclone (acts like cyclone to remove dust), gas cooler, water filter (scrubbing unit), gas distribution system, control system, cork filter, and storage tank.

No.	Object	Cost, Egyptian pounds
1	Gasifier with control system	30,000
2	Gas cleaner	10,000
3	Belt conveyor	15,000
4	Milling and pelletization of biomass	50,000
5	Gas holder	15,000
6	Gas distribution system 100 × 500	60,000
7	Gas stoves 100 ×350	35,000
8	Gas meters 100 ×400	40,000
9	Erection	10,000
10	Contingency (10%)	26,000
	Total	2,90, 000

$ = 8.5 Egyptian pounds.
Assume that:
Cost of material = 300 L.E/ton
Cost of preparation = 100 L.E/ton
Cost of raw materials = 0.25 ×400 × 360 = 36 000 L.E/year
Labors required = 2 ×1500 = 3000 L.E/month = 3000 ×12 = 36000 L.E/year
Income gas production = 500 m^3/day
≈60 kg L.P.G/day
= 60 × 360 = 21,600 Kg/year
= 21,600 × 7 = 151,200 L.E/year
Profit = income − raw material − depreciation
Profit = 151,200 − 36,000 − 36,000 − 29,000 = 49,400
Return on Investment = profit/initial cost
Return on Investment = (49,400/290,000)×100 = 17%
Therefore, the payback period is about 6 years based on the international prices of L.P.G.

Cost of equipment for gasification system.

The cost of land required for erection of the plant is not included in this calculation of the feasibility study.

7. Conclusions

Gasification represents a viable solution to overcome the energy shortage by developing commercial gasification units in rural and off-grid areas. An integrated system comprising a gasification-electrical generation method represents an ideal solution from the technical and economical points of view.

However, due to the wide varieties of available biomass feedstock, it is recommended to manipulate different systems in each location depending on the feedstock, produced syngas, and energy demands. Downdraft gasifier is recommended for small-scale applications in rural areas. The co-current nature of air and biomass flow reduces the tar content and increases CO and H_2 in the produced syngas. The syngas produced from downdraft gasifier can be used after a simple purification process in thermal applications. From a cost study, the payback period of a gasification system is around 5 years.

Acknowledgements

The authors would like to express deepest thanks for the Science and Technology Development Fund (STDF Egypt) for financing and supporting the development of the thermal gasification of biomass.

Author details

Mohammed Abed Fattah Hamad*, Aly Moustafa Radwan and Ashraf Amin

*Address all correspondence to: hamadnrc@hotmail.com

National Research Centre, Cairo, Egypt

References

[1] Sheth PN, Babu B. Experimental studies on producer gas generation from wood waste in a downdraft biomass gasifier. Bioresour Technol. 2009;100:3127–33.

[2] Khater EMH, El-Ibiary NN, Ki-Iattab IA, Hamad MA. Gasification of rice hulls. Biomass Bioenerg. 1992;3:329–33.

[3] Zabaniotou AA, Skoulou VK, Mertzis DP, Koufodimos GS, Samaras ZC. Mobile gasification units for sustainable electricity production in rural areas: The Smart-CHP project. Ind Eng Chem Res. 2011;50:602–8.

[4] Buragohain B, Mahanta P, Moholkar VS. Biomass gasification for decentralized power generation: The Indian perspective. Renew Sustainable Energy Rev. 2010;14:73–92.

[5] Karellas S, Karl J. Analysis of the product gas from biomass gasification by means of laser spectroscopy. Opt Las Eng. 2007;45:935–46.

[6] Ni M, Leung DYC, Leung MKH, Sumathy K. An overview of hydrogen production from biomass. Fuel Process Technol. 2006;87:461–72.

[7] Kirubakaran V, Sivaramakrishnan V, Nalini R, Sekar T, Premalatha M, Subramanian P. A review on gasification of biomass. Renew Sustainable Energy Rev. 2009;13:179–86.

[8] Radwan AM, Hamad MA, Singedy AM. Synthesis gas production from catalytic gasification of saw dust. Life Sci J. 2015;12:104–18.

[9] Yoon SJ, Choi YC, Lee JG. Hydrogen production from biomass tar by catalytic steam reforming. Energy Convers Manage. 2010;51:42–7.

[10] Hamad MA, Radwan AM, Heggo DA, Moustafa T. Hydrogen rich gas production from catalytic gasification of biomass. Renew Energy. 2016;85:1290–300.

[11] Klass DL. Biomass for Renewable Energy, Fuels, and Chemicals. Academic Press. San Diego, CA., 1998.

[12] Asadullah M. Barriers of commercial power generation using biomass gasification gas: A review. Renew Sustainable Energy Rev. 2014;29:201–15.

[13] Jain A, Chopra S. A review of fixed bed gasification systems for biomass. Agricultural Engineering International Journal. 2007; IX (5).

[14] Begum S, Rasul M, Cork D, Akbar D. An experimental investigation of solid waste gasification using a large pilot scale waste to energy plant. Proc Eng. 2014;90:718–24.

[15] Zhang W. Automotive fuels from biomass via gasification. Fuel Process Technol. 2010;91:866–76.

[16] Moharkar SP, Padole P. Design and development of downdraft gasifier for rural area. IJCA Proceedings on International Conference on Emerging Frontiers in Technology for Rural Area (EFITRA-2012): Foundation of Computer Science (FCS); 2012. p. 24.

[17] Reed T, Reed TB, Das A, Das A. Handbook of Biomass Downdraft Gasifier Engine Systems. Biomass Energy Foundation. Golden, Colorado,1988.

[18] Brown D, Gassner M, Fuchino T, Marechal F. Thermo-economic analysis for the optimal conceptual design of biomass gasification energy conversion systems. Appl Therm Eng. 2009;29:2137–52.

[19] Radwan AM. An overview on gasification of biomass for production of hydrogen rich gas. Der Chem Sin. 2012;3:323–35.

[20] Goyal HB, Seal D, Saxena RC. Bio-fuels from thermochemical conversion of renewable resources: A review. Renew Sustainable Energy Rev. 2008;12:504–17.

[21] Tavasoli A, Ahangari MG, Soni C, Dalai AK. Production of hydrogen and syngas via gasification of the corn and wheat dry distiller grains (DDGS) in a fixed-bed micro reactor. Fuel Process Technol. 2009;90:472–82.

[22] Tasaka K, Furusawa T, Tsutsumi A. Biomass gasification in fluidized bed reactor with Co catalyst. Chem Eng Sci. 2007;62:5558–63.

[23] Gil J, Aznar MP, Caballero MA, Frances E, Corella J. Biomass gasification in fluidized bed at pilot scale with steam-oxygen mixtures. Product distribution for very different operating conditions. Energy Fuels. 1997;11:1109–18.

[24] Bhavanam A, Sastry R. Biomass gasification processes in downdraft fixed bed reactors: A review. Int J Chem Eng Appl. 2011;2:425–33.

[25] Leung DY, Yin X, Wu C. A review on the development and commercialization of biomass gasification technologies in China. Renew Sustainable Energy Rev. 2004;8:565–80.

[26] Guangul FM, Sulaiman SA, Ramli A. Gasifier selection, design and gasification of oil palm fronds with preheated and unheated gasifying air. Bioresour Technol. 2012;126:224–32.

[27] Faaij AP. Bio-energy in Europe: Changing technology choices. Energy Policy. 2006;34:322–42.

[28] Schmidt DD, Gunderson JR. Opportunities for hydrogen: An analysis of the application of biomass gasification to farming operations using microturbines and fuel cells. 2000 US DOE Hydrogen Program Review. San Ramon, California, US: DOE; 2000.

[29] Wander PR, Altafini CR, Barreto RM. Assessment of a small sawdust gasification unit. Biomass Bioenerg. 2004;27:467–76.

[30] Zainal Z, Rifau A, Quadir G, Seetharamu K. Experimental investigation of a downdraft biomass gasifier. Biomass Bioenerg. 2002;23:283–9.

[31] Panwar NL, Rathore NS, Kurchania AK. Experimental investigation of open core downdraft biomass gasifier for food processing industry. Mitig Adapt Strategies Glob Chang. 2009;14:547–56.

[32] Vervaeke P, Tack F, Navez F, Martin J, Verloo M, Lust Nl. Fate of heavy metals during fixed bed downdraft gasification of willow wood harvested from contaminated sites. Biomass Bioenerg. 2006;30:58–65.

[33] Lv P, Yuan Z, Ma L, Wu C, Chen Y, Zhu J. Hydrogen-rich gas production from biomass air and oxygen/steam gasification in a downdraft gasifier. Renew Energy. 2007;32:2173–85.

[34] Zhou Z, Yin X, Xu J, Ma L. The development situation of biomass gasification power generation in China. Energy Policy. 2012;51:52–7.

[35] Tijmensen MJ, Faaij AP, Hamelinck CN, van Hardeveld MR. Exploration of the possibilities for production of Fischer Tropsch liquids and power via biomass gasification. Biomass Bioenerg. 2002;23:129–52.

[36] Guo F, Dong Y, Zhang T, Dong L, Guo C, Rao Z. Experimental study on herb residue gasification in an air-blown circulating fluidized bed gasifier. Ind Eng Chem Res. 2014;53:13264–73.